14th European Architecture Envisioning Conference (EAEA14 2019)

SHS Web of Conferences Volume 64 (2019)

Nantes, France
3 - 6 September 2019

Editors:

L. Lescop
A. Kepczynska-Walczak

ISBN: 978-1-5108-9233-0

Printed from e-media with permission by:

Curran Associates, Inc.
57 Morehouse Lane
Red Hook, NY 12571

Some format issues inherent in the e-media version may also appear in this print version.

For additional information, please contact EDP Sciences – Web of Conferences at the address below.

EDP Sciences – Web of Conferences
17, Avenue du Hoggar
Parc d'Activité de Courtabœuf
BP 112
F-91944 Les Ulis Cedex A
France

Phone: +33 (0) 1 69 18 75 75
Fax: +33 (0) 1 69 28 84 91

contact-edps@webofconferences.org

Additional copies of this publication are available from:

Curran Associates, Inc.
57 Morehouse Lane
Red Hook, NY 12571 USA
Phone: 845-758-0400
Fax: 845-758-2633
Email: curran@proceedings.com
Web: www.proceedings.com

TABLE OF CONTENTS

edp sciences

SHS Web of Conferences

Statement of Peer review

In submitting conference proceedings to SHS Web of Conferences, I certify to the Publisher that I adhere to the **Policy on Publishing Integrity** of the journal in order to safeguard good scientific practice in publishing.

1. All articles have been subjected to peer review administered by the proceedings editors.
2. Reviews have been conducted by expert referees, who have been requested to provide unbiased and constructive comments aimed, whenever possible, at improving the work.
3. Proceedings editors have taken all reasonable steps to ensure the quality of the materials they publish and their decision to accept or reject a paper for publication has been based only on the merits of the work and the relevance to the journal.

Title, date and place of the conference

EAEA14 - European Architectural Envisioning Association Conference
3-6 September 2019, Nantes

Proceedings editor(s):

Laurent Lescop
Anetta Kepczynska-Walczak

Date and editor's signature

02/05/2019

Foreword

Introduction

This monograph is the result of the 14th European Architectural Envisioning Association Conference **Envisioning Architecture: Image, Perception and Communication of Heritage** hosted by the Graduate School of Architecture of Nantes, France, 3-6 September 2019.

Following the mission of the European Architectural Envisioning Association, the conference was intended as a platform for communication and exchange of experience, experimentation, research and collaboration in the field of envisioning built heritage, with a special focus on ambiances.

Taking into consideration that contemporary culture is largely based on the visual perception, heritage envisioning is considered in particular as a process of representing knowledge about space, time, behaviour, light, and other elements that constitute cultural environments. Therefore, it was expected that the conference contributions would explore the following subjects:
• Visualization of the ambiance-related knowledge;
• A cultural approach to climatic, light and sound phenomena;
• Methods of communicating the ambiance issues to the society;
• Enhancing and stimulating sensitive approaches to the built environment in education;
• Envisioning all aspects of ambiance perception.

The EAEA14 theme

The concept of "ambiance" has been formed over the years by questioning the interactions between three attractors: architecture and the city, climatic and sound phenomena, uses and perception. Studied in pairs, each of these attractors refers to very different disciplinary fields, architecture and phenomena concern the physics of the city, architecture and uses interest sociology and uses and phenomena are rather turned to comfort.

Studies concerning ambiances are therefore highly interdisciplinary and open to many questions: living spaces, urban renewal and heritage, urban prospective and the city as a stage. For this, many conceptual and technical tools are mobilized: digital tools for simulation and immersion, investigation, surveys and storytelling, prototyping, field action. What may be new in the field of academic studies is the awareness of artistic creation as a resource for the use of digital tools, storytelling and the representation of complexity through original means.

Considering the broad and multifaceted problematics, the EAEA14 Conference sessions and research papers were organised in three tracks, namely:

1. Ambiance, storytelling and immersion (representation and communication of medium including analogue, digital and immersive environment). *Keywords: analogue digital media, immersive devices, representation.*
2. Ambiance and education (tools, techniques, and strategy of design teaching). *Keywords: parametric design, conceptual studies, digital world.*
3. Ambiance and Design (process, analysis, observation and execution of design for objects, space, built-form and urban environment). *Keywords: conceptual design, perceptual design, analytical design, performative design, visual design, parametric design, simulation design.*

The monograph

The layout of the book has been designed following the requirements for a multi-authored monograph. The monograph is composed of three parts reflecting the three main tracks of the EAEA14 Conference. Accepted papers have been adapted to chapters.

The papers published in this monograph were carefully selected through the process of two-stage reviews. With the great help of the EAEA14 International Scientific Review Committee, each submission was double-blind reviewed by three members independently. Then, submitted papers went through the second stage of the assessing process, and finally, the book was reviewed by two independent reviewers.

Acknowledgements

We would particularly like to thank Véronique Dom for the quality of her work, Sophie Provost who translated and reviewed all the contributions and Marine Bouquet who managed the entire financial part.

We also thank Thomas Leduc and Anne Bossé, the director of CRENAU, our beautiful research laboratory, and of course Daniel Siret, the director of AAU. We would like to thank the Director of the Ecole Nationale Supérieure de Nantes, Christian Dautel, and the administrative teams for their help in organizing the conference. We also thank our funding agencies. We would like to thank the Ambiance Network, which has been an important support, as well as the DCA and eCAADe associations for the dissemination of information. We thank the reviewers for the care taken in selecting the articles from among the 80 received.

It was an honour to have as confirmed keynote speakers at the conference

Keynote Speakers

Dr. Yan Breuleux is a professor at the École des arts numériques, de l'animation et du design à Montréal (NAD), he's also a researcher and practitioner in the field of visual music for immersive display. For twenty years, he has collaborated with musicians and composers to create multi-screen, panoramic, architectural projection and Fulldome pieces. Since 1998, in the PURFORM duo with the composer Alain Thibault, he created strong sensorial A/V performances. His most recent contribution in explore the problematics of Immersif Storytelling for FullDome display with the projects Les Planètes (2018), Re-Génération (2015), Nuée | Swarm (2015), VjGraph (2014) et Engima (2015-18).

Dr.-Ing. Anke Jurleit After her studies in urban planning at the universities in Hannover, ETH in Zürich and UC Berkeley, Anke worked at several small to large-scale architecture and urban planning offices in San Francisco. Her project involvement ranges from major planning and urban design assignments globally to large-scale community planning and design through projects to academic campuses and private homes. As assistant and project manager, Anke maintained an active teaching and engagement profile which included teaching at UC Berkeley Extension and being actively involved in San Francisco's ULI chapter. During her time in California, Anke's ever arising topic was to design for water scarcity. She was active participant in developing Aecom's SSIM (Sustainable Systems Integration Model), a tool that models key environmental, economic and social indicators of large scale projects. Since 2010, Anke is member of the REAP research and teaching group. Her main task was to develop and carry through the REAP Project studios. Her research activities involved the documentation of watersensitive community Jenfelder Au and the 'Hamburg Water Cycle' (http://www.jenfelderau-info.de/). Anke also assisted as technical advisor in developing the water-related topics and Indicators for the newly developed certification system NSQ 'Sustainable communities' at the German sustainable building Council (DGNB).

Ing. Guillaume Thibault Guillaume Thibault is senior researcher at EDF R&D (Saclay) and is pioneering in the fields of laser scanning, virtual and augmented reality, spatial cognition, to help maintenance operations in power stations as well as archaeological studies.

About EAEA

The EAEA was founded in 1993 in Tampere, Finland, and has reconvened every two years since then. What had originally started as a platform for European academic institutes making active use of optical endoscopy instrumentation, gradually but steadily evolved into a wider range of design visualisation and simulation interests.

The founding meeting, hosted by the department of Architecture of Tampere University of Technology in Finland, was the first international meeting of experts in the field of architectural endoscopy, coming from fifteen universities.

The association was intended to become "a platform for communication and exchange of experiences, experimentation, research and collaboration in the field of endoscopy and environmental simulation." Initially, the focus of the European Architectural Endoscopy Association lay exclusively upon the visual simulation of the effects of environmental interventions using optical instruments: 'capturing' photographic or analogue (video) images using physical scale models, generally using a viewing pipe.

Essentially, the first meeting was a gathering of academic professionals in this field, with the delegates representing institutes with some form of 'endoscopic' apparatus. During the conference the participants took part in a workshop session, using the facilities of the Tampere laboratory.

From the first session onward the exclusive focus on optical endoscopy began to shift, first gradually, then more and more steadily towards other environmental visualization opportunities, notably using digital media.

This clearly proved to be the case during the presentations of the second EAEA conference in 1995, hosted by the department of Spatial Simulation at the Vienna University of Technology. In particular, the interdisciplinary conference workshop – 'the (in)visible city' – stimulated the integration and comparison of analogue and emerging digital technologies.

For this workshop initiative participating institutes were sent a study model via the post and asked to prepare environmental simulations using their institute's facilities. The varied results were presented and evaluated during the conference.

Similarly, an important element of the third meeting, held at the Architecture faculty at Delft University of Technology in 1997, was formed by a creative study initiative: the 'Imaging Imagination' workshop. Essentially, conceived as a professional confrontation between 'Optical' and 'Digital' Endoscopy. In this case study, the participants were free to choose between a physical modelling package and a digital file, incorporating texture mapped 'facades'. Some fifteen visualisation proposals were prepared, brought to the conference and viewed and discussed during a special Imaging Imagination conference session.

Apart from the quality and content of visualization, the aspect of the Modelling as such also became a recurring theme. This was particularly the case during the fourth conference, at the Architecture faculty of the Dresden Technical University of 1999, whereby participants took part in an impromptu hands-on modelling exercise using an interior-scale model.

During the subsequent conferences (the 5th conference at the Institute of Urban Design and Planning at the University of Essen, the 6th conference at the faculty of Architecture at the Slovak University of Technology in Bratislava, the 7th conference at the faculty of Architecture at the University of Applied Sciences Dortmund and the 8th conference at the Moscow Institute of Architecture) the shift from 'straightforward' optical endoscopy towards new techniques and topical issues became more and more evident. Noteworthy developments included the increasingly adaptable, distinctive and indeed elegant modes of digital representation, but also the use of digital photography and film, the opportunities of combined media and graphics, but also the introduction of disciplines such as Experimental Aesthetics and Virtual Archaeology.

This led to recurring discussions concerning the association's name. To what extent should endoscopy be considered a fitting 'identity' for the increasingly diverse enterprises of architectural imaging and environmental visualization addressed at the meetings?

Generally, the sentiment tended to be to uphold the established 'label' and to keep the EAEA fraternity relatively exclusive and small-scale in comparison to other, more computer-oriented academic and professional platforms.

During the 2009 Cottbus conference, the thematic differentiation of architectural visualisation approaches and interests once again became manifest during the varied presentations, leading to renewed discussions concerning the EAEA's meaning and role.

What might be an appropriate name that would do justice to the reputation and tradition of (optical and digital) Endoscopy, whilst at the same time giving expression to the steadily unfolding of fields of interest?

Rather than Endoscopy, Envisioning was eventually agreed upon, as it was felt that this fittingly evokes the shared ambitions for a dynamic architectural visualisation practice and the continued exchange of ideas concerning the imaginative conception of future environments.

The EAEA – the European Architectural Envisioning Association

It was hoped that this small, but significant, name change would broaden the appeal of the association on an international level, amongst academics involved with architectural visualisation in the broadest sense, researchers and teachers, whilst at the same time stimulating the deepening of the intellectual discourse.

Previous conferences

13 EAEA Conference | 2017 | Glasgow | Glasgow School of Art
12 EAEA Conference | 2015 | Lodz | Lodz University of Technology
11 EAEA Conference | 2013 | Milan | Politecnico di Milano
10 EAEA Conference | 2011 | Delft | Delft University of Technology
09 EAEA Conference | 2009 | Cottbus | Brandenburg University of Technology
08 EAEA Conference | 2007 | Moscow | Moscow Institute of Architecture (MARCHI)
07 EAEA Conference | 2005 | Dortmund | University of Applied Sciences
06 EAEA Conference | 2003 | Bratislava | Slovak University of Technology
05 EAEA Conference | 2001 | Essen | University of Essen
04 EAEA Conference | 1999 | Dresden | Dresden University of Technology
03 EAEA Conference | 1997 | Delft | Delft University of Technology
02 EAEA Conference | 1995 | Vienna | Vienna University of Technology
01 EAEA Conference | 1993 | Tampere | Tampere University of Technology

Laurent Lescop & Anetta Kepczynska-Walczak
EAEA14 Conference Chair and Book Editor 2019

Envisioning ambiances: Representing (past, present and future) atmospheres for architecture and the built environment

Laurent Lescop[1], *Anetta* Kepczynska-Walczak[2]

[1]ENSA Nantes - CRENAU/AAU UMR-CNRS1563, France
[2]Department of Digital Technologies in Architecture and Urban Planning, Institute of Architecture and Urban Planning, Lodz University of Technology, Lodz (Łódź), Poland

Abstract. The concept of "ambiance" has been shaped over the years by questioning the interactions between three attractors: architecture and the city, climatic and sound phenomena, uses and perception. Studied in pairs, each of these attractors refers to very different disciplinary fields; architecture and phenomena concern the physics of the city, architecture and uses interest sociology and uses and phenomena are rather turned to comfort. Studies concerning ambiances are therefore highly interdisciplinary and raise many questions: living spaces, urban renewal and heritage, urban prospective and the city as a stage. For this, many conceptual and technical tools are mobilized: digital tools for simulation and immersion, investigation, surveys and storytelling, prototyping, field action. What may be new in the field of academic studies is the awareness of artistic creation as a resource for the use of digital tools, storytelling and the representation of complexity through original means.

1 Ambiance as reference

The term of "ambiance" is a relatively recent one. It appeared for the first time in *Les Nouveaux Contes cruels de Villiers* by Villiers de l'Isle-Adam, in 1885. In the *amour sublime* short story, the author invents this word as he plays with sounds, as it was often the case at this time. Therefore, nouns like *regrettance, reluctance, vibrance, tremblance* were created. In the original text, the quotation appears as follows: "*Évariste, — nous l'avons donné à entendre, — était un de ces hommes qui s'expliquent les choses sans jamais s'emporter, son calme énergique lui conférant le don d'étiqueter toujours, d'une manière sérielle, un fait quelconque, sans l'isoler de son ambiance, — et, par conséquent, de le dominer, en l'utilisant même, s'il se pouvait, — dans la mesure du convenable, bien entendu.*" *Ambiance*, coming from literature, was transferred to common language via the technical vocabulary used in painting to describe something phenomenological, then cinema, to identify a sound or a light, which helps to characterise a place. After the creation of the CERMA laboratory at the beginning of the 1970s, then in collaboration with the CRESSON laboratory within a dedicated UMR, the term of "ambiance", sometimes used in the plural, was used to describe the complex relationships between a place designated by its architecture, atmospheric and sound phenomena and the users' perception. By associating many competences for the description and simulation of the physics of phenomena and the understanding of urban practices, research works on ambiances aim to associate quantitative and qualitative dimensions in order to improve the design process of the architectural project.

The substantive "ambiance" does not have a direct translation in foreign languages, and the outlines of its definition remain blurry. The first English occurrence of the term came shortly after Villiers de l'Isle-Adam's invention. Theodore Child used the term in an article for Harper's called "*American Artists at the Paris Exhibition*". He stated, "*Thus, before we realize what the picture is, we already receive an impression of something rich, rare, and precious, and at the same time of something exquisite and quintessential, for the form which we discern in the dreamy ambience is of supreme elegance and fearless purity of line*" (p. 500). The spelling of *ambience* should be noted, while the *ambiance* version also existed, but was more seldom used. Indeed, in many languages, *ambiance* is often translated as *atmosphere.* Therefore, *ambiance* can be found as *atmosphäre* and *ambiente* in German. *Ambiente* is also the Italian translation, with a broader meaning. If the concept is studied phenomenologically, the term *stimmung* is more often used to refer to ambiances as sensitive particularities of a space. *Stimmung* is a musical term, with the same root as *stimme,* the voice, and it means tuning, tonality or effect. It also refers to a mood. In music, the emergence of what I would call a secret emotion, as its origin and explanation are not easy to pinpoint, takes place through a work of skilful variation around repetitive themes or the sound of the same tune. Such is the case of Stockhausen, who, by playing with duration, throws his audience into a hypnotising and meditative universe.

2 Ambiance as method

The contributions to the fourteenth EAEA conference prove that, beyond its actual definition, *ambiance* brings the research community together around questions regarding intangibility, perception mechanisms, and more broadly around the search for a non-demonstrative form of quality. It is thus a *keyword,* which we use as much to open a phenomenological approach of ecosystems, whether natural or built where humans interact, as to describe in a more specific way a prominent phenomenon, i.e. the sound, aeraulic or thermal ambiance of a place. It is also, and perhaps more importantly, a *grammatical word* that is required to understand and design places. Indeed, it entails taking into account the physical and symbolic interactions of atmospheric, sound, use phenomena in the built space, within the analysis and the conception.

If we consider the year 1995 as our starting point to locate the first definitions of ambiances, it appears that the concept has become richer, twenty years later, and has spread across the world, in turn making it harder to understand. Jean-Paul Thibaud wrote that "the ambiance escapes", which implies both the enrichment of the concept and the complexity to fully grasp it. The ultimate risk and the crisis to come would be to state that everything is *ambiance*, therefore preventing any attempts at delineating a proper subject. For now, let us go back to one of the most frequent uses: the perspective of ambiance.

The notion of architectural design presupposes that a representation phase will be needed. According to the canonical diagram, the design process goes is accompanied by sketches, which gradually transform as the project is defined, from a cursive and personal expression to a standardised layout. Within this process, the ambiance drawing accounts for qualitative aspects by assuming a perspective that is close to the future user's point of view. Therefore, there lies the first conceptual stake: the representation of ambiances, i.e. the graphical transcription of the perception of space and, consequently, the representation of intangibility. The creation of a corpus of graphical figurations of climatic and sound phenomena, of transcription of perceptions of places, is an essential step. The contributions hereinafter propose many ideas to enrich our graphical and instrumental universe. The experiences and realisations that are presented in this volume show how we can combine the project with precise indications in terms of intentions for control and regulation of lighting or sound, for example. It allows to go to the next step, which is the interaction between phenomena and perception mechanisms. It would be too hasty to consider that the construction of perception only involves the five senses, with touch dealing with what is the closest to us and sight and hearing with what is the furthest away. Effectively, the information is organised in batches, regardless of distance. We could call the intelligibility "background", and use the term "salience" for the sensitive qualification of an environment, with the appreciation of one or the other of these aspects. The *background* is a presence in which the information is perceptible and undefined, similar to a landscape from afar, where it would be hard to distinguish the components in detail, but still representing a coherent whole. Another example would be the urban rumour, which we can perceive, although we cannot separate its origin and the nature of the different sounds. The *intelligibility* is a whole, of which we can pinpoint the components and specify them. Lastly, the *salience* is a phenomenon that stems from an ensemble, which can sometimes characterise it, similar to a light beam that would mean that a room is sunlit, or an airplane, which would imply a sound nuisance.

Therefore, the conceptual enrichment gives the project not only a shape, but also a phenomenological aspect that belongs to a context.

3 Ambiance and education

The contributions that the reader will discover hereafter highlight the fact that education around a project is mainly based on the mastery of means of representation. This project-based education relies on three pillars: notional, referential and projectual. The notional pillar involves the broad fields of knowledge to identify schools of thought and conceptual tools, in order to embed them into the evolution of ideas. Architecture, intrinsically pluridisciplinary, nurtures itself with different sources, both theoretical and technical. The referential element represents the identification of trajectories within the architectural field, which can be recognised in the history of architectural styles, doctrines, building techniques or ideologies supporting realisations. Lastly, the projectual aspect spans the processes and techniques allowed to produce a built space. Researchers working on ambiances can find themselves in at least two of these notions.

The representation stage adds to the cognitive aspects supporting the perception of a place by connecting it to a context, a temporal dynamic, the identification of sensitive figures, symbolic values, or a certain attention. In the corpus of works on ambiances, and in some articles of this book, it is possible to find propositions to organise some of those registers into partitions to generate a synoptic reading. Beyond the academic world and within scenography and classical painting, many effective solutions exist, which associate sensitive analysis and technical implementation, similar to the architectural project. The result serves the storytelling: the ambiance becomes an actualised, and denotated, space. It is characterised by elements that connect it to a distant, fictional or not, "elsewhere", creating imitation or resemblance. We can distinguish three groups of space facilitating denotation. Firstly, the topical, i.e. a configuration associated with a sort of platitude, but whose elements and their articulation are found everywhere, with slight differences. For example, we can mention the village square. Secondly, we can find the archetypical, which refers to extraordinary forms that are easily identified and identical, e.g. restaurants and hotel chains. Lastly, the synecdoche of ambiance represents spaces where only one component is enough to identify all the others. Therefore, an armchair, a lamp and a television are enough to signify an indoor environment. These three groups represent what scenography proposes in terms of décor, and would be recognised in everyday spatial experiments, even highlighted when it comes to the augmented city.

For students, the validation of their hypotheses inevitably involves the project and the implementation of different configurations to perceive the influence and the variability of each factor, while still ensuring a phenomenological approach. Indeed, we should avoid the fragmentation of the analysis, which would produce reversed or biased results. For obvious practical matters, we use virtual technologies and immersion to start understanding how phenomena work and test their intensity on the audience. From then on, a new question arises to characterise the research protocol. From a neuronal point of view, we now know that a virtual experience is lived in similar ways to a real one. However, an in-depth reflection still needs to be carried out, on what we consider as "immersion", and on the nature of the mobilised devices. In other words, we need to work on both the form and the content.

4 Ambiance and digital instruments

One of the stakes mentioned in some contributions is the role of digital tools in education, and, thus in the design community. Two main dimensions characterise ongoing studies: immersion and parametric conception. They are both paving the way, which the field of ambiances is only starting to explore. The articles presented at EAEA14 on this theme will notably represent progress in that vein.

Immersive instruments have a very long history already, if we consider the fact that they represent the association of an illusory technical device, an ambiance setting and storytelling. Panoramas, created in the 18th century and made popular in the 19th century, already addressed most characteristics by proposing virtual journeys to a faraway "elsewhere". Ambiances have since long been one of the features of panoramas, trying to control lighting, sound, then atmosphere, sometimes even smells as the image became more real and the staging more spectacular. Similarly, theatre, seeking to abandon the constraining central perspective and frontal organisation, tried with Apia and Gropius to conceive a total show. Over this aesthetic and conceptual research, three movements started appearing, which would characterise immersion until today. The first is the envelopment of the audience as their visual space overflows. The second is the distance from the real world by stupefaction, when it overloads them with new sensory information. The third is the commitment to the virtual world, thanks to interaction interfaces, whether real or not. The combination of these three aspects gives the illusion of stepping into a diegetic space, of walking on the other side of the screen. If sight and hearing are easily mobilised, it is possible to find systems that reproduce airflows, smells, thermal engineering and kinaesthesia. Generally, the devices rely on the users' synaesthesic abilities, leaving them to mentally re-develop the missing sensory information. Designing sensory stimulations often goes with the question of simulating light, sound, smells – and we are thus back to the ambiances.

5 Ambiance and physical phenomenology

Immersion is often described as the alternative to reality. However, it also represents a constitution, as a set of new, unprecedented perceptions that can be experienced freely. The evolution introduced in the contributions shows how to virtually experiment physical phenomena by associating simulation and lived experience. Connecting the phenomenology of perceptions with the physical understanding of thermal, aeraulic, sound and light phenomena develops a global approach that is theorised by research works on *cognitive architecture* and studies on architecture and neurosciences. By the characterisation of the stories that accompany the sensitive experience, the ambiances highlight the fact that we do not perceive passively, but that we actively select within our *Umwelt*, even to the point of making us believe that we can modify or act on it. "*The privilege of humans is, in a way, to create worlds, to have at least the illusion that they can escape their Umwelt!*" wrote Alain Berthoz about simplexity. This implies that it is less about exploring the process of direct perception and more the formulation mechanism of these worlds.

This is the reason why we can be very sensitive to objects, elements or events that put our everyday perception off-centre, which we can see appear in urban scenographical experiences. Digital tools greatly help in the teaching of these aspects and in the design within projects. The latter do not have to be realisations for the contemporary world. They can be mobilised to convey historical ambiances. They can also offer an archaeology of ambiances as to no longer work on the object or the form only, but also on the context, the use and the symbolic representations. These themes, sometimes highly speculative, still renew some questions. They also help to understand what is permanent on a given territory.

6 Ambiance and heritage

To commemorate the centenary of the Bauhaus, while wars and accidents lead to the destruction of architectural tradition, questions on the knowledge and preservation of heritage are even more relevant. Studies on Łódź's industrial heritage, military heritage, as well as all sorts of knowledge and vernacular history prove that the consideration of ambiances allows both to enrich hypotheses with restitutions and with fine-tuned knowledge by distancing us from a pure object-based approach.

In Dessau, the answer given by Bruno Fioretti Marquez Architects, based in Berlin, is so provocative that it invites us to rethink the question of heritage in terms of ambiances. Inspired by writers, such as Thomas Demand, Hiroshi Sugimoto and Jorge Luis Borges, architect José Gutierrez Marquez stated during the inauguration of his project that "Our memory lives off blurriness and imprecision". The idea of a blurry memory is formally translated as a very light grey colour, uniform, inside and outside. The windows are transparent. From inside, the landscape appears uncertain and far away. From outside, the interior does not enable introspection. The built volume no longer houses a home; it is an empty shell, barely sculpted by a few lines in relief. This is not architecture anymore, but a geometric outline, where form does not follow function – it represents a sort of mould of the memory of the old building. This reinterpretation is iconoclastic, and it runs away from the picturesque, historical fetishization and subservience to the image. To mention the question from the first chapter, we can ask – what does this construction denotate? Certainly not the Bauhaus, since we cannot find spatial articulations, games of openings or even technical details that preserve buildings from time. It perhaps rather denotes Lyonel Feininger's work on colours and László Moholy-Nagy's greys and light. Either way, it sets up an outside environment that does not predict the inside one – the house tells a story that plays on discontinuities. Should we follow that path, should we stay on that of imitation as it is the case nowadays, should we freeze movement to maintain the memory?

7 Envisioning ambiances

The undetectable imitation, which would deceive our senses enough to be unable to spot what the misleading artifice, is the fantasy of perfect reproduction of phenomena or objects, of undetectable immersion. This would imply a type of imitation that would address each of our senses through mechanical devices so that no trace of context would disrupt the experience. Therefore, it differentiates an imitating space, the one that is not real, that we recognise as a diegetic space, from an imitated space, which is the lived space. The question of the model is thus raised, as it includes two reversed typologies: one corresponds to the simplified sample of a phenomenon. For example, those are our models of physical simulation. The other typology is a mathematical and computational construction, idealised to "*understand what we observe in nature*", as Henri Atlan states. Therefore, architecture completely immerses us into the diegetic space of our senses. The reproduction of this experience, for the sake of study or research, should not forget the immense synaesthesic capacities of our perceptions, which, thanks to compensation and substitution, are able to find and give meaning from the information of another, calling on memory. The representation operates on choices. If we want to imitate, we do not do so about a whole, but about prominent, characteristic elements that create meanings or sensations. The construction of an immersive space works in the same vein, so that the characterisation of the diegetic space happens through the mobilisation of references, the first being the ambiance.

The World Building Framework for Immersive Storytelling Projects

Yan Breuleux[1], *Bruno* de Coninck[1], *Simon* Therrien[1]

[1]Université du Québec à Chicoutimi, NAD-UQAC, 2566 Rue de Bleury, 7e étage Montréal (Québec) H3A 2J2, Canada

Abstract. This article explores issues associated with immersive storytelling in order to examine how the field of World Building can constitute a theoretical framework for practice in the context of VR-based and Full Dome artistic projects. With respect to immersion, the intent will be to interpret the concept of storytelling in relation with the recent formulation of the concept of extended reality (XR). The very concept of World Building is transauthor and transmedia by nature. The transauthor dimension of World Building resides in the idea of subcreation, i.e., designing environments and interaction rules that help create a storytelling basis for generating multiple stories. Once the universe has been conceived, stories written by different authors take shape through transmedia processes across multiple distribution media (film, video games, web, etc.). The question then arises: How can the World Building approach shape the construction of immersive experiences? The article sets out to answer this question, and in doing so, to contribute to the research on environmental storytelling.

1 Introduction

This article explores issues associated with immersive storytelling in order to examine how World Building can constitute a theoretical framework for practice in the context of virtual reality and Full Dome artistic projects. Our articulation of this theoretical framework is inspired by the conceptions of researchers and practitioners from the film and video game domains. This explains why the diverse sources for this article include books, articles, specialized websites, presentation excerpts, and interviews available online. We respectively present the conceptions of several figureheads of the narratological current of game studies such as Mark J.P. Wolf, Henry Jenkins and Marie Laure Ryan. As for practitioners, we discuss three designers: 1- Alex McDowell, film production designer and researcher; 2- Monika Bielskyte, futurist & future designer; and 3- Don Carson, game and theme park designer. These three profiles will allow for describing a particular conception of World

Building and environmental storytelling that is focused on practice. After sketching a broad picture of Digital Storytelling, we explore the World Building concept in relation with the notion of environmental storytelling. This lens will then be used to examine research-creation projects.

The aim of this article is thus to provide an outline of a World Building practice for the field of Full Dome and virtual reality experience design. To do so, it is important to begin by establishing a field of research for teaching and transmission in the field of the creative industries (i.e., digital art, visual effects and video games). Next, we attempt to shed light on how immersive installations shape the *mise en scène* of narrative dimensions. After proposing a typology of immersive environments, we examine several research-creation projects carried out at the École des Arts Numériques de l'Animation et du Design / School of Digital Technology and Animation (NAD-UQAC) and at the Société des Arts Technologiques de Montréal / Montreal Society for Arts and Technology (SAT).

2 Digital, Virtual, Interactive Storytelling

Digital Storytelling, as a field of study, emerged at the crossroads of common interests in the research fields of digital art, literature, communication, and computer sciences in conjunction with the development of new media and entertainment industries. These industries include Massively Multiplayer Online Role-Playing Games (MMORPGs), interactive and immersive experiences of all kinds, smart toys, etc. [1] Given the vast array of applications and practices arising from this field, it should be pointed out that digital components and storytelling are cross-cutting concepts that allow for much crossover between disciplines. In light of the applications and installations produced by researchers in the field, Digital Storytelling embodies the promises of the confluence of the new digital economy and technoculture.

As a brief historical reminder, storytelling dimensions have always been central to the development of new apparatuses and installations. When patenting the Telesphere Mask on October 4, 1960, Morton Helig referred to a "Stereoscopic-television appartus for individual use" [2]. Hence, he was already establishing himself as articulating a new entertainment apparatus. Subsequently, in 1966, Ivan Sutherland created the first real-time 3D animation experience with his virtual reality headset, named the "Sword of Damocles"—a name that aptly conveyed how heavy the device was. In his article "The Ultimate Display," Sutherland interestingly mentions the metaphor of *Alice in Wonderland* [3]. Virtual reality indeed emerged as a technological embodiment of the famed journey through the looking-glass that plunges Alice into her fantastic world where anything is possible. By making this parallel with a literary classic, Sutherland concretely describes the fantastic nature of this new medium, which, without this metaphor, may have struck some as rather abstract and therefore harder to grasp. In doing so, he also successfully anchors this new medium in the timeline of history by assigning it a specific and significant role from the standpoint of the public. Both of these virtual reality apparatuses—mask and headset—anticipated, or rather concretely realized (but without today's powerful calculation capabilities) the fundamental configuration of modern virtual reality devices.

If we were to look for a pivotal moment for digital storytelling in terms of a diversification of practices and the active assertion of the term, it could be located from the beginning to the mid-1990s [4]. With the development, democratization of the computer and the multimedia web browser (the Mosaic browser in 1994), new digital media supports and advances in 3D simulation, a new form of interactivity gained ground, based on real-time connectivity to databases accessible locally and/or online. The utopian period of the 1990s was characterized by the hypothesis of the advent of a new language borne by digital technologies, particularly immersive ones. Janet Horowitz Murray, with his book *Hamlet on the Hollodeck* initially published in 1997, became a leading figure on this topic, and continues to be a key reference to this day [5]. The research in this field also includes the proceedings of a series of computer science conferences that started in 2001 on the overarching theme of Virtual Storytelling (ICVS). As mentioned in the introduction to the first conference in this series, "The first International Conference on Virtual Storytelling gathers researchers from the scientific, artistic, and industrial communities to demonstrate new methods and techniques, show the latest results, and to exchange concepts and ideas for the use of Virtual Reality technologies for creating, scripting, populating, rendering, and interacting with stories, whatever their form, be it theatre, movie, cartoon, advertisement, puppet show, multimedia work, video-games..." [6]. In 2008, the term "virtual" in the conference titles was replaced with "interactive." It is worth noting that Janet Horowitz Murray, whose works are often cited as a theoretical framework, was the keynote speaker for the 2018 conference [7]. The utopia of creating a 360° literary remediation of a translation of Hamlet was realized in 2019 by the Commonwealth Shakespeare company in collaboration with Tech Giant Google [8].

Currently, to explore recent technological innovations in interactive storytelling, the community of practice associated with the Future of Storytelling (FoST) current has been taking an interest in the promotion of artists, theoreticians, researchers, entrepreneurs, designers, directors, etc., but also in research on new experiential practices [9]. Immersion issues hold a prominent place in this community. To emphasize the pluridisciplinarity of FoST contributions, it should be pointed out that this new field of creation also extends to documentary practices. For example, the laboratory of researcher William Uricchio, of the MIT Open Documentary Lab, explores, among other things, the experiential and informative potential of the virtual reality documentary [10] According to Émilie Ropert-Dupont, the practice of virtual journalism raises interesting questions about the role and purpose of emotions in transmitting information via immersive media [11]. Quebec is home to an exhibit series on the theme of Sensory Studies that partakes in the FoST current. This series is of cultural significance, given that it is presented at Montreal's Centre PHI every year and is affiliated with the founder of the Future of Storytelling, Charles Melcher [12]. The most recent conference in this series, entitled "The New Storytellers," featured Douglas Trumbull as its keynote speaker. The remarks made by Trumbull, who was responsible for the visual effects in the film *2001: A Space Odyssey* (produced by Stanley Kubrick), [13] addressed the practice and language issues surrounding immersive cinema [14]. To clarify the subject of this article in relation with Trumbull's thinking, we are witnessing the advent of a new medium shaped by the viewer's special position in a situation of immersion. The first-person point of view participates in co-constructing the story of the immersive experience. Indeed, immersing the viewer in the story makes him or her a full-fledged player within it. Michael Gödde et al. describe this practice as "VR film." They also mention that producing a 360° story requires guiding viewers' attention to the important elements in the scene, determining whether the "viewer" is an active participant or passive observer, choosing the right point of view for the immersive camera in order to determine the anchoring point in the environment, and thinking through the relationship between the temporality of the story, on one hand, and its relation to space, on the other. Finally, the concepts of framing and montage also need to be rethought in this context. The issue of the point of view and the language of immersion is also addressed in reflections by authors such as Celine Tricart and John Bucher [15].

These aspects show how the introduction of immersive cameras turns the cinematographic narrative structure on its head. In sum, the fundamental question of digital storytelling remains: How do digital technologies transform and shape the construction of the storytelling dimensions of the experience?

2.1 Storytelling and technological environments

The recent rise and proliferation of virtual technologies has led to the need to update the concept of virtual-reality continuum as proposed by Paul Milgram and his team in 1995 [16]. The concept of extended reality or cross reality (XR) has recently emerged in response to this need [17]. The concept is interesting to consider in that it allows for merging mixed reality technologies, including virtual reality, augmented reality and new forms of spatial virtualization emerging from ubiquitous computing. In this context, and with the increase in cell phones' recording, analysis and calculation capabilities, a hybridization of virtual and ubiquitous technologies may reasonably be expected in the (relative) long term. Without getting into the statistical details, it is very likely that virtual reality will be democratized primarily via the Internet of Things. This progression of technologies makes it possible to capture and model spaces and bodies in real time. For example, Google's previous smartphone research program Tango was able to project point cloudes in order to capture volumetric data in the user's environment. The program's spatial modelling allowed for designing augmented reality applications and games adapted to the particularities of the specific architectural space at hand. In addition, being able to digitally capture faces suggests the future possibility of adapting data to the specific profile of each individual. This means that our connected devices would transmit their actual real-world anchor point in space but also the topological data captured in each location. In turn, this would mean that any real-life environment could become the basis for a personalized immersive experience.

To better understand how such technologies shape storytelling dimensions, we have divided immersive environments into five types. The goal behind this typology is to define the conditions under which storytelling is developed and the extent to which it is possible to adapt and/or create a narrative and sensory experience. More specifically, we drew up our typology of immersive virtual environments (IVEs) following the logic of extended environments (XR). This typology is based on five forms of technological environments: 1- virtual reality environments, 2- mixed reality (MR) environments, 3- ubiquitous environments, 4- recorded environments, and 5- projected environments (outside the computers). Given that virtual reality (VR) environments are calculated and projected within the computer, we include video games, metaverses, simulators, etc. in this category. Mixed reality (MR) refers to Augmented Reality environments created by the synchronization and simulation of virtual objects that are adapted to the real world. Conversely, Augmented Virtuality Environments (AV) synchronize the presence of real-world objects in the virtual world. For example, this includes Location Based Entertainment (LBE), which leverages different forms of synchronizing the real world within the virtual world. LBE incorporates captured movement and modelling of physical space in synchronization with objects, people and visual textures. The third category in our typology is Ubiquitous Environments, which include physical computing, smart clothes and the Internet of Things. In this case, usage becomes the prime focus, the interface becomes invisible, and the object integrates into the real-world environment using various forms of geolocation in relation to the individual. For their part, Recorded Environments are based on 360° live surround capturing and recording of action in video format and point clouds. To avoid confusion, it is important to note that

Recorded Environments cover all forms of data capturing, from 360° video to Lidar technologies and up to photogrammetry and videometry [18]. The final category is Projected Environments (outside of computers), which include CAVE / FullDomes / Multi-screen and video mapping. These projected environments are created by synchronizing multiple projectors. This is referred to as Spatial Augmented Reality (SAR) [19]. In other words, a building, dome, sculpture or any other type of medium is augmented using various technologies and video projections. All of these environments help structure the narrative dimensions of immersive storytelling. The principal hypothesis of this article is that the World Building framework can contribute to research-creation in immersive storytelling.

2.2 World Building

The World Building approach upholds an industrial practice of transmedia storytelling that makes it possible to provide franchises, television shows, films and video games with a degree of unity with respect to the narrative, aesthetics and formal aspects, for distribution across a wide array of platforms [20]. However, in light of an examination of the books written on this topic, World Building can serve as a model for many diverse practices, from literary practice and futuristic exploration of architectural scenarios to the production of expanded reality (or cross reality) (XR) projects.

To properly circumscribe the World Building approach, it is worth relating a very telling anecdote. During a conference entitled "World Building the Marvel Way," Bill Rosemann, Director of Marvel Games, told the story of a Captain America fan who, during the 2003 Boston Marathon, decided to turn back to help out the rescuers; Roseman then precisely detailed his vision of the term: "When Marvel does it job right, we all do our job right, we create fictional world that can then impact the real world" [21]. Although there are many ideological elements to take into account in the context of such an assertion, it does highlight the foundations of the World Building conceptual framework, i.e., the complex relationship between the primary world (real world) and the secondary world (imaginary world). The imaginary world exists in parallel with the real world. Similarly, as Bielskyte noted in a critique of traditional media, "We have to see the world, not something that we see in a frame, but really as a world that would actually exist as parallel world to our physical world" [22].

In a similar vein, according to Mark J. P. Wolf, World Building consists in designing secondary worlds inspired by the real world and which establish the underlying conditions that enable stories to emerge. Wolf uses the term "subcreation," which he himself borrowed from J. R. R. Tolkien [23], and notes that the "world" can be imagined in a geographical as well as in an experiential form. A world can contain anything that can be experienced by characters within its fictional realm, and the same goes for the events that can give direction to their lives. Interestingly, Wolf makes reference to the etymological meaning of *weorld*, an old German term meaning "all that concerns humans" [24].

Wolf also describes World Building as being organized by three broad rules of design: invention, completeness, and consistency [25]. The first rule is invention. Invention grows out of, but also distances itself from, the real world by creating new hybrid figures in the design of technologies, cultures, artifacts, institutions and new forms of ecosystems. In our view, Bielskyte gives a fairly good description of the relationship that the principle of invention establishes between the secondary and primary worlds. To help bring this fictional world into being, Bielskyte asks a series of questions: "[…] what are the landscapes we do not know? What other plants, what are the animals. What is the contemporary art? What are the sort of scientific breaktroughs or scientific locations? And sort of really looking at the richness of diversity, plurality of our world and our culture and the environment that we live in [...] looking at it, at the shapes of theses colors, of theses stories and theses gestures and transforming them. Taking the best of reality, remix in it in a new way, changing the scale of things and adding a layer of magic and imagination" [26].

The second rule, completeness, has to do with the internal limits, or boundaries, that the world sets up. In other words, it refers to the quantity of information available to the viewer, visitor or user exploring the world. The idea is to enable this individual to find and verify the information they need to guide their actions. In other words, the question to ask is whether the explanations provided by the established systems inform the individual of the rules that organize the world, while also keeping its boundaries concealed. How can the world be contained, both in terms of territory and imagination? What strategies are employed to bound the world's geography while keeping its borders invisible to the world's "guests"? These two components must support the third rule, which is consistency. In other words, invention and completeness obey fictional rules which must, in the context of the subcreation, give the world a high degree of internal plausibility. Importantly, plausibility does not reside in the world's actual relationship to the real world, but rather in the world's consistency via the functioning of rules and fictional characters. Completeness refers to the degree of consistency structuring the world's ecosystem. A more complex world will make completeness progressively more complex. As such, the details are just as important as the whole. Indeed, the details must be in keeping with the general structure of the pre-established world. It is therefore important to meticulously follow the general structure of the system of rules that were put in place. Going along with imaginary systems, inventions and rules requires a temporary suspension of disbelief. For example, in the recent TV series "West World" (2016), the participants in a Far West simulation can break many actual laws of the real world. We accept the actions of the characters because we know that this fictional world is populated with androids. Consistency dictates even the smallest details, such as clothing texture, weapon design, and the use of certain transportation modes. All these elements must, by their presence, reinforce the logical and formal unity of the constructed world. It is thus no coincidence if many of these worlds are geographically located on islands or spaceships. The spatial confinement of the players in the story helps mark out the realm of action; indeed, isolation justifies the world's boundaries but also its social structures, architecture, costumes, etc. The three rules of World Building thus allow for devising an imaginary world that abides by its own internal rules and boundaries.

In 2008, Alex McDowell, Director of the World Building Institute (WBI) at the University of South California Department of Cinematic Arts, summed up the approach as follows: "World Building designates a narrative practice in which the design of a world precedes the telling of a story; the richly detailed world becomes a container for narrative, producing stories that emerge logically and organically from its well-designed core" [27]. Given this statement, it should come as no surprise that McDowell was Production Designer for the film *Minority Report* (2002). When it came out, the film foreshadowed the future use of dematerialized screens and multitouch technology. According to Wolf and McDowell, this approach makes it possible to generate stories in accordance with a transauthorial and transmedial logic [28]. Building fictional worlds also provides direction for urban planners and architecture designers. In an article entitled "World Building and the Future of Media. A case Study – Makoko 2036," Laura Cechanowicz, Brian Cantrell and Alex McDowell relate an experiment conducted by a hodgepodge group of students, architects, interactive media designers, musicians, engineers, urban planners, animators, directors and artists. Drawing inspiration from a specific location and culture, specifically the Lagos Lagoon in Makoko,

Nigeria, the group explored the articulation of different fictional scenarios, from both a story and technology standpoint. In this interdisciplinary World Building context, the authors of this last article wrote, "Any storyteller can weave a compelling narrative, but world builders create storyworlds that support myriad stories by multiple storytellers across disparate platforms — including those platforms that may not yet be named" [29]. It thus appears that the world exists in itself, outside of any sort of media embodiment, and thus serves as a basis for the infinite generation or proliferation of stories and substories. Whether transdisciplinary, transmedia or transauthor, any form of subcreation constitutes a world that guides the expressive use of digital technologies. Clearly, given its all-encompassing nature, beyond its commercial and industrial applications, the World Building approach is particularly fertile ground for designing immersive experiences. Without getting into details, Wolf distinguishes between three perceptual states: immersion, absorption and saturation. It can be postulated that any current or future world, whether existing or yet to be designed, can provide the basis for building a world obeying specific perceptual states. The visitor will thus be plunged into the world (immersion) and captivated by the elements that make up the storyworld (absorption). The world's complexity makes it impossible for the individual to experience all of its elements (saturation). Outside of any particular narrative structure, such as a story arc, being immersed in a fictional creation as a witness or participant brings into play the prime characteristic of digital storytelling, which is interactivity. A world existing and functioning according to its own specific organizational rules allows the person immersed in it to experience the fictional world according to their own point of view. It is worth reiterating that the world determines the preconditions for the emergence of multiple stories. Like a virtual world persisting in the cloud of digital data and evolving according to its own unique logic, by existing as a fiction, a world, an environment, a culture, etc., World Building is an immersive construct. On this topic McDowell, who primarily expresses this idea in talks and presentations, mentioned in a radio interview: "I think VR is just an opportunity to think about a completely different medium. Once you take that step and you get technology that allows you to immerse yourself completely in an unfamiliar environment, my sense is that this environment will provoke the stories that it requires" [30]. Hence, the environment engenders its own narrativity, by the mere fact of its existence. What is at play is therefore no longer a narrative structure, but an invented world that exists within certain boundaries and is organized according to its own specific rules.

From a practice perspective, there are many design rules and points of entry when producing a (story)world. For example, one can use the language of the inhabitants or the geography of the story as a starting point. Moreover, it is essential to determine the historical period (be it real or fictional), the planet, the city, architecture, the building, the institutions, etc., just as it is necessary to determine the characters' emotional characteristics, interaction rules, personal histories, powers, and so forth. Indeed, it is important for all these elements to be determined since they lay the groundwork for viewers to experience a degree of expectation within the context of the story. There is no specific methodology or recipe for designing a "storyworld"; rather, an infinite number of points of entry can be considered. The same is true of games, i.e., there are no specific rules in designing a storygame [31]. However, among all the available choices (characters, languages, cultures, institutions, etc.), one design variable is paramount: spatiality, which includes geography, architecture, location and space. In other words, spatiality, whether imagined in a literary or virtual form, is at the core of the practice of World Building. In fact, the practice of environmental storytelling is based precisely upon this affirmation.

2.3 Environmental storytelling

Inspired by the storytelling strategies used in the architectural design of theme park attractions, the field of environmental storytelling is premised on setting up the conditions for stories to emerge by first constructing the space. In the words of Don Carson, former theme park experience designer, "One of the trade secrets behind the design of entertaining themed environments is that the story element is infused into the physical space a guest walks or rides through" [32]. In this context, when it comes to both real-world and virtual environments, the practice comes down to articulation strategies. Carson also discusses the conditions for dialogue between real-world and virtual spaces. Based on an analysis of his design experience, he defines how the specific narrative structure of theme parks' architectural design helps better understand the field of environmental storytelling. Along similar lines, Henry Jenkins mentions in his text "Game Design as a Narrative Architecture" that spatiality is central to designing games, whether tabletop or virtual: "Game designers don't simply tell stories; they design worlds and sculpt spaces" [33]. Jenkins goes a step further by maintaining that the field of "environmental storytelling creates the preconditions for an immersive narrative experience in at least one of four ways: spatial stories can evoke pre-existing narrative associations; they can provide a staging ground where narrative events are enacted; they may embed narrative information within their mise-en-scene, or they provide resources for emergent narratives" [34]. Hence, the imbrication of places, be they open or closed, affords the interacting player or actor some agency enabling them to create their own narrativity. On the website "Gamasutra," Bart Stewart states that "Environmental storytelling is the art of arranging a careful selection of the objects available in a game world so that they suggest a story to the player who sees them," adding that, "Environmental storytelling is less direct. Instead of explicitly describing events, environmental storytelling shows the final outcome of a sequence of events, then it invites players to make up their own stories about what happened to cause that outcome" [35]. This goes to show that the player's agency conditions their narrative experience of the space. On this topic, Mary Laure Ryan distinguishes between different relationships to space from a strategic and emotional point of view. The space is emotional in that "the experience of that space is associated with emotional values," and Ryan goes on to say, "Whereas the emotional design of the space constructs the subject by the relationship he or she has with their ambient environment, strategic design constructs the subject as possessing agency" [36]. Hence, storytelling is also based on space as perceived from an aesthetic as well as functional standpoint. This means it is important to determine whether the subject will recognize a familiar (emotionally resonant) environment or a strategic one that serves functional goals. More specifally, the strategic space can be in the form of a sandbox, i.e., it can be viewed as pure indeterminacy, or in the form of level design that responds to the strategic requirement of a particular affordance of space according to the construction of the player's experience.

3 World Building: Full Dome research-creation projects

The World Building hypothesis emerged progressively and spontaneously as an inescapable solution in the context of creating a number of research-creation projects in the Full Dome format. First, the Environmental Storytelling hypothesis structured the scripting of the projects *Re-Génération* (2015) and *Les Planètes* (2018). The *Illumination Frankenstein* (2018) project was guided by the World Building hypothesis throughout, from ideation all the way up to performance in November of 2018 [37].

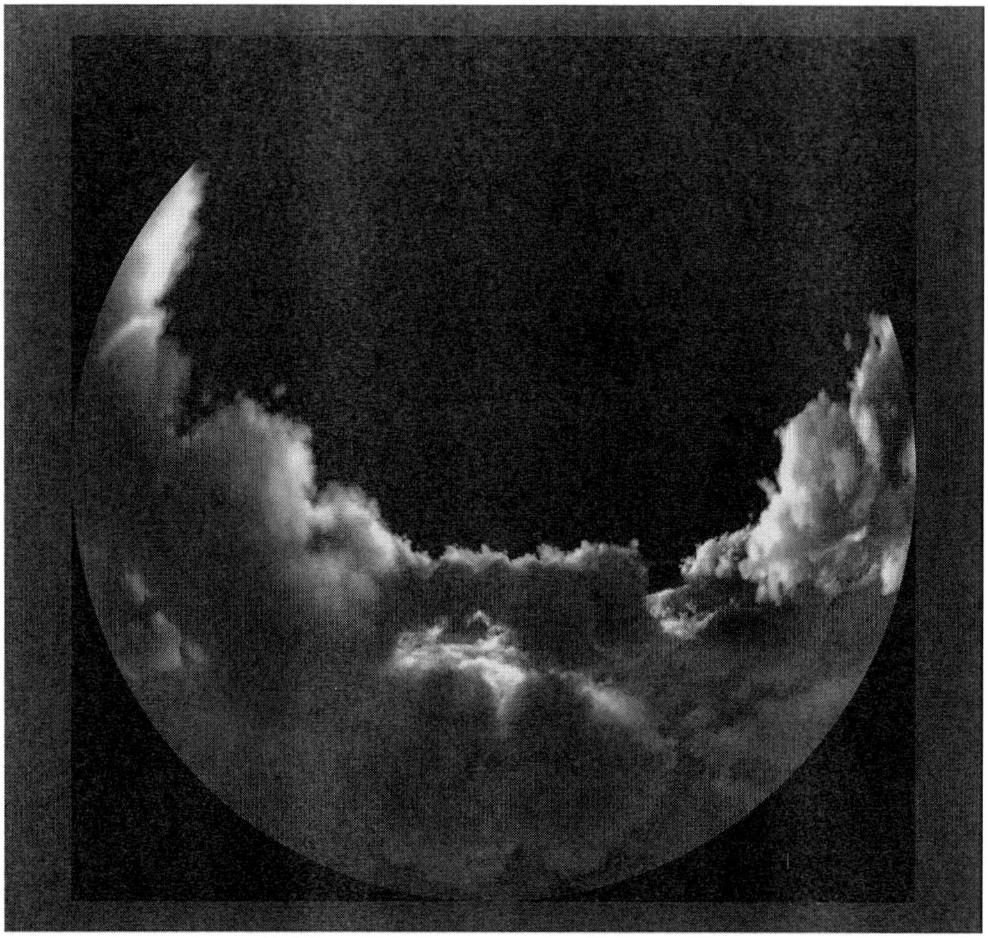

Fig. 1. *Re-Génération*. Société des Arts Technologiques. Cité des sciences et de l'industrie (2015)

To properly understand the evolution of the hypothesis of using the World Building theoretical framework, it is worth providing an overview of key moments in this evolution. In 2008, as part of its cultural activities on the heels of Cop21 in Paris, the Société des Arts technologiques was commissioned with creating an immersive work on the theme of climate change. David McConville, then Director of the Buckminster Fuller Institute, responded by imagining a new form of transmission based on relationships between the human body and the environment via embodiment approaches. His idea was to use environmental visualizations from NASA, simulations of living organisms and synthetic environments to convey a certain vision of climate change through the exploration of the sensory dimensions of the immersive medium that was put in the service of the narrative. The intent was to arrange multiple metaphors in order to create a sensory experience demonstrating that human beings are not separate or isolated from the ecosystems they are changing and transforming. The project, entitled *Re-Génération,* observed the natural world from the standpoint of the time sequence of day and night, the tides, and land movements, all viewed as one giant breath extending across the globe [38].

The narrative scenario, inspired by the Gaia hypothesis, consisted in identifying different levels of knowledge emerging from studies on the broad theme of regeneration. The second level involved writing a narrative presented in the form of a meditation on climate. The story behind the project, written by David McConville, was scripted on the basis of a succession of metaphors drawing links between the balance of ecosystems and the human body. The narration, or storytelling, was thus built on a succession of continually shifting environments. The structure was guided by the observation that narrativity, in a Full Dome situation, no longer plays out through images but through the environment. Once conceptualized, each environment in this scenario was produced by several teams composed of artists and directors. It is important to point out that while it was performed in the immersive format of a Full Dome projection, the project was created using a transauthor approach.

Fig. 2. *Les Planètes*. Société des Arts Technologiques (2018). **Photo:** Nathanael Corre

The second experience where the environmental storytelling conceptual framework was used was entitled *Les Planètes* (2018) [39]. Globally speaking, this project consisted in a series of environments that responded to music and that could, in keeping with a Transmedia aesthetic, be transposed to single-screen, multi-screen, panoramic, virtual-reality and Full Dome formats. The first version of the project was presented as an immersive Full Dome concert generated entirely in real time. The project involved creating a light-piano that entertained a visual dialogue with pianist Louise Bessette regarding a new interpretation of *Les Planètes* by composer Walter Boudreau. The light-piano enabled the projection of dynamic constellations of small spheres scattered across the dome's projection space. The visual shapes were based on analysis of different sections of the musical score. Each section was considered an abstract subworld, as it was divided into as many planets as exist in our solar system. Using environmental storytelling directly resulted in generating the immersive experience through a series of interactional rules between virtual objects. Thus, each environment explored an array of music and sound visualization strategies, and the succession of visualizations yielded a form of abstract storytelling or sensory storytelling. In this context, environments of abstract shapes constituted spaces of mental projection (similar to huge Rorsach ink blots) in which viewers drew their own meaning from the aesthetic shapes as they allowed their attention to wander. Along the same lines, this approach fundamentally relying on transmedia had already been explored in two previous projects carried out in collaboration with the composer Alain Thibault, namely *White Box* (2011) and *Enigma* (2017-2019) [40].

Fig. 3. *Illumination Frankenstein*. Société des Arts Technologiques (2018). Photo Yan Breuleux

The third Full Dome World Building experience that used environmental storytelling was based on the fictional world of Frankenstein as imagined by Marie Shelley, whose novel came to popularity in movie theatres. The project consisted in an immersive lecture by sociologist Céline Lafontaine [41]. In this case, we adopted the opposite approach by presenting the elements which, in the actual romantic era, shaped the world of Dr. Frankenstein and his creature. Thus, *Illumination Frankenstein* takes the shape of a spatialized story to present a succession of environments composed of archives from the complex world that gave rise to Shelley's scientific vision. This brings up the question: What was the real world that inspired the novel? For example, a study of the novel reveals that Frankenstein was put together from human and animal corpses, and was vegetarian. An interpretation using a World Building approach could allow for a new modelling of the creature, using details provided by the author. Indeed, studying the novel and comparing its different versions would allow for composing a coherent virtual world in keeping with a World Building logic, if we examine the story in its minute details. The immersive lecture helped expose the actual world that give birth to the novel *Frankenstein*. Following a narrative structure of environmental storytelling, composed as a journey within a myriad of immersive environments, the project was, fundamentally, underpinned by an understanding of different ways of applying the World Building model. The idea was also to show how, "Two hundred years after its literary birth, the prescient light that this prophetic novel shed on the beginnings of modernity still shines today and invites a reflection on the conquest, manipulation and devastation of the living world" [42]. In other words, by its very structure, the project attempted to draw links between the primary and secondary worlds of Frankenstein, which are at once romantic and modern.

3.1 World building: virtual reality research-creation projects

Fig. 4. *PeterPan360°*, Bruno de Coninck, NAD - UQAC (2019).

The theoretical framework of World Building has helped guide the research and structure the practice of a number of students at the NAD school pursuing research-creation in the course of their master's studies (Master 2). Bruno de Coninck has used different media to model a virtual environment of the world of *Peter Pan* by blending a number of different interpretations of this world imagined by James Matthew Barrie in the early 20th century (1904, 1911). Once the prototype was completed, it gave rise to a series of hypotheses for transcoding the world in different formats. The last version consisted in generating a virtual-reality world. To achieve this, Coninck compared different descriptions of this world and then modelled a 3D narrative environment that served as a medium for various game mechanics. Semantic blending of countless interpretations of the tale were used to add to, complement and compose a complete virtual environment.

Fig. 5. Hybrid environment. Simon Therrien (2019).

For his part, Simon Therrien has imagined a hybrid world based on a cross between the world of Franz Kafka's novel *Métamorphose* and Jim Henson's *Labyrinth*. His environmental storytelling project has taken the shape of designing a new video game depicting a number of fantastic environments as settings for action. Therrien's research has primarily involved defining a series of conceptual tools to nourish reflection on practice.

Fig. 6. Cités-Mondes. Samuel David (2019).

Samuel David, with his project *Cités-Mondes*, has worked on environmental storytelling by observing the creative dimensions of 3D modelling while adopting a reflexive practitioner approach. Thus, David has created an animated universe determined by fictional rules for creating animated ecosystems. In reality, he modelled a complete world that functions according to its own internal rules. Objects conceived in this way could enable writers to imagine stories, intrigues, tensions, etc. This is because the environment serves as the medium for an imaginary narrative space. In a different vein, Eric Filion (also known by the pseudonym *Vj Nokami*) has drawn inspiration from the study of the specific environment of Iceland to make an immersive film on the unique sense of *Solastagia* described by Glen Albrecht [43]. The goal of Filion's project was to help the viewer visualize this unique emotion, characterized as a sort of anguish provoked by the gradual shifting and transformation of our natural environment.

Conclusion

The theoretical framework of World Building offers a springboard for analyzing the practice of environmental storytelling in the field of Spatial Augmented Reality (SAR), and more specifically in the Full Dome format with its succession of environments. In sum, our analysis of projects upholds that the World Building approach structures the creation of storyworlds and storygames that can be adapted to multiple apparatuses and installations. As such, and viewed from an art and design standpoint, this approach makes it possible to structure storytelling projects while taking into account a diversity of projection platforms. To illustrate this assertion, one could mention for example the fact that the overarching theme of a museum exhibit can constitute a storyworld. In this context, the different artifacts are produced using a transmedia approach. After conducting our analysis, we have also noted that the world's organization is based on collaboration between players in the academic and institutional milieus, and with professionals from multiple disciplinary backgrounds. In our view, the transauthor dimension is especially relevant as it helps structure the contributions of all the players involved. Another interesting fact is that environmental storytelling can be embodied both in the exhibit's scenography and in the media design of parallel worlds, virtual reality experiences, etc. In this context, the experiential design approach adapts particularly well to the storyworld formulated by all of the players, especially since all historical periods obey rules pertaining to story structure and construction. Hence, the characteristics of World Building, as formulated by Wolf, provide a theoretical framework for organizing the relationship between art, design, technology and storytelling, by following a certain logic. Whether in museology, documentaries, films or digital art projects, this relationship helps define a new form of transmedia storytelling. In all cases, the process is based on a series of critical steps. The first is to immerse oneself in the historical world, and the second, to imagine a map, a place, an environment. However, the elements entailed by each step vary depending on the project; depending on the case, one may need to include interactions with the characters, the era, technologies, and common objects. From this world, this subcreation, all media choices become possible.

In our case, in the context of our recent research and that of our students, the worlds have, globally speaking, been based on the use of an existing story. The underlying research makes it possible to imagine a certain historical environment and to convey knowledge through a given storyworld, thus giving visibility to the paradigm on which the knowledge is based. In this general context, the theoretical framework of World Building also transforms the paradigm from which the project is created. The story plays out through the environment in which the action occurs, but partakes in what Bielskyte describes as a storyworld experience. The storyworld experience is developed in the field of VR films via close collaboration between four key individuals: the director, the production manager, the engineer (or technical artist) and the interaction designer. Each has a specific role to play:

1- the director is in charge of the story and the emotional design, 2- the production manager carries out environment-related tasks, whether in terms of architecture or narrative, 3- the engineer fleshes out how the story will be conveyed, and 4- the interaction designer determines mechanics and logics of interacting with the world. This collaboration somewhat diminishes the director's status as a demiurge. As the designer of the experience, the director adapts to and collaborates with an existing world and the other team members. The engineer now participates in storytelling, and the interaction designer (as a specialist in relationships with the world) determines the strategies the user will use to interact with the world. This said, it is important to understand that this structure brings into play not just skills, but also a space of horizontal communication. Indeed, each player in the process observes the issues surrounding the story, the environments, the interaction and the technological means, all from the characteristic angle of their particular profession. Thus, in a World Building approach, each role involves exploring the conditions for bringing the story and environment into existence (from both a modelling and technological standpoint) by connecting them with different forms of interaction.

The project *Dans la lune* (2019), currently being developed under the direction of Marc Caro in collaboration with a group of students and teachers from the École Supérieure d'architecture de Nantes, consists in a VR film experience inspired by a hybridization of the world of the film magician Georges Méliès and the retrofuturistic aesthetics of Jules Verne, whose Machines de Nantes are an outstanding example of this aesthetic [44]. The construction of the film's immersive décor is a central part of articulating the experience. Consistent with the World Building framework, constructing this film involves the combination of a world and the building of a storytelling environment where the action will take place. In this context, the theoretical framework of World Building serves as a foundation to analyze the project's narrative dimensions as well as its creation process.

In light of current trends in immersive technologies, it appears evident that the transmedia premise of storyworlds allows for adapting stories across different forms of technological environments. One avenue to explore would be to look at how the model applies to the four worlds of storytelling as set forth by Scott A. Lukas in 2013 [45], which was linked to the theory of possible worlds as studied by Wolf and Ryan. [46], [47]

According to Lukas, the primary world is the realm of reality (a), the second is imaginary (b), the third is experiential (c) and the fourth is the world of guests (d). These four perspectives constitute as many trades and complementary endeavours and approaches. The first world consists in the materiality of built structures, from an urban planning, architectural and industrial standpoint. The second world, that of the imagination, stems from literature, fiction and all forms of visualization (maps, models, patterns, language, etc.). The third world, experiential, comprises technological environments such as those described in this article, and the design of visitor interactions. The fourth world, or visitor world, corresponds to the perspective of the receiver. Clearly, our field is located in the humanities and social sciences. How can the unique perspective of the individual be understood in the context of the interrelationships between all these worlds? Moreover, what knowledge would help to include guests in the storytelling process? In our case, it is clear that we are in an interdisciplinary context. The next step of the present study would be to establish a design methodology making it possible to create a narrative continuum between these different worlds, beyond their technologies. This brings into play the hypothesis of World Building as a continuum of mixed reality or cross-reality—not technological, but rather narrative and experiential, in nature.

References

1. Miller, Carolyn Handler. *Digital storytelling: A creator's guide to interactive entertainment* (Focal Press, 2014), p.14.
2. Patent of the *Telesphere Mask* invention: [https://patents.google.com/patent/US2955156A/en], page accessed on February 6, 2019.
3. Sutherland, Ivan E. "The ultimate display." *Multimedia: From Wagner to virtual reality* (1965): 506-508.
4. A detailed chronology is available in chapter 1 of the following book: Alexander, Bryan. *The New Digital Storytelling: Creating Narratives with New Media--Revised and Updated Edition*. (Abc-clio, 2017).
5. Murray, Janet Horowitz. *Hamlet on the Holodeck : The Future of Narrative in Cyberspace* (New York: Free Press, 1997).
6. Balet, Olivier, and Gérard Subsol, Torguet, Patrice. *Virtual Storytelling; Using Virtual Reality Technologies for Storytelling*: International Conference, ICVS 2001, Avignon, France, September 27-28, 2001, Proceedings. Vol. 1. (Springer, 2001). P.2.
7. *International Conference for Interactive Digital Storytelling* : [https://icids2018.scss.tcd.ie/], page accessed on February 6, 2019.
8. New York Times: [https://www.nytimes.com/2019/01/25/theater/hamlet-virtual-reality-google.html], page accessed on February 6, 2019.
9. Future of Storytelling: [https://futureofstorytelling.org/], page accessed on February 6, 2019.
10. MIT Open Documentary Lab: [http://opendoclab.mit.edu/], page accessed on February 6, 2019.
11. Ropert-Dupont, Émilie. *Journalisme et réalité virtuelle: Émotion ou information?*. (Éditions L'Harmattan, 2017).
12. Centre PHI: [https://phi-centre.com/en/event/fost-2015-en/], page accessed on February 6, 2019.
13. *The new storytellers*: [https://phi-centre.com/en/event/the-new-storytellers-5-en/]
14. Douglas Trumbull : [https://www.youtube.com/watch?v=47wO-Az22v0], page accessed on February 6, 2019.
15. Gödde, Michael, Frank Gabler, Dirk Siegmund, and Andreas Braun. "*Cinematic Narration in VR–Rethinking Film Conventions for 360 Degrees*." In International Conference on Virtual, Augmented and Mixed Reality, pp. 184-201. Springer, Cham, 2018. See: Tricart, Celine. *Virtual Reality Filmmaking: Techniques & Best Practices for VR Filmmakers*. Taylor & Francis, 2017. Also worth mentioning is: Bucher, John. *Storytelling for virtual reality: methods and principles for crafting immersive narratives*. Focal Press, 2017.
16. Milgram, P., Takemura, H., Utsumi, A., & Kishino, F. (1995, December). *Augmented reality: A class of displays on the reality-virtuality continuum. In Telemanipulator and telepresence technologies* (Vol. 2351, pp. 282-293). International Society for Optics and Photonics.).
17. The New Zealand VR/AR Association (NZVRARA). *Virtual gets real. The Explosion of Cross Reality in New Zealand, report one*, September 2017.
18. Breuleux, Y. *Narrativité immersive pour Vr et FullDome*. Art & Technologie. Effects Mtl, 2017. Palais des congrès de Montréal.

19. Bimber, Oliver, and Ramesh Raskar. *Spatial augmented reality: merging real and virtual worlds*. AK Peters/CRC Press, 2005.
20. Boni, Marta, ed. *World Building: Transmedia, Fans, Industries*. Amsterdam University Press, 2017.
21. *World Building the Marvel Way*, Bill Rosemann, Marvel Games (2017):[https://youtu.be/p3J2ZuOmFWw?t=120], pages consulted on February 6, 2019. Also worth mentioning is an article on the links between the Marvel empire and the World Building approach: Jeffries, Dru. " *The Worlds Align*." World Building Transmedia, Fans, Industries (2017): 287.
22. Monika Bielskyte, *The New Storytellers II: World-building: Best Practices for Creating Story Worlds*: [https://youtu.be/jtr0AZK64Do?t=1093], page accessed on February 6, 2019.
23. Wolf, Mark JP. *Building imaginary worlds: The theory and history of subcreation*. Routledge, 2014. P.19.
24. Wolf, Mark JP. *Building imaginary worlds: The theory and history of subcreation*. Routledge, 2014. P.39.
25. Wolf, Mark JP. *Building imaginary worlds: The theory and history of subcreation*. Routledge, 2014. P.44.
26. Monika Bielskyte, *The New Storytellers II: World-building: Best Practices for Creating Story Worlds*: see (24:42): [https://youtu.be/jtr0AZK64Do?t=1093], page accessed on February 6, 2019.
27. *The World Building Institute*: [http://worldbuilding.institute/about], page accessed on February 6, 2019.
28. Wolf, Mark JP. *Building imaginary worlds: The theory and history of subcreation*. Routledge, 2014. P.28.
29. Cechanowicz, Laura, Brian Cantrell, and Alex McDowell. "*World Building and the Future of Media: A Case Study-Makoko 2036*." IEEE Technology and Society Magazine 35, no. 4 (2016): 28-38.
30. Voices of VR: *Alex McDowell on World Building in Storytelling*, [voicesofvr.com, podcast 309] (2016), page accessed on February 6, 2019.
31. Montfort, Nick. "*Interactive Fiction as 'Story,' 'Game,' 'Storygame,' 'Novel,' 'World,' 'Literature,' 'Puzzle,' 'Problem,' 'Riddle,' and 'Machine.'*." First Person: New Media as Story, Game, and Performance (2004).
32. Don Carson: [https://www.gamasutra.com/view/feature/131594/environmental_storytelling_.php], page accessed on February 6, 2019.
33. Jenkins, Henry. "Game design as narrative." Computer 44, no. 53 (2004): 118-130.
34. Jenkins, Henry. "Game design as narrative." Computer 44, no. 53 (2004): 118-130.
35. Bart Stewart, Gamasutra: [https://www.gamasutra.com/blogs/BartStewart/20151112/259159/Environmental_Storytelling.php], page accessed on February 6, 2019.
36. Ryan, Marie-Laure. "L'expérience de l'espace dans les jeux vidéo et les récits numériques." Cahiers de Narratologie. Analyse et théorie narratives 27 (2014). P.2.
37. *Re-Génération*: [http://www.uqac.ca/portfolio/yanbreuleux/recherche-creation/], page accessed on February 6, 2019.
38. See: [http://sat.qc.ca/cop21], page accessed on February 20, 2019.

39. See: [http://sat.qc.ca/fr/planetes], page accessed on February 20, 2019.
40. See: [https://vimeo.com/purform/enigma], page accessed on February 20, 2019.
41. See: [https://socio.umontreal.ca/repertoire-departement/vue/lafontaine-celine/], page accessed on February 20, 2019.
42. *Illumination Frankenstein*: [https://sat.qc.ca/fr/evenements/illumination-frankenstein], page accessed on February 20, 2019.
43. The theses of Bruno de Koninck, Simon Therrien, Samuel David and Eric Filion will be made available during the summer of 2019 on the following website: [https://constellation.uqac.ca/], page accessed on February 20, 2019.
44. The project is currently under production: [https://www.instagram.com/architecture.representation/], [https://www.instagram.com/architecture.representation/] and [https://architecturerepres.wixsite.com/decor360], page accessed on February 20, 2019.
45. Lukas, Scott. *The immersive worlds handbook: designing theme parks and consumer spaces*. Focal Press, 2012. P.230.
46. Bell, Alice, and Marie-Laure Ryan, eds. *Possible Worlds Theory and Contemporary Narratology*. University of Nebraska Press, 2019.
47. Wolf, Mark JP. *Building imaginary worlds: The theory and history of subcreation.* Routledge, 2014. P.17.

Psychological, physiological, and phenomenological effects of colored light

Jennifer Scott[1,*] and *Judy* Theodorson[1]

[1]Washington State University, School of Design and Construction, Pullman, Washington, USA

Abstract. This research explores colored light from historical, scientific, and technological vantage points. Effects of colored light on the human condition are examined on multiple levels, comparing theories in physiological, psychological, and phenomenological areas of study, addressed across a broad scope of literature. The focus of this research is to gain an understanding of the effects of colored light on the human condition, and how designing with it can provide an engaging phenomenological experience which marries centuries' old design techniques with modern technological innovations. The stages of research include a literature review, followed by a series of creative explorations with light, texture and form through small-scale models. Data gathered is then analyzed, categorized, and used to create a taxonomy for designing with colored light. Research is then explored on a larger scale through a pilot study conducted in a 5' x 9' room designed to create an experience of being bathed in colored light with user-controlled variability. Each research method may then be used to inform the design of interior environments, affecting the participant's physiological, psychological, and phenomenological experiences.

1 Background

Colored light has been used by architects and designers to shape our built environments for centuries. Daylight through colored glass panes adds a sense of transcendence and awe to sacred spaces. The advent of light-emitting diodes (LEDs) has enabled colored light to become a more ubiquitous presence in the designer's toolbox; a single fixture can produce every color in the visible spectrum. Research indicates that exposure to colored light has layered effects on the human condition – from physiological responses such as changes in blood pressure and eye-blink frequency [1-3] to psychological responses to specific *hues* (colors), *values* (lightness or darkness), and *chroma* (saturation) [4-6]. Given the potential for such a range of responses to colored light and given that technological advances have made this tool so readily available to designers, it is important to understand existing literature on the subject in order to make informed and strategic design decisions. The application of this research in a day spa provides an opportunity to design for the human condition in all three categories – physiologically – designing spaces which are bathed in

* Corresponding author: jennalynnscott@gmail.com

colored light, inducing physiological responses; psychologically – tapping into color theory which carries through cultural, experiential, and personal references; and phenomenologically – designing experiences which provide tangible engagement with colored light, providing momentary suspension of reality through extraordinary experiences.

1.1 Theoretical Framework

This research on the effects of colored light on the human condition can be traced back to an ever-evolving body of knowledge in the field of color theory – specifically considering chromotherapy. Scientists, artists, designers and writers have attempted to explain our perceptions of color. Linguistics, though able to offer poetic descriptions of color, is also limited in is variety of descriptors. According to Elliot and Maier, there are an estimated 2.3 million discernable colors [6]. Why then do we limit our identification of the colors of the rainbow to only seven? Our language is unable to adequately describe what our eyes perceive.

Color representation systems have dramatically evolved since the 1600s. Sir Isaac Newton (1643-1727), who discovered that the white light that surrounds us is actually made up of all of the colors in the visible spectrum, identified seven unique colors through his prism experiment in 1666 [7]. Johann Wolfgang von Goethe (1749-1832) published his *Theory of Colors* in 1810, in which the botanist and poet identified six colors, which he displayed as a color wheel. Unlike Newton, Goethe's interest was in our perception of color, and not in its physiological origins [8]. In the late 1800s, Michael-Eugene Chevreul (17861889), a French chemist, developed a 72-hue color circle, and began experiments with simultaneous contrast [7]. Perhaps the most influential to how we currently evaluate and analyze color perception is Albert Henry Munsell (1858-1915). Munsell categorized color as hue, value, and chroma. His 3D color system has been used in various case studies to determine psychological associations with color [4,5]. Though much research exists in the area of color theory, there is little available research specific to color theory as it pertains to psychological effects of colored light. A pilot study in this area was an initial step in bridging this gap, in the "User-Controlled Immersive Light Room Experience" discussed later in this paper.

1.2 Literature Review

A literature review was an initial step in the research process – collecting, analyzing, and categorizing existing literature which addresses the effects of colored light. Information gained from this review provided a framework through which to more thoughtfully apply colored light in interior design solutions – taking into account its empirically-proven as well as its perceived effects on the human condition. The literature covers a broad range of fields of study: medical, holistic nursing, psychology, sociology, and cultural anthropology, as well as interior design and architecture journals, to name a few. Colored light is of interest to many different fields of study for its impact on our physiology, psyche, and its influence on how we experience our built environment. Through the research strategy of logical argumentation, existing research and case studies in the realms of medicine, psychology, and design were examined. It was then sorted into three categories: psychological, physiological, and phenomenological effects of colored light.

1.2.1 Physiology

According to Peter K. Kaiser, Psychology Professor at York University Downsview in Ontario Canada, color has a physiological effect on us, whether we perceive it visually or not. Kaiser asserts that "physiological human responses to color as evidenced by the electroencephalogram, galvanic skin response, blood pressure, heart rate, respiration rate, and eyeblink frequency, and suggests that some are indirect effects mediated by cognitive responses to color." [9] Color is perceived visually, and changes with the quality of light. The eye perceives color based on the light that is reflects off of a surface – different wavelengths result in different perceived colors. A particular shade of blue will look lighter when the sun shines directly upon it than it does when cast in shadow. When in a dark space, or when we suffer from severe visually impairment, colors all become shades of black. Color does not exist without the presence of light.

Our perception of color is simply our minds interpreting wavelengths – or vibrations – of light. Since color would not exist without light, it makes sense that cultures have for centuries connected the two. According to Day and Rich, it was believed in ancient Egypt and Greece that healing was accomplished through a treatment of the body, mind, and soul – something which was addressed with the use of colored light [2]. In ancient Egypt, the "sick were diagnosed by color and then laced in rooms that radiated the particular color prescribed." These rooms were lit with colored light produced by positioning colored gems such that daylight would pass through [1,2]. Today, with the advent of modern technologies and advancement in medical treatments, colored light is still used in to treat chronic physical conditions. Colored light is used to cure dental fillings; red light has been found to prevent scar tissue; different colors of light are used in optometry to improve eyesight and balance; monochrome light domes are used to treat phobias, depression and stress; blue light is used to treat jaundice in infants; the list goes on [2]. As science advances our understanding of the human body, we are drawn back to ancient healing techniques, but in more precise, empirically-proven ways.

1.2.2 Psychology

Color is an integral part of our lives in many ways, both culturally and psychologically. Research indicates that color has an effect on our mood [9-11]. Research has correlated emotional reactions to and psychological associations with color hue, saturation, and brightness. [4, 5, 10, 11). The color red has been associated with excitement, anger, passion; orange has been perceived as distressing and upsetting; purple as dignified and stately; bright yellow as cheerful, blue has been associated with comfort; and green as having a calming effect [10]. However, these psychological associations, though shown to be somewhat consistent across cultures due to shared experiences (i.e. we are all familiar with the blue of a warm summer sky, the green of freshly mown grass, and the warm hues of a sunset), are also affected by personal experiences [5]. According to Elliot and Maier, psychological responses to color are biologically based (i.e. red is associated with strong, highly stimulating emotions, possibly because the wavelengths of red light are longer, penetrate the skin more deeply, and produce more intense physiological reactions), but they are also based on learned associations of color meaning [6]. There is substantial empirical evidence regarding the physiological effects of color and colored light, but much of the research related to the psychological effects of color are, by the nature of the subject matter, somewhat subjective.

1.2.3 Phenomenology

Research in phenomenology began long before it was applied to the built environment. It is an exploration of the essence of being human, the essence of existence. Considering the physiological and psychological effects of color, it follows that color itself plays a strong role in our experience of place – in the essence of our existence. According to Seamon and Mugerauer, "the phenomenologist works to circumvent the natural attitude and to undertake a fresh, exhaustive examination of consciousness and experience" [12]. It is through this fresh examination that architects and interior designers conceive spaces which provide engaging phenomenological experiences. In his book *The eyes of the skin: Architecture and the senses*, Pallasmaa questions "why, when there are five senses, has one single sense – sight – become so predominant in architectural culture and design?" [13]. He argues that by suppressing the other four sense, the experience of the built environment is less rich, and less engaging that it could be – than it should be. Research indicates that architecture is more than just visual – it is place, which through all of the senses, and through memory, provides a deeper meaning to what it is to be human, and is ingrained in our self-identity [14]. Since color is tied to memory, and is connected to our sense of place, of self-identity, and is a trigger for recollection of experience, color itself is a gateway to a multisensory experience, and can be used in strategic ways in design.

1.3 Primary research questions

- How can we design with colored light in deliberate, controlled ways?
- What is a person's response to being bathed in colored light, and how can we provide a positive phenomenological experience?
- How can this research be applied effectively in an interior design project?

2 Research methodology

Three phases of creative research inform the design of the day spa. The first is a series of light model studies which lead to a categorization of design output as well as a taxonomy of designing with colored light. The second phase is an evaluation of design precedents by six designers who incorporate the categories identified in Research Phase 1 in order to create a phenomenological experience through interior design or art installations. The third and final phase is a pilot study which builds on the work of James Turrell – specifically his Perception Cells. This study involves the design and installation of an immersive user-controlled light room experience. Participants spend time in the light room, controlling and changing the colors and intensity of the light, and then answer a series of open-ended interview questions to determine their perception of the experience. This feedback is later applied to the hypothetical design of the day spa to explore incorporating these ideas into an interior environment.

2.1 Creative Research

This phase of research began with a series of study models developed by members of the Fall 2016 Graduate-level Interdisciplinary Design Studio. We each created physical models with which to study daylight as it interacts with different materials and forms. I took these models and applied colored light through various techniques including white light through colored filters, RGB LCD light fixtures, and a combination of both. (Fig. 1)

Methods were documented, and then the models were photographed, resulting in a library of images which were then categorized. Three categories of design outputs were identified, which were labeled Painting with Light, Immersion in Light, and Capturing Light. (Fig. 2)

The findings of this research phase include the categorization of design outputs as well as a taxonomy of designing with colored light (Fig. 3).

Fig. 1. Photos of light study models, models photographed by Jennifer Scott

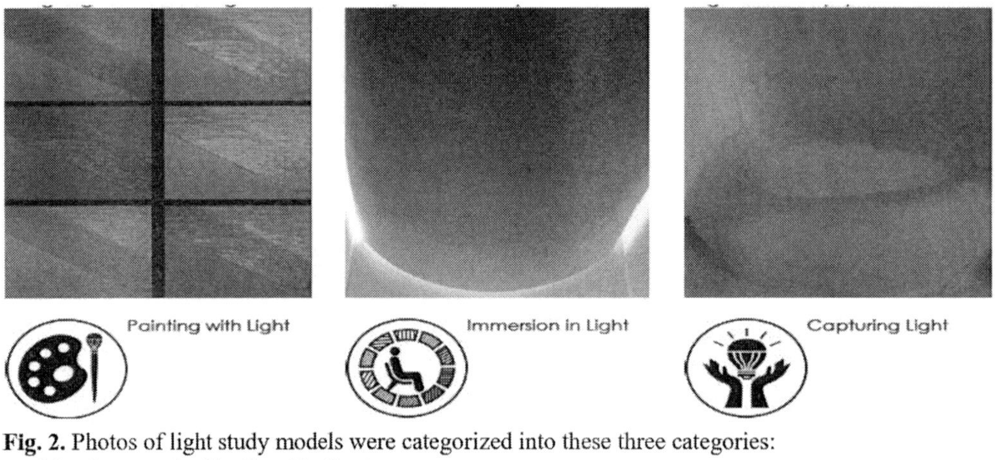

Fig. 2. Photos of light study models were categorized into these three categories: models photographed by Jennifer Scott

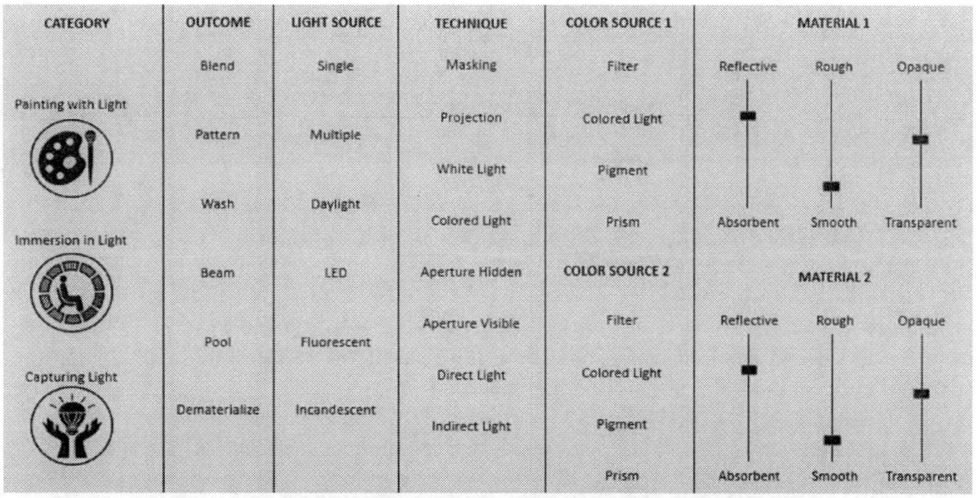

Fig. 3. Taxonomy of designing with colored light, developed by Jennifer Scott

2.2 Design Precedents

Identification of design precedents was based on the three categories identified through the creative research: Painting with Light, Immersion in Light, and Capturing Light. Six designers were selected and evaluated to determine how that work can inform the design of an interior space

2.2.1 James Turrell: Immersion in light: manipulating perception

Turrell explored the idea of manipulating perception through the use of colored light. His Perception Cells (images on right) created personalized phenomenological experiences through immersion in colored light and helped to inspire the Pilot Study discussed in Research Phase 3.

2.2.2 Rem Koolhaas: Immersion in light: immersive experience

Koolhaas designs color into his interiors in a powerful, unapologetic way. Through strategic materiality and lighting techniques, he creates spaces which immerse the visitor in colored light. His use of color is referenced in the day spa in the colored glass atriums, though in a less intense way, allowing for more universal comfort

2.2.3 Architects of Air: Immersion in light: suspension of reality

These inflatable installations are bathed in colored light with variations in internal temperature adding another dimension to the experience.

2.2.4 Regine Shumann: Capturing light: creating walls of light

Shumann's art installations sandwich colored LED light between clear panes, resulting in translucent walls which appear to be made exclusively of light. Her work is referenced in the spa in an indirect way – colored light grazes waterfalls which act as partitions between the changing rooms and the pool area. These light "walls" incorporate water for a multisensory experience.

2.2.5 James Carpenter: Painting with light: daylight mutability

Carpenter's work utilizes daylight as an integral element in his designs. The colors produced by his windows move around the room as the sun moves across the sky. Carpenter's work is referenced in the lobby of the spa, where daylight is refracted off of prismatic columns, casting mutable rainbows around the room and across the visitors, providing an engaging experience.

2.2.6 Steven Holl: Painting with light: vessels of light, reflection

Steven Holl creates the effect of colored light by applying colored pigment to surfaces which are hidden from view, but which reflect daylight, filling the room with blocks of color. Holl's work inspired me to hide the light source where possible, and to keep the viewer's eyes on the light, and not on the lighting fixture.

2.3 User-Controlled Immersive Light Room Experience

This research phase is a pilot study of a user-controlled immersive colored light room experience. An LED light installation was installed in a 5' x 9' room with no natural light. The LED lights consisted of two 30' long RGB tape lights which were installed around the room at heights of 2' and 6'. Prior to installing the tape lights, the walls were covered in white paper to create a consistent and reflective backdrop for the colored lighting. Each strand of lights was controlled with a remote, allowing the participant to control the intensity and color of the lights. Participants were recruited from design studios within the School of Design + Construction at Washington State University. A total of 22 subjects participated in the experience. They spent five minutes alone in the room, using the remote to explore different lighting conditions. Afterward, they were asked a series of open-ended interview questions, and their responses were recorded, transcribed, and analyzed.

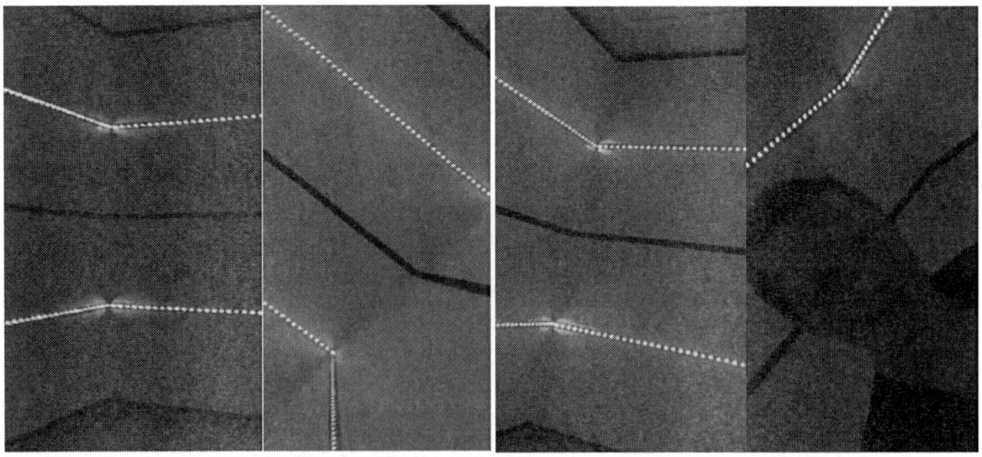

Fig. 4. Immersive Light Room Experience, developed and photographed by Jennifer Scott of the 22 participants, 100% claimed to enjoy the experience. The ability to change the color and intensity of the lights was cited by many as a positive aspect to study. When asked which color felt calming and which color felt stimulating, their answers varied dramatically. Some felt calm in a white room, since they are used to that color of light. Certain colors, such as red, elicited strong reactions – some positive and some negative.

Blue also had varying responses. This study informed that people do enjoy being bathed in colored light but need to be in control of the color and intensity. Also, nobody remained in the room for longer than five minutes, so it is undetermined if a longer exposure would lessen their enjoyment.

3 Application to a conceptual design project

The overall design objective is to utilize techniques for designing with colored light in the programming, wayfinding, and phenomenological experiences of a day spa – incorporating colored light as a tangible and interactive design element.

Three distinct design strategies for designing with colored light were identified in the literature review: first, the historical technique of designing with colored glass panels though which daylight or white artificial light is passes; second, the scientific technique of refracting light through a prism, resulting in the full spectrum of colors broken down into individual hues; and third, using light emitting diodes (LEDs), a modern technological technique. The first and second options offer the possibility of mutability as daylight shifts, though, each technique may be highly controlled if the light source is artificial, such as with LEDs. The design of this Day Spa incorporates all three techniques in strategic ways.

Historical: Used in the exterior glass atrium spaces, six in total and each with a different color (red, orange, yellow, green, blue, and purple). These spaces are lit with colored glass panels which illuminate the cavern-like interior space, providing a mutable softly-lit experience.

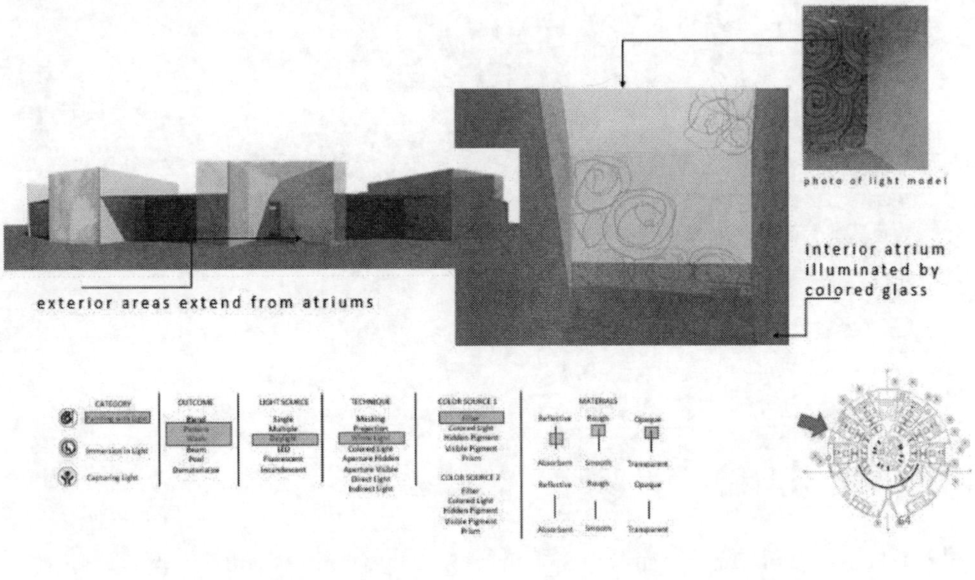

Fig. 5. Atrium Spaces of Conceptual Day Spa, developed by Jennifer Scott

Scientific: The lobby is filled with a forest of prismatic glass columns which refract daylight from the exterior glazing, and are also lit with white LED angled downlights, allowing for rainbows of light to paint the walls, floor, and visitors as they pass through the lobby to the main corridor. These rainbows will shift as the sun crosses the sky, and will stabilize as daylight ceases, and the lobby is lit purely with artificial light.

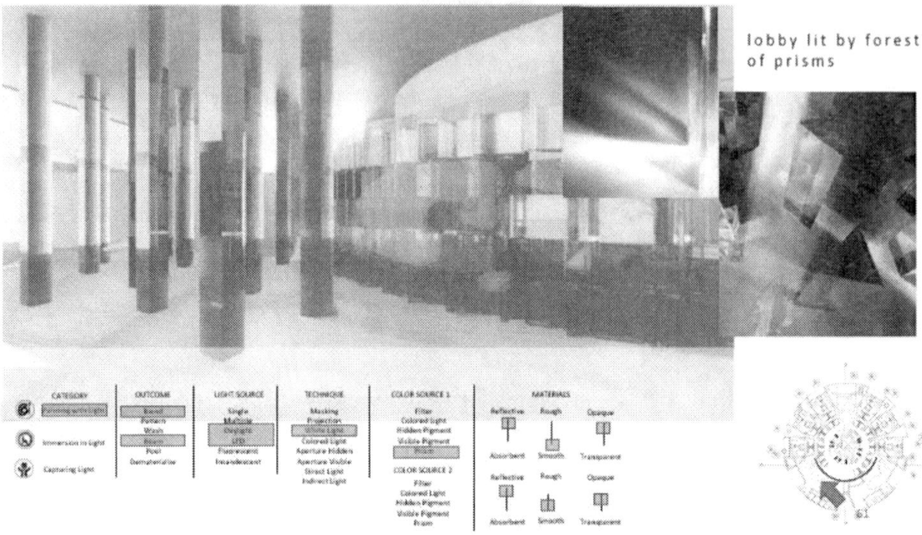

Fig. 6. Lobby of Conceptual Day Spa, developed by Jennifer Scott

Technological: LED lights provide stable, consistent color in areas which would not be possible with the first two methods: LEDs illuminate the underground stream which functions as a primary source of wayfinding. LEDs also fill the wall alcoves, allowing for controlled colored light along the corridor. LED fixtures also cast colored light which grazes the water at the entryway, and along the border between the changing rooms and the pool area.

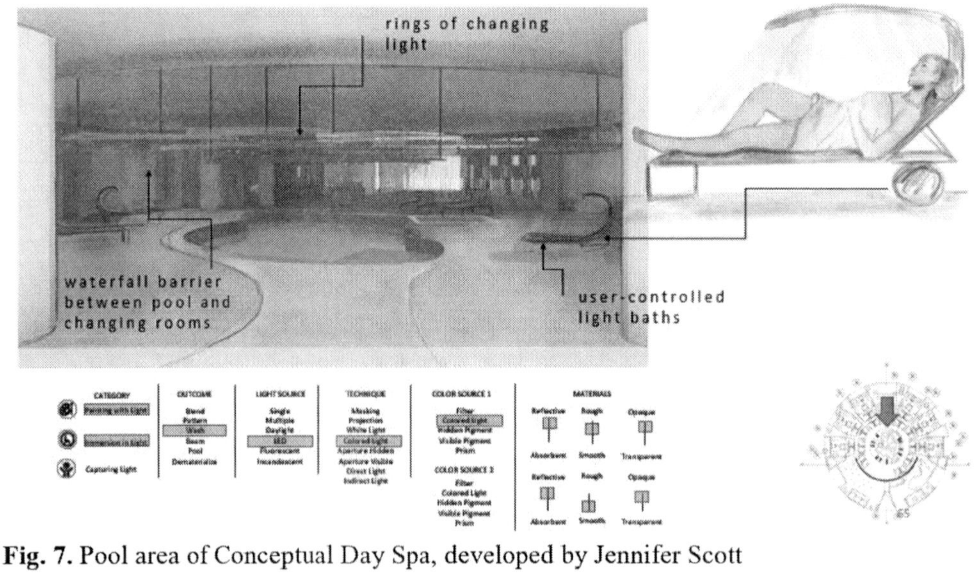

Fig. 7. Pool area of Conceptual Day Spa, developed by Jennifer Scott

4 Reflections

This project began with research into the effects of colored light on the human condition in order to better understand how this design tool engages our bodies, minds, and experience of place. A series of light model studies followed, which resulted in a taxonomy of design techniques from which to draw ideas for designing with colored light. A pilot study examining the psychological responses to being bathed in colored light in a user-controlled environment added to the body of knowledge in this project.

It is possible to imagine how one might feel when moving through this series of phenomenological experiences. It is not, however, possible to confirm that the spaces designed do in fact have physiological effects on the visitors. For this data, a more complete and controlled study would need to be conducted. We can infer, based on the research, that there would be some level of physiological response, but to what degree is uncertain. We can also assume, based on the research and the pilot study conducted in Research Phase 3, that the proposed design solutions would elicit psychological responses as well, and that the visitors would enjoy this more with a level of control over their environment.

As designers, it is important to understand the far-reaching impacts of our design choices. Since research indicates effects of colored light on so many different levels. This research aims to add to the body of knowledge in the field of color theory. It addresses color theory in relation to colored light, as opposed to applied color. While there is a substantial amount of empirical research regarding physiological effects of colored light, its psychological effects are not clearly known. This is a gap in the research that I lightly addressed in my pilot study, and plan to further address as a PhD thesis project.

References

1. S. T. Azeemi. M. Raza, A Critical Analysis of Chromotherapy and Its Scientific Evolution. Evidence-Based Complementary and Alternative Medicine,**2**(4), 481-488 (2005) doi:10.1093/ecam/neh137
2. T. D. Day, T. D., C. Rich, A Theoretical Model for Transforming the Design of Healing Spas: Color and Platonic Solids. HERD, **2**(3), 84-107 (2009) doi:10.1177/193758670900200307
3. A. Hankey, E. Ewing, New Light on Chromotherapy: Grakov's 'Virtual Scanning' System of Medical Assessment and Treatment. Evidence-Based Complementary and Alternative Medicine, **4**(2), 139144 (2007) doi:10.1093/ecam/nel060
4. F. M. Adams, C. E. Osgood, A Cross-Cultural Study of the Affective Meanings of Color. Journal of Cross-Cultural Psychology, **4**(2), 135-156 (1973) doi:10.1177/002202217300400201
5. R. D'Andrade, & M.Egan, The Colors of Emotion. American Ethnologist, **1**(1), 49-63 (1974) Retrieved from http://www.jstor.org/stable/643802
6. A. J. Elliot, M. A. Maier, Color Psychology: Effects of Perceiving Color on Psychological Functioning in Humans. Annual Review of Psychology, **65**(1), 95-120 (2014) doi:10.1146/annurevpsych-010213-115035
7. J. Eckstut, A. Eckstut, *The secret language of color: science, nature, history, culture, beauty of red, orange, yellow, green, blue & violet.* (New York, NY: Black Dog & Leventhal, 2013)
8. "Goethe's Theory of Colors," The Journal of Speculative Philosophy, **1**(1), 63-64 (1867) Retrieved from http://www.jstor.org/stable/25665594
9. P. K. Kaiser, Physiological response to color: A critical review. Color research & application, **9**(1), 29-36 (1984) doi:10.1002/col.5080090106
10. N. Kaya, H. H. Epps, Relationship between color and emotion: A study of college students. College Student Journal, **38**.3, 396-405 (2004)
11. P. Valdez, A. Mehrabian, Effects of color on emotions. *Journal of Experimental Psychology: General,* **123**(4), 394-409 (1994) Retrieved from https://ntserver1.wsulibs.wsu.edu:6117/docview/614327058? accountid=14902
12. D. Seamon, R. Mugerauer, *Dwelling, place, and environment.* (Dordrecht: Nijhoff, 1985)
13. J. Pallasmaa, *The eyes of the skin: architecture and the senses.* (Chichester: Wiley. 2004)
14. M. Paterson, More-than visual approaches to architecture. Vision, touch, technique. Social & Cultural Geography,**12**(3), 263-281, (2011) doi:10.1080/14649365.2011.564733

Experimenting intermediate lived spaces in residential subdivisions: Action, interaction and storytelling

Teresa Palmieri[1] and *Marion* Serre[2]

[1]PhD student, Faculteit Architectuur & Kunst - Faculty of Architecture & Arts, Hasselt University, Belgium
[2]Architect, PhD in architecture, Project[s] research laboratory – ENSA-Marseille, France

Abstract. Suburbanization and more precisely the development of residential subdivisions in Europe are at the core of many current debates. Over the years, they have been strongly criticized for the waste of land, increased infrastructure and network costs, car-dependency and the lack of urbanity. In this context, two processes are politically-driven in Europe: the densification of residential areas and the revitalization of the city centers. These political guidelines interrogate the transformation – and possible deterioration – of living environments and ambiances of residential subdivisions. Indeed, beyond the negative perspectives, other investigations have revealed the potentials of residential subdivisions from different points of view: ecology, social, architectural and urban. In order to contribute to these studies, this paper proposes to investigate the intermediate lived spaces in the specific context of residential subdivisions.

1 Introduction

Suburbanization and more precisely the increasing development of residential subdivisions in Europe are at the core of many current debates, on broad scientific and political levels. Over the years, they have been strongly criticized for the waste of land, increased infrastructure and network costs, the intensification of motorized transportation, individualism and the lack of urbanity ([1], [2]). In this context, two processes are politically-driven in Europe: the densification of residential areas ([3], [4]) and the revitalization of the city centers[1]. These political guidelines interrogate the transformation – and possible deterioration or decrease – of living environments and ambiances of residential subdivisions. Indeed, beyond the negative perspectives previously mentioned, other investigations have revealed the potentials of residential subdivisions from different points of view: ecological ([5]), social ([6], [7]), architectural and urban ([8]).

[1] For example: Seminar "Urban degrowth and land declines" ("décroissance urbaine et déprise foncière"), PACT Laboratory, Grenoble, April 2018.

In order to contribute to these studies, this paper proposes to investigate the intermediate lived spaces in the specific context of residential subdivisions. What are their forms, perceptions, social relations? What kind of interaction between public, private, individual and collective dimension do they reveal? Furthermore, what kind of interaction between institutional and residential strategies do they enlighten? What capabilities do they unveil? Could they trigger projects of transformation to emerge contextually enhancing citizen capabilities and help to develop more sustainable futures for these living environments[2]?

To provide answers to these questions, we will first define the concept of intermediate lived spaces by relying on founding theories and show how it enlightens living spaces and ambiances in residential subdivisions. Secondly, we will give some examples of existing intermediate lived spaces, describing 1) how they have emerged (which action?), 2) what kind of socio-spatial situation they create (what interaction?), and 3) what narratives and meanings they produce (which storytelling?). Hence, through experimenting *intermediate lived spaces* in residential subdivisions, we will mainly discuss the relation between two attractors of the concept of "ambiance": on the one hand, architecture and the city and on the other hand, uses and perception.

2 Intermediate lived spaces: between private, public, collective and individual dimensions

Over the years, studies in the field of lived spaces have explored various forms of "in-between", investing different concepts and notions, such as intermediate spaces ([9], [10]), third spaces ([11]), interval spaces ([12]), urban interstices ([13]), common spaces ([14]). Each of these concepts has been used to focus observations and reflections on certain characteristics. For example, third spaces defined by Soja (inspired by Lefèbvre and Foucault) are mobilized to reveal the interactions between "real and physical environment" (First space) and "imagined representation" (Second space). Another example, common spaces are defined by Stavrides as those that, by escaping the control of prevailing authorities and by distinguishing themselves from public and private spaces, renegotiate spatial rules and roles toward the creation of new forms of spatial sharing and common life. Based on these various definitions: **seeing residential subdivisions through the lenses of intermediate lived space could help us relate them to future-making processes ([15]) and the development of more sustainable alternatives for these environments.**

In design and ethnography research fields, Pink advances the definition of homes as ongoing projects that are being made and remade continuously by inhabitants materially and imaginatively: the home "is never completely realised materially, but instead exists partially in one's own imagination as a series of constantly developing dreams or plans" ([16]). Pink proposes the home as a possible entry point for future-making processes and design interventions that aim at developing more sustainable everyday living. However, as she explains, when researchers and practitioners engage with domestic environments as sites for change, they cannot avoid considering how people, "as everyday designers", are constantly transforming their homes through their material and imaginative engagement with everyday

[2] These questions are explored within the framework of the CAPA.CITY research program (JPI Europe 2017-2020), which questions the capabilities that need to be mobilized and associated in order to transform existing residential subdivisions. The originality of this approach lies in the interrogation of creating negotiation spaces between individual, collective and common interests, so that collective capabilities could be built between the different actors (inhabitants, professionals, institutions) implicated in the transformation of residential subdivisions.

environments. Looking together at perceived and imagined space can facilitate to understand this relation between the ways people live and imagine to live in their homes, and how it might already shape inhabitants' everyday lives and the materiality of their environments.

Like the home in Pink's definition, intermediate lived spaces of residential subdivisions can be attended to as 'projects'. By overcoming the strict separation between private and public spaces and individual and collective uses, which most commonly defines suburban monofunctional neighbourhoods, these spaces are the result of ongoing negotiation and imagination that can unveil possibilities for questioning and envisioning future alternatives for residential subdivisions in the present together with residents. In this perspective, we propose to define and identify them in residential subdivisions through the following model (Figure 1), in which:

- Public and private dimensions are related on the ownership status of the space: who does it belong to?
- Individual and collective dimensions are related on the uses and perception of the space: who can use it, who does actually use it, why and how?

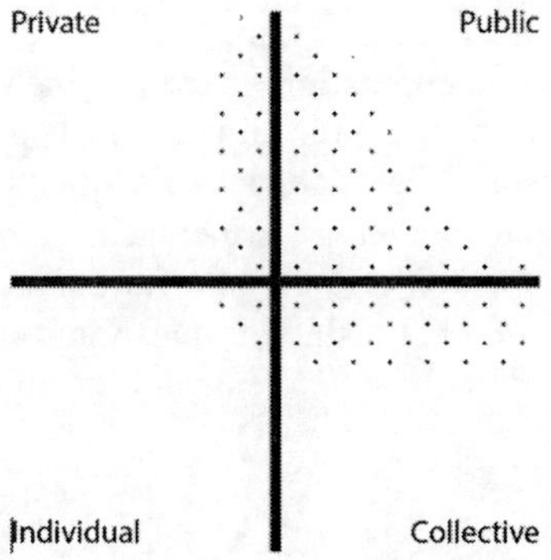

Fig. 1. Intermediate lived spaces analysis model © Marion Serre and Teresa Palmieri

This theoretical framework, put into perspective with the case studies, notably illustrates the negotiation between public and private spaces and between individual and collective uses, creating specific lived spaces and ambiances. Furthermore, it will reveal appropriation processes, strategies and tactics[3] developed by the inhabitants to adapt their living environment and ambiances to their needs.

3 Intermediate lived space as day-to-day city stages

The paper presents two case studies as part of the European Research project CAPA.CITY, one residential subdivision in Belgium (Lanaken, figure 2) and one in France (Aix-en-Provence, figure 3). They have been chosen as being typical suburban neighbourhoods, presenting common local challenges such as the lack of services and public spaces, the ageing of the population, space underuse, and the lack of connection with its surroundings. The

[3] A strategy refers to a set of coordinated actions, clever operations and maneuvers to achieve a specific goal. While a tactic refers to several actions carried out one by one, caught depending on opportunities (de Certeau, 1980).

residents were interviewed in their homes and were facilitated to explain and visualise how they live individually and collectively in the neighbourhood by using paper models of an average house of the area and/or maps.

Fig. 2. Residential subdivision in Lanaken

Lanaken is a small town in Flanders on the border with the Netherlands. The residential subdivision examined was developed by the municipality in the nineties as a reaction to the local trend that was seeing young people leaving the area. New plots were offered for affordable prices for young local citizens to buy and develop their own housing project. If Lanaken experienced a phase of population growth in the half of the last century with the development of housing projects, today the situation has changed and Lanaken has recently been confronted with a real-estate crisis due to the inability of selling large and old properties present on its territory and space underuse is increasing. To face local socio-economic and demographic trends, such as the aging of the population, lack of local economy, decreasing of household size, the municipality started questioning the possibilities to transform residential subdivisions into more diverse, green and liveable environments.

Fig. 3. Residential subdivision in Aix-en-Provence

In Aix-en-Provence, the residential subdivision was built in the 1960s. Recently, the municipal plan has been revised in order to increase the density of the outskirts already built. In this context, the district will have to support many urban transformations in a few years, such as (1) densification of residential subdivision villages such as La Clairnande, (2) urbanization of selected areas, especially old farmland, (3) creation of a new university with student accommodation, (4) creation of new infrastructures such as new streets and missing sideways. The openness of the rights to build has attracted several real estate developers to the residential subdivision. One of the residents tried to sell his house in 2017. As a consequence, the homeowners have filed a lawsuit against the city arguing for the inappropriateness of the allowed densification in the municipal plan claiming that their internal regulations could not allow such transformations. This example, in which, private interests conflicts with the public one, is very characteristic in France and is often used to block densification processes.

Both cases are currently facing future transformations, due to the upgrading of regulations and new planning strategies. In these contexts, the field surveys reveal that the inhabitants are not passive and that their actions on their own environment interrogate several characteristic stereotypes of residential subdivisions, such as homogeneity, individualism, car-dependency. Hence, by their differences and their similarities, these two cases can speak for European residential subdivisions, showing inhabitants' initiatives to adapt their environment. Among them, about fifteen intermediate lived spaces, created by the inhabitants, have been identified through the field surveys and classified in three categories:

1. **Porous limits**: private spaces transformed by collective and individual uses
2. **Transition spaces**: public spaces appropriated by collective uses
3. **Street interaction**: public and private interfaces used by individuals

These fifteen situations show a wide range of interactions between private, public, collective, individual dimensions and the capabilities of the inhabitants to modify their environments. To illustrate this, we present two examples for each of the mentioned categories.

3.1 Porous limits

We defined porous limits as the private spaces that are transformed by inhabitants together or individually to accommodate collective uses, confront individual and collective needs and that promote practices of sharing between the inhabitants of residential subdivision.

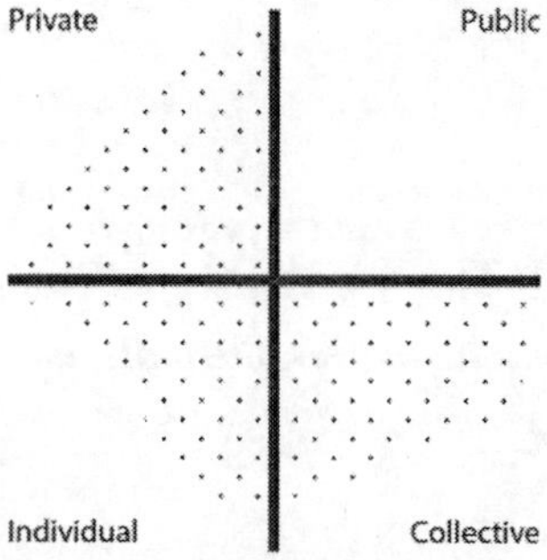

Fig. 4. Intermediate lived spaces analysis model: porous limit © Marion Serre and Teresa Palmieri

https://doi.org/10.1051/shsconf/20196401002

3.1.1 Sharing private gardens in Aix-en-Provence:

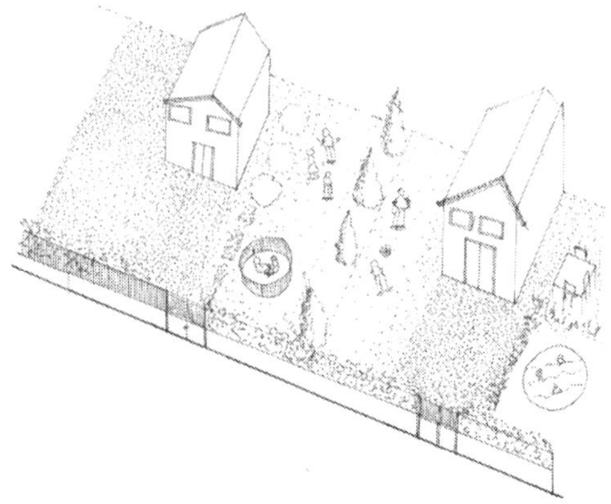

Fig. 5. Sharing private garden in Aix-en-Provence © Safa Ben Kheder

Action: Two families decided to remove the fence between their gardens. They kept a natural delimitation with existing trees and shrubs. This action reveals the capacity of the inhabitants to divert dwelling models (unlike Nordic or US models, fences are an integral part of the French model of suburban housing) and legal rules. Indeed, the internal regulations of their residential subdivisions have specific rules for fences:

"The fences separate the lots by a wall of one meter maximum, surmounted by an iron grating or any other device submitted for approval to the Departmental Service of Urbanism."

Interaction: Due to this transformation, these two families share a swimming pool, a trampoline, but also guest rooms, kitchen and gardening tools. The ambiance of the residential subdivision also changes by introducing porosities in generally closed spaces. This part of the residential subdivision is closer to the model of garden city[4] than the model of sector urban planning ([17]).

Storytelling:
"*There is something important in our lives, for our two households: we have opened our two gardens for a very long time. The gardens are...more than adjoining. They are open and we have lots of things in common: the mower, the tools, and the swimming pool!"*

[4] A utopian model which has strongly influenced the creation of residential subdivisions.

3.1.2. Sharing private basement in Lanaken

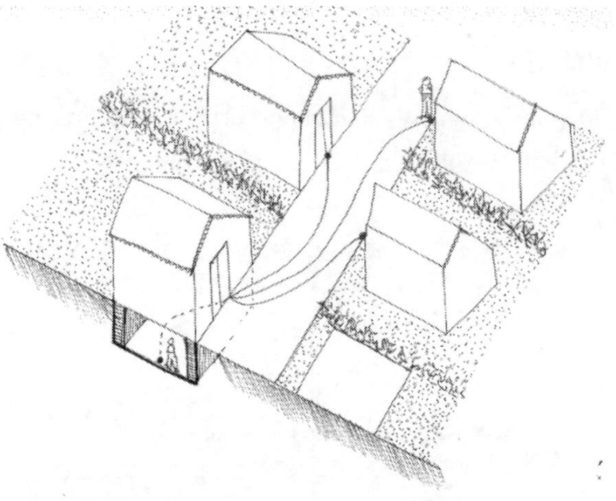

Fig. 6. Sharing private basement in Lanaken © Safa Ben Kheder

Action: After a change in the family situation, a resident started living alone in a large house. As a result of the change of the household size, the inhabitant was left with a lot of underused space in the house, especially the basement. The resident offered to share the basement with larger families in the neighbourhood in need of extra space to temporarily store things. This action showed the capacity of the inhabitant to modify the very strict plot-by-plot development of the neighbourhood that favour private life over a more collective dimension.

Interaction: As a result of this transformation, some families in the neighbourhood are using a private space for a common purpose. This way, the ambiance of the neighbourhood also changes and introduces the possibility for collective use not only in public spaces but also in private ones.

Storytelling: *"Under the house there is a large basement. It is really a lot of space. This space is now a storage for people, for my brother but also for other neighbours who need it. Somebody for instance bought a couch and needed space where to store the old one, so I told them: no problem, you can put it here. And if someone needs space they can use mine."*

3.1.3 Conclusion

As the comparison shows, porous limits grow from private initiatives and negotiations to transform private spaces to make a better use of them, as shown in the examples, by overcoming the underuse of space or by sharing facilities. They entail a reconfiguration of the space (e.g. elimination of the fences) and/or a rearrangement of its rules (who can use the private space, how, for how long etc.). As these intermediate lived spaces happen in inhabitants' private property, they are facilitated both by social relations as by spatial proximity (e.g. close neighbours) and necessary involve the building of trust between actors. These spaces are subjected to a constant negotiation between people. Therefore, they are temporary and last as long as inhabitants agree. These intermediate lived spaces highlight the opportunities to develop retrofitting strategies that are based on the capacities of inhabitants to negotiate and deviate from dwelling models.

3.2 Transition spaces

Transition spaces are public spaces that are appropriated over time by collective uses. These spaces are transformed and adapted by the inhabitants in order to facilitate their mobility.

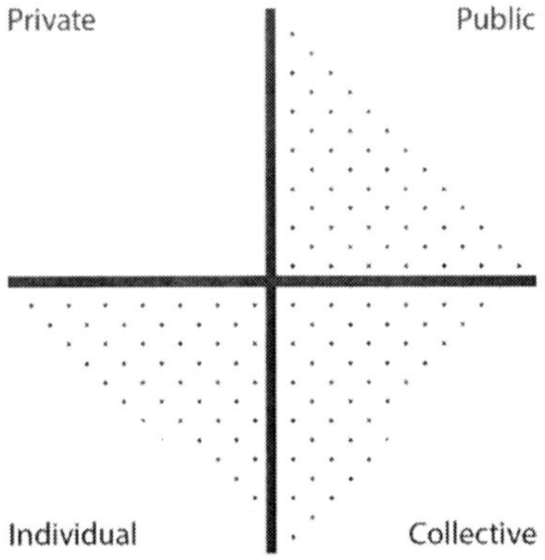

Fig. 7. Intermediate lived spaces analysis model: transition spaces © Marion Serre and Teresa Palmieri

3.2.1. Collective paths for alternative mobility in Lanaken

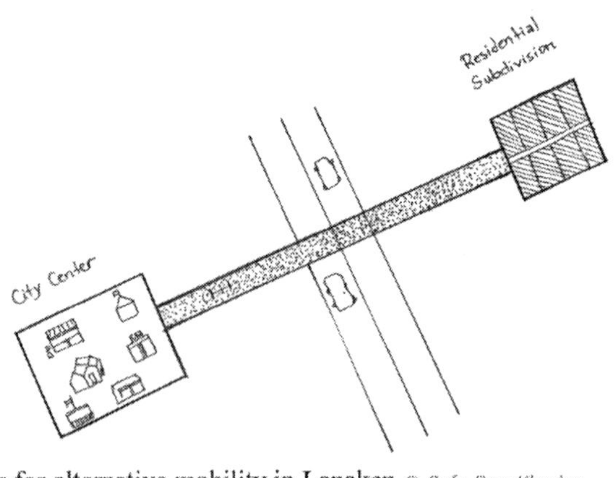

Fig. 8. Collective paths for alternative mobility in Lanaken © Safa Ben Kheder

Action: Shortly after the development of the residential subdivision, a large regional double lane road was set up near the neighborhood. To restore the fast connection to the center on foot and by bike, a number of inhabitants reclaimed a part of a residual space in the neighbourhood to build their own shortcut. In this way, they demonstrated their collective capacities to organise collectively to react to macro-scale changes locally and to transform their environment according to their needs at the micro-scale.

Interaction: The shortcut creates interactions on different levels. It creates a new interaction between residents and residual public spaces, which acquire new meanings and functions as a result of the collective action of the inhabitants. For example, they become spaces for inhabitants' collective projects and spaces which connect instead of separating. Furthermore, the action triggers a different interaction between the residential subdivision and other close areas (e.g., other neighbourhoods, the town center, etc.). Due to the implementation of the shortcut they become reachable by foot and by bike. Finally, the collective creation of the slow road creates a possible different ambiance for the residential subdivision, not only a car-oriented area, but one where other mobility systems are possible.

Storytelling:
"From this part of the neighbourhood it was not possible to quickly go to the center by bike. People made their own road. Although it is not supported by the municipality, people use it anyway. There should be a better connection between the neighbourhood and the center of the town and facilities, for instance by improving the bike mobility".

3.2.2. Collective stairs to improve pedestrian mobility in Aix-en-Provence

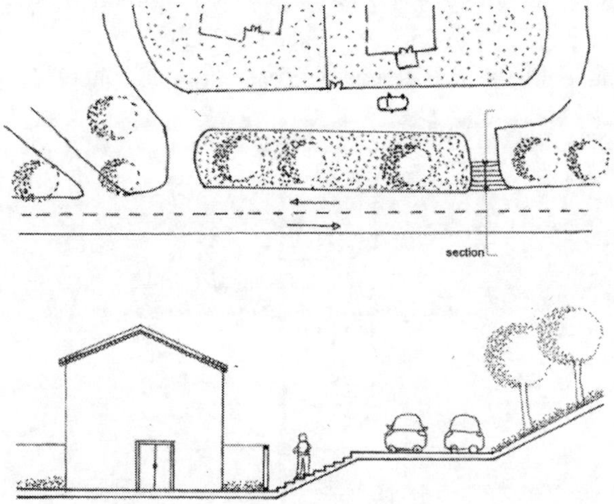

Fig. 9. Collective stairs to improve pedestrian mobility in Aix-en-Provence © Safa Ben Kheder

Action: As in many residential subdivisions, public spaces have been conceived for cars. Sidewalks have not been planned and this increases the car-dependency of the inhabitants. Starting from this, the inhabitants of the residential subdivision decided twenty years ago to build these collective stairs.

Interaction: Today, the stairs still work. They are used by all inhabitants: to go shopping in the neighbourhood, for a daily walk or for jogging. Although the stairs have improved pedestrian mobility, residents still face a major problem: there is no sidewalk to get from the stairs to the other neighbourhood.

Storytelling: *"If we wanted to go to the other neighbourhood before the stairs, we had to go around to go up and down to the shops. So, we created this staircase to make it easier for us."*

3.2.3 Conclusion

Residential subdivisions are most commonly characterised by car-dependency and lack of diversified mobility possibilities. This contributes to the unsustainability of these suburban neighbourhoods, not only from an economic and ecological point of view (e.g. high cost of infrastructures, high-energy demand, pollution, etc.), but also from a societal perspective. For example the increasing ageing residents cannot always rely on cars to move to perform their daily activities. The cases show the capacities of residents to act collectively to diversify the mobility system. They reclaim part of the car-oriented public space and residual public spaces to build slow roads, developing alternative mobility systems that facilitate residents to reach close facilities by bike or on foot. Hence, investigating 'transition spaces' in residential subdivisions can contribute to unveil possibilities for retrofitting strategies that consider slow roads as integral part of a more diversified and less car-oriented mobility system, contributing to transform them into more sustainable urban environments.

3.3 Street interactions

We define street interactions as private spaces that are directly in relation with the street and that change the function of the private space (by the opening of a shop for instance) or the atmosphere of the public space (defensible space or open space).

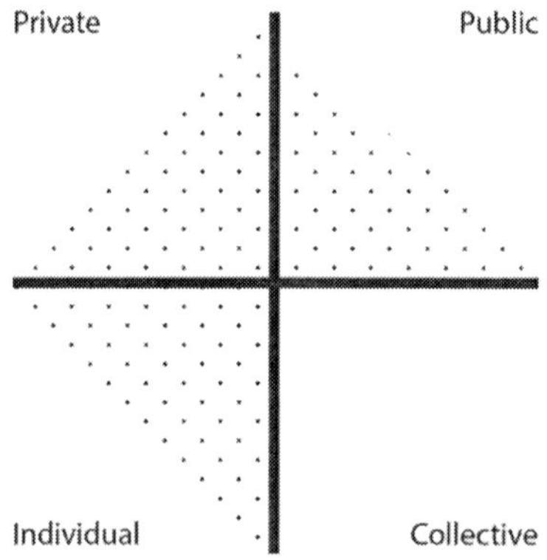

Fig. 10. Intermediate lived spaces analysis model: street interactions © Marion Serre and Teresa Palmieri

3.3.1 Neighbourhood pop-up shop in a private garage in Lanaken

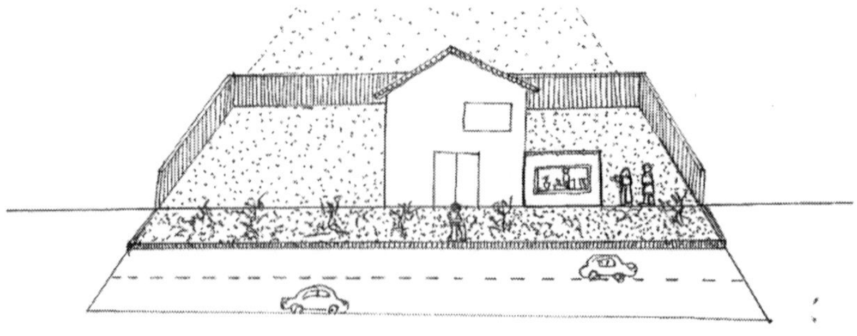

Fig. 11. Neighbourhood pop-up shop in a private garage in Lanaken © Safa Ben Kheder

Action: One of the inhabitants had transformed his garage into a small neighbourhood bakery open to the public, selling bread and other small goods.

Interaction: This transformation temporarily created a new ambiance for the residential subdivision as a more diversified environment, where functions, other than the residential one, were possible.

Storytelling: *"The bakery only lasted one year because for the municipality, these activities are not allowed. It was in a resident's garage and it was great because if I wanted some bread or a drink I didn't have to go to the center, but for some inhabitants the bakery was creating too much traffic."*

3.3.2 Open entrance gates in Aix-en-Provence

Fig. 12. Open entrance gates in Aix-en-Provence © Safa Ben Kheder

Action: In this residential subdivision, many people leave their entrance gates open, because it facilitates their mobility in it. Here, they feel safe and trust their neighbours.

Interaction: This spatial practice creates a special atmosphere less defined by closure and withdrawal than in other residential subdivisions.

Storytelling: *"I usually leave my entrance gate open. I have noticed that those who do not do so are more likely to park in front of their homes and not in their own garden. And cars in public space are a problem, both aesthetic and practical for the pedestrians."*

3.3.3 Conclusion

Residential subdivisions are most often characterised by the clear separation between private and public spaces and by mono-functionality, and, besides a few exceptions, residential subdivisions are made of single-family detached houses that cannot host shops or other facilities. This contributes to residential subdivisions being closed environments. The examples have shown the capabilities of residents to overcome the clear separation between private and public spaces in residential subdivisions by intervening in their private spaces to modify their interaction with the street. By so doing, they create new and more open atmospheres for these environments that may lead also to new functions, such as in the example of the pop-up shop in Lanaken. If on one side residential subdivisions are today often criticised for their mono-functionality, on the other side residents are often reluctant to discuss the reconfiguration of the separation between private and public spaces as perceived to disrupt the quiet and private lifestyle that is most common and appreciated in these residential areas. By relying on people's existing capabilities, street interactions of residential subdivisions can open the dialogue toward retrofitting strategies able to transform residential subdivisions into more open environments.

Conclusion

A proper investigation of intermediate lived spaces not only from the perspective of their spatial and physical configuration but by looking also at how they are conceived, produced, agreed upon and maintained, can enable a deeper understanding of residential subdivisions in their complexity against too simplistic and stereotypical definitions of residential subdivisions as homogeneous environments characterised by individualism. Intermediate lived spaces recognize inhabitants as key actors in processes of retrofitting who themselves constantly reconfigure and improve their neighbourhoods to make it more sustainable for themselves. Inhabitants are thus co-producers of their dwelling spaces that, through their capacities, might transform the ambiances of residential subdivisions. Intermediate lived spaces can become spaces for collective learning about and in urban environments especially because, as hybrid spaces, between private and public, individual and collective spaces, force us to reconsider the binary modes through which space is often perceived, described and designed. In this perspective, intermediate lived spaces invite thinking differently the forms, the ambiances and the meanings of spatiality and interrogate the roles of actors (i.e. inhabitants, local authorities, and spatial practitioners) in retrofitting processes.

The paper has presented a comparison between two case studies, one residential subdivision in France and one in Belgium. The analysis model (Figure 1) has been advanced as a tool to facilitate the comparison of contextual spaces and spatial practices by focusing on the interaction between private and public spaces and individual and collective uses that they present. The model therefore facilitates the emergence of different types of intermediate lived spaces. In this study, we describe three types of intermediate lived spaces we found, namely, porous limits, transition spaces and street interaction. By providing different interactions between private and public spaces and individual and collective uses, they reveal inhabitants capacities of reflecting and acting upon their built up environment and renegotiate spatial relations individually and collectively to improve their neighbourhood according to their needs and values. For instance, the porous limits analysed show the capacity of people to renegotiate private spaces by introducing collective uses and confronting in this way several challenges of residential subdivisions such as space underuse or the high resources waste by facilitating sharing practices. Further, the transition spaces described show the capacity of people to act collectively to re-appropriate the public space for diversify it for instance to develop alternative mobility systems. Finally, street interaction shows the capacities of people to question individualism and mono-functionality of residential subdivisions to develop more diversified, connected and open environments.

As shown in the case studies, participatory forms of urban planning and design can use the provided analysis model not only for having a deeper understanding of residential subdivisions and make more informed and situated retrofitting proposals, but, especially, for revealing the existing capacities of people to transform the spaces of residential subdivisions and renegotiate their purposes (i.e. private, public) and uses (i.e. individual and collective). We therefore propose intermediate lived spaces as possible entry points in processes of participatory urban planning and future-making to help to recognise citizens as full participants in the development of urban environments and enhance their engagement in transformation processes such as the retrofitting of residential subdivisions.

References

1. P. Newman, J.R. Kenworthy, *Sustainability and Cities. Overcoming Automobile Dependence*, Washington: Island Press (1999)
2. D. Pinson, *Le pavillon n'est pas la maison*, in SociologieS, Toulouse: Association internationale des sociologues de langue française (2017)
3. A. Touati, J. Crozy, *La densification résidentielle au service du renouvellement urbain: filières, stratégies et outil.* Paris: La documentation française (2015)
4. White paper for spatial policy plan Flanders, 2016
5. P. Frileux, *Le bocage pavillonnaire, une ethnologie de la haie.* Grane: Créaphis (2013)
6. R. Dodier, *Habiter les espaces périurbains.* Rennes: PU (2012)
7. A. Lambert, *Tous propriétaires! L'envers du décor pavillonnaire*, Paris: Le Seuil (2015)
8. F. Bonnet, *Aménager les territoires périurbains et ruraux.* Rapport remis au ministère du Logement, de l'Egalité des territoires et de la Ruralité (2016)
9. C. Moley, *Les abords du chez-soi.* Paris: La Vilette (2006)
10. A. Flamand, *L'invention des espaces intermédiaires.* Thèse: Paris Est (2008)
11. E. Soja, *Thirdspace: Journey to Los Angeles and Other Real- and-Imagined Places*, Oxford: Blackwell (1996)
12. L. Roulleau-Berger, *La ville intervalle. Jeunes entre centre et banlieue.* Paris: ed. Meridiens Klincksieck (1991)
13. S. Tonnelat, *Interstices urbains. Entre contrôles et mobilités, quatre espaces résiduels de l'aménagement.* Thèse: Université Paris XII et City University of New York (2003)
14. S. Stavrides, *Common Space: The City as Commons.* London: Wed Books (2016)
15. P. Ehn, E. M. Nilsson, R. Topgaard, *Making Futures.* The MIT Press, pp.1-13 (2014)
16. S. Pink, *Making Homes*, UK: Bloomsbury Publishing (2017)
17. D. Mangin, *La ville franchisée*, Paris: La Vilette (2004)
18. M. Elbakidze, L. Dawson, K. Andersson, R. Axelsson, P. Angelstam, I. Stjerniquist, S. Teitelbaum, P. Schlyer, C. Thellbro, *Is spatial planning a collaborative learning process? A case study from a rural-urban gradient in Sweden.* Land Use Policy, **48**, 270-285 (2015)

Augmenting Psychological Restorativeness towards Immersive Design of Underwater Ambiance

Nermin Sena Ozger[1,*], *Gözde Damla* Turhan[2], and *Guzden* Varinlioglu[2]

[1]Izmir University of Economics, Department of Visual Communication Design, 35330 Izmir, Turkey
[2]Izmir University of Economics, Department of Architecture, 35330 Izmir, Turkey

Abstract. Discussions of the relationship between humanity and built environment usually focus on the human's relationships with the earth's surface. Less attention has been given to underwater environments, a relatively underexplored research area at the intersection of user experience design and psychology. This paper shows the initial findings of experiments on divers, which can guide the design of augmented reality tools for underwater environment. The comparative evaluation process demonstrates an increase in psychological restorativeness of users in relation to ambiance design, indicating a difference between a specifically designed environment and a naturally vacant area for first time divers. Future studies will explore immersive design of underwater ambiance through augmented reality tools, proposed as a product design and framework, in order to improve psychological restorativeness.

* Corresponding author: nermin.ozger@iue.edu.tr

1 Introduction

During the course of history, humanity has intuitively explored different environments for living. Underwater habitats are environments for which today's technology has potential to extend the ability of humanity to adapt itself to extreme conditions. Although there are various studies, ambiance of underwater architecture and its effects on users are still areas or exploration through alternative methods, due to the complexity involved in such environments. This research focuses on a comparative experiment that measures psychological restorativeness (PR) in relation to underwater ambiance design, and displays a number of findings for furthering the research towards designing a product equipped with augmented reality tools.

The attempts in the literature show that the studies allow users to experience the immersive environment using virtuality tools by recreating a virtual environment in a 3D platform. However, these completely altered ambiance experiences are considerably different from actual environment itself. In contrast to these findings in the literature, we propose to use the real environment with enhanced its immersiveness through the deployment of 3D artifacts and, an augmented reality (AR) embedded diving mask.

1.1 Definition of Terms

Over the last two decades, many studies have shown that restorative environments help replenish psychological well-being, promote recovery from mental fatigue [1], and have beneficial effects on attention restoration [1,2,3]. The interaction between the built environment and psychological restoration was described by Kaplan and Kaplan in 1989 within the context of ART. Attention restoration theory (ART) builds on assumptions about the evolution of human cognitive capabilities in natural environments in relation to the concept of presence. Psychologically, there are three aspects of presence; the user's response to an immersive system; the sense of being there and the extent to which the virtual environment takes precedence over the real one; and finally, the way users refer to their experience as having been to a place [4]. Furthermore, the level of fascination, and stimulation of the cognitive skills are also related to the user's experience, and the ambiance [5]. Therefore, the design of immersive environments can improve the user experience in terms of psychological restorativeness [1].

Facilities that enable visitors to recover from everyday stress, so-called restorative environments, are those where personal adaptive resources and ability to focus are renewed [6]. It is stated that highly populated urban centers create stress on human psychological and physical resources [1]. On the other hand, psychological restoration is supported by natural environments [1,7]; in particular, riversides and seashores have higher levels of impact in comparison with urban spaces, and even green areas [1]. However, in most the cases, the environmental influences of underwater on human psychology are understudied, and therefore, the recreational element is the specific focus of this study.

Diving is a profound physical activity that provides a physically and mentally immersive user experience, and potentially contributes to the improvement of psychological restorativeness. In terms of physical impacts, it can take place in natural reefs, as well as artificial environments created by deploying ships, planes, and other large-scale structures. These wrecks become touristic attractions, but, as seen in the Mediterranean and the Red Sea, artificial reefs can damage the sensitive ecosystems when metal wrecks are inappropriately deployed.

On the other hand, as an immersive embodied experience, diving mobilizes a complex of human senses that are tactile, audial, odoral, taste, and visual. Even though these senses interact to create material engagement and feeling positionality in space, visual sense is the focus of this paper, since the use of other senses is limited due to water having a denser material characteristic than the air surrounding us. Although visual senses are also distorted by making objects seem 33% bigger and closer [8], diving masks overcome this problem. The distortion in audio is a greater problem because there is no satisfactory equipment designed to minimize the effects of sound, traveling four times faster underwater than on the surface, making it difficult to detect the direction and the decibel level of the sound. Tactile senses present similar challenges, as water speeds up heat-loss by 25 times compared to air.

To sum up, underwater diving is an activity in which human senses are stimulated and psychological restorativeness is provided by nature, so that human abilities such as learning, paying attention increase. These features are the basis for re-creating an underwater experience with augmented reality tools for divers.

1.2 History of underwater explorations

In order to understand the possible scenarios for underwater habitations, it is important to review the pioneers in the field. In the 15th century, legend has it that Alexander the Great ordered a glass barrel to be dropped into the sea. The first records of such efforts in Greece date back to 1538 [9]. These so-called "diving bells" were lowered beneath the sea while the pressure was equalized by creating a vacuum with the air inside. This principle allowed divers to breathe under the surface.

An alternative diving system was not designed until 1828, when John Deane and his brother modified a suit made for firefighters to enter smoke-filled buildings. The new suit consisted of a copper helmet big enough to completely cover a diver's head, attached to a cloth body suit to protect from cold water, and weighted shoes to counteract the positive buoyancy created by the air in the suit. The air was supplied through a pump on the surface, and the helmet was not sealed, so it flooded when the head was turned at an angle. Later, Augustus Seibe developed the diving suit by connecting the helmet via seal to prevent it from flooding [10]. Without the technology to compress air into a container (and on-demand regulator as the fundamental equipment) at this time, it was impossible to develop a self-contained underwater breathing apparatus (scuba). It was Cousteau, the first scuba diver, who first used independent air supply with the invention of the aqualung regulator, freeing him from the need to be connected to those above the surface.

All of these early attempts led humanity to remain under the sea for longer periods while observing the marine life. In general, the explorations underwater started with attempts to create a space through glass barrels and diving suits, and eventually developed into scuba. The development can be reviewed from spatial designs to product design. Nevertheless, diving equipment has not been sufficiently integrated with digital technologies for the purpose of enhancing the psychological restoration. Therefore, an experiment was conducted to explore an alternate immersive underwater experience.

2 Experiment on the relationship between PR and design of immersive underwater ambiance

It is observed that functioning of daily routines typically demands our capacity to direct attention, and that this capacity may weaken over time, causing direct attention fatigue. This, in turn, results in negative emotions, irritability or decreased sensitivity to interpersonal issues [1]. This depletion of directed attention can be restored by resting in such environments where this mental fatigue can be reduced [7], although the occurrence and intensity of restorative experiences is unpredictable [6].

A literature research revealed three types of restorative impacts on humans: Emotional, physiological, and cognitive impacts evaluated by the psychological restorativeness scales that measure the relationship between diver (user) and underwater ambiance design [1,5], and by the Stroop Effect test to measure the cognitive abilities before and after the activity. According to Pasini et al [11], restorative environments encompass a wide range of settings from wilderness to indoors, and a variety of different scales. In the literature, there are four common identified properties of restorative experience settings: Being away, extent, fascination, and compatibility characteristics of both the imagined/mental domain and the actual/physical domain [5]. When developing the Psychological Restorativeness scale (PRs), two components of a restorative experience were consistently identified as present in each analysis: "being away" and "fascination", and these are the core features of the ART [7].

In relation to these two components of psychological restorativeness scales, recreational diving is defined as a recreational activity to which ART applies, since it provides sense of being away and fascination in terms of both mental and physical domains. At this point, the integration of immersive digital media to designed underwater ambiance can potentially enhance the user experience while decreasing the environmental concerns. Immersive media, such as augmented reality (AR) tools, are being widely explored in relation to user experience. In discussing AR tools, the visual aspect of the interaction with the imposition of three-dimensional computer graphics is often central to discussions; however, any media within a synthetic space, provided by, for instance, sound or haptic features, can increase the users' sense of reality for various explorations [12].

Digitization of the ambiance of the built-up environment has a potential to provide alternative solutions to a wider spread of knowledge and experience. Additionally, the use of immersive reality tools could make the experience more appealing and innovative. Through this research, users will be able to experience a completely innovative underwater ambiance without the harm to the built-up environment caused by current applications.

2.1 Methodology

This study starts by proving how psychological restorativeness (PR) is enhanced through an immersive underwater ambiance design. Designed space has been shown to improve PR more compared to the naturally vacant space. Furthermore, it is known that immersive media improves underwater user experience [13]. These results led us to explore alternative of digital applications, i.e. augmented reality (AR) to experience underwater ambiance, while reducing the need for physical and artificial attraction points. For this, an AR embedded diving mask design is envisioned. For acquisition and dissemination of knowledge, augmented reality (AR) are often used for making physical content digitally accessible, especially when the actual content is rather inaccessible. However, in this study, the concern was over issues of physical impacts rather than accessibility, and users' psychological responses were studied to gain insights into the outcomes of an immersive underwater experience.

Research on restorative environments shows the precise outcomes for experience of nature, generally by comparing an attractive natural ambiance to the less attractive built environment [14]. In parallel, this study compares the psychological effects of a built underwater ambiance and vacant underwater ambiance with the aim of identifying whether differences in design have a critical role in the level of psychological restorativeness in terms of ambiance experience. Moreover, psychological tests are critical to measuring the level of changes in psychological restorativeness, but these have rarely been conducted within the framework of user experience design, especially for underwater studies.

In order to promote and integrate psychological restorativeness (PR) as one of the criteria in the design of underwater ambiance, the PR and Stroop Effect tests were conducted with 30 divers in shallow waters. PR scales were designed according to studies by Han and Hartig [1,5]. The cognitive response durations were measured by the Stroop Effect tests. The initial results have shown that users diving for the first time in an artificially created built environment showed a greater amount of psychological restorativeness than those in a naturally vacant area. Moreover, through The Stroop Effect tests, an increase in participants' cognitive abilities was recorded.

To elaborate, in the Izmir/Karaburun area, August 2018, the corresponding author designed the dive site, deploying clay-made archaeological replicas (i.e. columns and amphoras) at shallow levels (between 5-7 mt depth) on the seafloor, composed of sand, seagrass, and rocks (Fig.1). Variables such as the users' physical features, e.g. weight, were not taken into consideration. The environmental conditions where the tests conducted were pegged. Each trial dive lasted for 20 minutes (-/+3 mins), and the average water temperature was 21°C.

The ambiance design was planned according to the trial dive route underwater. The artifacts were carefully placed in three groups; first group consists of a small group of amphoras as a starting attraction point; second part is a group of amphoras at a deeper level, and third group involves columns, amphoras, and other clay artifacts, as the final stop for the trial divers.

For the pre- and post-dive, participants were asked to indicate the colors of rectangles, to ensure that they could clearly see the colors. Second, they were asked to read aloud the words that were written in a different color from the word they signified, e.g. the word "red" written in blue. Finally, they were asked to identify the color each word was written in, ignoring the word itself, for instance, the word "red" written in blue, should be identified as blue. The time taken for reading and the number of errors were recorded for each phase. After the trial dive, divers evaluated their experience by the self-rating psychological restorativeness scales.

Fig. 1. Deployed columns and amphoras, Cagri Guven (2018).

2.2 Findings

The experiment showed a direct relationship between the ambiance design and psychological restorativeness. According to the Stroop Effect test results, Group A, as experiment group, shows greater progress than Group B, as the control group, in all three phases. In indicating the colors of rectangles, the average response time fell by 14% for Group A, and 8.50% for Group B after the trial dive. The improvement in recorded time average for reading words is 11.70% for Group A, and 6% for Group B. The improvement for indicating the colors of words for Group A is 13.32%, and 10.22% for Group B.

The self-rating psychological tests were conducted post-dive. They aim to obtain an insight about the divers' psychological and physical reaction to the underwater. The results of the post-dive self-rating psychological tests state that the divers were more likely to be curious for further exploration of underwater, especially when they noticed that the ambiance was specifically designed. Moreover, the test group indicated that they would spend more time underwater to explore the area after they started to be familiarized with their surroundings. This proven point can be also interpreted as the desire to be around artifacts they know of.

These findings are complementary to the hypothesis that the design has a critical role in the level of psychological restorativeness in terms of underwater ambiance experience. Diving as a physical activity has an effect on the users, however, the ambiance design has a greater impact on the users' psychological restorativeness in terms of providing a challenging immersive experience.

Although artificial environments created by deploying ships, planes, and other large scale structures are attractive to many divers, an integration of augmented reality (AR) technologies can further facilitate this experience, along with a virtual ambiance design. In order to understand the potential for AR to enhance PR within an underwater ambiance design, an AR-embedded mask design is proposed.

3 Proposal for an immersive underwater ambiance

For underwater ambiance, having contact with water elements is essential. In this sense, the "immersion" into the water helps us to dissolve categorizations of water. The very essential activities of swimming and diving, as the very basic forms of immersion, allow the body to interact with the environment. Thus, we did not try to duplicate the underwater environment, as the immersion is not easily reproduced and transformed into the virtual world [16].

We consider augmenting the real environment, rather than re-creating a virtual one by adding a visual concept. As indicated before, diving uses all the senses, but our focus has been on the visual effects. Thus a diving mask (Fig.2), which is used for a clear vision underwater, will be designed using augmented reality (AR) technologies. The mask will enable divers to experience the ambiance through AR and to see both the real environment and the embedded 3D artifacts.

In order to explain the design of a 3D immersive environment, an insight from the existing literature is briefly reviewed. The approaches might vary from non-realistic re-composition of space as in the case of Char Davies' VR work from 1995, to recent mobile apps, i.e. BBC Civilizations. The level of reality/virtuality depends on the equipment and embodiment. In her work entitled Osmose, Davies [17] explored recreation of underwater ambiance from a fictional viewpoint. However, interaction of the user with the virtual environment copies the movements of a diver. The user's movement in virtuality is through her bodily movements, and breathing rhythm allows the user to move up and down. Thus, the ambiance and movements of the underwater environments are recreated.

Another study exemplifies the use of visual graphics and navigation in order to experience the underwater ambiance. The system works through kinesthetic sensors that enable divers to interact with the underwater space through their own motions. Their aim is to recreate the sensations and physical requirements of scuba through design, without intruding into the space [13].

 https://doi.org/10.1051/shsconf/20196401003

In our design proposal, we do not attempt to recreate a virtual underwater environment and movements, but, instead, the ambiance will be augmented through a transparent glass visor/screen, without changing the overall diving experience. The diving mask, a mandatory equipment for the divers, will also have additional info-graphics displayed in the screen while the diver/user experiences the actual underwater environment. We envision AR-embedded diving mask that will allow divers to see the objects through markers, thus improving the underwater space experience and psychological restorativeness.

Fig.2 Conceptual AR-embedded diving mask design to experience underwater ambiance showing info-graphics, by the corresponding author (2018).

Conclusion and future research

The genesis of this study is to create a link between ambiance design, and user's psychological restoration from a designerly perspective. Once the relation is established between the ambiance as an experience, and psychology, the inclusion of the innovative technologies of AR will potentially improve the potential for design to play a role in psychological restorativeness. The self-rating psychological restorativeness tests are to measure the improvement, if any, caused by the immersive ambiance experience. The tests are the first step to understanding the relation between three key concepts. User experience, ambiance design, and psychological restorativeness.

During the test phase, the goal was to enhance psychological restorativeness through an immersive underwater ambiance design. From a designerly perspective, the influence of underwater ambiances on the user's psychological restorativeness was measured, taking into account environmental concerns. Regarding underwater ambiance design, the findings showed a direct correlation between the built environment and psychological restorativeness, and cognitive abilities. More research is needed to validate the changes in restorativeness, and to create a greater level of interaction between the user and objects. The immersive characteristics of AR provide attention restoration through providing sense of being away, thus, digital technology can be integrated as a tool to meet the challenge of providing an immersive underwater ambiance.

The Restorativeness scales and Stroop Effect tests proved that the immersive design has a positive impact on user's psychological restorativeness. Additionally, even though the structure of immersive design has been shown to have the power to affect the outcomes, there is a need to study the input of the virtual elements such as AR.

Future studies have the potential to enhance the understanding of immersive ambiance design and the psychological restorativeness. The results of the tests will shape the virtual immersive ambiance design and help to eliminate exaggerated physical interference in the underwater environments. Although AR can never replace reality, it can be integrated or overlapped into the real world experience through design; as such, it has a role to play in improving psychological restorativeness and the quality of the experience through the design of underwater ambiance.

We would like to thank Efes Diving School at Karaburun and the diving instructors Cengiz Arslan, Erkan Ayvazoglu and Burak Bambul for their support in deployments; Taner Yilmaz for supplying the clay artifacts; Asst. Prof. Dr. Burak Erdeniz for his academic contributions in Psychology; Cagri Guven for underwater photography; and the significant part of this research, the participants.

References

1. T. Hartig, K. G. Korpela, W. Evans and T. Gärling, A measure of restorative quality in environments, Scandinavian Housing and Planning Research, **14**:4 (1997)

2. C.A. Lewis, Gardening as a healing process, in M. Francis, and R.T.J Hester (ed), The Meaning of Gardens: Idea, Place and Action. The MIT Press, Cambridge, pp.244–251 (1990)

3. C.M. Tennessen, B. Cimprich, Views to nature: Effects on attention. J Environ Psychol **15**:77–85, doi: 10.1016/0272-4944(95)90016-0 (1995)

4. D. Jain, M. Sra, J. Guo, R. Marques, R. W-u, J. Chiu, and C. Schmandt, *Immersive Terrestrial Scuba Diving Using Virtual Reality*, Proceedings of the 2016 CHI Conference Extended Abstracts on Human Factors in Computing Systems - CHI EA '16 (2016)

5. K.T. Han, *A reliable and valid self-rating measure of the restorative quality of natural environments* (2003)

6. R. Kaplan, S. Kaplan, R.L. Ryan, *With People in Mind: Design and Management of Everyday Nature*, Island Press, Washington, DC (1998)

7. S. Kaplan, The restorative benefits of nature: toward an integrative framework. J. Environ. Psychol. **15**, 169–182 (1995)

8. https://www2.padi.com/blog/2017/12/20/scuba-diving-makes-smart/

9. F.X. Ross, Undersea Vehicles and Habitats: The Peaceful Uses of the Ocean. New York: Thomas Y. Crowell (1970)

10. https://www.deepblu.com/post/81641b20107c11e7aa9ad181bc785596

11. M. Pasini, R. Berto, M. Brondino, R. Hall, C. Ortner, *How to Measure The Restorative Quality of Environments: The PRS-11* (2014)

12. T. Shute. Is it OMG finally for augmented reality? interview with Robert Rice. Ugotrade: Virtual Realities in world 2.0. (2009)

13. Jain, D., Sra, M., Guo, J., Marques, R., Wu, R., Chiu, J., & Schmandt, C.. Immersive Terrestrial Scuba Diving Using Virtual Reality (pp. 1563–1569). Association for Computing Machinery (ACM). https://doi.org/10.1145/2851581.2892503 (2016)

14. B. Erdeniz, E. Serin, M. Aslı, D. Şenol, and G. Alankuş, Virtual Reality - The Effects of Simulated Virtual Reality Environments on Psychological Restoration (2016) 8

15. T. Shute. Is it OMG finally for augmented reality? interview with Robert Rice. Ugotrade:

Virtual Realities in world 2.0. (2009)

16. Varinlioglu, G. The FluidExperience of Space: Physical Body in Virtual Spaces Over an Analysis of Osmose.Master Thesis. Bilkent University (2013)

17. Straughan, E. R. (2012). Touched by water: The body in scuba diving. Emotion, Space and Society, **5**(1), 19–26.doi:10.1016/j.emospa.2010.10.003

Traditional and Digital Representations of Atmosphere in Architecture. A first systematisation

Danilo Di Mascio[1,*] and *Yvette* Putra[2]

[1]Department of Architecture and 3D Design, School of Art, Design & Architecture, The University of Huddersfield, United Kingdom
[2]Melbourne School of Design, Faculty of Architecture, Building & Planning, University of Melbourne, Melbourne, Australia

Abstract. In architecture, atmosphere and its many implications have, in recent years, received increased attention. This paper considers atmosphere in architectural representation, and begins with an exploration of different definitions of these atmospheric representations. This paper then identifies and summarises representations of atmosphere in architecture and their key aspects, and proposes, in the form of a timeline, a preliminary systematisation of these aspects. This paper considers both traditional and digital representations of atmosphere in architecture, and focuses on aesthetic and emotional qualities of atmosphere. Thus, this paper is not limited to realistic and scientific approaches of atmosphere as meteorological conditions, but extends to atmosphere as emotions and mood. This paper also suggests cross-fertilisation, in the representation of atmosphere, between architecture and other fields. Therefore, while this paper explores atmospheric representations in architecture, it establishes its discussions by crossing disciplinary boundaries. It is anticipated that this paper will encourage further research, including recommendations for negotiating the apparent incompatibility of traditional and digital representations, to maintain qualities characterising well-known examples from the past.

1 Introduction

This paper considers atmosphere in architectural representation, and begins with an exploration of different definitions of these atmospheric representations. This paper then identifies and summarises key aspects relevant to representing atmosphere in architecture, and proposes, in the form of a timeline, a preliminary systematisation of these aspects.

In architecture, atmosphere and its many implications have, in recent years, received increased attention. The word 'atmosphere' offers several interpretations, as it is linked to a number of concepts. These concepts range from objective meteorological conditions to more subjective moods, in which the latter are prompted by tangible and intangible characteristics of buildings and places.

* Corresponding author: D.DiMascio@hud.ac.uk

There is a great deal of research around the meaning and concepts of atmosphere in architecture, with most of these works investigating atmosphere relative to real-world buildings and places [1]. In contrast, there is a lack of scientific study exploring atmosphere in architectural representation.

Atmosphere in architectural representation is particularly relevant in the present, when considerations for architecture's functional and quantitative aspects appear to have overtaken considerations for its qualitative aspects. In terms of atmosphere in architectural representation, its qualitative aspects include emotions, triggered by architecture and architecture's attendant characteristics.

Architecture is one of many disciplines contributing to atmosphere creation. For centuries, representational techniques and tools have been used to depict and interrogate atmosphere in architecture. Yet, despite the significance of atmosphere in architecture, there appears to be no known critical investigation or systematisation of atmospheric representations in architecture.

This paper considers both traditional and digital representations of atmosphere in architecture. It focuses on aesthetic and emotional qualities of atmosphere, as suggested by Zumthor in *Atmospheres*: "We perceive atmosphere through our emotional sensibility [...] We are capable of immediate appreciation, of a spontaneous emotional response, of rejecting things in a flash. That is very different from linear thought" [2][†]. Thus, this paper is not limited to realistic and scientific approaches of atmosphere as meteorological conditions, but extends to atmosphere as emotions and mood.

2 Methodology

A methodology of four main steps has been used to systematise traditional and digital representations of atmosphere in architecture, and their key aspects (Fig. 1). These steps constitute this paper's main structure.

This paper begins with a brief analysis of definitions of atmosphere in architecture. As 'atmosphere' carries different meanings, it is critical to explore its origins and different interpretations, and consider how they support this research. This paper then examines atmosphere's role in representation, by establishing the role's importance and highlighting two possibilities for representing atmosphere. This paper's subsequent section describes the techniques and content of atmospheric representations, both historically and in the present. The final section proposes, through a timeline, a first visual systematisation of atmospheric representation in architecture, and key aspects of these representations.

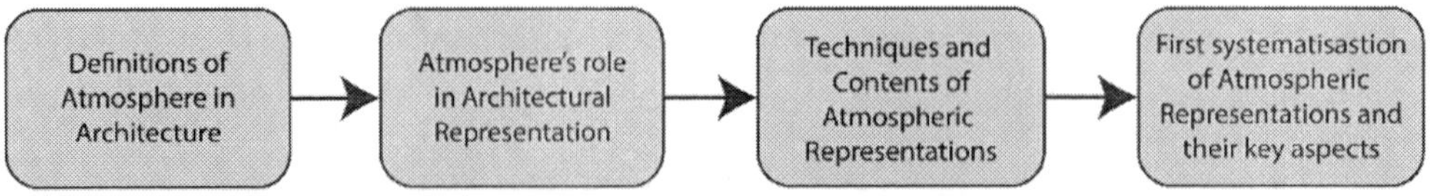

Fig. 1. Main methodological steps to systematise traditional and digital representations of atmosphere in architecture, and their key aspects.

[†] (Zumthor 2006: 13)

3 Definitions of Atmosphere

The word 'atmosphere' has its origins in the mid-seventeenth century, from the modern Latin *atmosphaera*, and from the Greek *atmos*, meaning 'vapour', and *sphaira*, meaning 'ball' or 'globe' [3]. Earliest attestations of the word describe gaseous envelopes around, most commonly, the Earth and other heavenly bodies: "That there is an Atmo-sphaera, or an orbe of grosse vaporous aire, immediately encompassing the body of the Moone." Over time, atmosphere's meaning extended to "mental and moral environment", "pervading tone or mood", and even "fascinating or beguiling associations or effects".

For the meanings "layers of gas surrounding a planet" and "spatial ambience and mood", Vignjević uses the shorthand "meteorological notion" (or "technical notion") and "aesthetic notion" respectively [4]. Yet, the distinction between these two positions is not altogether clear. This is seen, for instance, in how mood is affected by weather, which is an established correlation in psychology [5-6]. In the context of architectural representation, Wigley expresses this connection thus: "The weather conditions represented in traditional drawings have always contributed greatly to the mood of the project" [7][‡].

To return to atmosphere's semantics, Ford suggests that the interdependence of its two definitions is traceable to the late seventeenth and early eighteenth centuries [8]. This coincides with Romanticism's flourishing, when "atmosphere was used both to articulate the divergent autonomous forms taken by the poetic imagination and scientific disciplines, but also to mediate their continued communication" [8][§]. Such practices are evident in the era's literary and visual arts, and, as this paper discusses later, constitute a shift in Western aesthetics.

Current architectural discourse reflects the challenges of atmosphere's definitions. According to Wigley, atmosphere "is some kind of sensuous emission of sound, light, heat, smell, and moisture", but is, at the same time, "a swirling climate of intangible effects" [7][**].

In representation, there is an unclassifiable variety of contexts and media. As such, each of atmosphere's definitions shows a likewise extreme variety of iterations in its representations. But, as representations of meteorological phenomena must be recognised by viewers, they are necessarily limited by widely accepted conceptions of how such phenomena appear. On the other hand, representations of tone, mood, and effect are more complex. This is because their meanings are inherently subjective, and, therefore, more varied. This paper acknowledges atmosphere's dual meanings, yet underscores the meanings' inseparability. In subsequent sections, this paper does not make any distinction between these meanings.

4 The Role of Atmosphere in Representation

The first possibility of atmospheric representation is 'literal'. In literal representations, viewers are, either through realism or suggestion, drawn into visualised worlds. The main aim of literal representations is to increase the viewers' direct involvement with these representations, in which viewers are either participants or spectators in specific moments. The viewers' involvement with a precise moment is, as Bois writes of perspectives, "the

‡ (Wigley 1998: 19)

§ (Ford 2018: 5)

** (Wigley 1998: 18)

petrification of the spectator, similar to the Medusa's head" [9][††]. Meanwhile, the potential for viewers' immersion is described by Deleuze in this way: "As a spectator, I experience the sensation only by entering the painting, by reaching a unity of the sensing and the sensed" [10][‡‡].

The second possibility in representing atmosphere is 'symbolic', in which symbolic representations function largely to convince viewers of particular ideas. Symbolic representations include iconography and allegory [11]. In symbolic representations, the sky may suggest universal order, and a tree may link the celestial and terrestrial spheres. Symbolic representations require their viewers to grasp their symbols, so these representations encourage less immediate immersion than literal representations.

Across both literal and symbolic representations, atmosphere is used to increase the viewers' emotional engagement, and to, ultimately, evoke emotions such as admiration, curiosity, and wonder. Emotional qualities of atmospheric representation are described in a note, in Powell and Leatherbarrow's *Masterpieces of Architectural Drawings*, about Piranesi's 'Prisons' (1745-50): "The nightmare emotions [...] were transformed into a visual aesthetic experience through pictorial means" [12][§§]. Powell and Leatherbarrow then mention how Piranesi's use of double diagonal perspective evokes "a feeling [...] of immensity" [ibid.]. These descriptions support that representations of atmosphere in architecture are, primarily, to elicit emotions, and, accordingly, to allow viewers to experience certain moments.

In architectural representation, emotions were particularly emphasised in the Picturesque, an English aesthetic category of the eighteenth century [13][***]. In the Picturesque, there was a preference for architecture which was "more visual than intellectual", leading to "architectural drawings less geometrical and abstract and more concerned with colour, light and atmosphere than they had ever been before" [ibid.].

5 Techniques and Content of Atmospheric Representations

This section examines, through an analysis of selected works, some of the main techniques and content of atmospheric representations in architecture. This research recognises the interconnectedness of architecture and other fields, so the selected examples are not limited to those produced in an architectural context. These examples have been chosen chiefly for their frequent references in the literature. This section begins with a discussion of historical representations of atmosphere, before focusing on more current examples.

5.1 Historical representations

Shifts in atmospheric representation over time are evinced in historical architectural representations. This is seen in works of Piranesi (1720-78), Boullée (1728-99), Gandy (1771-1843), Ruskin (1819-1900), Garas (1866-1925), and Ferriss (1889-1962). In Gandy's 'Architectural Ruins, a Vision' (1798) (Fig. 2 - left), atmosphere is conveyed both literally and symbolically. The rotunda at the Bank of England, although completed the same year as this painting, is depicted as a ruin. It is illuminated by sunlight in an overcast sky, conveying

[††] (Bois 1981: 46)

[‡‡] (Deleuze 2004: 25)

[§§] (Powell and Leatherbarrow 1982: 35)

[***] (Curl & Wilson 2015: 578)

decline but leaving traces of hope for the future. Garas' 'Artist's Interior' series (1896) comprises imaginative views which "strive to express ideas, sensations or musical rhythms in architecture" [14][†††]. This series, much like van Gogh's 'Café Terrace at Night' (1888), demonstrates how intimate scenes communicate evocative atmospheres.

In art, examples of atmospheric representations of architecture are found in works of Romanticism, such as those by Turner (1775-1852), and Impressionism, such as those by Monet (1840-1926). In the 'Rouen Cathedral' series (c. 1892-94) and 'Houses of Parliament' series (c. 1899-1901), Monet investigated and represented changing light conditions, and qualities of light itself on architecture. These series show how light influences the atmosphere of their architectural subjects. The overall effect is more literal than symbolic, as viewers find themselves caught in exact moments of architectural and climactic collision. In identifying representations of atmosphere in art, it is necessary to also mention occasional earlier proponents such as Claude (1600-82). Claude's use of light, typically in expanses of sky and over bodies of water, communicates idealised places, as seen in 'The Embarkation of the Queen of Sheba' (1648) (Fig. 2 – right).

These examples allow for the identification of elements characteristic of representations of atmosphere, such as strategic use of points of view, and depictions of light and shadow. Architectural aspects, such as shape and composition, can be included as important components of some atmospheric representations. Further, this historical survey establishes a framework for discussing the current state-of-play of atmospheric representation, in which digital technologies appear to have all but superseded traditional methods of representation.

Fig. 2. (left) Joseph Michael Gandy, "Architectural Ruins, a Vision" (1798) (Source: Photo © Sir John Soane's Museum, London); (right) Claude Lorrain "The Embarkation of the Queen of Sheba" (1648) (Source: Image © The National Gallery, London 2019).

††† (Mathieu 2003: 64)

5.2 Modern and contemporary representations

The analysis of drawings in *100 Years of Architectural Drawings, 1900-2000* [15] demonstrates that World War II, and the ensuing Modern movement, precipitated an important change in the style and content of architectural representation. Representations of the preceding generation are more imaginative, and show influences from artistic movements such as Romanticism, Futurism, and Expressionism. Modern representations, in contrast, show greater simplicity and functionality. Typical drawings of the Modern period have clearer line-work, with light and shadow almost absent.[‡‡‡] Modern representations are, it may be suggested, often as functional and aseptic as the buildings and spaces they represent. These representations usually do not depict light and shadow in any Expressionist manner. However, there are some exceptions from the Modern era, with architects whose representations display atmospheric qualities akin to those in historical examples. Wright (1867-1959) and Kahn (1901-74) were two such Modernists, with Wright's rendering of Fallingwater (1935) described as having a sky "drawn as a series of parallel wavy blue lines that echo the shape of the building and trees. It appears to be an atmospheric aura produced by the design." [7][§§§]

[‡‡‡] The collages of Alison (1928-93) and Peter Smithson (1923-2003), for the Golden Lane Estate competition entry (1953), are prominent examples of this.

[§§§] (Wigley 1998: 20)

Fig. 3. (top left) View of the foyer of the Kiasma Museum by Stephen Holl (1993. Source: © Steven Holl); (top right) View by Darcstudio of Pingus Winery by Peter Zumthor (2018. Source: © Darcstudio); (bottom) View of the National Center for Theodor Kittelsen by Snøhetta (2017. Source: © Snøhetta).

At present, one of the main objectives of digital visualisations in architecture is, arguably, to market projects to clients and other parties. Again, however, there are exceptions to this condition, as in works of architects and practices such as Holl (b. 1947) (Fig. 3 – top right), Zumthor (b. 1943) (Fig. 3 – top left), and Snøhetta (Fig. 3 – bottom). Holl's restrained use of traditional media draws viewers into atmospheres of contemplation and controlled movement. Zumthor's design is represented more vividly, to convince viewers of the space's illuminative, material, and scalar possibilities, and of the moods which follow. Snøhetta's vision of an interior at the National Center for Theodor Kittelsen communicates, through its view towards the adjacent forest, a feeling of discovery and wonder. This is, more precisely, represented in two children looking out at an 'enchanted' view, reminiscent of Kittelsen's drawings and in an apparent homage. Here, the window opens onto a real landscape depicted fantastically, emphasised further by the light's ephemerality and the presence of two faintly rendered animals.

For most of history, representations of atmosphere in architecture have been largely visual through, for example, painting and film. In these instances, viewers occupy a relatively passive role. But with the advent of virtual reality and related technologies, atmospheric representations increasingly include the other senses.

6 A First Systematisation

A timeline, spanning the seventeenth century until today, may be established with examples from this paper's previous sections and elsewhere. This timeline provides a first visual systematisation of traditional and digital representations of atmosphere, in architecture and related fields. This diagram (Fig. 4.) shows a portion of this timeline, including the examples' titles, along with names of their authors (such as architects or artists), and key aspects defining the examples' atmospheric qualities. Each coloured line suggests a different visual art such as painting, film, and video games. This timeline, as it appears below, overviews some of the main examples. It may be further developed with more examples, for more complex analyses. This systematisation allows for reflections on different representations of atmosphere.

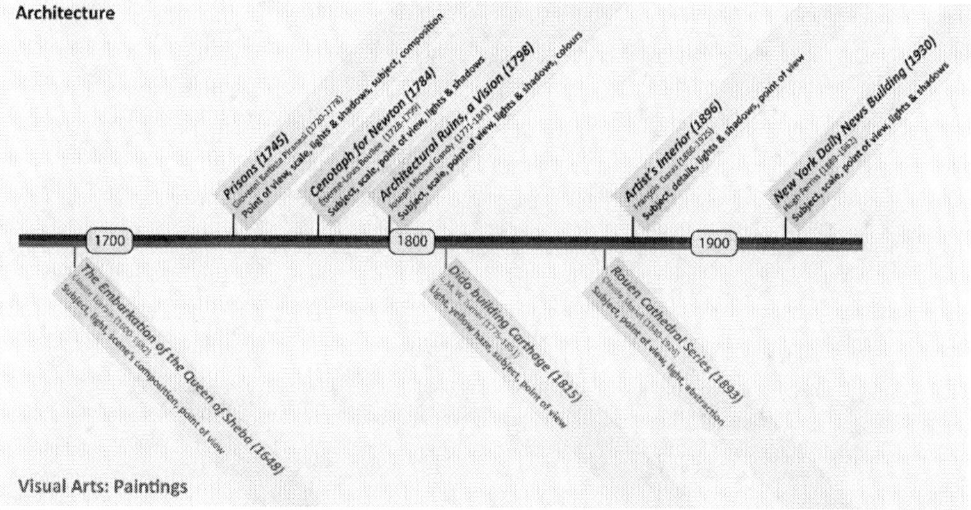

Fig. 4. A portion of the timeline showing the examples' titles, authors, and key aspects defining the examples' atmospheric qualities.

Conclusion and Future Developments

This paper identifies and describes traditional and digital representations of atmosphere in architecture, and their key aspects, and proposes their first visual systematisation. This systematisation is achieved through a timeline, in which a line indicating architecture is matched with lines indicating painting, film, video games, and other visual arts. In this way, this paper suggests cross-fertilisation, in the representation of atmosphere, between architecture and other fields. Therefore, while this paper explores atmospheric representations in architecture, it establishes its discussions by crossing disciplinary boundaries.

It is anticipated that this paper will encourage further research, including recommendations for negotiating the apparent incompatibility of traditional and digital representations, to maintain qualities characterising well-known examples from the past.

Atmosphere is an essential element of architecture. With the growing orientation towards architecture's quantitative aspects, this paper proposes that atmosphere, as a subject of discussion, will, paradoxically, find greater relevance.

The authors wish to thank Steven Holl Architects (www.stevenholl.com), Snøhetta (https://snohetta.com/), and Darcstudio (http://darcstudio.co.uk/), for granting permission to include their images in this paper.

References

1. G. Böhme, C. Borch, *Architectural atmospheres: On the experience and politics of architecture* (Birkhäuser, Basel, 2014)
2. P. Zumthor, *Atmospheres* (Birkhäuser, Basel, 2006)
3. OED Online, <www.oed.com/viewdictionaryentry/Entry/12552>
4. A. Vignjević, *AM Journal* **0** (2017)
5. E. Howarth, M.S. Hoffman, *British Journal of Psychology*, **75** (1984)
6. M.C. Keller, et al., *Psychological Science*, **16** (2005)
7. M. Wigley, *Daidalos,* **68** (1998)
8. T. H. Ford, *Wordsworth and the poetics of air* (Cambridge University, Cambridge, 2018)
9. Y.-A. Bois, *Daidalos,* **1** (1981)
10. G. Deleuze, *Francis Bacon* (University of Minnesota, Minneapolis, 2005)
11. M. Battistini, *Symbols and allegories in art* (J. Paul Getty Museum, Los Angeles, 2005)
12. H. Powell, D. Leatherbarrow, *Masterpieces of architectural drawing* (Orbis, London, 1982)
13. J.S. Curl, S. Wilson, *The Oxford dictionary of architecture* (Oxford University, Oxford, 2015)
14. C. Mathieu, *Orsay architecture* (Paris: Editions Scala, Paris, 2003)
15. N. Bingham, *100 years of architectural drawing, 1900-2000* (Laurence King, London, 2013)

SHS Web of Conferences **64**, 01005 (2019) https://doi.org/10.1051/shsconf/20196401005
EAEA14 2019

Post-flânerie: How do mobile devices shape the experience of the city?

Gorsev Argin[1,2,*], *Burak* Pak[2], and *Handan* Turkoglu[1]

[1]Istanbul Technical University, Department of Urban and Regional Planning, 34367 Istanbul, Turkey
[2]Altering Practices for Urban Inclusion Research Group, KU Leuven Sint-Lucas Brussels and Ghent Campuses, Department of Architecture, 9000 Ghent, Belgium

Abstract. In the last decade, the advances in mobile technologies and location-based applications reshaped our mutual relationship with the urban environment. These technologies, as both a mean and barrier to the engagement between humans and their environment, have transformed the urban experience in profound ways. Urban experience is a relatively new concept introduced with the rise of modern cities in the nineteenth century. Its loss due to the rapid urbanization has been a subject of debate since then. Among the discussions that take place in this debate, the figure of 'flâneur' plays an extensive role. The flâneur is a figure who wanders through and appropriates the metropolitan city in pursuit of urban experience and *reaps aesthetic meaning from the spectacle of the teeming crowds*. Flânerie, or the act of wandering, and its implications for our understanding of urban life have been profound. Today, mobile technologies create a new kind of urban wanderer which is described as "post-flâneur". In this paper, by examining the altering concept of flânerie, we discuss the effects of mobile technologies on urban experience. Based on an informed study of a wide range of theories, we make reflections on the impact of mobile devices on the mutual relationship between humans and their environment, introduce key concepts for understanding the emergent phenomenon of post-flânerie and elaborate on its interconnections with the phenomena of cyber and hybrid-flânerie.

* Corresponding author: gorsevargin@gmail.com

1 Introduction

The modern city of the nineteenth century introduced the figure of flâneur and the act of flânerie. Benjamin's [1] 'flâneur', a nineteenth-century character firstly portrayed by Baudelaire, walks in the city and observes rapidly transforming urban spaces of the modern city. This flâneur uses walking as a way of experiencing the city and uses this experience as a latent protest against the rapid alteration of the modern city. Although the flâneur was appropriating the rapidly evolving modern city by walking in it, Benjamin predicted that flâneur could not escape the fate of modern-self and the "loss of experience" was inevitable. In the 1950s, the Situationist movement brought back the concept of flâneur. This new the flâneur was walking through the city in search of an unplanned urban experience. Eventually, the urban experience, which was perceived as in favor of the modern city by Baudelaire, became something to fight for in the twentieth century. With the rise of consumer capitalism, a new kind of 'urbanian self' emerged, which has been described by Jameson [2] as the 'society of the image'. The Ghent Urban Studies Team [3] has identified this spectatorial consumer as 'neo-flâneur', which is the heir of Benjamin's middle-class flâneur.

Today, new technologies create a new kind of urban wanderer which can be named 'post-flâneur'. Whether it is a smartphone, a tablet or a wearable technology, by using these mobile devices and location-based applications installed to these devices, people move through and experience the urban space in a whole new way. Furthermore, by using social media applications, people can record and share their memories along with their routes by attaching them to the specific geographic locations and re-appropriate the urban space by strolling, recording, and sharing it.

Within this context, the aim of this study is to understand how these mobile technologies affect the mutual relationship between humans and their environment and the act of post-flânerie. The questions to be addressed are:

- How did the concept of flâneur evolve through time?
- What are the key concepts to understand and discuss the post-flâneur and differentiate it from other figures of flânerie?
- Is post-flânerie a loss or a rediscovery of the urban experience? And how is it related to the phenomena of cyber- and hybrid-flânerie?

To answer these questions, this paper examines various theories on the concept of flânerie that has been redefined and cited over time, and redefines the act of post-flânerie by providing critical reflections on these theories.

In this respect, the first part of the study will focus on the emergence and transformation of the concept of urban experience and the figure of flâneur (Section 2). Following this, the second part will examine urban experience in the era of mobile technologies and, by examining the evolution of flâneur, introduce the figure of the post-flâneur (Section 3). In the third part, the emergence of the post-flânerie will be discussed through four main alterations in the concept of flânerie; alterations in (re)construction of the 'city', rhythm, instruments of experience and instrumentalization of experience, and lastly, the concept of privacy (Section 4). The paper will conclude with final remarks on post-flânerie and future directions of the study (Section 5).

2 Urban experience and the flâneur

The industrial revolution and the concomitant technological innovations of the nineteenth century had revolutionized the traditional city and constituted what we call the 'modern city'. Not only had the urban space dramatically changed, but also people who live in, their behaviors, everyday practices and the ways they imagine and experience these urban spaces [4]. At this point, flâneur takes the stage both as a product of and an attempt to appropriate modern city [5].

Benjamin, in his article on Baudelaire, revives the figure of flâneur and uses it in his study called "The Arcades Project" [1] (continued between 1927 and 1940) to investigate and examine the 'new' experience of an individual, or flâneur, with his terms, who walks in the city, observes rapidly changing urban space of the modern city and *reaps aesthetic meaning from the spectacle of the teeming crowds* [6]. According to him, a flâneur does not only walk but also tries to determine everything abolished and captured by capitalism while decoding the meaning of this new way of life [7]. This act of walking as a way of experiencing the city was a sort of a latent protest against the rapid alteration of the modern city and a mean to appropriate this changing urban phenomenon. Connerton [8] in his book "How Modernity Forgets", reminds us of De Certeau's [9] analogy, which suggests "act of walking is to the urban system what the act of speaking is to the language system", and describes walking as "a space of utterance; as speech is to language, so walking is to the appropriation of urban space". Although the flâneur was appropriating the altering modern city by walking in it, Benjamin predicted that the flâneur could not escape the fate of modern-self. According to him, 'loss of experience' was inevitable [10].

It is important to clarify what Benjamin means while using the word 'experience'. Although it is expressed with one word in English, the discussions on experience indicate two different understandings on experience. In German, these two different concepts are defined as *erfahrung* and *erlebnis*. *Erfahrung* means a more collective experience, such as a knowledge developed over time, while *erlebnis* indicates a more personal experience, that is to say, the act of experiencing. Contradictory to the thoughts of the theorists that state 'experience is no longer possible' and 'we are living in a society of experience' born from this duality, the former one means *erfahrung* while the latter one means *erlebnis* [11]. In this sense, the inevitable fate of loss declared by Benjamin should belong to the former one: the *erfahrung*.

It can be claimed that many criticisms on the modern city have struggled against this fate. In the second half of the 1950s, the Situationist movement defined a new discipline, which they called 'Psychogeography' to study "the effects of the geographical environment, consciously organized or not, on the emotions and behavior of individuals" [12]. The act of walking was the main tool of these scholars since in the act of walking, the body can trace a new map, one that escapes the rigid hierarchies of an imposed order [13]. Hereby, the concept of flâneur was back on stage again. This new flâneur was not walking the major roads and touristic districts determined by authorities but instead, by using methods such as '*dérive*' went into every nook and cranny streets/districts of the city in a search for an unplanned urban experience.

Subsequently, at the beginning of the 1960s, postmodern criticism of the modern city was on the rise [14, 15]. This time, the main criticism was the loss of human-centered urban experience in modern cities. According to these critical voices, as a result of modern planning implementations, there had been an increasing hegemony of automobiles in urban space and walking-based urban experience had been interrupted. Two decades later, these arguments bring along New Urbanism, which advocates "livable" spaces at the human scale [16]. Thus, determining spatial organizations which provide walking and perceiving the city from a human perspective became the main interest of the postmodern urban scholars [17, 18].

In the 1980s, with the rise of consumer capitalism, the urban space itself became the subject of commodification [19]. One of the visible aspects of this commodification is the increasing instrumentalization of public spaces for the advertisements, which, in turn, leads to the loss of self-control over visual perception during a daily walk. McCormick [20] defines his walking experience in Melbourne streets as a comprehension of the loss of public space. He adds that we still have streets, shops, parks, however, there is no urban space that we do not hear or see messages that force us to buy something [20]. Jameson [2] defines the new society which has been in a rise in last decades as "a new 'society of the image' in which consumerism and market frenzy are not the issue so much as consumption by the eyes". Based upon this definition, The Ghent Urban Studies Team [3] introduced the phenomenon of 'neo-flâneur' who is "today's spectatorial consumer" that "should be seen as a direct heir to the popular middle-class flâneur described by Benjamin" [3]. The main goal of the neo-flâneur is not necessarily drawing alternative aesthetic meanings and experiences but purely market-driven consumption.

3 Urban experience in the era of mobile devices and the post-flâneur

Urban experience is a relatively new concept which has been a subject of debate following the rise of modern cities. Furthermore, loss of urban experience due to the rapid alteration of the urban spaces seemed to be the key concern of further critics of the modern city within these discussions. Today, urban experience has reached a new level due to the advances in mobile technologies that have rapidly penetrated into and shaped our daily lives. As Castells [21] explains, today, we are walking in a built environment "that is increasingly incorporating electronic communication devices everywhere" and we live in "a new urban form in which we constantly interact, deliberately or automatically, with online information systems, increasingly in the wireless mode".

Today, the urban experience is generated by digital information mediated by mobile devices in ways that were not possible just a decade ago [22]. One of the most current examples of these mobile devices are smartphones, which have been adopted more rapidly than any other technologies in the history [23] and has an extraordinary rising rate all over the world [24].

These mobile technologies offer new perspectives to individuals but also emerge as a mean and barrier to novel forms of urban life and its experience. Today, it is the question that whether people walking on the streets with their heads down with their mobile phones are still a part of the street life or not [25]. Urban life consists of everyday mobile encounters in the streets and visual, or face to face interaction, is a key feature of these encounters [16]. As a result of mobile technologies, people are no longer automatically part of these encounters, half of them "there in body but not in any other way" [25, 26]. This also affected the figure of flâneur profoundly (Figure 1).

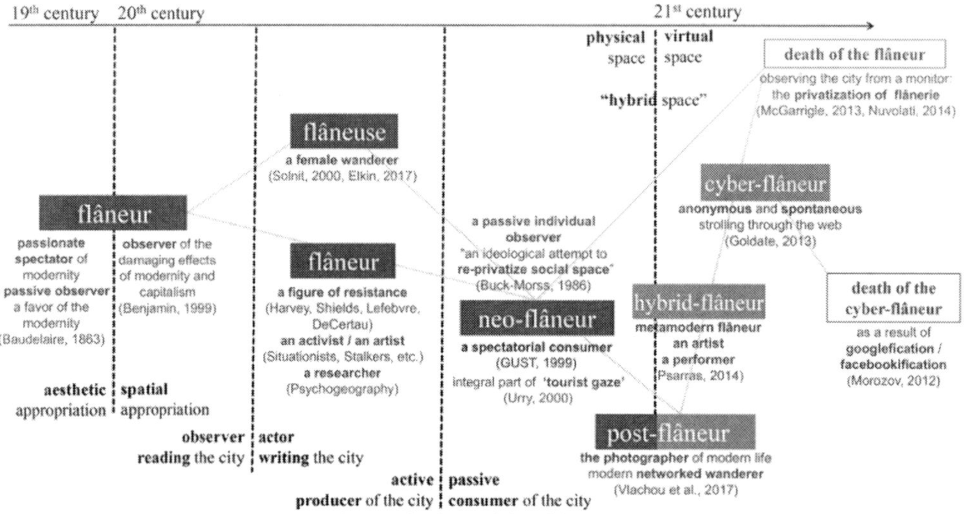

Fig. 1. The evolution of the flâneur.

3.1 Death and resurrection of the flâneur

Baudelaire's and Benjamin's flâneur are both the observers of the modernity. The former walks through the city for an aesthetic appropriation while the latter slides towards the spatial appropriation of the altering urban space. According to Benjamin, flâneur is the person who observes the damaging effects of capitalism and his extinction is inevitable as the loss of experience. When flâneur takes the stage again in the late 1950s and 1960s, he – and this time also she – is not only an observer who reads the city but instead is an actor who writes the city. He/she appears as a figure of resistance, an activist, an artist or a researcher. With the rising commercialism in 1980s, the figure of flâneur comes to the threshold of death but doesn't die. However, Benjamin is right about the fate of the urban experience. Flâneur who turns into 'neo-flâneur' [3], a "puppet in the hands of capitalism" [5], a "passive individual observer" [27] or a tourist [28] who perceives the urban environment from pre-scripted perspectives and becomes the consumer of the city and the urban experience that appears only as an *erlebnis*.

In the twenty-first century, with the advances in technologies, the virtual space has become a new wandering ground for the flâneur. In-between the death announcements of the "old" flâneur [29, 30], a new wanderer who strolls in the endless possibilities of the web was born and called as 'cyber-flâneur' [31]. Cyber-flâneur has some similarities with the classical figure of flâneur. Both of them spontaneously wander their environment which is a virtual space for the former and a physical space for the latter. Cyber-flâneur also has a similar fate to his/her ancestors. With the rise of guiding search engines and social networking websites, the guides became the directors of the surfing experience and anonymous and spontaneous nature of strolling came to an end [32].

3.2 Emergence of hybrid space and the post-flâneur

With the advances in mobile technologies, a new 'hybrid' space was born in between the physical and virtual spaces. This hybrid space brings the flâneur back to the stage one more time. It is redefined with respect to the different challenges that came with mobile technologies and called as hybrid-flâneur [33] or post-flâneur [34]. Psarras [33] defines the hybrid-flâneur as the "metamodern flâneur" who acts as an artist and a performer. Post-flâneur, on the other hand, can be anybody who walks the city with his/her smartphone, takes photographs and shares them in the social media, who can be called the "photographer of modern life" or the "modern networked wanderer" [34].

This paper, on the other hand, uses the term of post-flâneur in a wider context, to define a wanderer with a mobile device who shuttles between the virtual and physical space in pursuit of an urban experience. Whether he/she takes photographs or navigates through his/her mobile device, the aim of the walk should be the experience itself.

4 Key concepts for understanding the post-flâneur

In this part of the paper, post-flânerie will be examined through the key concepts reflect behavioral and perceptual similarities and differences between various states of flânerie which are alterations in the (re)construction of the 'city', rhythm, instruments of experience and instrumentalization of experience, and lastly, the concept of privacy.

4.1 Walking and (re)construction of the 'city'

Flânerie is a way of wandering through the parts of the city and seizing its details. Moreover, it also means to provide a perception and narrative about the 'city' as a whole. As mentioned before, De Certeau [9] suggests that the act of walking means for the city in the same way as the act of speaking means for the language. Perceiving the city by walking was much easier in the traditional city, which was a unity. However, that unity has become fragmented in the course of time due to the rapid urbanization. The citizens have been alienated from the traditional city and become unable to mentally map their own positions and the urban as a unity [14].

In the years when Debord suggested the act of *dérive* as a version of a flânerie, Paris was in a somewhat rapid urbanization process. Perceiving the city as a whole – which can exist only in a virtual state – was becoming more and more difficult for the citizens [35]. The aim of the *dérive* was to make the city a unity again, a unity which is in motion and transformation [35]. With this purpose, Situationists were simultaneously walking in the streets of physically separated parts of the city and sharing their experiences concurrently by using all kinds of mobile communication tools such as walkie-talkies [35]. This pursuit of unity had been continued until the 1960s when it became impractical due to the undeniable mass urbanism.

Today, we have collective digital maps reachable from mobile and online devices developed by large communities. A post-flâneur can walk through the city by both using and contributing these maps. It should be underlined that *dérive* maps show significant distinctness from the current collective digital maps in many ways (i.e. aims, ideologies, representations, etc.). Nonetheless, the unity of the city that can only exist in the virtual state is becoming more visible both in the virtual and also in the hybrid space as a result of the advances in mobile technologies and especially location-based mapping tools. Today, people who do not know each other can walk on the streets and share their comments and experiences simultaneously. The form of these narratives can differ depending on the sharing tools. In this way, a collective map of the city defines a 'new' whole which is constantly in motion and transformation. The people who make this map are not members of a small group like the Situationists, however, they have to be a member of a wider group (perhaps this one is much easier to enter) to exist in these platforms.

It should also be added that flânerie, in the classical sense, does not need any guidance of a map. Psychogeographers, in some of their flâneries, go one step further and use maps contradictory to its *raison d'être* by using maps of different cities [13]. For a post-flâneur, on the other hand, maps become a default instrument comes along with mobile devices. Today, as a result of the advances in mobile devices and location-based applications, post-flâneurs wander through the representation of the city as a network of coordinated places, data, and etc. The "blue dot", or the avatar of the post-flâneur, wanders through this virtual representation of the city, at the same time, the post-flâneur strolls in the actual physical space. In one sense, this is a reflection of an egocentric walking experience into an exocentric representation of the experience. For a post-flâneur who uses the navigation of the mobile device or takes and shares photographs by adding location information, the representation of the city in his/her mind should be different from the previous flâneurs.

4.2 Altering rhythm: walking the turtle to walking with a smartphone

According to De Certeau [9], if we think about "the memorable here and there", which is often incited by walking in the city, memory is kind of a spatial practice [23]. Since spatial memory requires a certain measure of stability, walking – by establishing oriented things as identical things and hence constituting a "steady system of places" – contributes to the construction of spatial memory and "a particular pace of assimilation" makes a place and what is experienced there, memorable [8].

The city has always been in a rush compared to the rural areas [36]. However, by the turn of the twenty-first century, the tempo of the city has been increased from many perspectives. This phenomenon has been explained by certain theories such as "hyper-acceleration", "turbo-capitalism" and the "digital speed revolution" [37]. Respectively, the effect of this capitalism on urban space has been conceived as the altering tempo of everyday rhythm and therefore resulted in the change of walking speed in the urban environment. Plus, with the introduction of new and faster mobility forms, a new form of flow has captured the city. Castells [21] attributes this new urban phenomenon to the era of "space of flows rather than a space of places". This is generally related not only to the acceleration phenomenon of social life but also to urban life.

Benjamin refers to the walking men with their turtles while narrating the flâneur who walks through the nineteenth-century streets and arcades. In these walks, the turtle decelerates the pace of the flânerie, thus, the flâneur who keeps this pace experiences the environment from a different perspective. In his novel "Slowness", Kundera [38] expresses his longing for the flâneurs of the past and asks the question of "Why has the pleasure of slowness disappeared?". In fact, by giving the example of flâneurs with turtles, Benjamin himself emphasizes the tempo of a flânerie [27] and addresses this kind of pleasure of slowness. Kundera [38] also defines a correlation between slowness and memory and between speed and forgetting. This brings to mind Benjamin's use of the term 'experience' in the meaning of *erfahrung* which also can be associated with the memory.

If we imagine a post-flâneur walking with its smartphone in today's 'hyper-accelerated' world, we may conjure a mental picture of a person who frequently slows down and stops. This may also mean that the post-flâneur, that his/her ancestors have been flowing in the crowds, may interrupt or change the flow from time to time. An extreme example of this is called as the smartphone zombie, "who walks slowly and without attention to their surroundings because they are focused upon their smartphone" [39]. In this sense, this figure is quite far away from the post-flâneur, by moving physically in the crowd while mentally strolling in another world far from the crowd.

Whether used for taking pictures or finding directions, mobile devices often determine or at least influence the pace of the walker. A post-flâneur who wants to take a picture firstly slows down in seeking of a desired frame and mostly stop to press the shutter or a navigator post-flâneur's actions on the map reflect his/her walking actions [40]. However, the aim behind the slowness in a post-flânerie can be questionable. Is it a pursuit for physical and mental engagement with the environment or a way of using the mobile device as a mean of virtual engagement? It can also be claimed that the slowness of turtle stimulates the instant experience while slowness provided by smartphone is an arrangement for the next experience.

4.3 Instruments for experience and instrumentalization of experience

Flâneur, in its classical meaning, uses walking as an instrument for the urban experience. Besides, he/she uses a variety of instruments to accompany this walk. Benjamin's flâneur walks with a turtle while some of the situationists use walkie-talkies in their *dérive* experiments. Today, the widely-used instrument for the post-flâneur is smartphones and applications that help to capture, record, share, navigate, and etc.

All of these instruments change people's perceptions of their environments. The increasing use of photography in everyday practice due to the advances in smartphones and the rise of visual narratives is a case in point. Related to advances in mobile technologies and social media platforms, human beings are no longer interested only with the environment as itself, but also "in the representation, recording, management and monitoring of these environments", to put it another way, this is not only about the environment, but rather it's versions that "invented, idealized, perceived and experienced through countless filters" [41]. These also reflect the post-flâneur's relationship with the city. According to Young [42], it has been weakened and loosened as media representations of the city have evolved.

The post-flâneur, who is defined as the "photographer of the modern life" [34], walks in search of creating this kind of representations of the city. This figure can also be read as a version of Baudelaire's "painter of the modern life" who walks the city in a search for an aesthetic appropriation. However, this flâneur does this appropriation by photographing and uploading images to social media in which his/her attention is increasingly concentrated on their mobile phone screen and reactions to these images [34]. To put it another way, in some cases taking and sharing the photo become the main aim of the post-flânerie in which the experience itself became an instrument. Today, a photographer post-flâneur's flânerie can be under the pressure of finding the "perfect" photo whose criteria have been built by the influencers.

We can witness the instrumentalization of experience also in the other forms of flânerie. As mentioned before, Benjamin's flâneur, from time to time, walks through the city in an attempt to protest latently the transformation of the city due to the rapid urbanization. Furthermore, in the 1950s, the flâneur starts to walk in an ad hoc manner. For instance, the main aim of a flânerie can be creating situations for a Situationists, while it can be doing ethnographic research for a Psychogeographies. However, the significance of the "real" experience is still prominent in these examples.

As already mentioned earlier, the crisis of experience has been on the table every time the flâneur has been brought to the stage. Today, if we live in a world of experiences and hear and use the word 'experience' consistently, this may tell us that the 'experience' (*erfahrung*) which is in crisis has already come to an end and the act of 'experiencing' (*erlebnis*) has taken its place. The instrumentalization of experience today means precisely the instrumentalization and consumerism of this somewhat *erlebnis*.

4.4 Figure of privacy and the privacy of the figure

The alteration in the relationship between the flâneur and the concept of privacy can be discussed from two perspectives. The first one is transformation of the figure of post-flâneur into a figure of privacy [31] every time he/she gives his/her attention to the virtual space through the mobile device. The classical figure of flâneur, the man of the crowd [6], strolls within the flows of the modern city while engaging with the urban environment physically and mentally. On the other hand, post-flâneur who shuffles between the physical and the virtual space by using his/her mobile device, from time to time, focuses on the virtual and moves away mentally from the surrounding crowd. This isolation creates some limitations on the real-world engagement. Bull [43], in his article that announces "the end of flânerie", indicates the emergence of somewhat "mediated urban isolation" for the people who listen music with their mobile devices while walking. In light of these, it can be claimed that, today, as a result of mobile technologies people are no longer automatically part of the city and the walking experience. Golberger [31] defines the effect of this situation on post-flânerie by giving a cell phone example:

"...even when the phone does not ring at all, and is being used quietly and discreetly, it renders a public place less public. It turns the boulevardier into a sequestered individual, the flâneur into a figure of privacy" [31]

The second transformation in the concept of privacy can be seen in the privacy of the flâneur. From the very beginning, the major things that make possible the concept of flânerie has been the expanding understanding of the city and the growing crowd in it. That is to say, the classical figure of flâneur was trying to blend into this crowd in order to see without being seen [44], to stroll freely by the help of anonymity. The birth of cyber-flâneur was also a product of a somewhat anonymity derived by the early internet experiences, and its death was announced with the end of this anonymity.

Today, almost every act in the virtual space leaves footprints behind. This also questions the anonymity of the post-flâneur since every information about the physical space or its representation produced or consumed by post-flâneur in the virtual space can be traceable. One step further, countless applications track its users in the name of providing personalized experiences. In the case of post-flânerie, this also jeopardizes the spontaneity and anonymity of the strolling, firstly due to the incongruity of pre-scripted experience with the nature of flânerie, and secondly due to the genericization of their online 'personality' since they are mainly developed for target groups instead of individuals.

Conclusions and future directions

The flâneur of the twenty-first-century hybrid space; post-flâneur, is a wanderer who is walking through the city accompanied by his/her mobile device(s), wandering between the physical and virtual space in pursuit of an urban experience. In a way, these devices can be a helpful instrument for the post-flâneur since they can provide an enhanced urban experience by guiding individuals through an immersive and interactive environment with valuable information [45]. From this perspective, it can be claimed that post-flânerie still provides an experience and this means the experience has not been lost but transformed. This transformation occurred in the meaning of experience and it is not only the result of advances in mobile technologies but mostly related to the "spirit of the age" we are in. With this respect, today, when we are discussing the effects of mobile technologies on the experience of post-flâneur; firstly, we should underline the paradigm shift in the meaning of experience, from *erfahrung* to *erlebnis,* collective to personal experience.

Benjamin's flâneur was walking in an age of *erfahrung*. In today's world of high-speed society, *erfahrung* is in a crisis. *Erlebnis* insists that more personal experience substitutes this. Today, we are living in an era in which the state of 'experiencing' surround us on every side. Correspondingly, the twenty-first-century relationship between human and the city has mainly described by referring to a somewhat personal experience (*erlebnis*). At this point, it is important to remember that, Benjamin's friend Adorno [46] gives attention to the commercialism of these kinds of experiences (*erlebnis*) and puts a question mark on their 'personality'.

The rise of personal experience (*erlebnis*) and its commercialism have been in the front burner for a long time and their effects on flânerie can also be seen in the 'spectatorial consumer' of the late twentieth century: the neo-flâneur [3]. However, with the emergence of hybrid space mediated through mobile technologies, there has been another alteration within the *erlebnis* itself.

Hybrid space emerged as a result of the coexistence of and interactions between digital information, and communication flows and the physical space. This interaction was not on an ex parte basis. While the latter is affecting the physical environment and its perception, the former and its experience started to settle in the virtual space within the form of its representations and narrations. With the advances in mobile technologies, social networks, and location-based applications, these representations of the physical space and its experience also became a significant part of everyday life. Today we are living in a world of representations [47]. Especially, with the introduction of visual-based mass communication tools (i.e. Instagram), the experience of representation of the "real" has been substituted the "real" experience. This also accompanied by a paradigm shift from private experience to shared experience. In some cases, experience of sharing the experience can substitute the "real" experience. In this context, we can also claim that a new kind of *erfahrung* has reemerged in the virtual as an online social experience substituting the real.

In light of these, building on the reality–virtuality continuum of Milgram et al. [48], we suggest redefining the term of **post-flâneur as a continuum between the flâneur and cyber-flâneur**. Today, when a post-flâneur engages with physical space without any mediation of mobile technology and experience the 'reality', he/she approaches a flânerie

(Figure 2). This flânerie, naturally, has distinctions from the older ones due to the changes in the concept of experience through time. The anonymity of this figure can also be questioned in an era that face-recognition systems have been spreading towards the public space. However, it can also have some similarities in terms of the cognitive construction of the space and the rhythm of the pace. On the other hand, when a post-flâneur's attention is increasingly concentrated on the virtual, he/she approaches to the cyber-flâneur. In this situation, the experience is again open to discussion in terms of the degree of the relation of the experience with the reality.

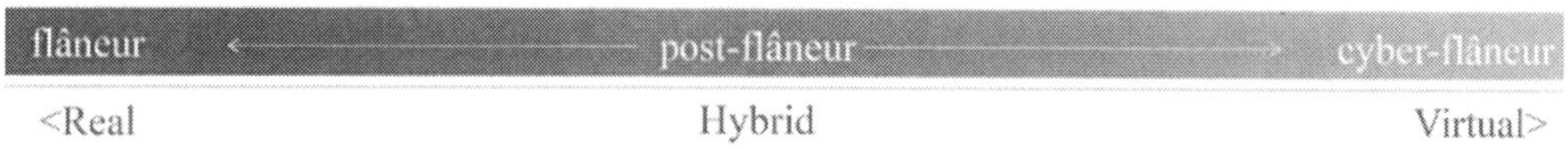

Fig. 2. Post-flâneur as a continuum between the flâneur and cyber-flâneur.

The post-flâneur can also be located in the middle of this continuum in which he/she experiences the "reality" through its representations in the virtuality. Here, it will be convenient to address Baudrillard's [47] four degrees of representation which are the representations that reflect directly the reality, masks and filters the reality, masks the absence of the reality, and lastly the simulation which has no relation with the reality. Accordingly, when the representation used or produced by the post-flâneur approaches to simulation, the act of post-flânerie moves away from the reality and the experience of this reality.

It is clear that, in the near future, humans are going to get more and more connected to the virtual world, and the hybrid space mediated through mobile technologies will be more dominant. This means that the perception of the urban space and its experience will continue to transform. At this point, the emergent phenomenon of post-flânerie gives us a chance to observe and discuss the effects of these mobile technologies on urban experience and offers a contextual framework towards the future of urban experience. In this context, the further step of this study will be to go on the field, observe post-flâneurs in the urban crowd and examine the key concepts introduced the degree of dominance of the virtual representations mediated by mobile technologies on the post-flânerie. This observation can be conducted at least at two levels: from a behavioral perspective of the "outsider" and through the eyes of the post-flâneur. Both of these can provide complementary knowledge which can help us understand the emergent phenomenon of post-flânerie deeper, lead to novel ways of rethinking and designing our future cities.

This paper is produced as a part of a Ph.D. dissertation on Post-flânerie by the first author supervised by the second and the third authors.

References

1. W. Benjamin, *The Arcades Project* (The Belknap Press of Harvard University Press, Cambridge, 1999)
2. F. Jameson, London Review of Books, **18**, 7 (1996)
3. GUST, The Urban Condition: Space, Community, and Self in the Contemporary Metropolis (010 Publishers, Rotterdam, 1999)
4. R. Solnit, *Wanderlust: A History of Walking* (Viking, New York, 2000)
5. C. Jenks and T. Neves, Cultural Values, **4**, 1(2000)
6. K. Tester, "Introduction," *The Flâneur* (Routledge, New York, 1994)
7. A. Oktay, *Metropol ve İmgelem* (Türkiye İş Bankası Kültür Yayınları, Istanbul, 2002)
8. P. Connerton, *How Modernity Forgets*, (Cambridge University Press, Cambridge, 2009)
9. M. De Certeau, *The Practice of Everyday Life* (University of California Press., Berkeley, CA, 1984)
10. M. Hansen, Public Culture, **5**, 2 (1993)
11. M. Jay, Songs of experience: Modern American and European variations on a universal theme (University of California Press., Berkeley and Los Angeles, 2005)
12. G. Debord, Les Lèvres Nues, **6** (1955)
13. T. Richardson, *Walking inside out: Contemporary British psychogeography* (Rowman & Littlefield International, Ltd., Lanham, Maryland, 2015)
14. K. Lynch, *The image of the city* (Technology Press., Cambridge Mass, 1960)
15. J. Jacobs, The death and life of great American cities (Random House, New York, 1961)
16. J. Middleton, Urban Stud., **55**, 2 (2018)
17. W. H. Whyte, *The social life of small urban spaces* (Conservation Foundation, Washington, D.C., 1980)
18. J. Gehl, *Life between buildings: Using public space* (Van Nostrand Reinhold, New York, 1987)
19. D. Harvey, The condition of postmodernity: an enquiry into the origins of cultural change (BasilBlackwell, London, 1989)
20. M. McCormick, Dialogue, **4** (1996)
21. M. Castells, "Space of Flows, Space of Places: Materials for a Theory of Urbanism in the Information Age," *The Cybercities Reader* (Routledge, London, New York, 2004)
22. M. J. Kelley, GeoJournal, **79**, 1 (2014)
23. J. Kalin and J. Frith, Rhetor. Soc. Q., **46**, 3 (2016)
24. P. R. Center, "Smartphone Ownership Is Growing Rapidly Around the World, but Not Always Equally" (2019)
25. P. Goldberger, "Our Cell Phones, Our Disconnected Selves" (2013) [online, accessed 9.2.2019] Available at: https://www.metropolismag.com/ideas/paul-goldberger-cell-phones-disconnected-selves/
26. M. De Waal, The City as Interface: How new media are changing the city (nai010 Publishers, Rotterdam, 2014
27. S. Buck-Morss, New Ger. Crit., **39**, 99-140 (1986)
28. J. Urry, Sociology Beyond Societies: Mobilities for the twenty-first century (Routledge, New York, 2000)
29. C. McGarrigle, 19th International Symposium of Electronic Art (2013)

30. G. Nuvolati, "The flâneur: A way of walking, exploring and interpreting the city," *Walking in the European City: Quotidian Mobility and Urban Ethnography* (Routledge, 2014)
31. P. Goldberger, "Paul Goldberger on Our Cell Phones, Our Disconnected Selves" (2013) [Online, accessed 9.2.2019] Available at: https://www.metropolismag.com/ideas/paul-goldberger-cell-phones-disconnected-selves
32. E. Morozov, "The Death of the Cyberflâneur" (2012) [online, accessed 9.2.2019] Available at: https://www.nytimes.com/2012/02/05/opinion/sunday/the-death-of-the-cyberflaneur.html
33. B. Psarras, "Initial Reflections on a 'Metamodern' Flâneur" (2015) [online, accessed 9.2.2019] Available at: https://hybridflaneur.wordpress.com/2015/06/09/on-a-metamodern-flaneur-pt-i/
34. C. Vlachou, P. Ganouri and M. Skayannis, "Κατασκευάζοντας τον post-flaneur" (2017) [online, accessed 9.2.2019] Available at: http://gavagaiphilosophy.weebly.com/uploads/8/2/0/0/82000672/1.pdf
35. H. Lefebvre, Henri Lefebvre on the Situationist International [Interview] (1983)
36. D. Harvey, *Paris, Capital of Modernity* (Routledge, New York, 2003)
37. H. Rosa, Constellations, **10**, 1 (2003)
38. M. Kundera, *Slowness* (HarperCollins Publishers, New York, 1995)
39. J. Mwakalonge, S. Siuhi and J. White, J. Traffic Transp. Eng. Eng. Ed. **2**, 5 (2015)
40. E. Laurier, B. Brown and M. McGregor, Mobilities, **11**, 1 (2016)
41. R. Coyne, Interactions, **21**, 5 (2014)
42. S. Young, Scan: Journal of Media Arts Culture, **2**, 1 (2005)
43. M. Bull, "The End of Flânerie: iPods, Aesthetics, and Urban Experience," *Throughout* (The MIT Press, Cambridge, Massachusetts, London, 2012)
44. Z. Bauman, "Desert spectacular," *The Flâneur* (Routledge, New York, 1994)
45. O. Guntarik, J. E. Garcia, S. R. Howard and A. G. Dyer, Leonardo, **51**, 2 (2018)
46. T. W. Adorno, *Aesthetic theory* (Routledge and Kegan Paul, London, 1984)
47. J. Baudrillard, *Simulacra and simulation* (University of Michigan press., Ann Arbor, 1994)
48. P. Milgram, H. Takemura, A. Utsumi and F. Kishino, *Telemanipulator and Telepresence Technologies* (1994)

Esquis'Sons! Sketching architecture by listening

Théo Marchal[1,*], *Nicolas* Rémy[1,2]

[1]Grenoble School of Architecture Univ. Grenoble Alpes, Research unit AAU-CRESSON, France
[2]Department of Architecture, Univ. of Thessaly, Greece

Abstract: The Esquis'Sons! application can be used to envision sound atmospheres and prior to designing spaces. The Esquis'Sons application is a compiled module written in the MAX/MSP musical and visual programming language which generates stereophonic sound tracks in relation to geometric parameters extracted from the CAD software Rhinoceros 3D and its plugin Grasshopper. Rhinoceros and Esquis'Sons communicate and sound is played in real time depending on the listening point chosen by the designer and on the architectural scene created. This paper compiles several training periods during which Esquis'Sons has been used in workshop exercises. It also offers material for an architectural study in the rehabilitation of a district. The paper shows that the Esquis'Sons app is a pedagogical tool for designers of space and brings to light several dimensions that are hidden most of the time in the architectural design process: ***the distance of creation*** is discussed through the sound immersion offered by Esquis'Sons app. ***Scales*** of the architectural project are jostled by the fuzzy limits of sound phenomena and ***time*** brings layers of complexity and sensibility in global design.

1 Design principles behind the Esquis'Sons! application

Since 2015, the Esquis'Sons! application has been developed and released in order to help designers to create new building facades by taking into account the sound qualities of their project. The application works as an immersion tool for the evaluation of potential soundscapes designs via listening. Esquis'Sons! started as an exploratory research in which six sustainable neighbourhoods in Europe (Germany, Spain, France and Sweden) have been studied with the objective of describing the sound qualities experimented by inhabitants when they use intermediate spaces located on the facades of building such as balconies, loggias, terraces and external corridors. A cross analysis of the built space's physical dimensions, of the sound environment and of the users' perceptions led to as description of the minimal conditions of existence of these particular sound situations. By comparing

* Corresponding author: marchal.t@grenoble.archi.fr

results from these six *Eco-districts*, an application using parametric conception tools was developed and shared within the designers' community [1].

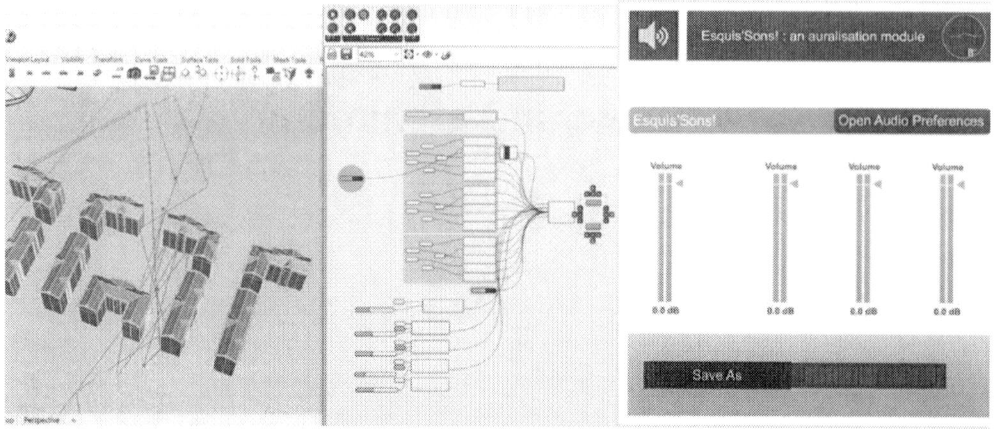

Fig. 1. Esquis'Sons' triple interface. From left to right: rhinoceros visualization, grasshopper module and auralization app.

After three years of development and several uses of this tool with designers, this paper presents how the Esquis'Sons! application can be used in order to envision sound atmospheres and assist the designing process. The application is a compiled module written in the musical and visual programming language MAX/MSP [2] that generates stereo soundtracks related to the geometrical parameters of an architectural project designed with Rhinoceros and extracted with the help of the Grasshopper plugin. The communication between Rhinoceros and Esquis'Sons leads to sound being played in real time depending on the listening point chosen by the designer in the virtual scene created.

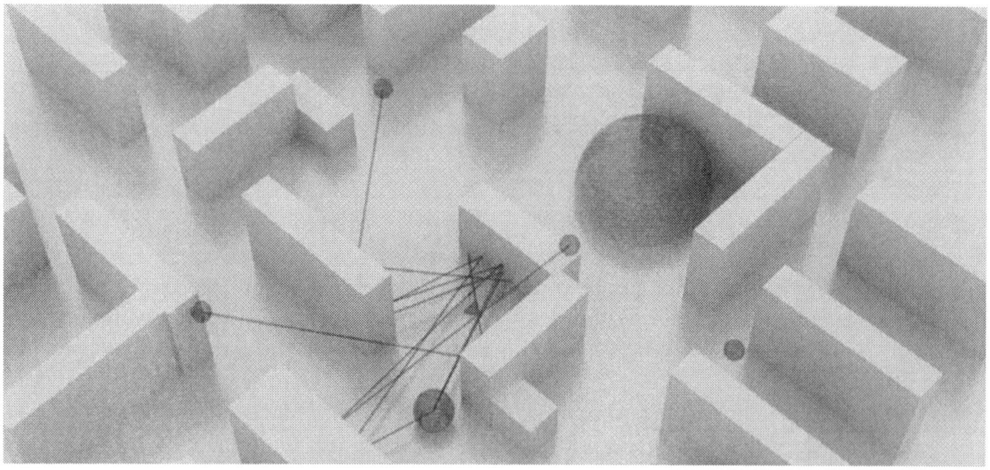

Fig. 2. Esquis'Sons' scene: first reflection visualization (rays), sources (blue spheres) built environment (gray volumes) and listener (red triangle) pointing the "direction of listening".

Esquis'Sons! was designed to allow parallel listening to co-exist: the aim is to make audible – by comparison of these sound atmospheres – sonic consequences of architectural choices made by the designer. A scene is created with the help of the spatial 3D modeler. Then, the scene is sonorized with the Esquis'Sons! application and users can select several listening points that take into account geometrical parameters of the space model.

These parameters are then "translated" into sound mixing properties into the Esquis'Sons! application. Any point of the modelled space can be used as a listening point, distances between the position of the microphone and the sources are automatically calculated and transmitted to the mixing console. The distance between buildings, the openness and closure of the islet, the type of facade, the opacity of the balconies are also evaluated in the geometrical model and generate several filters able to affect the sound sources. Several sources of various nature can be declared in the interface, such as environmental sources (traffic, transports, nature), people's activity (playgrounds, people walking and talking, markets), which can be mixed by trimming levels in the application's console [3]. The resulting sound sketch constitutes a means to assess the design in a qualitative way with sounds while it is still in progress. Esquis'Sons is not a computational tool in the sense of a software that modelizes and calculates a sound scene with a waiting time to be able to visualize or to hear the result (the sound scene). The space environment is sonorized "live" and placing a listening point in the scene is placing a virtual microphone in an environment that never stops sounding. Accordingly, by selecting several listening points in the scene, designers can switch from one point to the other. By comparing and by listening, designers can hear how their project sounds depending on their position in the spatial model, and, of course, also depending on the properties of their spatial design.

Sound propagation and sound behaviour in urban scenes have been translated into Esquis'Sons by using several mixing effects performed by MAX/MSP software and informed by geometrical parameters calculated by Grasshopper from the spatial 3D model [4]. The *auralization* module is informed by the geometric characteristics of the spatial pattern and vice versa. In other words, this application allows you to sketch a space by listening to it.

After several different releases (the last one was the recoding of the application with Python for grasshopper [5] and a future version compatible with rhino6 windows/mac is coming [6]), Esquis'Sons application has been used in several architectural workshops gathering students and professionals from different fields such as architecture, urbanism, music and arts. The main focus of the workshops was introducing the idea of "designing with sound" with an easy sketching sound tool related to the space we are designing. What assessment can we draw from these experiences and how does Esquis'Sons! help working on the sound ambiances that emerge?

2 The notion of sound sketch

As Esquis'Sons is a simple immersion tool that uses stereo sound tracks listened through headphones connected to a computer (or loudspeakers), it is an easy tool to introduce "sound" in the architectural or urban project: the sound scenes created are made of short sounds arranged in loops, selected in the first phase of this research. Users can also add their own recordings. Mixing effects used are here to express mainly geometrical parameters linked with sound propagation and built form effects (distance, elevation, mask, etc.). The qualities of the sound scene could thus be evaluated through realistic sound tracks.

In French, the name "esquissons" has two meanings. It means "let's sketch", but it could also be read as "sketch + sounds". We are talking about sketching as a tool that makes the sensitive qualities in the design "audibles". It then tries to translate the sound in an understandable and accessible way without leading to an oversimplification and loss of information. That is made possible by the new digital tools used in architectural and urban design introducing calculation within the design process.

In French, Esquissons is also a grammatical form that uses the imperative form of the present and which, with a sense of humor, commits the designer to sketch a project thanks to the sound. It is a commitment because the tool does not validate or correct a project already defined in detail, it engages the designer to take risks and to consider for a time that the sound traces of his architectural choices. The engagement is therefore exploratory because the designer creates a scene, proposes one or more scenarios, listens, modifies, listens again, does some changes and so on until he is satisfied with what he is hearing. Then, he notices the spatial architectural consequences of these "sonic" choices.

So why sketch? One can say it is a noticeable part of the design process while its purpose is not to validate or invalidate something already finished. By positioning it as a sketching tool, it is a bet to position the sound as one of the project materials. It is also a bet and a goal to challenge sound in its potentiality to be a main vector to thinking and designing spaces in a more sensitive way. The inclusion of the sound directly in the formulation of architectural interrogations allows to question the formal, material and functional issues of a building or an urban project – those aspects are all likely to affect the propagation of the sounds. It also allows questions on the organization and distribution of activities, uses that can also produce very different sound environments. In other words, it is a way to question the directions and aesthetics of a project beyond its visual representations.

Esquis'Sons! was also developed to exploit the potential of new processes of design that emanate from the contemporary ways to think architecture. Digital tools allow an easier articulation between the different phases of a project. Designers no longer have to respect a logical and linear sequence from the sketch to the simulation validating or invalidating their choices. Exploiting digital tools makes easier to think the project as an iterative process and an interaction is needed between evaluation/simulation and design [7]. That logic is even true between tools and machines as it is possible to prototype inside the design process and not only at the end.

3 Esquis'Sons! in use

Since its first release, Esquis'Sons! has been used in several architectural workshops where, each time, students and professional experiment the use of the application in order to create the new sound atmosphere(s) of a place. As we described it before, Esquis'Sons! is not a simulation nor validation tool, but a tool to engage sounds and their composition in the architectural designing process. Esquis'Sons! has been tested and experimented through design exercises during the CRESSON winter school each year. More recently, it has also been used by an architectural office in Paris to support their architectural feasibility study for the renovation of the historic district (classified heritage of the 20th century) of the Abbey in Grenoble. Observing the designers allows us to discuss three architectural issues that arise from the sound material manipulated.

3.1 Design from inside: how does it change the way designers look at the project?

The first findings could be that the tool allows a more "prospective" way of thinking about sound, space and their links. Indeed, designers are not seeing the project from outside anymore (as it is handed sketched or drawn as a plan or section). Designers are "inside" their project in order to listen to it.

Users are really "using" or manipulating the sound, and not representing it visually or mathematically. It reveals its incredible immersive potential. Sound helps to "be inside" because when you are listening to something, there is no such thing as a zoom or an "unzoom" for example, you are listening to a precise point, inside what you drew.

Teaching in studios in architecture and in workshops reveals that that computer-assisted design tools are very occulocentric – not to say only. Sight is the main sense used to design, sketch, draw and validate any project. These tools tend to distance and objectivize the spaces drawn. Students are working in front of an "object" which is put away, on the other side of the screen. By using Esquis'Sons!, and therefore hearing, it brings users back "inside" their design. Simply with the medium of hearing, it offers a strong immersive potential that helps to design spaces in a more sensitive way: it quickly appeared in the various workshops that the participants positioned themselves in the project and that it gave them the possibility to entangle a set of components that conditions the space and the conditions of possibilities of its ambiances. Thinking sound in the project is not just a technical thought or just a sensitive thought and Esquis'Sons! helps to cross those different dimensions or different approaches of sound in space (see sound sketches of Cresson Winterschool, edition 2017 [8]).

3.2 Scales of design: what are the scales of sounds and what are the dimensions of the space that are heard and designed?

The use of Esquis'Sons! in those different workshops showed that sound can introduce a different way of thinking scales in the project. Three broad categories of "scales" can be described and emerge from participants' productions in the workshops, but one of the stakes observed is a "reimplantation" of the designer into the environment he conceives. The term "dimensions" should be here preferred: it characterizes better the size of the ambiances than the scales of the design. The three categories that came out of those workshops are:

- The ***global dimension***, which is basically about the "urban" scope of space: it focuses on the urban environment, the wide ambiances, the distance and the size of the environment, but also on the background noise and the emergences. It mobilizes the articulations of ambiances, and it can bring out notions of homogeneity, heterogeneity and singularities in the project ambiances. In order to illustrate this from the perspective of the architectural project, this dimension appears when the design is made transversely between the district, the block, the building and the building facade.
- The second one is the ***architectural dimension*** that raises several questions regarding the thresholds, the interfaces, the near and the far or even the displacement and the stay of the bodies in a given space. Working on the porosity of the ambiances, the singularities of the spaces and the singularities of the ambiances between these spaces becomes possible. The categories defined by Gregoire Chelkoff in his work "architectural sound prototypes" [9] could be mobilized here in order to understand the potentialities of Esquis'Sons tool: in another words, it gives the opportunities to test the *articulation*, the *limits* and the *inclusion* of sonic spaces between them and thus to test and work on the composition of architectural spaces.
- The last one is the ***particular dimension***. It is more about the perception and the action of the users inside the building's arrangements (as balconies, terraces, loggias, rooms, passages, etc.) located in a building or at a specific location of a public space (urban furniture for example). It questions the ambiances of proximity, the relation between inside and outside and actually the sound ambiance at the scale

of the human body in action. Again, the focus is on the local emergences but also on their link with the background (feelings of protection, exhibition, projection, feeling of being inside, or outside, etc.). What is the ratio between intimate or peripherical spaces, for example? These dimensions specifically question relations between local ambiances; the interaction with the body is very important here and refers to postures and gesture; can still be illustrated with Esquis'Sons! (listening orientation, uses, social practices, etc.). [10]

Working with Esquis'Sons! then encouraged to investigate and articulate these "dimensions" even when the initial focus was at a precise scale. "Urban scale" design investigates up to the body scale [11], building and architectural designs extend their target and sound intentions to urban design and details [12], and then, detail designs (furniture or facade projects for example) broaden to ambient anchors up to the neighborhood scale [13].

3.3 Sound is time: what does time bring to the project?

Because the basic manipulated material is the sound, Esquis'Sons also deals with another important and crucial parameter: time. Indeed, very often, a visual representation always has the problem of being a snapshot, space is represented by a very specific time. Unfortunately, temporal dimension of the lived space, which is one of the main factors of the ambiances [14] cannot be shown. Because the ambiances change in time, they transform themselves. Our perception is based on a certain "relativity" with the environment and with time. The sound in particular is a dimension of the ambiances and of our environment that can literally not exist without time and that is why it is so difficult to represent without using the sound itself.

Sound only exists in association with time, and listening to the sound sketches makes designers think about space and time together. It could bring different scales of time as for example, the same spot for five minutes or, more often, the desire to listen to space by practicing a path inside the scene (it is one of the first things that people do when they use Esquis'Sons!).

In consequence, using a sound sketching tool brings a very important and specific characteristic of space in the tools for architects or spatial designer: time. It encourages the consideration of the temporal issues associated with an architectural object whether they are on a long-time scale (as variability during the seasons) or on a short time scale (a day, or the time of a given experience). We are more precisely thinking about workshop projects focused on the experience of a walk and the resulting experiences according to temporalities by bringing uses and materiality rather than on a designed object, for example [15]. All these results, their analyses and interpretations are developed and put into perspective in a PhD work in progress [16].

4 Discussion

As seen previously, the Esquis'Sons app is a pedagogical tool for space designers, and it brings to light several dimensions that are hidden most of the time by the domination of the visual sense in the architectural design processes: the question of the distance of creation is discussed by the sound immersion of the designers, scales of the architectural project are jostled by the fuzzy limits of sound phenomena and time bring layers of complexity and sensibility in global design.

To illustrate this part, recently, Esquis'Sons has also been used in a feasibility and programmatic study of the neighborhood *L'Abbaye in Grenoble* (France). The Parisian architectural office Particules [17] used an Esquis'Sons! sound sketch to support their scenario for the rehabilitation of the existing buildings. The district of *L'Abbaye* is a working-class residential area built after the Second World War and currently suffering from a lack of renovation. The majority of apartments, managed by a social housing landlord, are empty because they do not comply with the rules of contemporary construction. The social housing landlord, along with the City Hall had obtained to classify these buildings as elements of the industrial heritage of the 20th century.

Fig. 3. 3D Satellite view (Google map/IGN) of Abbaye District, Grenoble, France, 2019.

All previous studies concluded that it was more advantageous to destroy the 15 buildings concerned and to rebuild new ones that would be better insulated, less energy consuming and more adapted to the uses of the 21st century. Faced with this decision, the former and current residents of the neighborhood have organised themselves in an association in order to defend the hypothesis of the district rehabilitation. The first reason was their attachment to their housing, the second was of course the fear of a very significant increase in rent because of new construction and finally, they argued that the atmosphere of the place is also part of the heritage and needed to be preserved. In this difficult context, the local counsel and the social housing landlord granted the architectural agency Particules the mission to study the feasibility of a rehabilitation plan. Particules worked over a year and has therefore decided to install several teams of architects and researchers in order to understand what makes this place so different or, in other words, to describe which elements of this district are part of the built, cultural and intangible heritage.

The Esquis'Sons! app has been used in order to assist the design and also convince the owners. After an *in-situ* campaign of acoustic measurements, sound recordings and several interviews in order to understand the sound atmospheres that people can experiment in this place, Esquis'Sons! created several sound scenes in order to help the architectural office Particules.

As part of the heritage that had to be "saved", was the semi-open urban islet form of the building blocks than creates an important cutting effect [18] when people get inside. One of the five existing buildings had to be destroyed (the one with the C shape form, *see figure 3*) but a new one is reconstructed with a similar size and footprint in order to keep the urban typology with three half-closed islets. It offers the district new spaces for public services and facilities. One goal is to rebuild almost in the same size the building islet in order to keep the specific sound atmospheres that might be created in the center. It is also the place for plenty of potential sonic interactions between the facades' users and the public spaces. It was also decided to reinforce several uses in the public spaces (as playgrounds for children, a public gardening space, increasing the size and the number of markets during the week, open the ground floor of the buildings to rent for local shops and for craftsmen or young companies). Finally, it also gave goals for the "acoustic" qualities of the facade of the new building. All these scenarios have been input in the application and the result can be listened by following the link below.

Fig. 4. Visualization and final Esquis'Sons! sound sketch for the Abbaye District, Grenoble – (Architectural Axonometry with the kind permission of Particules Office) France, video [19]

To conclude, these sound sketches were actually used as a tool for envisioning architecture and sound atmosphere, but also as a tool to create spaces and practices. As described before, it pushed the architectural office to think the global dimension of the district, which means not only the design of the new building but also its links with the local public spaces, the neighbourhood and with all its local environment – playgrounds, roads, parks, schools, and the city. It helped designers to be immersed in the sound of the public spaces and by selecting and listening from several spots it offered the opportunity to think and design the architectural limits of the project. And finally, by composing this scene, time has been reintroduced in the project and a proposal has been designed in order to offer several remarkable sound ambiances during the day, the week and the year.

Esquis'Sons! is a sketching tool, which means manipulating objects and hearing the results can be quite fast. Thus, we see Esquis'Sons! as a real opportunity – in opposition to the traditional and heavy simulation tools – to bring more attention to sound/acoustic and more generally to ambiances and sensitive issues during the design process. The Esquis'Sons! application offers interesting possibilities for the designers to produce sound and ideas for their designs: it welcomes sound in space creation.

References

1. Esquissons' website : www.esquissons.com [consulted on February 2019]
2. MAX/MSP Cycling '74 – Wrote by Miller Puckette /IRCAM. https://cycling74.com/products/max/ [Consulted on February 2019]
3. N. Rémy, T. Marchal, G. Chelkoff, H. Amini, J.L. Bardyn, and N. Gamal. *Euronoise 2018 proceedings,* Esquis'Sons ! Sketching soundscapes by using parametric tools: application to the design of balconies, loggias, terraces and corridors of building facades., May 2018, Hersonissos, Crete, Greece. (2018) – https://halshs.archives-ouvertes.fr/halshs-01807918 [consulted on February 2019]
4. G. Chelkoff, T. Marchal, N. Rémy. *Proceedings of the 13th French Congress on Acoustics and Vibrations,* Esquis'Sons! Outil d'aide à la conception d'environnements sonores durables, Le Mans, France. pp. 2143-2149 (2016). See http://hal.univ-grenoble-alpes.fr/hal-01341194 [consulted on February 2019]
5. Python interpreter addon for grasshopper: McNeel - Steve Baer & Giulio Piacentino https://www.food4rhino.com/app/ghpython [consulted on February 2019]
6. Esquissons grasshopper plugin and application: AAU Laboratory - Theo Marchal https://www.food4rhino.com/app/esquissons [Consulted on February 2019]
7. A. Picon, Architecture and the virtual : towards a new materiality? *Praxis*, 6, p. 114-121, (2004).
8. Théo Marchal, Noha Gamal Saïd. *Prospectives sonores avec l'outil Esquis'sons!* Article de blog. 2017, http://ehas.hypotheses.org/849 [consulted on February 2019]
9. G. Chelkoff, *Prototypes sonores architecturaux.,* CRESSON, 188p., (2003).
10. N. Remy, & al. *Esquis'sons ! Outils d'aide à la conception d'environnements sonores durables*, Grenoble: CRESSON/ADEME Direction Villes et territoires durables. http://hal.univ-grenoble-alpes.fr/hal-01274959 [Consulted on February 2019]
11. Student work : « Amers sonores » S.B. Romdhane / L. Contramestre / B. de Baudinière. *Winterschool 2017* http://cressound.grenoble.archi.fr/son/winterschool/2017_04_Amers_Sonores.mp3 [Consulted on February 2019]
12. Student work : « Variations en suspension » S. Benhis / Z. Chalaux / J. Di Stefano / F. Kubenka. *Winterschool 2017* http://cressound.grenoble.archi.fr/son/winterschool/2017_05_Variations_en_suspension.mp3 [Consulted on February 2019]
13. Student work « place aux arbres » S. Robert / A. Sayah. *Master 2 workshop Architecture acoustique et espace sonore* ; G. Chelkoff, T. Marchal
14. J.F. Augoyard, L'environnement sensible et les ambiances architecturales. *L'espace géographique,* 4, p. 302-318, (1995).
15. Student work : « inTERREaction » A. Barkani / S. de Pertat / B. Lambert / U. Ozturk *Winterschool 2017* http://cressound.grenoble.archi.fr/son/winterschool/2017_03b_esquissons_in_TERRE_action.mp3 [Consulted on February 2019]
16. T. Marchal, dir. G. Chelkoff, PhD in progress : "Sound Ambiances and Digital Architectures, towards the creation of a sensitive space design tool." – EDSHPT – ENSAG/UGA Grenoble Alps University, http://www.theses.fr/s147276 [Consulted on February 2019]
17. Particules architectural office in Paris and Berlin, Co-founded by Fabienne Boudon et Lou Bellegarde, http://www.particule-s.eu [Consulted on February 2019]

18. J.F. Augoyard and H. Torgue, *Sonic Experience: A Guide to Everyday Sounds (*Andra McCartney & David Paquette translators), Mc-Gill-Queen's University Press, 230 p., (2005)
19. T. Marchal, sound sketch video: http://www.esquissons.fr/tutoriels/1099-quartier-de-labbaye-en-projet/?lang=en [Consulted on February 2019] *we thank Fabienne Boudon and Lou Bellegarde from particules agency for the axonometric picture used.*

Digital Immersion - case studies in virtual reality

Christopher Welty[1] and *Arief* Setiawan[1]

[1]Kennesaw State University, Department of Architecture, Atlanta, Georgia

Abstract. Digital tools have transformed the ways we generate designs, pushing the boundaries in formal explorations, as well as the ways we represent them. In this vein, the impacts of the digital revolution affect both design practice and design education. We would like to investigate these impacts on our teaching of design studios, focusing on the virtual reality. Our premise is that the use of first-person immersion in a virtual environment is a means for experiencing space. We are interested in ways in which the abilities of virtual reality to simulate ranges of sensorial information could inform design processes. This paper will report in our research in our teaching that speculated about the values and application of these techniques. This paper aims to discuss our learning processes and experiences as well as to reflect on possibilities of digital means effective design pedagogy.

1 Introduction

> *"What is taking place is a radical extension of the body and a reframing of its sensorium," he suggested, as computer technologies offer a "multi-layered reality" for redirecting our senses. Instead of "the heroic individual," what will emerge is "a multiplication of sub-selves inside a networked individual."* – Antoine Picon [1]

As in every aspect of life, digital tools have revolutionized the design world. It has transformed the ways we generate designs, as well as represent them. In terms of generative design, their capabilities for investigations of complex shapes allows for explorations of non-Euclidean geometries, in contrast to the limited formal vocabularies of conventional tools. Not only do these tools offer means to generate those geometries, they also allowed for geometric modeling techniques that is very intuitive, such as in parametric designs. Digital tools have also opened up new possibilities and opportunities in realizing design ideas into real structures. Digital fabrication techniques have offered ways to fabricate those formal explorations at the scale of a building. Along this line, building information modeling has helped in constructing buildings in much more integrated and sophisticated ways. In this line of thought, this paper will report in our research in our teaching that speculated about the values and application of these techniques within architectural designs. How can we develop methodologies to use virtual reality to contribute in the design process, not only in terms of representations, but also in generating design? Through this study we document and explore ways in which techniques in virtual reality cover aspects of creating, documenting and delivering decisions about space and professional communications. This paper aims to discuss our learning processes and experiences as well as to reflect on possibilities of digital means effective design pedagogy.

2 Background Thoughts

Digital tools pushed the boundaries of the geometry in generating designs. In the past, using manual tools, designers had to rely on descriptive geometries to arrive at satisfying forms. The tools and the techniques limited the possible explorations of geometric shapes within the realm of Euclidian geometry, limiting the geometry of design to basic shapes and their three-dimensional derivatives. The computational capacities of the digital world changed the situations. With powerful engines, contemporary computers were able to perform much more complex computations in shorter time. Software keeps evolving to facilitate, taking advantages of the advances in the hardware. This phenomenon led to the design explorations in complex geometry such as folding, for example, with examples in practice in the work of Diller Scofidio + Renfro used folding techniques in their design. Even folding in design has been discussed in relations to the philosophy of Leibniz [2]. Another foray into formal investigations opened up by digital tools was the emergence of blob architecture, or Binary Large Object [3]. If Euclidian geometry featured angled intersections of line and planes, blob offered continuous and sinuous geometry, which could resemble organic entities. Some architectural work, such as those by Greg Lynn and Future System, exemplified the use of blob. As Picon has pointed out, in terms of software, the advance in Wire-frame, solid modeling, NURBS (NonUniform Rational B-Splines) have facilitated designers to explore geometry of design in much more intuitive and complex [4]. These innovative geometries understandably led to the fascinations with forms. This indicated shifts from Euclidean geometries to topologies, as discussed by McGrath and Garden, among others [5]. Architecture, however, is not only a matter of form, but also, about space. How the digital revolution in design deal with spatial aspect of design is an interesting point. One way related to this issue pointed to the depictions of space generated by design. A type of computer-generated three-dimensional representations is virtual reality. As computing power increases and software broadens, the use of virtual reality environments and immersion simulations are becoming more accessible. With virtual reality, visualization is no longer dependent on conventional architectural projection drawings such as plans, elevations, sections and perspectives. Instead, it allows for immersion that offered potential advantages of virtual interaction, an experience that is not only isolated with the designer but also open to participant. Designers and participants could engage in the discovery, understanding, and experience of space.

The impacts of the digital revolution affect both design practice and design education. In this vein, we would like to investigate these impacts on our teaching of design studios. Salama has pointed to four aspects of architectural education [6]. The academic aspect dealt with compositional theories and formal design while craftsmanship instilled in students' proficiencies in design trades, including aptitude in digital tools. The latter related to the technological aspect, including building technology as well as technology for various aspects of design processes, including that of digital representations. Along this line, we believe that technical understanding could stimulate imaginations. We are interested in the relationship between design representations through digital media and design thinking, especially in the use of virtual reality in the design studios. We set up our research in the design studio by pairing virtual reality with conventional design processes.

3 Student Case Studies

3.1 Case Study 01: Jose Rodriguez

In our early case study, entitled "*Constructed Reality, A Study in Spatial Perception through Virtual Reality*", Jose Rodriguez was interested in investigating the nature between drawing and space while harnessing the malleability of virtual space. At the beginning of his thesis he asked a simple question, with VR can we experience a drawing? His research indicated that perception and visualization remain the human system responsible for organization, identification and interpretation of sensory information. He argued that VR could bridge the modes of representation between the human eye and the digital lens to create immersive spatial experiences. His research started by taking a historical look at representation and creating a graphical timeline the tracked how representation has changed over time. He concluded that representation had evolved from static one- and two-dimensional representations to dynamic three-point perspectives and collaging. As time passed and the development of how we reproduce improved, complex ideas were better represented and communicated. The study revealed an emphasis of perspectival representation heavily influenced by Brunelleschi's perspective method of tracing reality.

Fig. 1. Rodriguez, Timeline of representation.

Examples from the timeline were analyzed for how they were constructed, identifying parameters of perspective and categorized into the types of experiences and feelings they invoked. The findings of these exercises allowed the project to begin understanding the correlation between representations and the categories of experiences. Further analysis focused on perceptions of space and visualization of space. Findings from these perspective drawing studies served as an influence for understanding static vs dynamic representation. In this light, perspective was used as a way to model spatial perceptions. As the analysis of spatial perceptions advanced Jose, sought out other mediums that communicate feeling or ideas. He concluded that static perspective was created by drawings and photos while movies created moving perspectives. This led to looking at techniques in moving perspectives from cinema. The study then focused on the connection that architectural representation and film have in the way a scene is constructed in order to communicate a feeling or an idea. Film has always immersed us into representations to communicate an expression. Using cinematic principles, he questioned whether we can gain an understanding on how to construct a spatial experience in architectural representation. A questionnaire was developed based on comparing manual perspectives and digital perspectives. It was interested in capturing feedback about perception of space for both architecture and non-design participants. This led to the development of two modules to test ideas learning about the perception of space now in a virtual world.

The first module of exploration in the virtual environment was a study of Le Corbusier's Villa Savoye and how to further understand Le Corbusier's Five Points of Architecture. By using software to model and then porting into a gaming engine, an immersion environment was created to illustrate a dynamic understanding of what Corbusier wanted the user to experience. By using the gaming engine's capabilities, scripts were incorporated to make the environment even more informative with the advantage of being able to tweak the environment itself and allowed the user to filter information to further represent the five points. This case study helped to fully learn the capabilities and disadvantages of the workflow as well as an understanding where the project could head towards in the next stage. The conclusion gained from this exploration was that this was an immersion through input.

The second module in the study of perception in virtual space, used Francesco Borromini's gallery at Palazzo Spada as a precedent. Building on the first module this project looked to replicate an environment of immersion and experience. Again, virtual reality would be employed to study perception and user experience. This time, a user was asked to question the actual size of the space. The gallery columns, vaults, and floors were angled so that a space on a floor plan would normally read at 60 feet long, actually is only 28 feet long. The user's vision is challenged to comprehend the size of the space. It is not until you reach the interior that it becomes apparent the space has been modified. The project led to design strategies based on scale that transformed using physical alterations to manipulate a space.

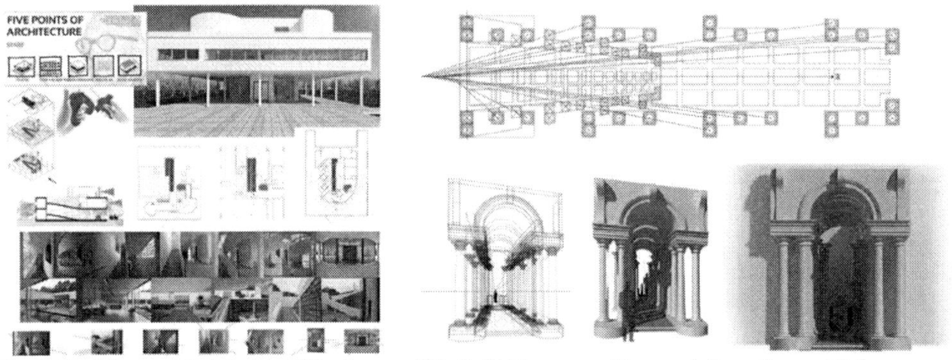

Fig. 2. Rodriguez, Module 01, Villa Savoye VR (left) Francesco Borromini's gallery at Palazzo Spada VR (right)

The final outcome for the thesis was a digital environment entitled "*Immersion Lab*", that encouraged spatial exploration for immersive and dynamic experience. Drawing on earlier modules the project was envisioned as an immersive space based on the synthesis of cinematic representation that allowed a user to experience different types of feelings. The lab was divided into portal exploring two parts of perception, happiness and fear. Villa Savoye, from module one provided the context. First happy, where it is placed in a bright and sunny forest setting complete with a light breeze and birds chirping as you travel down a path, and secondly fear, situates the building in a desolate, post-apocalyptic and damaged environment dark with the occasional rain drop and you hear the wind. A haunting soundtrack quietly plays as the user passes through the shadows. Just like mise-en-scene, the components of the environment could be altered independently so to provide different variations of experience in the desire to create new and unique experiences.

He concluded that the stock experiences presented in the portals helped perception but stayed in the realm of representation and not experience. With virtual reality, the immersion and interactions happened at true scale perhaps illustrating the connection to body and space.

The representation of happiness and fear were significant by themselves but because portals were separate environments there was now link between the two and could be more powerful if they were combined allowing greater interaction between one another.

3.2 Case Study 02: Jay Joyner

In our second case studio, entitled, "*Deciphering Sacred Space, Testing Virtual Reality Through Exploration of Sacred Spaces,*" student Jay Joyner was interested in light phenomenon and immersive experiences as a design tool. His question was whether virtual reality technologies could be used as more than a representational tool. Specifically, he was testing whether VR could immerse the body and allow a designer to test and experience parameters within a virtual space. The research developed in two parallel tracks, understanding light in architecture and developing the technical skills required for application in the virtual environment. The former started with an interest in the essays of Luis Kahn which outlined his design process that begins with the immeasurable and then built through the measurable. Once the architecture is built, the immeasurable starts living again. To further understand the impact of light on architecture the research lead to Marietta Millet and her book, "Light Revealing Architecture," which explores principles of lighting in architecture and how specific lighting effects are realized architecturally. Through orthographic projections, diagrams and renderings Millet looks at contemporary and historical architecture through the lenses of experience, form, space, and meaning. This line of inquiry resulted in a series of cases studies where the student analyzed three case studies known for their use of light, Church of light by Tadao Ando, Kimbell Art Museum by Louis Khan and Chapel of Notre-Dame-Du-Haut by Le Corbusier. In some sense, the research was investigating an analog approach to understanding digital media.

One of the challenges of VR is the required knowledge and technology to stitch all the pieces together. Beginning research focused on developing the skills with the digital tools and techniques of modeling digital light. Small artefact models were used to create and understand different lighting methods and their effects. For the project, the students chose to use Unity software and Oculus Rift VR eyewear. Early digital experiments allowed interaction with light at a human scale with the Oculus Rift. Precedent studies were modeled and evaluated in VR. The analog drawings from the case studies were created digitally with a focus on transforming light that was changed through a series of apertures, filters, and openings. In order to evaluate the digital environments, attention was given to changing the context of each precedent. The outcome was to produce a taxonomy of parameters for light phenomena.

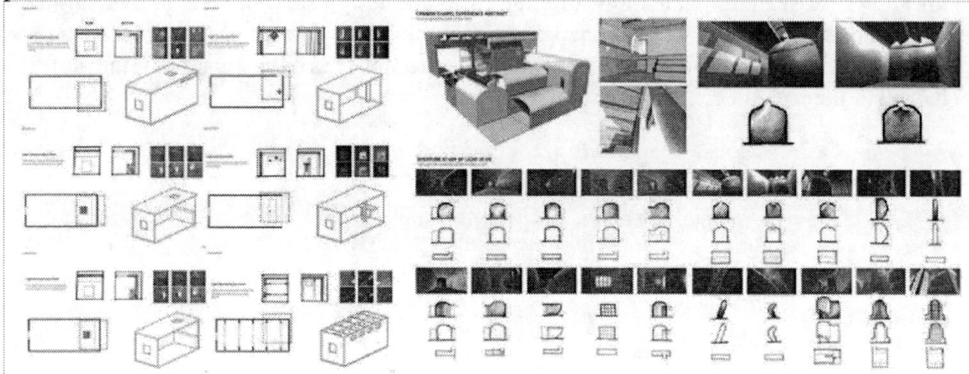

Fig. 3. Joyner, Analog light and digital light studies (left), Cannon Chapel artifact VR (right).

In order for this the thesis to be used as a tool, the "measurable" need to be defined. Again, a combination of analog and digital experience was used. The student set out to visit Cannon Chapel by Michael Graves on the Emory campus. This sacred space is known for its use of light and space. A combination of digital studies and narrative experiences were used to analyze the space resulting experience matrix that documented light, depth, and form through a narrative passage as well as a series of drawings diagrams and sketches.

After experiencing Cannon Chapel at Emory University, he created an artifact of his analysis. The resulting VR model was interested in linking what we had experienced with the virtual world and the digital realm. The arrangement of the artifact was configured based on the analysis of light and space. Materials were not included in the were study so the investigation would focus on light, space, and form.

Empowered by the success of capturing the "measurable" in the prototype digital artifact, Jay proposed creating a VR game to test his hypothesis of digital immersion. Utilizing the taxonomy developed in the research of sacred spaces the game focused on five aspects of understanding architectural spaces, form, space, order, tectonics, and light. Linking to his earlier research, Jay envisioned developing the game to explore the relationship between light, order and materiality. The strategy allowed users to choose design parameters while making design decisions within the different categories to create new sacred spaces and experiences.

3.3 Case Study 03: Holden Holley

In our third case study, entitled "*Beyond Flat Space: Representing Architecture in Virtual Reality*", Holden Holly explores the use of virtual reality as a communication device. Of particular interest was the question of how could VR be used to communicate with clients to influence design intentions and client understanding. Holden speculated if he could harness the information present in a BIM model to efficiently create virtual reality.

Holden argued that drawings once done on a drafting board are now being constructed in BIM programs like Revit and believed that constructing a BIM model represented a majority of the work of moving to VR and would easily dispel clients concerns with cost.

Early research focused on the available of hardware and software. Of particular interest in the modeling process was the integration of virtual software within these programs. He researched the interoperability of design and production programs to work together. For instance, Revit can use almost any BIM file type and convert to almost any file type as well. This allows for seamless transitions through different design software and allows for the use of virtual design methods without changing too much of our current design process. It can be as simple as a plugin for our current software or as complex as exporting to another software. This led to the use of Enscape with Revit.

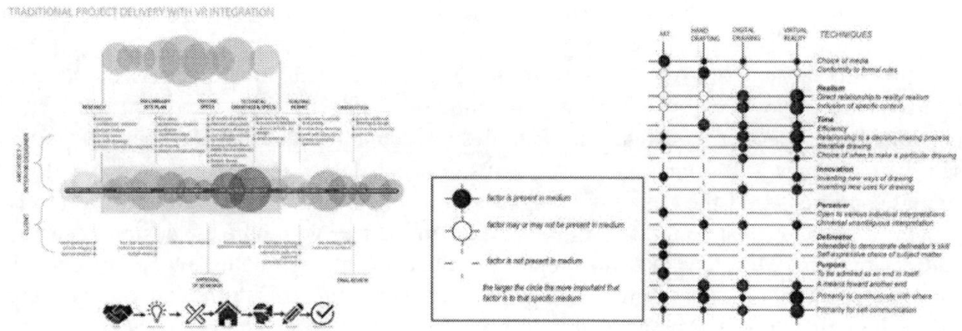

Fig. 4. Holley, Tradition design workflow integrating virtual reality (left), Representational style analysis including virtual reality (right).

Parallel to software and hardware analysis Holden speculated how this technology could be integrated into a traditional project delivery method. He concluded that virtual reality could be used anywhere from late stage schematic design to construction documentation, due to its integration within our current processes and software. For his research Holden developed a matrix of virtual reality representational techniques and their definitions. Current reality as defined as “what is”, “what already exists”, or “the norm”. Augmented Reality as a digital overlay on real scenes. A mix of computer-generated images and real-world photos. A new category called augmented virtuality utilizes virtual imagery with a layer of realism. It is more in depth and more virtual than augmented reality. Virtual reality, the area of his research, is the category of “what could be”, completely generated by digital means. This study led to the developed a comparison of representational techniques in with virtual reality which illustrated different types of design media and the strengths and weaknesses of each based on measurable attributes. He analyzed that virtual reality is very good at simulating realism and can be used for both self-communication and communication with others. He concluded that virtual reality not only helps us as designers see realistically how these spaces come together, but it also is equally as good at showing clients the same information about these spaces. He cited five reasons to use VR in architecture, the ability to make changes on the go, an added billable value to the project, increased attention at presentations, the media allows for long-distance presentations and the ability to persuade a client based on visuals. His final project tested these ideas with a collaborative effort to create a VR presentation of a fellow student’s urban intervention project which called for series of spatial experiences.

4 Discussions and Reflections

In a way, virtual world could serve as a new prosthetic, an extension of the human body to experience spaces. Technological advances not only expanded possibilities for visual experiences of space, but also allowed integrations of the use of other senses in experiencing space, hence, simulating both visual and haptic experiences of space. We are interested in these abilities to simulate ranges of sensorial information could inform the design thinking. They gave new ways to simulate space and evaluate beyond the formal aspects of spaces, its shapes and dimensions, into the ability to feel spaces. Indeed, these simulations allowed designers to assess factors including the effects of colors, textures, materiality, and lighting conditions, both natural and artificial. Thus, they allowed for the increasing capacities to assess and evaluate the performance of space that underlined efficacy of design explorations. In this line of thought, Picon pointed to emerging notions of affect and elegance [7]. The latter referred to the aesthetic aspect of formal explorations; a way to assess and appreciate the formal results of design explorations that accentuated the richness and complexity of

https://doi.org/10.1051/shsconf/20196401007

geometric shapes, yet maintained the coherence. Picon cited the description of elegance by Ali Rahim and Hina Jamelle as a quality obtained :

"through the creation of a family of formal features that are distinctive, yet remain interrelated as the transform form one to another" [8]. The former, affect, covered the performative aspects of design, the way design elements modulated space and created sensory experiences. It is concerned with ways in which design moves informed and affected spatial conditions and the behavior of spaces, informing spatial experiences beyond given programs. A factor that could determine spatial experiences would be the design of surfaces. Along this line, Moussavi has developed arguments that ornaments and other design of surfaces were more than creating visual interests, but creating sensations and affects in space [9]. The notion of affect, with regards to both the foray into design of surfaces and to the impacts on spatial experiences, underlined what Picon called as the spatial turn and spatialised-intelligence, in which knowledge embedded in space [10]. Hence, our perceptions and experience of space carried with them some forms of knowledge, most importantly the knowledge on spatial organizations and manipulation. Simulations of spatial experiences through digital simulations offer the possibilities to uncover and explore this knowledge. Hence, digital tools foregrounded the primary role of space, as they were a tool not only to generate design, but also to feel space. These investigations attempted to understand the relationship between the human body and one's spatial context.

In this research in design teaching, we are interested in the notion of spatialized intelligence, that space itself carried with it a range of knowledge that designers and users could unearth. In this line of thought, digital simulation produced virtual spaces that designers and users could experience, even though they were still in the realm of ideas. The question for us was to find a model to understand this thought process, in which we turned to the notion of cybernetics or feedback loops. McGrath and Gardner identified several features in cybernetic [11]. The cybernetic process looked at the whole context, hence, architecture in relationship to all factors that informed it, including contexts and users, instead of architecture as a free-standing object. Because of this, the inquiries turned to explore relationships and mapping those. The findings of this would be the patterns of these relationship and their qualitative aspects. The thought process itself worked from continues feedback loop instead of in a linear fashion. Our work with students demonstrated that constructing virtual models allowed them to immediately assess spatial conditions as a result of design iterations. These virtual models allowed them to bring together changes and alterations of various factors, including changes in plans, sections, elevations, openings, and surface conditions. They also allowed for modification with regards to the site and site conditions. Hence, their virtual models provided a means to integrate relationship between various space-defining elements, between space and the environment, and between space and users inside and outside the space. Further, the models provided the results of these integration immediately. In this sense, the models served to map relationship that emerged in the design process. Each design iteration proceeded with feedback generated by these virtual models.

This finding related to the nature of design thinking itself. With regards to the generative process of work of arts, Nelson Goodman has distinguished the types of ways in which the work emerged into autographic and allographic works [12]. The former was a way in which the work of art was produced directly by the hands of the artists, such as paintings and sculptures. On the other hand, allographic was a way in which artists worked indirectly in creating work of art. It required media through which the artist worked to realize the work. Architecture fell within this category, since architects produced drawings instead of buildings. Obviously, in our inquiry, the production of architectural representations, both drawings and models, happened digitally. As we documented in our research, students had to work with multiple software and hardware, each of which required a certain level of competency. The implications were the need on the aptitude in working with software, as well as in connecting between software.

In reflections, our experience with these projects demonstrated that virtual models and digital immersion are less of a design tool. Instead, they remain at the level of tools for architectural representations. The projects that we worked with tended to focus more on technical issues, mainly learning in mastering software and hardware. As a consequence, the process of those projects was less on exploring spatial and formal aspects of design, but more on constructing representations of design. Engaging this mode of architectural production, in terms of design pedagogy, seems to result in developing the skills and ability and less in understanding of architectural design. Hence, it is more on the vocational side and less on the academic dimensions. A distance emerged between the digital fluency and the spatial knowledge. On the other hand, it led to a knowledge in connecting different digital tools. In a way, the problems during the process seemed to shift away from simply dealing with ordering space and form to figuring out ways in which software and hardware work and, more importantly, strategies to connect those digital tools.

Their value, however, lays in their capacities to widen the possibilities of understanding the impacts of design decisions. Especially, it lays in their efficacy in providing that information. This strategy allowed an engagement with design results, as it allowed viewers to be inside a space, and highlighted relationship between various design variables. This mode allowed to integrate changes and map the result. Further, it allows for developing ways to measure changes. Hence, it maps changes and results from that changes. The case with case 1, Joyner developed tools to change surface properties, of which the models allowed us to assess the impacts; while with case 2, Rodriguez developed tools to control variables in perspective. In this line of thought, this mode allowed for controlling changes in a measurable manner. In a way, this mode provided a means to quantify atmosphere. This would be the potential in which digital immersion could serve as a means in activating the feedback loop processes.

References

1. A. Picon, "Vital Signs: Is drawing dead?," *Constructs*, p. 9 (Fall 2012).
2. See for example G. Lynn, *Folding in Architecture: Architectural Design Profile 102*, Academy Editions (London: 1993).
3. See the discussions in A. Picon, *Digital Culture in Architecture: An Introduction for the Design Professions*, Birkhäuser Architecture (Basel: 2010), p. 62-65.
4. Ibid., p. 70.
5. B. McGrath, J. Garden, *Cinemetrics: Architectural Drawing Today*, Wiley (New York: 2007), p. 15.
6. Salama, Wilkinson, "Introduction: Legacies for the Future of Design Studio Pedagogy," in Salama, Wilkinson, ed. *Design Studio Pedagogy: Horizons for Future*, Gateshead, UK: The Urban International Press, 2007, p. 4-5; A. Salama, *New Trends in Architectural Education: Designing the design studio*, Tailored Text & Unlimited Potential Publishing, (London: 1995), p. 40-41.
7. A. Picon, *Digital Culture in Architecture: An Introduction for the Design Professions*, Birkhäuser Architecture (Basel: 2010), p. 100-107.

8. ibid., p. 100.
9. F. Moussavi, *The Function of Ornament*, ACTAR, Harvard Graduate School of Design (New York: 2006), p. 5-12.
10. A. Picon, *Smart Cities: A Spatialised Intelligence*, Wiley (New York: 2015), p. 11-14.
11. B. McGrath, J. Garden, *Cinemetrics: Architectural Drawing Today*, Wiley (New York: 2007), p. 24.
12. N. Goodman, *Languages of Art*, 2nd.ed., Hackett (Indianapolis: 1976), p. 113.

Mixed Robotic Interface Γ: Searching for a hybrid cyber-physical design/experience interface using virtual/actual robots

Ebrahim Poustinchi[1,*]

[1]College of Architecture and Environmental Design, Kent State University, Kent, Ohio, USA

Abstract. Mixed Robotic Interface is a project-based design-research investigation, studying new ways of creating hybridized cyber-physical design and experience interfaces, at the intersection of robotics—as its core component, and augmented reality, game design, projection mapping, and digital fabrication. Mixed Robotic Interface Γ—as part of Mixed Robotic Interface series of research projects, focuses on using "actual" and "virtual" robot arms as a possible creative medium and extensions of design/gaming environment creating immersive atmospheres for "experiencing" design. This research questions the possibilities of creating an architectural/spatial atmosphere through digitally enhanced experiences. Different from some of the current experiments with augmented reality (AR), virtual reality (VR) and projection-mapping in architecture, Mixed Robotic Interface Γ is not looking into "immersive" experience as a way to "blur" the boundaries of digital and physical—similar to virtual reality experience with headsets. Instead, Mixed Robotic Interface Γ creates a recognizable gap between real and virtual to open up a creative space for the user/audience to be involved between these two mediums. Mixed Robotic Interface Γ uses para-fictional storytelling as a way to engage the audience with the experience and to create continues atmospheric qualities.

1 Introduction

Almost immediately after the introduction of digital tools for design in the early 1990s, the question of bridging this medium to the physical word has been one of the central demands in the field of digital design. Introduction of computer numerical control (CNC) machines, followed by the presence of robotic arms as part of the fabrication facility in design set-ups, made this communication between the digital environment and physical world happened. This bridging brought up the question of the role of technology—specifically robot arms, not only as a fabricator (as they have been intended to be used in the past couple of decades), but as possible physical extensions for digital design/experience platforms.

* Corresponding author: mpoustin@kent.edu

On the other hand, the rapid development of immersive user experience (UX) platforms such as XR (Augmented, Virtual and Mixed reality) and interactive projection mapping, made it possible to "project"/overlay the digital content directly onto the physical world to create hybrid curated atmosphere.

Mixed Robotic Interface Γ research aims to study user experience through storytelling and environmental ambiance, and uses both a robotic arm (both virtual and actual) and immersive experience platforms—specifically AR and projection mapping. The "Impossible Objects" design series by Kruysman-Proto and Curime Batliner [1] and the "Aether Project" by Guvenc Ozel studio at University of California Los Angeles [2] are some examples of using a combination of projection mapping and robot arms to create "qualities" and ambiance. Another familiar example of the use of robotics in conjunction with projection mapping to curate an atmospheric spatial effect is the work of Bot and Dolly—creative robotic studio in designing and shooting special effects of the "BOX" [3].

2 Mixed Robotic Interface Γ

Mixed Robotic Interface Γ follows Sylvia Lavin's "Man vs. Machine" retooling creativity exhibition conversation, where Lavin asserts that increasingly larger amounts of creative resources are being put into producing new tools and concepts that are designed not to make things but, to amplify the creative capacities of others [4]. Similarly, in his article "Postarchitecture," Guvenc Ozel calls for a combination of virtual and actual, and argues that our surrounding "space" is a hybrid of both physical materials and electronic bits [5].

Mixed Robotic Interface Γ, as a design research investigation, looks into new technologies and their possibility to become mediums for experiencing atmospheric effects of design.

Enhanced with both augmented reality techniques and projection mapping methods, through video/animation making, storytelling, cellphone-application developments, digital and experimental fabrications, and "super-graphics," Mixed Robotic Interface Γ presents two projects: "Robotic Pool Party!" and "Child Robot". Both projects look closely at the issue of cutting-edge technology in conjunction with storytelling and "user" experience and focus on three central themes:

- Storytelling through para-fictional scenarios, where a technological icon—an industrial robot in this research, becomes one of the core storytellers.
- Creating curated ambiance/experiences through seamless (or purposely "seemed") cyber-physical interfaces.
- Revisiting the potential of projection-mapping and augmented reality—as existing techniques for immersive design, to create custom atmospheres considering current contemporary discourse in architecture through classical issues/conditions such as poche, corner, envelope and new concepts of part-to-part and part-to-whole relationship, considering flat architectural ontology as suggested by Tom Wiscombe [6].

2.1 Mixed Robotic Interface Γ project I: ROBOTIC POOL PARTY!

Designed and exhibited as part of the *3-Ways* show curated by *Anthony Morey*, *Ivan Bernal* and *Ryan Tyler Martinez*, at *the A+D Architecture and Design Museum* in Los Angeles, the "Robotic Pool Party!" project revisits the potential of virtual animated robots and projection mapping as a method for experiencing the space and creating atmosphere through story telling (Figure 1).

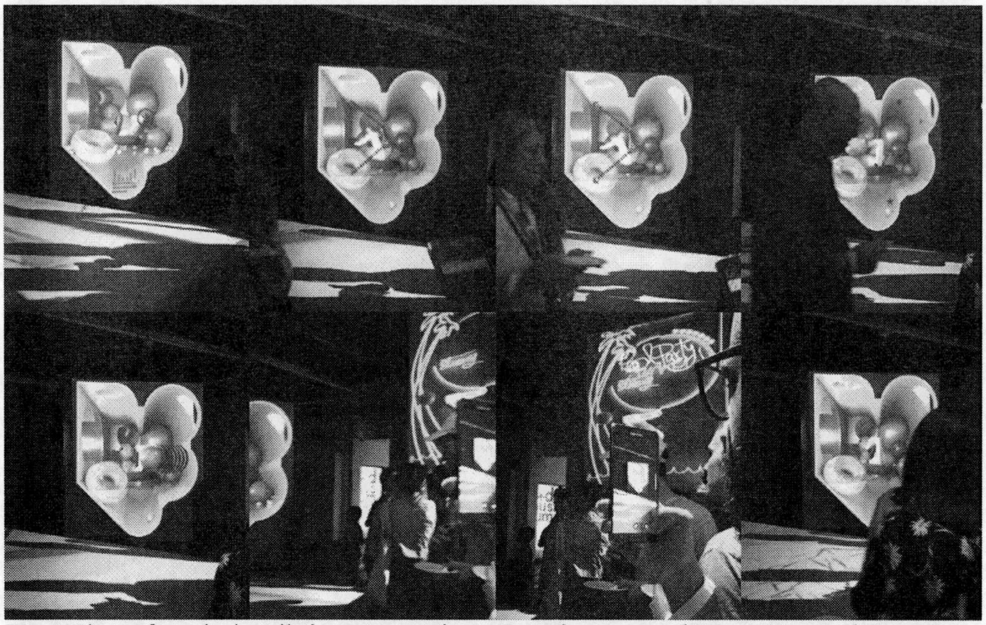

Fig. 1. Photos from the installation set-up at the *A+D Architecture and Design Museum* in Los Angeles. The animation is projected on a physical "canvas" provided by the museum and has become a window to a new ambiance.

Using animation as a story-telling medium, "Robotic Pool Party!" uses a combination of realistic and fake representational techniques to achieve a unique mixed atmosphere. Realistic physic simulations of Autodesk MAYA in conjunction with Inverse Kinematic (IK) techniques—to animate the robot on one hand, and highly saturated plastic/toy like qualities of the materials—even water, on the other hand, creates a blended yet divergent mixture of qualities to craft a unique story that lives between mediums (Figure 2).

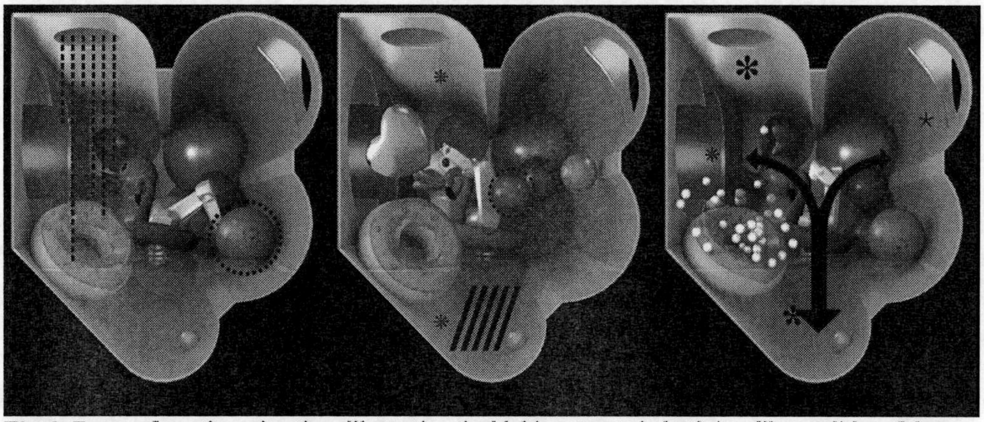

Fig. 2. Frames from the animation, illustrating the highly saturated plastic/toy like qualities of the materials in the scene.

Migrated from a factory assembly line, the robot at this pool party becomes a "humanoid," and the ultimate character of the "story". Celebrating the friction between actual and virtual, and real and projected, and interacting with its surroundings, the robot—in combination with its context, becomes a medium in which, the audiences, experience the atmosphere of the installation. The robot plays (interacts) with inflatables in the pool augmented with humanistic gestures and characteristics, as a way to engage the audience into the "performance" and story of the installation. In addition and through poche and sectional studies, "Robotic Pool Party!" also examines the potential of fake-depths created using projection mapping and in contrast with the project's three-dimensional canvas. Using non-realistic representational techniques, audience can clearly distinguish the projected/fake depth. However, this detachment is being used as part of the story to ultimately increase the engagement of the audience throughout the installation.

To architectural respond to the idea of celebrated representational detachment between the projected content (virtual) and the physical context of the show (actual), the project uses the notion of object-inside-an-object as Tom Wiscombe describes. "Robotic Pool Party!" aims to amplify the atmospheric contrast between the dark plain cubic physical canvas of the projection and the vivid, playful, dynamic and active projection content. The Section is dynamically changing to illustrate the difference between the outside and "inside" of the created depth as a way to engage the audience in the story/atmosphere that the project is aiming to develop (Figure 3).

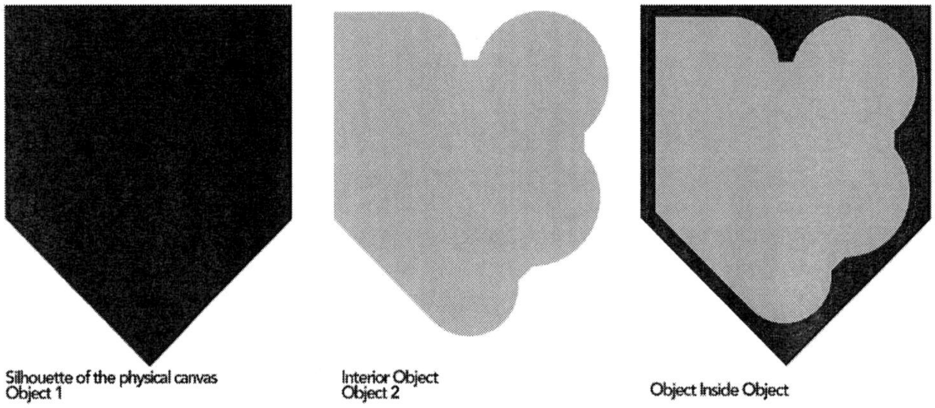

Fig. 3. The use of poche as a way to amplify the atmospheric contrast between the outside and inside.

2.2 Mixed Robotic Interface Γ project I: Child-Robot!

The Child-Robot project studies the relationship between actual and virtual—through storytelling as a bridge, and using a robotically animated hybrid—digital-physical scene/stage as the vehicle. Using multiple techniques and platforms ranging from custom augmented reality cell-phone applications, custom robotically camera controlling platform, green-screening techniques and super-graphics as spatial connectors, the Child-Robot project redefines the use of augmented reality techniques, not as a seamless "immersion" tool, but as a gateway to a story that is happing in two different worlds: the world of the physical robot and the world of the digital animation overlaid through AR (Figure 4).

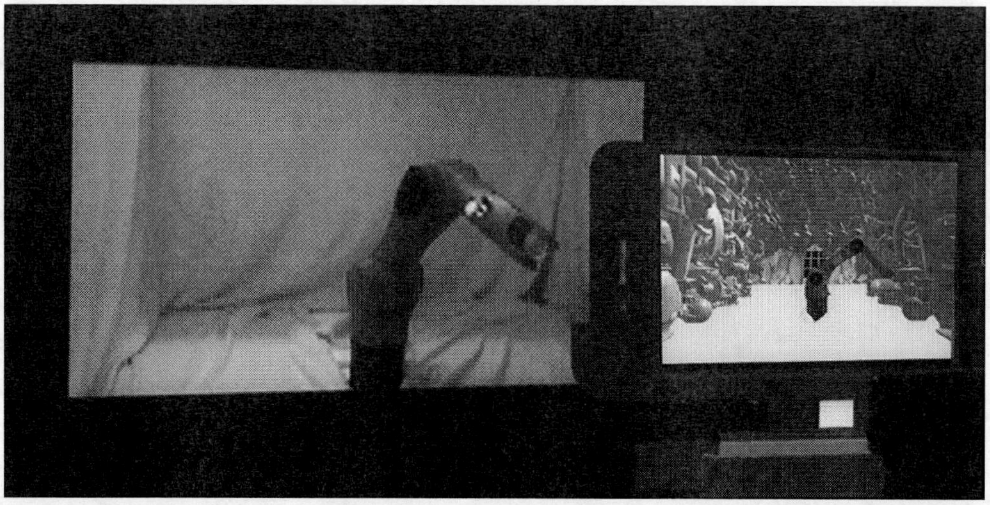

Fig. 4. The cell phone AR application creates a mixture of realities with combining the virtual room and the actual footage of the robot arm.

The story is set in a para-fictional scenario on the speculation of a future child's room where the robot arm plays the role of the child to illustrate a futuristic atmosphere and as way to initiate the story. The set-up is the result of a precise calibration of the physical robot arm, the virtual room and the virtual animated kit-bash scene (child-robot room) to enhance the experience. Using augmented reality techniques, Child-Robot enables the users to experience a mixture of reality, fantasy, actual and virtual through a custom-made software application and an actual scene. The audience can move around the robot in action, and through the lens of the device, overlay the customized context of the story to enrich the experience/performance of the robot as the actor (Figure 5).

Fig. 5. Screen captures from the robotic scene experience through regular and augmented cameras. The robot becomes the performer of the story and the AR environment, amplifies the atmospheric effects of the story/performance by introducing a toy-based.

The audience can move back and forth between the actual robot/physical set-up and the virtual/augmented environment, with which the robot is interacting. Using Oriole—custom-made robotic videography and controlling plug-in for Grasshopper 3D, Child-Robot also examine the potential of a hybrid digital-physical set-up to create, curate, tell and visualize a story from both robots' perspective and the audience/users (Figure 6).

Fig. 6. Screen captures from the robot perspective. Using Oriole plug-in for the grasshopper, the audience was able to also see through the eyes of the robot, as the leading "actor" in the science.

The Child-Robot project questions the potential of hybrid digital-physical platforms as not only design, representation, and simulation tools but ultimately as a source for spatial experience and storytelling where digital and physical, augmented and real, and artificial and natural blend into a distinguishable mixture.

3 Conclusion

The Mixed Robotic Interface Γ examined the potential of storytelling as a medium for immersion. Using robots as "programmable" actors that "think" digitally and "act" physically, and through AR and projection mapping techniques as transitional "windows" between actual and virtual, we observed that initial studies—Robotic Pool Party and Child Robot projects created engaging experiences for the audience through a hybrid experience medium. Going back and forth between multiple representational and atmospheric set-ups, and having the ability to immerse into the digital experience without losing the connection to the physical environment, we observed through informal conversations and questioners that our audience fund Mixed Robotic Interface Γ different from conventional AR or projection mapping. The main difference was that in this research projects, instead of the tool/device—virtual reality or mixed reality headsets, the story became the primary tool to blur the boundaries between actual and virtual.

4 Limitation | Delimitations, Future Research Plans

Mixed Robotic Interface research in general and Mixed Robotic Interface Γ in particular are very young research intonations that are seeking new mediums for design, communication, experience, and storytelling. One of the primary limitations of the current research is the limited number of users and limited feedback from them. Currently, we are looking into more proper data-collection techniques—such as human factor measurement or questioners, as a way to precisely compare the differences between Mixed Robotic Interface Γ and other existing methods. Another parameter of this research that we would like to develop further is the use of Mixed Robotic Interface method as an architectural design tool to generate from, space and compositions based on the interaction of the audience with the "mixed" story.

Currently, we are working on expanding the research, looking into set-ups with multiple robots where both the physical space and the digital "projection" are equally dynamic and performative.

References

1. B. Kruysman and J. Proto, "Impossible objects," Architectural Design, no **227**, ed. Bob Sheil, 106–111, (January/February 2014).
2. G. Ozel, "Case for an Architectural Singularity: Synchronization of Robotically Actuated Motion, Sense-Based Interaction and Computational Interface," in *Proceedings of the 34th Annual Conference of the Association for Computer Aided Design in Architecture (ACADIA)*, ed. David Gerber, Alvin Huang, and Jose Sanchez , 399–408, (2015)
3. Bot & Dolly Fuses 3D Animation and Industrial Automation, (2013, November 25),
4. Retrieved from https://www.roboticsbusinessreview.com/consumer/bot_dolly_fuses_3d_animation_and_industrial_automation/
5. S. Lavin, (2015, July 13). "Man vs. Machine: Sylvia Lavin Retools Creativity." Los Angeles Forum for Architecture and Urban Design. Retrieved from http://laforum.org/delirious/man-vs-machine-sylvia-lavin-retools-creativity/
6. G. Ozel, "Toward a postarchitecture," Log **36**, 99-105, (Winter 2016)
7. T. Wiscombe, Discreteness or towards a flat ontology of architecture. Project, (Issue **3**), 34-43, (2014).

"Characters in Search of an Author". Human figures and storytelling in architectural design communication

Fabio Colonnese[1]

[1]Sapienza University, Dept. of History, Drawing, and Restoration of Architecture, 00186 Rome, Italy

Abstract. In a presentation drawing, human figures have a natural attitude to work as an optical reference to measure the design space and to provide a sort of instructions to use it, but over the centuries, their agency has been multifaceted. The practice of photo-collages, which was fed by photography and cinema development, has recently spread thank to the digital techniques and human figures in architecture renderings seem to have become as fundamental as a top-modes for a fashion magazine cover. Besides providing a recognizable mark to the design entry, selecting peculiar figures can visually connect a design to a specific place and time, working as a cultural, situationist and sensorial agent. This seems to be true particularly for the cultural typologies. In renderings of museums, theatres or libraries, often ordinary people are integrated by figures of artworks and celebrities, like in Alberto Campo Baeza and Raphael Gabrion's design for a Louvre new building in Lievin, France, whose figures are placed in the renderings not only to explain the functions but also to remind the ambiguous threshold between representation and reality.

1 Introduction

In the context of the digital production by "CAD monkeys, rendering farms, and out-sourcers", hyper-rendered seductive scenes look often detached from real people and places [1]. An architectural "scene" is commonly intended as a mere picture of an imagined place for commercial goals, while it can also take into account the social context and agenda to depict a theatrical "scene", in which "action is carried on, business is being done, or events are happening" [2]. This specific attitude is currently practiced by those architectural visualizers who use to portray designed buildings immersed in peculiar environments, at night, under the rain falling or half-hidden in the mist. The people inhabiting the picture, whose agency seems to have become more and more central in any architectural communication, exploit much of the potential narrative of an architectural "scene". This seems to be particularly important for the culture-addressed typologies such as museums, theatres or libraries. Here, the diffusion of digital photo-collage in the renderings gives the architects the possibility to integrate figures of ordinary people with those of celebrities, artists and artworks, amplifying the ambiguity implicit in every visual representation and questioning the relationship between art and architecture. The analysis of the 2015 competition entry for the Louvre new building in Liévin by Alberto Campo Baeza and Raphael Gabrion provides the opportunity to frame and discuss the current development of human figures in the communication of architectural designs. In particular, the analysis of the digital renderings by disassembling the visual components and individuating the source of some of the human figures and artworks shown, compared with the specific quality of the proposed building, reveals not only the central role of figures in denoting it as a Louvre-related museum and the activities planned inside it, but also a critical approach toward the museum as institution, the relationship between true and fake, and the fragile boundary between virtual and real.

2 Human figures in architectural drawings

Renaissance architects used to associate the human body to architectural orders and overlaid it to church plans and palace facades to demonstrate the proportional relationship between human and architectural body [3]. Instead, "conceptions of the body of modern architects have their roots in the post-Galilean view, which conceives of the physical body as a machine and as a subject of mechanical laws" [4]. In particular, between the 19th and 20th century, the human body was gradually reduced to a series of physical parameters and performances to optimize production pipelines, machines, and urban spaces, generally idealizing movements and behaviours according to Taylorism. From half of last century, the presence of human figures in architectural drawings is being interpreted as symptom of a particular sensibility toward human scale and needs, not only according to the kinds of representation but to the different ideas on men and women.

Fig. 1. Luigi Rossini, Veduta del Cortile di Palazzo Silvestri, 1818 ca. On the left, the original print; on the right, a photomontage showing the actual size of people compared to building's size (Museo di Roma, MR23670; Elaboration by the author).

Human figures are conventionally used in architectural drawings as an optical reference to visually express the scale and size of design space, [5] like "scalies", as suggested by Waverly Lowell [6]. In contrast, they can be used to deceive the actual size of either a real place, like the little courtyard of the Palazzo Silvestri in Rome drawn by Piranesi's follower Luigi Rossini, or imagined spaces, like in artfully oversized perspective views of Boullée's metropolitan church. In this role, human figures may either look as consistent as the architecture is, like the black shadows drafted by Mies van de Rohe [7], or even become grey or transparent silhouettes to better reveal the buildings behind them.

Human figures may affect the reception of an architectural scene by contributing to defining its spatial and temporal context, but they are also used to provide the readers with further information on either the represented architecture or its designer's ideas. For example, they can provide a sort of instructions-to-use the design space or some of its components: an attitude inversely proportional to their graphic definition and detail, as the more a figure is detailed, the more it suggests an additional fictive role in the drawing and engage reader's body. Think of Le Corbusier's domestic interiors inhabited by people acting, reading, painting, cooking or playing, in order to describe the potentials of modern housing and to offer the viewer a new way of living with an unconventional sense of freedom. Somehow, Le Corbusier used people not only to explain his unconventional spaces but also to promoting and "marketing an aspirational lifestyle" [8].

Fig. 2. Bramante consulting a plan in the courtyard of Palazzo della Cancelleria; Baldassarre Peruzzi and Raphael observing the frescoes in the Villa Farnesina; Michelangelo inviting Raphael in his own house; Vignola in Villa Giulia's courtyard. From Paul-Marie Letarouilly, *Edifices de Rome Moderne*, 1868-74 (Elaboration by the author).

In the fundamental *Edifices de Rome Moderne*, Paul-Marie Letarouilly used to add the figures of the Renaissance architects he supposed to have designed the palaces and villas he had surveyed, "corrected" and illustrated. The figures of Bramante, Baldassarre Peruzzi, Raphael, Vignola or Michelangelo are not only measuring the space and reminding the designer, but also legitimizing the rectifications Letarouilly made on the effective building surveys to restore their original form, at least according to his own idea. Quite on the same wake, a century later, the American architect Richard Meier cut-and-pasted Karl Friedrich Schinkel's trees and copied Otto Wagner's old-fashioned figures to provide an historical environment to his museum designs [9]. While these apparent contradictions can be easily accepted in an architectural survey, they look bizarre in the rendering of an architectural design, which is expected to show an anticipation of an upcoming transformation. The existence of elements contesting the space-temporal reality of the design are used to make the beholder formulate conjectures about them, adding a mental dimension to the mere visual fruition of the picture. Although the final visual effect of the rendering can be quant or perturbing, those anachronistic elements may help the design be recognizable, distinguished and memorised by the commission of a competition.

3 From Manual to Digital Photo-montage

Adding people to an architectural scene requires a certain compatibility between the characteristics of the single figure (pose, clothes, light-and-shade) and that of the space representation (point of view, surfaces rendering, colors, shadows, etc.). Architecture visualizers can take staged perfect people from one of the several on-line digital photography stacks or can do it by themselves, by cutting-and-pasting figures of their own friends or other real people doing real things. Of course, one can also choose peculiar people, such as politics, celebrities, artists and even fantastic characters in order to convey specific meanings or political agendas [10]. In the early decades of the 20th century, when the practice of collage and photomontage spread, the photographs available, often offered by a few fashion magazines, were scarce and their fixed size strongly constrained their use in a perspective view, contributing to the surreal and subversive effects that have made the collages of Max Ernst or Bruno Munari famous.

Fig. 3. P. Bottoni, G.L. Giordani, M. Pucci, Design of Piazza del Duomo in Milan, 1937, Photo-reproduction detail of the perspective photomontage (Milan, Politecnico di Milano, DAStU, Archivio Piero Bottoni, op.155, Photographs, 18).

Inspired by early photography artists as well as graphic designers, architects began to enrich their drawings with photo-collage treatments, often with original consequences. Since 1930s, the Italian architect Piero Bottoni, one of the pioneers of this technique, has applied the cinematic practice of self-cameo, which was pioneered by Alfred Hitchcock, in his perspective photomontages, revealing how the growing influence of photography and cinema enhanced the fictive role of human figures in presenting architecture [11]. In general, the use of photographic figures in collage-drawings has the potential consequence of both relating the picture to other places and times and of including extra-architectural narratives useful to convey artistic, political, and social meanings, especially when figures of recognizable people are pasted. In the 1952 project for the Golden Lane reconstruction in East London, Alison and Peter Smithson cut-and-pasted the photographic figures of cinema and sport celebrities like Gerard Philipe, Claude Autant-Lara, Peter Ustinov, Marylin Monroe and Jo di Maggio onto their renderings, to express the subversive and ground-breaking character of their housing design and, at the same time, to superimpose a sort of new “as-found” reality [12].

After the political peak of architectural photomontages by Archigram and Superstudio in early 1970s, the advent of Xerox machines and then of digital technologies removed the aforementioned operational limits and today copying, cutting, adapting – in terms of size and colour – and pasting a figure taken from the internet massive archive into any digital rendering is easy, fast and for everyone equipped with a Photoshop-like software. At the same time, human figures in architectural renderings seem to have become as fundamental as top-models for a fashion magazine cover.

4 Figures in the Museum

Designing a museum is always dealing with the concepts of art and social institution, especially in the classical wake of architecture as a synthesis of the arts. Already Schinkel's perspective view of the main staircase gallery of Alte Museum in Berlin shows human figures "captured at the moment of being amazed by the paintings or immersed in a conversation while walking through the museum, act as characters in a montage that unleashes the imagination of the spectator." [13].

From the second half of 20th century onwards, architects such as James Stirling or Mansilla+Tunon took the habit of adding peculiar figures in the renderings of their museum designs. For example, the rooms and garden of Thyssen-Bornemisza Museum Stirling designed in 1987 are haunted by British artists Gilbert & George, De Chirico's silent muses and renaissance artists with their young apprentices, while black-and-white photographic figures of artists like Le Corbusier and Joseph Beuys appear like tutelar deities in most of Mansilla+Tunon's renderings [14]. At the same time, the photographic images of artworks can be directly reproduced in the renderings, even when they are made with manual techniques, as can be seen in Steven Holl's watercolour views for the Swiss Embassy Gallery in Washington, DC.

Fig. 4. Alberto Campo Baeza and Raphael Gabrion, Conservation and Storage Facility for the Louvre Museum in Liévin, 2015. Digital rendering of the building (Courtesy of Alberto Campo Baeza and Raphael Gabrion).

4.1 Campo Baeza and Gabrion's design of Louvre

In 2015, the Spanish architect Alberto Campo Baeza and the French architect Raphael Gabrion have proposed a further innovative step in this kind of architecture presentation. Together with Elena Jiménez, Tommaso Campiotti, María Pérez de Camino, Imanol Iparraguirre, Ignacio Aguirre, Alejandro Cervilla, as well as Raphaël Gabrion's collaborators in Paris, they have designed a rectangular building for the Conservation and Storage Facility for the Louvre Museum in Liévin, in an area in which two sturdy coal chimney stacks still mark the memory of a previous anthracite mining centre. The building is described as "A large, dark podium housing all the workshops and services, with the cubic part of this great warehouse emerging like the chimney of a great boat" [15]. Such a metaphorical reading is suggested by their motto *bateau ivre* (the drunken boat), echoing the title of the well-known Rimbaud poem and working as a primary metatext.

The project is presented by a number of different drawings and pictures: plans, sections, pictures after a *maquette*, exploded axonometric view and perspective section after a digital model, and no less than 15 black-and-white renderings: seven outdoor and eight indoor views. Conceived as a concrete cubic storage crowned by offices and laboratories around narrow patios, the building looks like neither a building dedicated to art nor an artwork in itself. Besides a huge *fenêtre-en-longueur* looking onto the coal pyramids, the renderings show the rooms are quite neutral, characterized by indirect light provided by the closed courtyards and a high sky-light. In the architectural context, most of the explaination of functional goal of spaces is trusted on the figures pasted onto the views.

Fig. 5. Alberto Campo Baeza and Raphael Gabrion, Conservation and Storage Facility for the Louvre Museum in Liévin, 2015. Digital rendering decomposed in main layers (Courtesy of Campo Baeza and Gabrion; elaboration by the author).

4.2 Artworks and Figures

Figures of artworks constitute a first category of visual elements denoting the building as a cultural institution. Moreover, the sculpture known as *The Marly Horse* or Eugène Delacroix's *La Liberté guidant le peuple* are commonly relatable to the Louvre, with Hubert Robert's *La Grande Galerie* directly showing its interior rooms. The figures of people looking at, moving or working on the artworks constitute a second category of visual elements identifying the museum as a place for ordering, conserving, restoring and exhibiting pieces. Some of them are not generic figures, like tourists and visitors looking at works, maybe taken in the rooms of Louvre itself, but specific figures. For example, a group of them is presumed to be pasted from a screenshot of Stan Neumann's *Leonardo da Vinci: The Restoration of the Century*, a documentary presenting the restoration of Leonardo's *St Anna and the Vergin* made between 2010 and 2012.

The rendering of the main entrance to the building shows a larger figure pasted on the right, absently welcoming the beholder. It is clearly recognizable as the Italian Neorealism director Vittorio De Sica who seems to have been chosen for some specific reason. The figure has been cut out of a 1950s picture taken in London and recently shown in the exhibition *Tutti De Sica* [16]. He might suggest an intimate relationship with the Pantheon, which Campo Baeza is very fond of, for a sequence of his 1952 movie *Umberto D.* was shot outside the monument and its light oculus is quoted in the Louvre design sketches, too. Eventually, De Sica could somehow validate the black-and-white quality of the renderings and convey both the idea of fiction, which is implicit in every representation, and the condemnation of the current trend of spectacularizing museum space.

Fig. 6. Alberto Campo Baeza and Raphael Gabrion, Conservation and Storage Facility for the Louvre Museum in Liévin, 2015. Digital rendering of the main entrance with Vittorio De Sica cut, mirrored and pasted from the 1950s picture on the left (Courtesy of Campo Baeza and Gabrion).

In the rendering of the "large format unpacking area," the sculptures and potteries among the wooden boxes are looked at by an elegant man seen from behind, which has been cut out of a screenshot of Giuseppe Tornatore's *La migliore offerta*, a movie released in 2013. The figure shows the Australian actor Geoffrey Rush as Virgil Oldman, a standoffish and aloof art auctioneer. Together with his friend and painter Billy, Virgil uses his expertise in forgery to manipulate the mind of the bidders and acquire artworks at a very low price. The figure is taken from the sequence in which he is revealing his secret collection of woman portraits to his beloved Claire, a fragile lady apparently affected by agoraphobia he had known for business but eventually revealing to be a member of thievery plan organized to steal Virgil's priceless collection. On the first impression, the figure of Virgil could simply recall the visual seduction of art as well as the excesses and risks of its trading, but there something more. The concepts of fake and authentic are central in the whole movie, as testified by the following passage:

"– Claire: In an old article of yours I found on the internet, you said: There's something authentic in every forgery. What did you mean?"

"– Virgil Oldman: When simulating another's work the forger can't resist the temptation to put in something of himself. Often, it's just a trifle, a detail of no interest. One unsuspected stroke, by which the forger inevitably ends up betraying himself, and revealing his own utterly authentic sensibilities".

This dialogue casts a different light on the story and the rendering itself. It reminds that every fiction, like an architecture scene apparently designed to sell a building, hides something deeply real and the figures chosen to inhabit the scene are likely to be that detail from the real life betraying the author and declaring his or her true intentions.

Fig. 7. Alberto Campo Baeza and Raphael Gabrion, Conservation and Storage Facility for the Louvre Museum in Liévin, 2015. Digital rendering with Geoffrey Rush from La migliore offerta, compared with the movie original shot on the right (Courtesy of Campo Baeza and Gabrion).

4.3 Representations of Representations

A third category of visual products denoting the building as a cultural institution is established by three supplementary photomontages conceived as "postcard images". More than just renderings, they show the museum design inserted into the background of Jean-François Millet's *L'Angélus,* Claude Monet's *Soleil Levant* and as a model onto the table of Jean-Baptiste-Siméon Chardin's *L'Enfant au toton*. All of them being property of Louvre, they constitute an original way of presenting a museum design by providing a sort of short-circuit between container – the museum – and content – the work.

Despite Leon Battista Alberti's and Raphael's condemnation of perspective views in architecture design communication, the use of painting is a long-term consolidated practice, from Hubert Robert and John Soane to Zaha Hadid and Pezo von Ellrichshausen, or the Italian architect Giancarlo Rosa, whose competition entries are usually presented by a De Chirico-like painting by his wife and artist Maria Carla Savi [17]. Nevertheless, Campo Baeza and Gabrion's postcards are something different. They are neither architecture renderings with special figures pasted onto nor paintings made on purpose for illustrating an architectural design but existing paintings turned into architecture design presentations. It is no more the human figure that contextualize the project but the project image that recontextualize the figures and the whole painting, parasitizing their story-telling and unlocking it to new meanings.

The intervention on *L'Angélus* has involved some propaedeutic alterations of the work. For example, the higher part of the sky with some ducks flying has been cut away and a wider distance has been created between the two figures in the foreground in order to better fit the building shape in the background as well the two pyramids of coal taken from a photograph.

Quite the same building shape has been placed in the upper part of *Soleil Levant* (it is quite difficult to speak of perspective or depth here) and textured with impressionistic brushstrokes captured from the painting itself, finally removing also the artist's signature.

The intervention on Chardin's *L'Enfant au toton* is by far the most interesting of all. Here, the design is presented in form of a little dark-grey model onto a desk replacing the spinning top the boy was playing with. This montage causes several consequences on the design reception. First of all, the model is connoted as a game in itself, a sort of innocent board-game or mysterious mathematical box explored by the boy's right hand. Second, this is no more a building in a landscape painted onto a canvas but a representation of a building into a room painted onto a canvas, furtherly mediating the sense of the picture and deceiving any clear boundary between reality and representation. Third, the spinning top removed cannot but recall Christopher Nolan's *Inception*. As known, in the movie released in 2010, Leonardo Di Caprio's character Cobb used a similar tool to understand if he was currently living in a dream or in the reality. Once again, this missing detail in the "postcard" could be a cinematic suggestion reminding that this picture is just a fiction, a sort of virtual reality with several levels of meaning but, at the same time, that it can be as important as the reality is.

Fig. 8. Alberto Campo Baeza and Raphael Gabrion, Conservation and Storage Facility for the Louvre Museum in Liévin, 2015. Digital rendering of the building pasted onto Jean-Baptiste-Siméon Chardin's *L'Enfant au toton* compared with the original work at right (Courtesy of Alberto Campo Baeza and Raphael Gabrion).

Conclusion

In the context of the massive production of anonymous renderings generously worldwide spread by internet, selecting peculiar figures may result an antidote against the impending process of homologation and provide an opportunity to orchestrate the architectural scene according to a sort of secret plot. This strategy can be used also to connect the design to a specific place and time, working as a cultural, situationist and sensorial agent, or to deny it, by perturbing the design space with anachronistic elements that require a conjectural extra-work, like an open work waiting for the reader's collaborative interpretation.

The analysis of the renderings of Alberto Campo Baeza and Raphael Gabrion's design of the Conservation and Storage Facility for the Louvre Museum in Liévin, resulted in a differentiated use of human figures and artworks in relationship with their perception by the beholder:

1. Measuring figures, dedicated to express the size of design space;
2. Functional figures, dedicated to express the activities of design space, here connected with the presence of artworks;
3. Fashioned figures, dedicated to contextualize the design in a specific place and age;
4. Recognizable figures, dedicated to provide cultural and social sense to design space, calling for subjective interpretations in the effort of deciphering the hypothetical intent of the designer. This effect is particular evident and layered when these "celebrities" are either involved in the visual arts or imagined characters from literature or cinema, for instance.

The three "postcard images" are an innovative format to present an architectural design that appears to fit perfectly the topic of the museum but would probably be inappropriate with other architectural typologies. They are also a litmus test of the opportunities offered today by the technique of digital collage in mixing-up and deceiving visual products from different sources, eventually contributing in tearing down every traditional wall between different artistic expressions and challenging any division between representation and reality.

References

1. B. McGrath, C.-L. Hsueh, P.C. Hoi Shan, Rendering the Social in the Architectural Scene: Digital Representation and Social Inclusion on Architectural Design, Thinking, and Education. *AMPS*, **8**, 4 (2016), p. 1
2. McGrath, Hsueh, Hoi Shan, Rendering the Social, p. 1
3. M. Frascari, The body and architecture in the drawings of Carlo Scarpa. *Res,* **14** (1987), pp. 123-142
4. R. Imrie, Architects' Conceptions of the Human Body, Environment and Planning. *D: Society and Space,* **21,** 1 (2003), p. 47
5. A.T. Anderson, On the Human Figure in Architectural Representation. *JAE*, **55,** 4 (2002), pp. 238-246
6. *Designing People,* 2015 exhibition at the Environmental Design Archives at the University of California Berkeley. Https://archives.ced.berkeley.edu/blog/upcoming-exhibit-designing-people
7. F. Espuelas, *El claro en el bosque: reflexiones sobre el vacío en arquitectura* (Barcelona: Fundación Caja de Arquitectos, 1999)
8. McGrath, Hsueh, Paul Shan: Rendering the Social, p.1
9. F. Colonnese, Human Figure as a Cultural Mediator in Architectural Drawings. *Cultural Influences on Architecture* (Hershey, PA: IGI Global, 2016), pp. 110-111
10. H. Melcher, From fireworks to kayaks, step inside the well-produced, world of renderings, https://archpaper.com/2014/07/from-fireworks-to-kayaks-step-inside-the-well-produced-world-of-renderings/
11. F. Colonnese, Collage and Photomontage in 1930s: Piero Bottoni's Architectural Designs. *Studies in Visual Arts and Communication*, **5**, 1
12. V. Iborra Pallarés, I. Capdevila Castellanos, J. L. Oliver Ramírez, Drawing a New Architectural Paradigm. *Graphic Imprints*. (Cham, Springer, 2018), pp. 274-284
13. J.M. Falcón Meraz, El espacio gráfico habitado: la figura humana en la comunicación visual arquitectónica. *Arquiteturarevista*, **11**, 1 (2015), pp. 36-37
14. Colonnese, Human Figure, pp. 116-117
15. https://www.campobaeza.com/es/conservation-storage-facility-louvre-museum-lievin/
16. *Tutti De Sica*, exhibition arranged by G. L. Farinelli, Museo dell'Ara Pacis, Roma, 8 febbraio – 28 aprile 2013
17. G. Rosa, *L'insegnamento della composizione architettonica* (Roma, Officina, 1998)

Exploring territories by documentary: The appropriation of public spaces by collective listening

Luciana Roça[1,*], *Maria J. S.* Martins[1], *Marcelo* Tramontano[1]

[1]Nomads.usp. University of São Paulo, Institute of Architecture and Urbanism, CEP13566-590, Av Trabalhador São-Carlense 400, São Carlos, SP, Brazil

Abstract. Sound has a relational interplay with space, influencing the arrangement of ephemeral and symbolic territories in public spaces. Within this perspective, this article aims to discuss the influence of collective listening in the appropriation of public spaces, discussing the events conducted spontaneously by the youth in the Coronel Salles Square, São Carlos, Brazil. Weekly on Fridays, this marginalised community builds an ephemeral territory in the square, which is also influenced by collective listening by powerful speakers in cars. As methods, we use documentary procedures to take a closer look at these events. Being an assertion of the historical world, a perspective is taken by a non-neutral part, a documentary performed as practice-as-research offers a useful set of methods to researchers to explore urban issues and matters and, it is a reflexive method which questions the researcher's posture.

1 Introduction

Sound plays an essential role in the apprehension of the city and corresponds to society's ways of living by reflecting its actions and dynamics of daily life. Sound is an element that has a political and aesthetical potential since it reconfigures and participates in the multisensory apprehension of the city. Within this perspective, this article considers an inter-relational approach between sound and space, arguing that collective listening influences dynamics and uses of public spaces regarding ownership, frontiers and refuge.

Therefore, the sounds play a role in sociopolitical relations that take place in urban environments, influencing or reassuring perspectives, and making pre-existing conflicts more remarkable or elaborating them.

To discuss these matters, we develop a reading on the occupation of Coronel Salles Square (São Carlos, São Paulo State, Brazil) by teenagers who play aloud Brazilian funk music to provide a meeting point, as a form of appropriation of public space. The context of these practices is discussed briefly in the following section of this article. This article has an

[*] Corresponding author: lucianaroca@usp.br

interdisciplinary approach by making a documentary in the research process to reach a specific issue in urban spaces [1].

The documentary tradition has complex and overlapping boundaries with fiction since documentarists, screenwriters and directors of fiction narratives are constantly exploring its limits [2 - 4]. In this article, we do not elaborate an exhaustive analysis on the film narrative or history of documentary film, since other authors have contributed significantly with this matter [2 - 5]. Instead, we argue that the means of a documentary video are useful to elaborate issues on daily life and, at the same time, being a non-neutral form of approach and representation, provides an immersion in which the researchers/documentarists may be aware of their influence on the addressed issues. This awareness raises questions on how to represent what we explore or discover during the process as well as it enabled a more in-depth approach of the researchers on the event.

The documentary film raises questions on the construction of the "self" and the "other", and it has been used as a form of field research by social scientists, anthropologists, media researchers and others. Within this methodological perspective, the documentary video is fruitful for urban studies, either through analysis of documentaries [6] or by producing a documentary with research purposes [7]. The members of Nomads.usp, Center of Interactive Living Studies at the University of São Paulo, and research partners have been exploring the making of documentary videos as a key to interpret the city [8 - 10].

2 "Let's go to the square!"

Fig. 1. View of Coronel Salles Square on Google Earth. The square's area is highlighted.

It is Friday night at São Carlos, in the countryside of São Paulo State, which is the State with the highest population density of Brazil. The Coronel Salles' Square is in the centre of the city of approximately 220 thousand inhabitants. The Coronel Salles' Square, also called "Pigeons' Square" or "Central Square", is in an area of historical, commercial and economic interest. Many adolescents come from the impoverished peripheries by walking, by bus or by hitchhiking, while some of the oldest people, above 18 years old, arrive at the square by cars with potent speakers. The "open-air party" gathers teenagers and young adults, listening and dancing to Brazilian funk in most of the time. Here, cars with powerful speakers, of which some are from dB-SPL competitions and others called "*paredão*" (we can translate literally as "big wall" since it is formed by speakers put side by side), play Brazilian funk and is possible to listen to it a few blocks away. The lyrics of the music try to expose the problems of daily life in the *favelas* and other marginalised communities, discuss criminality and sometimes has high sexual content (*funk putaria*). These events occur on Friday nights and persist until the last buses leave or until the arrival of the military police.

There have been registers accessible online on these events since 2010 by local television news [11]. According to members of tutelary council and the secretariat of childhood and youth, both from city hall government at 2010, this group tried to occupy other different areas in the city, a shopping mall and another square. Even though this group has continuously been changing due to their ageing or change of personal interests, they were expelled by police, inhabitants or security guards from both locations.

Throughout this time, an independent music collective gave structure to these events by offering speakers and a cultural program. Later, the city hall government did similarly, but imposing more restrictions, as they curated the music. The programs used to include dance competitions, such as b-boy and street dance, and rap battles. Nevertheless, there was already some conflicts concerning the neighbourhood, mainly because of the loud sound, and some people also argued that there were problems with drug trafficking. In 2013, the city hall stopped its support due to the change of government by the elections. During the time of this research, in 2016 and 2017, there was a complicated situation between the teenagers and military police, a process that is further aggravated by the presence of drug dealers. Despite this statement, this context still represents a lack of events and shared activities in public spaces destined for these teenagers.

Despite its specificity, this context is essential for discussing this event. We understand these events on Friday night at the square as an ephemeral and symbolic territory, formed by the appropriation of this public space by the teenagers. We understand that the concept of 'territory' lies beyond a macropolitical configuration and formalisation, so uses, dynamics, practices also form a territory. As the Brazilian geographer Rogério Haesbaert [12] argues, the relations of appropriation are in a continuum: from the political-economic domination to the subjective and 'cultural-symbolic' appropriation. The appropriation of public space is also a matter of territory, which is constructed with exercised power by groups, institutions and people, in many different scales. Surely the power relations in this symbolic territory are not simple, due to the complex relationships among law, teenagers, drug dealers and the sociopolitical context. The party is not open to everyone since this event builds up an ephemeral symbolic territory that is difficult to be permeated by people who does not fit this group. This appropriation is fostered using collective listening. To observe and understand these events, we understand the concept of territory as fundamental to understand the relationship between people and space and recognise the importance of sound and collective listening.

3 Methods: making a documentary, interpreting public spaces

To better understand these relationships, we rely on our process of making a documentary video as an immersion on this subject, to approach and cover these relations of sound and appropriation of public space. The complex process of creating a documentary video, including research, argument, script, interviews, field recording, sound and image montage and exhibition, can be performed as a method to read and represent the dynamics and sociopolitical issues of space.

The audiovisual product, and more precisely the film, fiction or documentary, is composed of a set of choices and stages of production. From the script to the final product, the audiovisual work, film or video, this process of non-neutral representation proposes many questions to the researcher. There is a historical line that separates fiction and documentary, notwithstanding both documentary and fiction filmmakers may tension the limits of this line.

At first, we must acknowledge that a documentary is a narrative form of audiovisual that is a creative representation of reality, an assertion, created by a non-neutral observer. The documentary video does not have ties to rules of narrative form or tools; i.e., it does not work with stable regulations or standards, as the film theoretician Bill Nichols [2] argues. Albeit it seems like a free form of work, the documentary may repeat a series of strategies in its process. Beyond that, a documentary has ties to a part of reality; it is an assertion, a point of view of a part of the historical world, that may be ethical or not, as Fernão Ramos [3] argues. A documentary shows an idea, an assertion on a specific matter. To consider a documentary as a statement also remarks that there is an oscillation between a historical reality's recognition and its representation by the observer, and the documentarist can also influence this reality, by acting in the given situation.

The documentary is a set of choices, stages of productions, raising questions, as well as highlighting to the researcher its choices, the point of view, ways of interpreting. The documentary itself is a proposition of field research and tool for revisiting the field during its post-production — furthermore, the documentary results in videos that can go beyond the academic community.

We had as a starting point to research tensions, limits and territories in urban space through documentary processes. This process from research to the final documentary took approximately five months. After that, we have made discussions on these processes and a reflexive analysis.

Regarding the documentary, we choose the following methods for our documentary on the events of the square: primary research on the overall context and the activities at the square; non-participant observation, in different schedules; participant observation with interviews with the occupants; video recording and audio recording. In addition to this, the post-production (i.e. video and audio editing) was also a moment to deepen the reflection on our role, as researchers, in this context.

Fig. 2. The teenagers in the square. Frame from the documentary video. Images by Maria J. S. Martins and Luciana S. Roça.

The field research conducted our choices for the documentary. First, the localisation of the square, surrounded by a sports club, commerce, banks, schools and the city council. The square without trees or comfortable seats is not so inviting, especially during the day due to the heat and sun. There are not so many places of permanence or high contact [13]. On the other side, when the events took place during our presence, this does not seem to be a problem to occupy the square. Groups usually occupy the stair steps while the cars are commonly in a narrow passage close to the city council. Since the beginning, it was already difficult for us, researchers, to be at the square even without cameras or audio recorders. During some interviews, the teenagers talked about the retaliation they suffer from the police, just for being at the square. Despite not being utilised in our documentary, the interviews highlight a strong opposition from the teenagers to the police. Thus, we took the ethical choice to not include any interviews on the documentary to save the identity of the teenagers and young adults, even with proper legal authorisation.

The young people interviewed by us reside peripheral neighbourhoods of different areas of the city. Most of them use public transportation to get to the central square. In some cases, their responsible adults bring them to the square. The meetings are publicised at social networking events and in the "word of mouth" between friends. What motivates them to go to the square is the possibility of meeting; some of them argue that in their neighbourhoods there are no adequate spaces to meet and that the square is the place where they can find young people from different regions, which makes possible to meet new people and make new friendships. This stage of the work was fundamental to understand the context of the occupation and decisive for the elaboration of the final argument of the film.

Fig. 3. The teenagers in the square. Frame from the documentary video. Images by Maria J. S. Martins and Luciana S. Roça.

During our field research, we notice shared isolation among the groups. The belonging and appropriation of the square denote both a moment of being together in small groups but, still, isolated from the others who are out of the group. The specific behaviour, clothes, music, walking in small groups, drinking, affective encounters are welcomed in this temporary territory, suspending established social rules. The territory is marked by its use and by who perform it. Admittedly, it generates conflicts by their presence, making the surrounding inhabitants uncomfortable. Then, this reinforces the argument in which the police act to protect the public space and the wellbeing of the surrounding inhabitants. The collective listening strengthens the ephemeral territory but is also part of the moral argument since the sound is a disturbance for those who do not fit in this territory.

In the post-production, we revisited the images and the audios. Trying to interpret this appropriation, from a perspective closer to the teenagers' point of view, we decided to assume that we are out of these groups, looking at them. Otherwise, more time would be needed to strengthen relationships with the people who go to the square weekly. The resulting video states a point of view in which we, the researchers and documentarists, are out of this group, by the position of the camera. We, as outsiders of this group, assume our difficult position in this territory by the camera.

The process of conception, accomplishment and finalisation of the documentary was a reflexive process, full of questionings. Acting on the border of this refuge established by the youth enabled us to comprehend the dynamics better and to question ourselves, the place of the researcher on making assertions. This blurry border is a locus of conflict and, at the same time, of communication. In performing the documentary, there was not only a process of looking at the situation and context; making documentary has also become a look at ourselves before the world as researchers. Then, it is not only a matter of gathering information on this event to analyse it, but it is also a matter on how to perceive forms of interpretation and perception, as in a reflexive practice, given that the documentary is an assertion of what happens in a particular situation.

4 Gathering together in public spaces: on collective listening and its ephemeral territories

The sound reproduction of music by loudspeakers, resulting in collective listening in public spaces, is not a novelty in many regions of the world. It may be on the sidewalk of a bar, the stores announcing special offers and playing music aloud, someone playing music on the mobile phone without earphones in the streets. Despite that, the context of our event is fruitful to bring into discussion the tensions and power relations reinforced by collective listening in public spaces, on how music allied to social practice and context can form temporary territories and expose its tensions and limits.

A common sound environment may characterise the collective listening for a group of people; it is the opposite of individual listening, such as the one enabled by headphones. Collective listening is a sound environment which establishes a common ground among the listeners, which can be listened attentively or not. It may invite people to participate within a determined event or not since it addresses cultural value to the activity.

Collective listening may be or not elaborated by sound reproductions and other technological mediation. Evidently, the more intense the sound, the broader is the area affected by these sounds and, consecutively, more difficult if for someone of not being part of the activities. In our case, collective listening is intentionally formed by a group using sound reproduction, the speakers in the cars. The collective listening in the events of the square is a transformation on the current sound environment, adding cultural and social values to the activity. By doing that, the music aids to build an ephemeral territory in the square.

It changes the sound environment of who have the intention to participate, dance and listening what is played, but also changes the sound environment of other communities, pervading physical barriers that mark a difference from the home/street, to what is a shared and collective space and a private one.

In our study, we understand that the matter of collective listening of music in public spaces is beyond that of leisure activities. It reinforces a symbolic territory and dominance over the geographical location and is one of the few opportunities of these people to play dominance over the city's space. On the other hand, this claim to take part in the city's public life and to occupy public spaces is temporary ownership that excludes who is different from the formed group. This point and counterpoint, between us and the others, appropriation and disturbance, is also a reflection on the socio-spatial segregation of the city and how it is reproduced in its public spaces.

Beyond that, Brazilian funk is marginalised, often regarded as a minor musical genre, and it is also frequently portrayed in media as a subculture permeated by drug trafficking [14]. For that, the listeners of this genre are also taken as uncultured by the dominant discourse. These stigmas reinforce a territorial relationship elaborated by the teenagers and their conflicts with surrounding inhabitants and police. The events provide a duality between a cycle of agency and invitation to join the party, the ones who gather in the square, and the disturbance and the pervasion of their personal space, i.e. the neighbourhood, is inherent of the characteristics of sound. Some legislation on sound and activities in cities are a reflection on these issues, also influencing on what we consider as noise [15].

The collective listening helps an appropriation of the public space by a group, forming borders, an inside, together with the group, and an outside, areas where the others are tolerated. The structures of physical space also reinforce these outside areas, such as the street. Nevertheless, in our case, collective listening invigorates the dualities between public and private space: on the one hand, the group takes temporary ownership of the square, transforming it in a temporary private space; on the other hand, the walls of the surrounding private residential properties are not sufficient to restrain sound. By considering sound as a phenomenon that influences our apprehension of the space, it may forms territories with porous borders and limits.

The collective listening is part of the significant transformation on the way of interacting in public spaces since it denotes a conviviality among equals, people and groups that share the same preferences or sense of belonging. Collective listening, as in our case study, also implies territories and power relations among people, groups, institutions, law, and economic and political forces.

The collective listening builds up an ephemeral territory with a blurry boundary, and by the very characteristics of sound, it is also a disturbance for those who do not participate or not want to take part of the activities, since sound can pervade spaces considered public or private. Moreover, the stigmas of funk also add cultural and social value to the activity, reinforcing the territorial aspect of the event.

5 Findings

The documentary methods are useful to approach a perspective, as immersion and as an investigation to comprehend other worldviews. As a non-neutral representation, an interpretation, the process of a documentary video invites the researcher to be in the thin line of the observed territory. The documentary can aid the researcher to approach a specific way of life or situation, as some urban anthropologists already did. Furthermore, the documentary, as an assertion of the historical world [3], invites the researcher to be continually questioning its perspectives and statements, demanding a posture. There are ethical questions; historical and social contextualization; the sensibilization towards to the approached issues; to comprehend and form the assertions by the combination of images and sounds.

The documentary proved to be a useful way of approximation and immersion on these issues to understand the formation of this practised territory. Furthermore, a documentary carries fruitful methods and proposes situations to discuss the urban problems, which may be convenient to many knowledge areas such as urbanism, architecture, history, cultural studies, social sciences, and others.

A symbolic territory and appropriation are given due to the conditions people give ephemerally to this public space. The square becomes a fleeting territory of shared isolation, suspending established rules and enabling their own, designating what is acceptable or not for the group. The actions and dynamics of the teenagers resignify the given built space of the square, but at the same time this marginalised community is appropriating a public space, paradoxically they exclude the others too, resulting in conflicts.

The music reproduction, resulting in a collective listening, influences this symbolic territory and tensions the limits of it: while the collective listening is a matter of a constructed identity and belonging to these groups, it also tensions the blurry borders of this territory since the neighbourhood understand it as a disturbance.

Collective listening is enabled by sound technologies of reproduction for almost a century, changing forms of appropriation of public spaces and influencing its dynamics. The study of different activities performing collective listening, the modalities of appropriation of public spaces by sound technologies compose a future subject of research.

The collective listening is a perspective to study the transformation on how people interact in public spaces since it denotes ephemeral territories and power relations between people, institutions, and other groups, and not only social forces but also political and economic ones. Beyond that, collective listening also denotes a conviviality among equals, people and groups that share the same preferences or sense of belonging.

We thank FAPESP, The São Paulo State Research Foundation, for the financial support of the research. We thank Prof Dr Arthur Autran, Federal University of São Carlos (UFSCar), for providing fruitful discussions on the relationship between documentary and city, and Varlete Benevente, for the contribution.

References

1. The documentary can be found at the link https://youtu.be/5xM3KuCiZoc.
2. B. Nichols, *Introduction to documentary*. (Indiana University Press, 2001).
3. F. P. Ramos, *Mas afinal, o que é mesmo documentário?*. 2ª Ed. (SenacSP, 2008).
4. S. Da-Rin, S, *Espelho partido: Tradição e transformação do documentário.* (Azougue Editorial, 2004).
5. N. Burch, *Praxis du cinéma*. (Gallimard, 1969).
6. S. Olivieri. *Quando o cinema vira urbanismo: o documentário como ferramenta de abordagem da cidade*. (SciELO – EDUFBA, 2011)
7. A. Vogel, M. A. da Silva Mello, O. Mollica. *Quando a rua vira casa – A apropriação de espaços de uso coletivo em um centro de bairro.* 4ª. Ed. (Eduff, 2017).
8. M. J. S. Martins, L. S. Roça, Documentary and the city: representing fragments, forming views. *V!RUS*, **13**, (2016).
9. M. J. S. Martins, L. S. Roça, N. Benatti, J. E. Calijuri, M. Tramontano. Frontier Zones: documentários para ler cidades. *Encontro Intern. Cinemídia* (2015)
[10] Tramontano, M., 2016. Species of spaces: notes on a potential exercise. *V!RUS,* [e-journal] 12. [online] Available at: <http://www.nomads.usp.br/virus/virus12/?sec=6&item=1&lang=en>. [Accessed: 21 January 2019].
11. Video available at https://youtu.be/DQYIsFuo0Og.
12. R. Haesbaert, Da desterritorialização à multiterritorialidade. Boletim Gaúcho de Geografia, **29**, 1 (2004)
13. J. Gehl, *Life between buildings* (Island Press, 2011).
14. A. Facina. Quem tem medo do 'proibidão'?. In: C. B. Batista (org.). *Tamborzão: olhares sobre a criminalização do funk.* (Revan, 2013).
15. L. D. Cardoso, L. D. *Sound-politics in São Paulo, Brazil* (University of Texas, doctoral dissertation, 2013).

Building the city as a stage, the example of Nantes

Emmanuelle Gangloff[1,*]

[1] Laboratoire Ambiances AAU – CRENAU, ENSA Nantes, 6 Quai François Mitterrand, BP 16202, 44262 Nantes CEDEX 2 Nantes, France

Abstract. At a time when cities try to stand out from one another, the construction of ambiances is brought back to the foreground of urban, social and cultural dynamics in metropolises. With this contribution, we aim at taking the example of Nantes and the Machines of the Isle of Nantes to show the transformations of an industrial territory into a cultural open stage with cultural practices. Indeed, following the shutdown of the shipyards in the 1990s, the city pursued a major cultural policy to make the industrial wastelands attractive. In successive steps, the temporary artistic practices changed urban ones, based on the question of ambiances and their narrative. Favouring the implementation of projects taking place in public spaces, the city became a 360°-urban scene. As opposed to a generic city, Nantes turned urban scenography into a way to single itself out, by accompanying the creation of plots and narratives around its territory. Between artistic experiments and narrative, Nantes thus appears as a stage-city that invites to include user experiences.

1 Revelation of the dramatic potential by the artists

With the birth of urbanism, the rise of metropolitan centres as economic actors without intermediaries, and the progressive structuring of street arts as a sector, functionalities coming from a dramatic culture are empowered, which re-activates an archaeological connection between theatre and the city. Indeed, theatre took place outdoors for more than fifteen centuries before being brought indoors in dedicated locations. Throughout the ages, scenographic functionalities were present to organise the connection between representation space and the city [1]. Supported by the emergence of the event, cultural or creative city, the scenographic dimension of artistic projects serves the spectacularisation of cities, the dramatization of public spaces, turning precarious spaces into artistic experiment locations. It allows to reinvent the relations between inhabitants and their city by building an urban imagination. In France, this shift intensified after the events of May 1968. Society needed publication places, and public spaces appeared as potential crossroads. From the 1970s, the artists turned public spaces into means of expression, and the city became a background for theatre. The twentieth century has undergone a remarkable growth in the theatre offer out of its walls, which was available in many forms, either the circus, the fun fair or urban forms. It

* Corresponding author: emmanuelle.gangloff@crenau.archi.fr

reconnected with the archaeological aspects of scenography. Stages multiplied, and the city as a whole became a 360° stage. The cultural democratisation and the desire to establish a new relation to the public encouraged open-air temporary artistic experiments. Simultaneously, outdoors, urban centres developed themselves and convened theatricalness to spaces to strengthen their attractivity. Scenography thus accompanies this evolution of the interface, between play, interactions and processes to make visible. Therefore, the concept of *urban scenography* can be considered as an answer to the responsibility of scenography in contemporary cities, which, in turn, becomes a space of representation, play, or elements of the décor. The artistic environment reinvents functions of theatrical scenography, which evolves with these practices that appear to be new, so that they can reveal the dramatic potential of the city.

In Nantes, street artists instigated the first scenographic arrangements in public space. In the 1990s, artists used the urban wastelands, set their experiments under abandoned hangars and turned public space into a one-time representation space. Theatre companies took the city as a background for artistic action, and the built environment became a piece of the décor that is used to unfold fiction. The agglomeration supported these practices to create new cultural dynamics. It was about reconciling the city with a painful past, by changing the image of the industrial heritage that used to make its reputation. The Isle of Nantes – in its representation – reminded the population of the shutdown of the shipyards. The choice was to accompany these urban scenographic practices. Supported by the city, the street theatre company Royal de Luxe based itself in Nantes and found a setting for its experimentations. With the decision of funding this company, the actors of the city discover an artistic proposition that rethinks the possible relation with the city, turning it into an open scene. On the other hand, artists who are funded can come up with spectacular propositions. Thanks to a financial and logistical support, the company can imagine shows that transform the creation of the city, playing on the monumentality of the created arrangements. After its installation in the city, the first production was *The True History of France*. This show, presented in 1990 at the Avignon festival, represented one more step towards institutional recognition. In France, street arts are increasingly considered as an artistic field in itself, and the companies are commissioned by cities and festivals.

Royal de Luxe then developed Theatrical Accident and introduced a change of scale with its *Giants*. The Giants are the main characters of a show based on an urban tale for the whole city. They are monumental puppets activated by members of the company, humanising them as they move. The shows of the Giants are no longer thought as static theatre, but rather as walking spectacles with a plethora of skits over several days.

Fig. 1. The titanic giant and the diver, Nantes, Royal de luxe (2009)

At this scale and time, as Jean-Luc Courcoult states, "*The viewers cannot see everything, because several events are taking place at the same time in different places*" [2]. This type of show transforms the relation between creation and city. First, on dramaturgy, there is an inclusion of the viewer within the construction of the narrative. The story is given in a sequenced way, and the audience must look for the missing pieces by using its own imagination and/or by talking to the other viewers who have seen other main scenes of the show. These walking forms call in the scenographic function in the aesthetic and spatial organisation of the connections between the different scenes. The aim is to organise the representation space in relation to its context, the city and its streets to predict the relation to the public, while maintaining some mystery, so that the audience can be active. A series of scenographic effects are mobilised, such as apparition, to trigger surprise. By following the steps of the *Giant who fell from the sky*[2], the audience sees the city as a décor for theatrical action. Theatre becomes an urban reality, and the city is its stage. It is no longer limited to a place, a building. Street artists build a dramatic relation with the location and question the representations and imaginations of the city. The city of Nantes, which housed the birth of the Giants, sees the possibility of using the artists' narrative that follows the characters, who, through the media and the inhabitants' memories, see Nantes in the background as *an operational theatre*. By addressing the relation between urban space, event and narrative, the scenographic function is democratised and relayed by the urban governance. Policies aim to maintain the dramaturgic action, in order to produce a city narrative based on the artists' stories.

2 Perpetuation of temporary elements to build and tell the city

With companies such as Royal de Luxe, as well as the development of art festivals like *Les Allumées* or *Les Rendez-vous de l'Erdre,* the city has gotten used to morph into a stage every once in a while. The artists have started a transformation in the practice and the view of urban space. Culture is thus considered as a tool to build a territory, according to Elsa Vivant's

[2]First show of the Giants.

"*global approach*" [3], and the municipal structures reflect upon art in the city [4]. The metropolis thus seeks to perpetuate the imagination that the street artists convey. The story brought by the artist wove itself with the construction of a singular metropolitan identity, organised around key urban moments [5]. Driven by the success of urban parades and a temporary architecture created by the Giants, a lasting architecture was born: the Machines of the Isle of Nantes.

The Machines project represents the beginning of the second phase of cultural development policies, in which the city of Nantes aims to perpetuate artistic presences on its territory. The idea was born in 2000, after a meeting[3] between François Delarozière and Pierre Orefice, when they learned that the first phase of transformation of the Isle of Nantes was launched. This proposition gradually stood out, thanks to the multiplication of temporary urban events that required artistic teams from street arts (Royal de Luxe, La Machine, Manaus) and the strong acceptance by the public. In 2003, the success of the *Grand Répertoire* show-exhibition, designed by La Machine in the old Alstom halls consolidated the validation of this proposition, underlying the culture of performing arts in Nantes.

Fig. 2. The Machines of the Isle of Nantes (2019)

In 2007, the city inaugurated the Machines equipment, located under the naves of an industrial shed that was previously dedicated to port industry (Dubigeon naves). It also represents the beginning of a new phase in the urban renewal project for the Isle of Nantes, with the simultaneous launch of the Estuaire contemporary art biennale. With François Delarozière and Pierre Orefice's fantastic bestiary, the project creators, the universe of this company and the involved street artists is projected in a lasting way on the territory. The Big Elephant, key element of the project, became the emblem of a city that aims to make its cultural identity evolve. It is the symbol of a collective memory around major urban parades, while gradually distancing itself from the fictional story told by the first company during the show. Indeed, an image of the animal appeared for the first time on a poster for *The Sultan of the Indies visit on his time travelling elephant* show, by Royal de Luxe. The mechanical pachyderm, city machine, follows *The Elephant,* the machine that was designed and built by La Machine for Royal de Luxe for *The Sultan of the Indies visit on his time travelling elephant* show (2005). This machine perpetuated this Theatrical Accident form. *The Elephant* was thought for the Machines project. However, at the time, the newly created Machine company was still collaborating with Royal de Luxe as builders. François Delarozière, artistic director for La Machine company, showed his drawings of the pachyderm to Royal de Luxe's stage

[3] Pierre Orefice and François Delarozière met in the 1980s, while they were respectively administrator (1985-1998) and constructor-scenographer (1984-2004) for Royal de Luxe.

director, who wanted to make it one of the animated automatons of the Giants. This elephant, dramaturgic element of a show, became a 1:1 prototype of the lasting version of the pachyderm.

Between the two elephants, differences in uses are noted. For Royal de Luxe, The Elephant serves fiction, whereas *The Big Elephant* reveals the city and is connected in a more poetic way to its past. The animal belongs to its context and echoes with the construction of an urban imagination that is turned to myths and major local figures (the construction sites filmed by Jacques Demy, references to Jules Verne or important machine inventors, like Léonard de Vinci). The visual representation of the elephant thus refers in turn to an element of a show and the architecture of a city.

With the launch of the Machines of the Isle of Nantes, a large urban tale started, layered over different levels. The myth built by the designers of the cultural equipment is disseminated by the artists themselves, as well as the communication services of the city. Based on this artistic proposition, the urban governance finds ways to largely disseminate an urban imagination that, "*with the use of emblems, symbols (...) allows to appropriate a space, transfer a territorial belonging that constitutes the collective and/or individual identity*" [6].

A Nantes *storytelling* starts with the implementation of a new narrative order that relies on the production of a cultural equipment, multiplying narratives around the universe deployed by the artists to touch on the progress of the city project. *The Big Elephant* is then used as an emblem – and no longer just as the symbol of an urban story from the theatre – to communicate around a creative city that "*placed its bet on culture*"[4]. The mechanical elephant becomes a symbol to promote local tourism with the *ohlala* promotional campaign[5], before several structures used this figure to connote creative dynamism.

Fig. 3. *Ohlala* promotional campaign, Loire-Atlantique (2009)

Companies and organisations integrated the symbol in their logo to recall that they are based in Nantes. The animal was also transformed by artists, such as Jean Jullien, who caricature

[4]On this topic, you can read M. Grandet, *Nantes la belle éveillée: le pari de la culture* (2010). Several newspapers and magazines address this cultural policy, in particular the *Parcours France* journal, which has a special issue on the subject in October 2015.
[5]*Ohlala* is a brand created in 2009, used until 2015. A campaign with five visuals was used, two of them using the elephant as their main emblem.

it, integrating it within the public space close to the equipment, participating in a spatialised narrative that physically delineates transition spaces between the Machines and the everyday urban flow. The pachyderm is *turned into a logo* for a city that wants to be described as creative and culturally dynamic. The inclusion of a temporary theatrical cultural also comes from this, and claims a singularity typical of Nantes.

Beyond the presence of the art piece on the territory, its image influences the construction of the urban imagination. Through the *pollination* of artistic motivation, the territory differentiates itself and its representations evolve. More broadly, this project impacts the identity of the metropolis through the lens of its urban ambiances. The Machines of the Isle of Nantes sets in a permanent way an immaterial heritage of images that aim at triggering emotions in the population. In an urban theatre culture, the territory is revealed in another light, and builds itself in synchronization with urban, architectural and artistic projects. They mobilise the sensory capital, deal with affects and capture the inhabitants' perceptions. Henceforth, this capital provides a new way of building the city from ambiances, and their narrative.

3 From ambiances to the construction of a city-stage

A scenographic culture is emerging, and it is part of the rise of a production regime of the city as a stage. The example of Nantes demonstrates that practices from the theatre join new practices in urban planning. What used to be about temporary moments is perpetuating itself, and the territory's dramatic construction influences the ways actors work with the construction of the city. Via theatre techniques applied on the city scale and by becoming a scenic city, Nantes can question indirectly the actors of the construction of city about the place of the sensory elements in urban development. The interventions done by artists, cultural operators and developers are valued in the urban construction, where each one is responsible, in his own way, of a scenographic dimension to establish a narrative, animate the territory or integrate the population. With street artists such as Royal de Luxe, it is the opportunity to scratch the surface and make urban asperities visible. With time, the artists managed to experiment with different modes of representation and to appreciate the space with a quirky attention. With a rebound effect, the population and the communication services mobilised this sensory capital, dealt with affects and captured the inhabitants' perceptions. As explained by Henri Bahrmann and Ho Van Mang, "*The appreciation of space can be taught, it even improves with experience*" [7]. Questions are raised around the presence of spectacular and monumental elements in the city, which compose a new aesthetical order.

The Machines project itself highlights a symbolic and physical embedding of culture in the city. Since the 1990s, the Loire metropolis has gradually opened up to all the forms of temporary and permanent scenographic equipment. As François Duconseille reminds us, "*The city is understood as an artistic territory, a space to write temporary or permanent expressions, the scenography becomes urban, the world becomes a public stage to use from inside, but also to displace*" [8]. The capital of the region became once again a stage for theatrical reasons, before it opened to all the arts. With the rise of festivals dedicated to contemporary art, in-situ artists have in turn integrated a scenographic dimension to socialize their pieces. Lastly, the multiplication of artistic installations available in public space calls in a scenographic dimension to create a network between the artworks and to represent the territory. Nantes integrated in its own way functions of urban scenography at the heart of it metropolitan project, between narrative, space/time management and relation to the public. When the city is superimposed to the stage, the scenography is revalued as a culture of space in itself, it escapes from the theatre and influences the way to imagine the construction of urban centres. The culture of temporary elements invites the production of experiences, and the integration of the population seems to be an essential component to avoid the production of a mock public space where everything would be regulated by the institution that would build the story of a factitious city with no meaning. In an agglomeration that uses culture as a lever for development, it consists in developing conditions for the interaction between a piece, its context and its public. The latter slowly exits its spot as a spectator to become an actor while the perpetuation of the scenographic culture operates.

Taking up the challenge of the inhabiting experience was made possible thanks to time management. By welcoming street performances around the Giants, the metropolis acclimated the population to the operation of prospective scenarios. Thanks to the presence of artists on the territory, the temporality is present in the shows, over a day, a week or several years. The municipal teams have gotten used to switch between seasonality and artistic production cycles. The sensory experience also operates over time and can play with it, by stopping the urban everyday time, with highlights or, on the contrary, by perpetuating fleeting moments in the implementation of a storytelling that relies on a temporary culture. The local authorities thus have grasped the abilities of artists to tell stories and establish common bases for the urban imagination, which can then be spread by the governance. Since then, these initiatives have been institutionalised, leaving physical marks in the city, with the installation of permanent pieces, elements that strike the memories, but also the trade of images, sensations and a dedicated language. This type of experience, from a culture of urban arts, infuses the room with an aesthetical dimension that is specific to it and completed by a set of temporary and urban artistic interventions.

While the convening of a long time is usual for development professionals, the institutionalization of a culture of temporary time highlights a new management for the temporalities of urban projects. On the Isle of Nantes, following the artistic teams, construction projects produce those events. They adjust depending on the uses and the room taken by the temporary artistic equipment in the story of the city. Gradually, the actors of the urban construction, through the temporary, give themselves some room to experiment and animate the present time, and prefigure uses. The perpetuation of the temporary works at different levels. The event has an influence on the time beyond its proper duration, resonating with the story of the city. The median time is used to connect what constitutes micro-events with the long time, of territorial transformations. Then, within the median time, the short time is thought to animate the present time and allow to experiment situations that could last for longer. This game of temporalities opens up possibilities to constantly experiment, adjust and sequence a territorial development project.

Moreover, the rise of the need to control the connection between artwork/viewer has a positive effect on the way we value the sensory and poetic dimension of the city. A new attention is given to the way the audience accepts the art piece and is integrated to the urban narrative. When public spaces are staged, the question of the audience's physical and symbolic implication within the story is asked. It translated in the city into the addition of signs around iconic elements, such as the Big Elephant. There is a need to experience once again the catharsis, to mitigate the tensions in urban centres and to make them pleasant to live via the production of festive urban moments. The metropolis is told like a *happy city* and promotes the dissemination of an image of the city where it is pleasant to live, generating nice feelings [9] and multiplying the conduct of *interactive artistic experiences*. Initiated by artists, the responsibility of interactivity in urban spaces on a short-term basis was relayed by the city services. However, the risk lies in turning scenography into a lever to thread the relationships between the inhabitants and their living spaces, where the institution would only show the right side of a spectacularisation of urban spaces. The attractivity appears to work through the production of unprecedented urban experiences that appeal to the emotions and the senses. Another risk is the gradual loss of the utopian dimension of the artistic project in a form of dependence to the city project, and the loss of autonomy for the artists in relation to the local cultural structures. It echoes with old practices, when in Roman times the theatre disappears, from scenic games – *Ludi Scaenici* – to circus games – *gladiatorial combats* – with a loss of confrontation to catharsis and a return to reality.

While the figure of the city-stage appeared to be embodied by the artists, the example of Nantes shows how metropolises tend to become prescriptive of this model. Indeed, between temporary artistic interventions and delivery of an equipment coming from a street arts culture, the multiplication of experiences generates learning phenomena for the city, which has gotten used to telling itself through a game of temporalities, multiple images and in a spectacular way. Nantes grew up with the Giants, and some urban transformations are taking into account the possibility of parades. The relation to the environment is anticipated: the constraint is a pretence to establish new propositions and it becomes a stage. With the deployment of the Machines of the Isle of Nantes, the narration of the city becomes a way to make the territory attractive. We thus have a power play that is inverted, with a city organiser and producer of events, and companies that are institutionalised, to the point where they leave lasting traces of their presence on the territory. The city as a stage tells itself to its inhabitants, but also becomes a space of visibility that is prone to phenomena of patrimonalization. Therefore, the operation of the dramaturgic potential of the city by the artists encourages the consideration of the inhabitant through the ambiance and the production of unprecedented urban experiences. Between represented and place spaces, when the city becomes a stage, the user comes back to the centre of attention and can be involved in the construction of his living environment. Facing the risks of patrimonialization and of instrumentalisation of the city-stage, the question is to find the ways to continue the inclusion of artists and inhabitants to make a model that perpetuate the connections between art, development and inhabitant.

References

1. E. Gangloff, *Quand la scénographie devient urbaine, Nantes comme cas exploratoire des fonctions du scénographe* (2017)
2. J.L. Courcoult, O.Quirot, *Royal de luxe: 1993-2001*, 37 (2001)
3. E. Vivant, *Qu'est-ce que la ville créative ?* (2009)
4. C. Ruby, *L'art public un art de vivre la ville* (2001)
5. M. Miranda, *L'événement urbain festif : vers une « gestion de site exploratoire » sur lespace public ? : les cas de Nantes et de Bordeaux* (2007)
6. E. Bonerandi, Géocarrefour, *Le recours au patrimoine, modèle culturel pour le territoire* ?, **80/2**, online (2005)
7. H. Bahrmann, H. Van Mang, *L'ambiance urbaine. Réflexions sur la ville et l'environnement sensible,* 49 (1972)
8. F. Duconseille, J.C. Lauquetin, *« Du scénographique » Porosités et extensions des pratiques et de l'enseignement de la scénographie* in L. Boucris and al., " *Qu'est-ce que la scénographie ?,* 243 (2012)
9. L. Matthey, *Building up stories. Sur l'action urbanistique à l'heure de la société du spectacle intégré,* (2014)

Urban visions: Back from the future

Olivier Chamel[1]

[1] Florida A&M University, School of Architecture and Engineering Technology, Tallahassee, Florida, USA

Abstract. The persistent growth of the human civilization, fueled in large part by technological progress has brought upon us a series of very serious challenges. The quality of our overall environment, energy and food supply are subjected to increased pressure, while access to decent employment, housing and medical care remains broadly unequal. According to the current trends most of the world's future population growth will occur in cities, therefore positioning the city as a key component to solving challenges associated with human development. Based on that assumption, it seems crucial to think about what the city of the future should be and look like. If we look for existing and graphically convincing representation of the city of the future, we are inevitably drawn to popular culture media such as movies and graphic novels. For almost a century, movies in particular have proposed realistic constructs of future urban settlements along with the life associated with them. Based on a number of ideas expressed in motion pictures over the years about urban life in the future, one can argue that both past and recent predictions tend to be technologically optimistic but socially and environmentally pessimistic. This paper proposes to identify and discuss a number of challenges as well as opportunities associated with urban development in the next 100 to 200 years and present a series of urban visions to illustrate both positive and negative trends.

In the context of a continued and unprecedented growth for the past 50 years in cities worldwide, the goal of this paper is to identify trends in urban development and offer urban visions of what the city of the future might be and look like. Considering that this considerable urban growth is likely to continue, it will bring a whole array of challenges, but also opportunities to improve upon the current quality of life in urban centers. To that end we will look at how society has already produced convincing visions of an urban future through popular media, and more specifically motion picture. We will also attempt to outline realistic trends and visions for the metropolis of the near future based on current technology, social issues and economic development already affecting our cities today.

Historically, and maybe especially since the Industrial Revolution, cities have been both a symbol of progress as well as alienation. In that context, just as urban centers have brought the hope of employment and higher wages, they have also been synonymous with an oppressive social and physical environment. In that regard, the 1927 movie Metropolis, by German director Fritz Lang, proposes a very pessimistic vision of the city of the future based on some of the changes brought by the industrial revolution. Within this construct, the city is conceived as a reflection of society. Most of the urban population is composed of an oppressed working class who lives and works in poor conditions, while a small elite enjoys a life of leisure and luxury. Metaphorically, the working masses work for the city to make it function so that it can be enjoyed by a select few. In that context, the city itself becomes the symbol of alienation with its massive buildings and machines. Most of its physical environment is dark, uncomfortable and depressing. Metropolis makes sense as a social criticism of its time as the society of the day is still very much engaged in the industrial revolution. The idea of the city as a symbol of alienation, but also as a physical environment that is oppressive can be traced to the urban planning associated with the industrial revolution of the late 19th century and the 20th century modernism. Large urban housing projects such as the Cité Radieuse and the Plan Voisin by architect Le Corbusier are in fact a testament to the urban planning ideas of that time with their repetitive and out of scale designs. With a similar message, the 1965 movie Alphaville by French director Jean Luc Godard also addresses issues of social disconnect and personal alienation. It also uses the ubiquitous aesthetic of the modern architecture and urban planning of the 1950s and 1960s of Paris and its suburbs to represent alienation. Both Metropolis and Alphaville can be described as movies using in one case an imaginary urban context and in the other a real one as a vehicle for social criticism. In both cases, the premise is that cities have been designed by technocrats with little to no concern for the well-being of most of their inhabitants. In contrast, the urban environment created in Blade Runner (Ridley Scott 1982) presents a futuristic city that appears to have grown incrementally and organically to form a seemingly unplanned maze. The city as a whole seems to be functioning, but again its physical environment does not seem very conducive to a harmonious life. In that context, the city of the future is a result of an unplanned growth characterized by a network of overcrowded and rundown buildings. That vision, though decidedly pessimistic, is based on the understanding that cities evolve incrementally and over relatively long periods of time. In the more recent Ready Player One (Steven Spielberg 2018), we discover an urban environment alike of a waste field where a derelict urban world has been replaced by a virtual alternative. Again, just like Metropolis, Ready Player One is presenting an urban environment, which has very little quality for human life.

Fig. 1. Metropolis by Fritz Lang. 1927.

Fig. 2. Blade Runner by Ridley Scott. 1982.

As we compare the ways in which the city of the future is represented in these films over a period of almost 100 years, it is interesting to note that the city, as an image of society, becomes a symbol of its bad deeds and failures. The urban visions presented in these movies, as though they were representative of different times, seem to share the idea that society is firmly engaged in technological progress but has neglected to provide adequate solutions with regards to the quality of the urban environment, affordable housing, access to nature, air quality and the overall human experience in the urban world. Or, to put it more accurately, a small portion of the urban population has access to quality housing and urban environments, while the vast majority of urban residents does not. These movies share the idea of the city of the future as a dystopia and a microcosm, which displays great inequalities. Using the physical environment of the city as an expression of society itself, these movies propose a pessimistic vision of the future as there are undoubtedly a number of serious challenges ahead.

Now if we consider for a moment the cities of our times, one can summarize a number of these challenges we can actually observe. As there continues to be global trends of positive migration towards cities both in developed and developing countries, the sheer size of cities has become an issue. In fact, once urban structures reach a certain size, providing a number of basic services turns into complex problems. For example, food and water supplies as well as garbage collection systems, become especially challenging for very large cities. Coordinating commercial, public and private transportation systems poses a number of problems, including environmental issues. The construction of new buildings requires considerable updates and improvements in terms of infrastructure, and so do transportation networks. As a very visible consequence of the growth and densification of large urban areas worldwide, transportation in these cities has turned into huge challenges. In that context, providing access to clean air, parks and meaningful spaces for urban residents remains a challenge in many cities around the world.

1 Housing affordability and urban density

Though decent and affordable housing is key to the well-being of people, cities are struggling to meet that need in both developing and advanced economies. If current trends in urbanization and income growth persist, the number of urban households that live in substandard housing could grow from 330 million to 440 million by 2025. This would mean that the global affordable housing gap would affect one in three urban dwellers or approximately 1.6 billion people. In that scenario, the disproportionate cost of housing relative to overall life cost would make other essential services, such as healthcare, unaffordable for urban households (Woetzel 2017).

Billions of people in cities around the world are faced with a great challenge whenever they attempt to find a new home. From Tokyo to Medellin, housing costs are creating financial stress for a vast majority of the world's urban residents. The cost of owning or renting a dwelling has risen far faster than incomes in most countries, especially in large cities, where many people want to live and where job opportunities are concentrated. This issue stems from the fact that there is a large disproportion between supply and demand. Population growth, a steady trend toward urbanization, and overall increase in global incomes are all contributing to a steady demand increase. The number of housing units in urban centers around the world has not significantly increased to keep up with this surge in demand. It would take a serious increase in the number of housing units produced to achieve greater affordability. But this is no easy task considering the sheer number of people moving to urban areas.

One of the first issues to tackle is to find vacant sites across these cities, and preferably along public transportation corridors. For example, sites that are underutilized can be identified as priorities for redevelopment. Incentives, such as investment in public parking and other types of infrastructure, can make these types of projects more attractive to developers. Another possible strategy, especially in very dense urban centers, involves building infill housing on vacant parcels. Even dense neighborhoods may have empty lots that could serve as viable sites. Other buildings, such as outdated sports facilities, military bases, or transit hubs, may also provide valuable land. Some cities may be able to convert derelict industrial sites and create large attractive mixed-use developments. Large unused industrial parcels can offer tremendous development potential. In addition to these potential solutions, a relatively new trend has emerged where cities can encourage the owners of single-family homes to add accessory dwelling units. Accessory dwelling units are by nature affordable because they use existing land, buildings, and infrastructure.

It is critical for congested cities to promote density along public transportation corridors rather than encourage sprawl and longer commutes. Transit-oriented development may involve redeveloping existing residential structures or encouraging new buildings by permitting higher floor-space ratios, loosening height restrictions, or allowing greater density in specific target zones. These zones can be selected to promote local objectives, such as reduced dependence on private vehicles or the development of mixed-use, pedestrian-friendly cityscapes. In many cases, cities may not even need to increase density thresholds. They can build out on residential parcels that are not taking advantage of currently allowed density (Woetzel 2017).

As cities can create incentives for new development and make land available, the whole design and construction industry needs to increase efficiency and streamline production. In fact, when compared to other sectors of the economy, such as services and industrial production, productivity in the construction industry has remained stagnant for years. A big boost in productivity could come with the concept of a manufacturing-inspired mass-production system. Prefabrication presents a huge potential for a broad array of building types and bring a series of improvements with regards to material waste, workers' safety, quality control and overall lower costs. This involves more standardized elements, panels manufactured and assembled off-site, and limited finishing work conducted on-site. Countries such as Sweden already produce a large portion of their housing stock using various degrees of prefabrication. Other recent technological developments, such as 3D printing using composite materials or concrete mixes, have seen exciting developments and are starting to be applied in large scale projects. Saudi Arabia, for example, is looking at building a large real estate project in Dubai using a 3D printing concept. Considering that, around the world, buildings' heating and cooling account for 35 to 60 percent of total energy demand and produce nearly 40 percent of emissions, the effort to reduce energy use and emissions, from buildings will continue to remain an important challenge to solve.

Fig. 3. 3D printed house by Apis Cor.

2 Transportation and mobility

A number of trends are influencing urban-mobility systems around the world, such as shared mobility, autonomous vehicles, electric vehicles and public transit. Trends, like vehicle electrification and the development of autonomous-driving technology, relate directly to mobility and are in turn pushing towards the decentralization of energy systems, as modes of transportation come to rely more and more on electricity as an energy source. Ride-hailing services have grown rapidly over the past few years and now compete not only with traditional car-sharing and car-pooling providers, but also with public transit and private vehicle ownership. Investments in ride-hailing companies, such as Uber and Lyft, have taken off, too, more than doubling to $11.3 billion in 2015 from $5.3 billion in 2014 (Gao 2016). Advances in autonomous-driving technology seems promising with regards to road-safety, transportation cost and access to mobility. Autonomous vehicles will turn driving time into free or more productive time. Autonomous vehicles could also lead to higher overall vehicle mileage, as people take advantage of their convenience by making more trips or even sending autonomous vehicles to run errands for them. Globally, electric-vehicle sales have risen quickly, from 50,000 in 2011 to nearly 450,000 in 2015. Purchase subsidies, falling battery costs, fuel-economy regulations, and product improvements have contributed to the increase. Bloomberg New Energy Finance estimates that battery costs will drop below $100 per kilowatt-hour in the next decade. If that happens, electric vehicles should achieve cost competitiveness with conventional vehicles (Gao 2016). On the other hand, the widespread of information technology applications into vehicles and infrastructure will generate data with a variety of uses. For commuters, software systems can facilitate trip planning and guide autonomous vehicles based on real-time conditions. Transit authorities could use the same data to analyze the movement of people and vehicles, identify bottlenecks, adjust services, and make long-term transit plans. Cities around the world are expanding and improving their public-transit networks. Adding autonomous features to transit vehicles may reduce operating costs, while new deployment models, such as fleets of shared vehicles, can make transit more flexible and accessible. The United Nations Population Division projects that the world's urban population will increase by more than two-thirds by 2050 (UN Report 2014). Such an increase of people in motion could put more strain on city roads that are already struggling to keep up with increases in vehicle miles. But infrastructure upgrades that favor public or shared transit and bicycling could reinforce a shift away from car ownership.

In a broader context, integrated mobility systems should improve the lives of city dwellers in a number of ways. One is environmental quality. As more commuters shift to electric vehicles, shared mobility services, and public transit, emissions of carbon dioxide, nitrogen oxides, and fine airborne particulates in cities should go down. This will contribute to reducing health problems associated with air pollution.

When it comes to urban commercial transport, a number of solutions have been identified, such as order grouping, route optimization, and night deliveries, which could be implemented more or less immediately. Others, such as the use of robots, and autonomous ground vehicles, are currently in research and development, but likely years away from large-scale applications (Bouton 2017). A large portion of the current transportation networks of our cities will most likely remain though the nature of the vehicles involved will change. The use of low and no-emission vehicles is a critical part of solving the transportation and mobility challenge while at the same time reducing greenhouse gas emissions and improving air quality for urban residents.

3 Sustainability and resilience

The unprecedented urbanization we have experienced in the last five decades has primarily taken place in developing countries and areas exposed to natural hazards, such as riverbanks and coast lines. The fact that a number of these urban development projects have consisted of informal and unplanned settlements exacerbates existing challenges and leaves cities more vulnerable to natural hazards. The idea of concentrating large numbers of people and assets geographically logically increases the potential impact of natural disasters, especially in the context of increasing extreme weather events. Large urban settlements can therefore suffer devastating events in terms of human lives lost, infrastructure and economic livelihood. The poor segments of the population are particularly exposed, as they tend to live in lesser quality housing and lack the necessary safety nets to recover from both economic and environmental disasters. Therefore, the major resilience challenges of our times, be it poverty reduction, natural disasters, climate change, environmental sustainability and social inclusion will take place in our cities.

In addition to the threats generated by climate change, the continuous population migration from precarious parts of the world, such as Africa and the Middle East, to more stable developed countries are creating additional strain to existing urban centers. The recent impact of Hurricane Michael in October 2018 on the Florida Panhandle shows how even a relatively well-developed urban area can suffer tremendous infrastructure and economic loss in the context of unexpected climatic events. The need for human settlements in general and urban centers in particular to gain a certain level of resilience has emerged as a crucial challenge in the last decade. The concept of urban resilience can be defined as the ability of an urban system, including its inhabitants, to maintain continuity through stress and disasters whether they are man-made, or weather-related. Discussions on urban resilience have focused primarily on three specific threats, climate change, natural disasters and terrorism (Coaffee 2008). One key component of addressing resilience lies in the ability of cities to become more sustainable. In fact, more and more cities are being evaluated in terms of their ability to adapt to changing conditions (Ayooh 2016) and against the notion of cities as highly complex adaptive systems. In that context, urban planning would move away from a traditional approach based on geometric plans to an approach informed by network science.

Climate-related challenges, such as flooding, heat waves and droughts, have been found to provide a threat to approximately 1.7 billion people, with a combined GDP of USD 35 trillion according to a recent study (UN Habitat 2016). Heat waves, which have in recent years affected a large number of urban areas across temperate climatic zones, have resulted in significant deaths along with spikes in electric demand, forest fires, air pollution and reduced biodiversity. Agricultural losses from heat and drought can in turn have severe consequences in terms of food access and social stability. In that context, the direct fatalities linked to heat waves are statistically concentrated in urban areas because of inadequate housing conditions and the urban heat island effect. The urban heat island effect has been known to generate daytime temperatures up to 10°C higher in a city and between 5 to 6°C warmer at night. The primary factor for this phenomenon is the presence in cities of a number of materials, which absorb and store heat much more effectively than in rural areas. Urban geometry also plays a role as tall buildings can reduce wind and air movement therefore limiting convection cooling (Oke 1982). With regards to urban heat island effect, a number of strategies has been identified and is currently being implemented. The idea is to increase the amount of natural cover within the city and increase surface refractivity. Trees are especially effective in that regard, as they provide shading to both buildings and ground's surfaces and contribute to creating pedestrian friendly spaces with cooler temperatures. Another important strategy aimed at reducing the urban heat island effect consists in increasing the albedo or light reflectiveness of urban surfaces especially horizontal planes such as roofs.

In terms of flooding, risk urban areas have seen increased occurrences of serious flooding due to the effects of climate change. The cities of Brisbane, Australia in the summer 2010-11, New Orleans, USA in 2005 and Nimes, France in 1998 have all experienced catastrophic floods. A broader approach consisting in replacing as many non-porous surfaces with more porous green spaces can certainly help. Typical grassed areas and gardens can be complemented by high absorption bioswales that replicate natural environments such as swamps. The deeper the soil, the more rainwater it can absorb. One effective strategy simply consists in creating enough space for the excess water generated by a storm. Flood plain clearance, which removes structures and pavement from flood plains and promotes a natural environment, is a logical and nature-inspired solution that has a lot of benefits. It provides large natural areas capable of absorbing large quantities of water in order to limit or avoid the flooding of adjacent urban areas (Tanner et al. 2009). As another potential strategy, green roofs have gained popularity in the past 15 years, though their impact is typically limited due to limitation of depth and weight. In certain situations where cities are on flood plains or along rivers and coasts, levees and other flood barriers may remain an indispensable tool. Unfortunately, this type of equipment and infrastructure is typically out of reach for some of the most vulnerable communities in the developing world.

4 Urban environment quality

As we mentioned the urban heat island effect and flooding as urban vulnerabilities, their occurrences present a number of opportunities aimed at improving the physical environment of cities. Historically, the physical growth of cities around the world has been primarily driven by economic development, therefore leading to the massive building structures we can experience today. One can argue that the development of cities has been a constant balance between economic pressure and livability. The two issues are very much tied, as livability is an important factor for making cities attractive from an economic standpoint. The greatest and most livable cities around the world are known for providing people with the ability to physically interact with their environment with such amenities as vibrant streets, parks, plazas and gardens. Being physically engaged with the environment where one lives and the ability to connect to nature and the outside is a basic human need. Therefore, and in order to remain competitive and attractive, cities have to commit to creating and nurturing positive, diverse and safe urban spaces, as well as a clean and healthy overall environment.

5 A few possible scenarios

The cities of the future, along with their suburbs, will be bigger and denser as land is a finite resource and more and more people will continue to migrate towards them. Real estate value will continue to grow along with speculation. Commute time will remain, but its quality may improve due to more efficient mass transit and the increased flexibility of shared and public transportation. Large high-tech buildings will continue to be built by public entities and large corporations, along with micro un-planned developments. Cities will likely remain less and less affordable. When it comes to the space of the city and more specifically that of the street, one can wonder who will prevail between vehicles and pedestrians. Hopefully, we will reach some type of status quo. The existing roads will most likely be used by automated electric vehicles to transport heavy things from food to construction materials and everything in between. Most deliveries will be scheduled at off-peak times and at night so as to improve flow and efficiency. Private and shared vehicles would use the same roads as commercial and utility vehicles, though at some point one could see additional roads being built above the network of existing streets. More utilitarian traffic would occur at ground level, as higher levels would offer a more enjoyable experience for commuters.

Fig. 4. Shenzhen master plan. Avoid obvious architects and Tetra.

At some point, the ground street level may not be available to pedestrians. A compromise would consist in creating dedicated multilevel transportation axes while dedicating other streets purely to pedestrian access. We are already seeing that trend when it comes to pedestrian streets worldwide. If the space of the streets at ground level was to change in nature and become unsafe, as suggested by a number of cinematographic representations, then the traditional street experience would be lost (e.g. 5th Element, Blade Runner, Next Gen). In that case, the basic relationship of people with the ground and all its important connotations would also be lost. The wealthy-enough urbanites would commute via drone-type vehicles on a system of carefully established elevations similar to the way commercial aviation currently operates. Similarly, small packages may be transported by drones along specifically determined aerial corridors. The use of lightweight private electric-powered drone-type vehicles will inevitably see an increase but will most certainly remain costly.

Buildings themselves will be built and operated more efficiently. They will be assembled on-site using pre-fabricated components and will have high-tech envelopes able to react to the exterior environment with regards to light, wind and water. They will be self-cleaning and every exterior surface will be designed to collect solar energy. In parallel to high-tech buildings, cities may see the re-introduction of nature-inspired ecosystems within their footprint in order to mitigate water run-off, heat waves and provide meaningful green spaces for urban residents. Overall, the physical structure of the city will harvest energy and water, and the shift towards everything electric will help with controls, efficiency and creating a much cleaner environment for urban dwellers. Urban planners will study urban morphology to offset negative weather patterns (heat gain, excessive wind loads) and enhance positive ones (cooling breezes, shading). Technological progress at every level of society will contribute to making things more affordable. Prefab technology should increase affordability for urban housing as long as adequate public policies continue to support it. Technologies such as digital fabrication and large-scale 3D printing along with advanced sustainable practices will lower construction and operating costs, therefore lowering overall costs. There is no doubt that technology and automation will have an effect on cost, but they will also influence job access. The hope is that overall life conditions will raise across various social levels of society so that cities can remain ethnically, socially, economically and culturally diverse. Social inequalities will most likely remain, but there is hope that with the advent of new technologies and efficiencies more people will be afforded an improved basic standard of living. As less and less physical tasks will require the need of human workers the emergence of a universal basic income seems inevitable. In the context of intense technological changes, the big question regarding the city of the future is whether it will be able to remain a vibrant and healthy place to live while continuing to be a crucial economic driver.

References

1. J. Woetzel, S. Ram, S. Peloquin, M. Limam, J. Mischke. *Housing affordability: a supply-side tool kit for cities*. Executive briefing, Mc Kinsey Global Institute (2017)

2. P. Gao, H.W. Kaas, D. Mohr, D.Wee. *Disruptive trends that will transform the auto industry*. Executive briefing, Mc Kinsey Global Institute (2016)

3. UN, Department of Economical Social Affairs, Population Division, esa.un.org. (2014)

4. S. Bouton, E. Hannon, S. Knupfer, S. Ramkumar. *The future of mobility: how cities can benefit*. (2017)

5. Coaffee. *Risk, resilience and environmentally sustainable cities*. Energy Policy, 36 (12). (2008)

6. UN Habitat. *World cities report 2016: emerging futures*. (2016)

7. T.R. Oke, The energetic basis of the urban heat island. Quarterly Journal of the royal meteorological society. (1982)

8. T. Tanner, T. Mitchell. *Urban governance for adaptation: assessing climate change resilience in ten Asian cities*. (2009)

The affective notes of represented space as motors of emotional and sensorial response

Elisa Morselli[1,*]

[1]Sapienza University of Rome, Department of Architecture and Design, Via Flaminia 359, Rome, Italy

Abstract. The following essay proposes to investigate the perceptual and emotional aspects related to the visualization of architectural images. The field of research is limited to a well-defined category: figurative representations as the photographic and digital images of contemporary architecture. In particular, two types will be analysed: the un-built architecture produced by Studio MIR and Bloomimages compared with the photographed built architecture. Using figurative images as a tool of reading, the aim of this work is to identify and classify three types of affective spaces capable of generating a specific kind of perception, producing a sensorial classification of atmosphere for architecture. The study of the Psychology of Art, as well as Aesthetics and Neuroaesthetics can be a valuable tool in understanding the phenomena of the present, considering the marked pictoriality of these images. The application of the analytic methodology, developed in these disciplines, can suggest a new way of "looking" at the project, paying attention to the representation of the atmospheres, which characterizes the experience of felt space.
Keywords: Affective Space, Perception, Representation, Aesthetics, Atmospheres, Design Research, Generators.

1 Introduction

The research moves from a consideration on the current condition: in the graphic apparatus of architectural design, today more than ever, the focus is shifted towards the processing of image-effects that aim to attribute "affective" qualities to the space. In other words, contemporary architectural images, like works of art, through the increment of painterly aspects are intended to stimulate in the observer a bodily and emotional response that influences the perceived subject regardless of the intrinsic qualities of the presented work.

Consistently with the current philosophical debate led by the German phenomenologist Gernot Böhme, it is possible to see how the aesthetic mark of contemporary images tries to evoke the "felt" atmosphere of lived space [1]. To produce this feeling in images, architecture is never shown in isolation, it is instead restored to the multiplicity and at the same time to the transience of the phenomena that characterise the reality of the world. This means to pay

*Corresponding author: elisa.morselli@uniroma1.it

attention to the flow of time, changing seasons and meteorological conditions, as well as the presence of people and objects that animate the scene.

In the last ten years, the need to excite the viewer emotionally and sensorially has led figurative representations towards a morphing from clear to indeterminate. The goal to pursue is no longer to depict architecture cleanly, making every detail readable. Both photographers and renderers are moving away from photorealistic references: through an overlap of filters they "dirty" the image, restoring the perception of lived space. In succession, I will describe three categories of affective space, focusing on the emotional and sensorial perception of the atmosphere presented.

2 The representation of Clear Space

To better understand the issue, let us start by analysing the first category with two images showing different architectural solutions (a staircase and a ramp) for exhibition spaces.

Both these architectural elements, if we limit ourselves to providing a purely technical description, can be described as "transit space" designed to overcome a difference in height. From an architectural point of view these elements could be realized with infinite solutions: the only permanent feature is given by the bodily sensation that can be experienced in this "space of movement". The vertical shift, obtained by means of ascent and descent, qualifies this element in every project, regardless of the type or intended use. By observing the two images of different projects by different authors, it is evident how this motion is central to both representations and how it is generated by means of different devices that can be compared to each other.

Fig. 1. Digital Image by Mir, Asif Khan Architect, Pavilion Expo Dubai 2020 (© Mir)

Fig. 2. Photo by Iwan Baan (2009), Zaha Hadid Architects, Maxxi Museum, Rome (© Iwan Baan)

The first expedient is identifiable in the perspective structure of the images. The simulated ocular direction proceeds from top to bottom, "sucking" the observer into the space funnel [2]. Due to the observer's location, the gaze opens up in correspondence with the great central void around which the staircase is articulated, thus generating a sensation of vertigo. Through the loss of balance, we actively act on observer's perception of movement: in both images, the eye, through the camera, is placed at the top level, projecting as far as possible from the safety balustrade. While in the second picture you can still see the border, in the first one the tension is greater: the area coinciding with the observer's eye is placed zenithally outside the parapet. By assuming an unrealistic position, the sensation of vertigo is increased by simulating a suspension on an inaccessible level. The effect generated by this type of perspective structure can be compared to the illusion produced by the perspective "from below up", widely used during the eighteenth century for ceilings and vaults frescoes. In the frescoes, a whirling combination of architectural elements and moving figures leads the observer's eye towards the centre in the space where the sky opens. Usually, this technique uses fake architectural structures that disappear in a naturalistic background (often a sky) so as to give the illusion – by means of the depth of the perspective system – of a dematerialized covering.

Fig. 3. Andrea Pozzo, Gloria di Sant'Ignazio fresco, Rome, (1691-1694) and characters details

Furthermore, the effect obtained by this perspective implant is amplified by the sensations generated in the observer by means of the embodied simulation mechanisms. Vittorio Gallese's and David Freedberg's studies demonstrated that these mechanisms are activated in spectators even by the vision of static images: "when the actions, emotions and sensations perceived are depicted". [3]

The representations of sensorimotor activity and emotions have therefore great importance to describe an *ambiance*. In the first image of the ramp (Fig. 1), the feeling of vertigo is generated by the characters leaning forward and looking down, capturing the depth of the space in height, while the character in the foreground is looking up towards an indefinite point, extending the depth of the space towards the sky. Through this performance of corporeal imbalance, the observer is pushed to sensorially grasp the depth of the space. Furthermore, there is a direct ocular contact between the eye of the first man at the top and the eye of the observer over the screen, just like the figures depicted in the Baroque vaults. Differently, in the second image examined (Fig. 2), the motion is generated by people continuously walking through the stairs space.

The images show numerous affinities producing a precise narration of space. In both of these (un-built and built architecture), the need of generating the bodily and sensorial experience of lived space is central. However, again observing the first one image, it is possible to grasp another important detail: the perception of vertical movement is accentuated by the introduction of an artificial "fog" that crosses the great void of the ramp, invading its entire height and overcoming the top edge of the photo. Expanding and contracting in an uncontrolled manner the fog generates a "widespread" movement that "touches" characters and architecture at the same time, dirtying the image. With the introduction of this particular the representation begins to break away from the photo-realistic contacts. The first image is a rendering elaborated by the Norwegian studio MIR for Asif Khan, taking part in 2020 Expo

Competition in Dubai. The second image shows the interior stair of the Maxxi Museum in Rome designed by Zaha Hadid Architects, photographed by Iwan Baan in 2009.

3 The representation of Ambiguous Space

The second category concerns the representation of ambiguous space. Starting from this category, the analogies between built and un-built architecture become more blurred. This splitting is more evident for the outdoor stage, where the description of meteorological phenomena becomes artificial to increase the observer's involvement. The manipulations of colour and light effects are in line with the current trends of urban design projects. The urban planner and sociologist Jean-Paul Thibaud highlights how nowadays the aim of designers is no longer limited to the formal definition of the spatial elements, but rather extends to install an "atmospheric" space capable of transforming the perception of architecture by acting on its inhabitants [4]. The goal is to offer a sensory and emotional experience. This need has led representation techniques to acquire strong effects of light and colour, to simulate extreme weather conditions, stimulating the observer's emotional response. In digital representations, by exploiting the software's potential, the natural atmospheric conditions are altered to evoke extreme and extended meteorological phenomena, while they generally continue to be sought in photography. Most photographers work using ideal lighting conditions – clear and sunny skies – that highlight architectural features.

The outdoor images of Hamburg's *Elbphilharmonie* by architects Herzog & de Meuron are particularly appropriate to illustrate this aspect. The two photos below depict the main front of the building taken from a dock port. The building project is part of a large-scale urban redevelopment operation (HafenCity), launched to develop the waterfront of Hamburg. The new *Elbphilharmonie* is grafted onto a pre-existence – the deposit of Werner Kallmorgen – and occupies a privileged position: it stands on the terminal end of the tongue of land between the Sandtorhafen and the Grasbrookhafen. This location coincides with HafenCity's sea entrance and, therefore, the Philharmonic becomes the landmark of the new district.

Fig. 4. Photo by Thies Rätzke (2017), Herzog & de Meuron Architects, *Elbphilharmonie* of Hamburg, Elbphilharmonie ©Thies Rätzke

Fig. 5 Rendering by Bloomimages (2007), Herzog & de Meuron Architects, *Elbphilharmonie* of Hamburg, ©Bloomimages

Reading from left to right, the first image (Fig. 4) is a shot taken by Thies Rätzke in 2017, while Fig. 5 is rendering made by Bloomimages ten years earlier. Starting from the description of the image's structure it is possible to note many affinities: both the photographer and Bloomimages adopt the same position, simulating a shot from the dock to the side of the Philharmonic. In the first image – Rätzke's photo – the building envelope reflects "normal" lighting conditions, while in the digital image the background undergoes increasing changes. In the digital representation (Fig. 5), the sky becomes cloudier assuming dark blue colours. In that night setting, a sky full of clouds is reflected on the building and the sea. The gloaming cloudy sky generates a feeling of excitement on the viewer, and increases the tension induced by the formal solution of the building. The taut curves of the envelope's upper part are enhanced by the colour contrast; the Philharmonic is framed by a white glow. By this artifice, there is a simulation of movement that interacts with the observer's corporal perception. In an article on colour perception, neuroscientist Vittorio Gallese states that when there is "a sudden chromatic jump [it] stimulates a sensation of movement and physical change of position. [...] It is, in fact, known that one of the visual features of low level that influence the perception of movement is colour" [5]. Moreover, the apparent movement induced by the colour's nuances accentuates the emotional character of the image. German psychologist Rudolf Arnheim, has already identified this perceptual mechanism in the late 1960s. In *Art and Illusion,* he writes that "all colour gradients have the power to create depth, but among them the gradients of brightness are the most efficient" [6]. This explains again how the perception of movement is activated in the observer. Regarding this expedient, Arnheim reminds us that this kind of illusion has been widely used by the artists of the Nineteenth century. Romantic painters "represented the sum of local brightness, local colour, and the brightness and colour of the illumination through a single shade of pigment. [...] This pictorial procedure also defined the individual as being partly the creature of its environment, subject to influences that cannot simply be shed like veils" [7].

Fig. 6. Caspar David Friedrich, Wanderer above the Sea of Fog (1818), Kunsthalle Hamburg

Today these concepts are more current than ever. Since as Jean Paul Thibaud argues, the landscape representation is no longer limited to a "natural" analysis of the context but focuses on an artistic representation. Likewise, Thibaud quotes Romantic painters, in this case to underline the feeling involved in this artistic production: "is the idea of *Stimmung* that neatly expresses the intrinsic relationship between the perceptible unity introduced by the landscape and its constitutive feeling" [8].

4 The representation of Indeterminate Space

In the final category, the artificial manipulation on the environment is dominant. The alterations on colour and light describing the *ambiance* are not limited to the background;

instead they invade the entire space of the image, overlapping the architecture. In this way, interpreting Thibaud's thinking, making all the components present in the space work together and "integrating them by giving the same tonality to all that appears" [9], the true essence of *ambiance* becomes visual reality. "As we may learn from the art world, working on sensory matter consists in giving the world a singular face, a certain overall physiognomy, a particular colour scheme." [10]

In the digital image analysed below, the need to homogeneously characterize the ambiance translates into a visual effect of rarefaction with a predominant tone of colour. For the observer, due to this effect it is difficult to recognize every single element and its spatial arrangement. As in William Turner's work *Norham Castle* (Fig. 7), where the blinding pure colours express the light while the historic building and landscape merge together; in the digital representations, the background, the work, and its inhabitants are all given the same shade and therefore affective tone. Using a definition formulated by the historian Robert J. Pepperell: "visual indeterminacy can be defined, as the perceptual experience occurring in response to an image that suggests the presence of objects but denies easy or immediate recognition" [11].

Fig. 7 J. M. William Turner, Norham Castle Sunrise (1845), Northumberland

Fig. 8 Moreau Kusunoki Architects, Digital image (2014), Guggenheim Helsinki Museum, © Moreau Kusunoki Architects

According to Pepperell, the indeterminate category can be effectively described by means of a single digital image: an external view of the winning project for the Guggenheim Helsinki Design Competition in 2014, by Moreau Kusunoki Architects (Fig. 8).

Beginning with the analysis, let us imagine asking an observer to describe what they see without giving them any information about that representation. The observer in question would certainly feel lost, experiencing difficulties in recognizing both the elements of the composition and their position. The landscape appears, indeed, entirely wrapped in a dense grey-blue fog; the presence of some light points guides the eye in the distance, allowing the part of inhabited space to be sensed. The cold colours together with the widespread fog could suggest a Nordic landscape. The elements of the context – the surrounding buildings, the sky and the sea – are hinted, but not clearly distinguishable from each other. Some elements allow the observer to recognize the presence of a large building: it is possible to perceive the height differences of the volume and the curves on the roof, but it is equally difficult to add anything else.

In this image, the only readable element is a variation of chromatic tonality that extends spatially in every direction: it describes an affective tone. To explain this effect, we can use the definition of *actual fact*, formulated by Josef Albers at the end of the 1960s [12]. Gernot Böhme – who reminds us that the goal in Albers' colour research was not to represent anything, but to put on stage a chromatic fact – took up this concept. Albers, following the experiments conducted in *Homage in the square*, distinguished the *factual fact*, that is the colour physically spread on the canvas, from the *actual fact* that is the perceived colour, the thing in act: "which cannot be located on the canvas or on the single places, being rather in a certain sense an excitation of the perceptual space in which the observer comes in when dealing with the painting" [13].

The tone of this image generates various perceptual aspects. Firstly, the grey-blue fog is able to generate a bodily sensation, by means of synesthesia it is possible to perceive cold and humidity on the skin. Secondly, the warm lights and bright colours catch the attention of the observer, generating a corporeal sensation related to heat. Furthermore, the image as a whole generates movement. The represented environment thus acquires a strong dynamism because the observer, unable to recognise the elements, is forced to explore it visually. In this image, it is indeed difficult to clearly understand the museum's spatial position: the ground is not perceptible, just as it is not possible to understand the connection with the sea since the pier is immersed in the fog. The museum is depicted from a considerable distance, and due to the positioning of the camera the observer is placed at the centre of an indeterminate space. In this way, a growing desire to reach the only inhabited place – identifiable in the image thanks to the light – is produced. The observer is doubly stimulated: the subject's imagination makes an effort to analyse space – performing an analytic function on the recognition of the elements present – and at the same time bodily sensations are produced.

Conclusions

The goal of these images is to excite the observer's sensorial experience: the representation of fogs and particular weather conditions stimulates hearing, smell and touch as well as the representation of air movement stimulates the tactile sense of humidity and temperature. These *ambiance* qualities are generated in observatories not only with the insertion of people and objects, but primarily with slight shades of colour blurring the image. In this sense, it is possible to argue that the representation of lived space is undergoing a *morphing* from clear to indeterminate.

To underline the atmosphere of the lived space, these images propose to simulate not only architectural shapes, but rather the contingency of its appearance. The aesthetic features are not limited to the anticipation of the perception of a solid body in space; but instead to the stimulation of an emotional response. Consequently, the narration of *ambiance* is the true representation's subject. Architecture is a vital part in this context; it is a living character that vibrates together with its inhabitants, the changing of the seasons and the passing of the day. Welcoming the contingency of life means to strip the architecture of its idealism, to bring denser aspects potentially more imprecise and therefore truer. It means making architecture more hypothetical, pictorial, imaginative and paradoxically more real, enabling it to evoke the affective and sensory experience that characterizes every day urban life.

References

1. G. Böhme, *Atmospheric architectures: the aesthetics of felt spaces*, Bloomsbury Publishing (2017)

2, 6. R. Arnheim, *Art and visual perception*, Univ. of California Press (1974)
3. V. Gallese, *Arte, corpo, cervello: per un'estetica sperimentale*, Micromega, **2**, 49-67 (2014)
4, 8-10. J. P. Thibaud, *The backstage of urban ambiances: when atmospheres pervade everyday experience*, Emotion, space and society, **15**, 39-46 (2015)
5. V. Gallese, M. Ardizzi, *Il senso del colore. Tra mondo, corpo e cervello*, Colori: l'emozione dei colori nell'arte, Milano: Silvana Editoriale, 23-35 (2017)
11. R. Pepperell, *Connecting art and the brain: an artist's perspective on visual indeterminacy*, Frontiers in human neuroscience, **5**, 84 (2011)
12. J. Albers, *Interaction of color*, Yale University Press (2006)
13. G. Böhme, *Atmosfere estasi e messe in scena: l'estetica come teoria generale della percezione*, Milano: Marinotti (2010)

Bibliography

R. Arnheim, *Art and visual perception*. Univ. of California Press (1974)
J. Albers, *Interaction of color*. Yale University Press (1971)
G. Böhme, *The aesthetics of atmospheres* (J.-P. Thibaud, Ed.) (1 ed.). London; New York: Routledge (2016)
G. Böhme, *Atmosphere as the fundamental concept of a new aesthetics*. Thesis eleven, 36(1), 113-126 (1993)
A. Damasio, *Descartes' error: emotion, reason, and the human brain*. New York: G. P. Putnam (1994)
D. Freedberg, V. Gallese, *Motion, emotion and empathy in aesthetic experience*. Trends in cognitive sciences, **11**, 5, 197-203 (2017)
D. Freedberg, *The power of images: Studies in the history and theory of response*, Chicago: University of Chicago Press (1989)
V. Gallese, V. (2014). *Arte, corpo, cervello: per un'estetica sperimentale*. Micromega, **2**, 49-67 (2014)
V. Gallese, *Bodily selves in relation: embodied simulation as second-person perspective on intersubjectivity*. Phil. Trans. R. Soc. B, 369(1644), 20130177 (2014)
T. Griffero, *Atmospheres: aesthetics of emotional spaces*. Routledge (2016)
D. Leatherbarrow, *Atmospheric condition*. Phenomenologies of the city: studies in the history and philosophy of architecture, M. Sternberg & H. Steiner Eds., 85-99 (2015)
J. P. Thibaud, *The backstage of urban ambiances: When atmospheres pervade everyday experience*. Emotion, space and society, **15**, 39-46 (2015)
J. P. Thibaud, *The sensory fabric of urban ambiances*. The senses and society, **6**, 2, 203-215 (2011).
D. Vesely, *Architecture in the age of divided representation: the question of creativity in the shadow of production*, MIT press (2004).
S, Vizzardelli, *Verso una nuova estetica: categorie in movimento*, Milano: Mondadori (2010)
S. Zeki, J. Nash, *Inner vision: an exploration of art and the brain*, **415**, Oxford: Oxford University Press (1999)

Acknowledgments

I would like to thank the following for their kind permission to reproduce their images: Mir for the rendering of the Pavilion Expo in Dubai (Asif Khan Architect), Iwan Baan for the interior's photograph of Maxxi Museum, Herzog & De Meuron for the images of

Elbphilharmonie of Hamburg and Moreau Kusunoki Architects for the rendering of Guggenheim Helsinki Museum Competition.

Temporal Sections Conceptual Tool: Articulating Space and Time in Representing Urban Ambiances

Noha Gamal Said[1,*]

[1] Assistant prof. Depart. of Urban Design, Faculty of Engineering, Ain Shams Univ. Cairo, Egypt
Associate researcher, School of Architecture Univ. Grenoble Alpes, AAU-CRESSON, France

Abstract. This article aims to present a new methodological protocol and a representation tool that articulates both space and time in understanding urban ambiances: temporal sections. The elaboration of this conceptual tool is based on a theoretical background dealing with the palimpsest of urban ambiances. The urban palimpsest is a metaphor underlying the dynamic of the formation process of the territory resulting from a temporal, spatial and social stratification that manifests in the lived experience. Combining urban transect, a geographical tool, and stratigraphy, a geological one, temporal sections represent the sensory experience as a palimpsest that integrates the different pasts and projects on the future. This tool provides a very specific way of immersion and opens new realm in constructing a storytelling; it proposes a rereading and writing the ambiance in depth.

1 Palimpsest, ambiance, representation of time

The construction of cities presupposes a complexity of time. In fact, time, as much as space, are both mediums in which a city changes. In this optic, a city is seen as an evolving organism, *un-être-en-devenir-permanent* [1], exposed to incessant transformation and subject to the order of *in process*. Cities reflect the work of time in space, as time is incarnated in the different modes of construction, in various architectural styles, in distinct urban fabrics and in mixed social practices. The coexistence of all diverse spatial and social configurations that represent different temporalities gives rise to a historical amalgam and reveals temporal complexity.

Several authors have tried to nominate this temporal complexity that characterizes the contemporary landscape: palimpsest [2] or metissage [3]. Others have associated several qualifications to time: stratified time [2], consolidated time [4], sedimented time [5]. These are conceptions of time that underline a state of change, *une pensée de devenir*, in which time is perceived as a flux, movement and transformation.

* Corresponding author: said.n@grenoble.archi.fr

This article tackles the evolving nature of cities in terms of urban ambiance. Despite its fragile nature, space ambiance incorporates a certain temporal depth and unfolds the historical dimension of a territory. Moreover, an ambiance evokes a subjective experience and reactivates individual memories and internalized experience by its anamnestic quality. It has a strong presence in composing a collective memory and in creating a particular sense of belonging and place attachment. The incorporation of time in terms of ambiances marks a deviation from materiality to immateriality, where sounds, odors, texture, motions and emotions, become the main substance of traces and souvenirs.

Ambiance-palimpsest constitutes the main theoretical base of this article [6]. A palimpsest is defined in dictionaries as a two-dimensional writing support, which, by a process of erasing and rewriting, thickens in three dimensions; it is also defined as something having usually diverse layers or aspects apparent beneath the surface. The term palimpsest is introduced as a metaphor in urbanism by André Corboz [2] and revisited by Sébastien Marot [7-8] and others [9] in order to point out the temporal dimension in which a territory is composed.

The association of two notions: palimpsest and ambiance underlies the formation process of an ambiance resulting from a temporal, spatial and social stratification that manifests in the sensory experience [6]. Far beyond the immediate perception of places, sensory phenomena carry ambient traces and configure a sensory heritage: a palimpsest of urban atmospheres. As much as the space, an ambiance embodies a temporal depth, *épaisseur temporelle,* where the present time is enriched by recomposing values coming from the past.

How to represent the coexistence of diverse sensory phenomena in the lived experience that come from different temporalities? What sort of appropriate tools can be created to capture and embody time; tools that are capable of incorporating, saving and representing urban temporality in a graphical transcription as a shared language between architects and urban planners? These questions point out the necessity to create a new tool that helps to reread the intangible layers composing the urban palimpsest and to suggest new possibilities for designing spaces by considering the evolution of cities.

In this perspective, this article responds to the theme: "Ambiance, storytelling and immersion". Based on a comprehensive methodological protocol, the article introduces a conceptual representation tool that articulates both time and space in understanding and analyzing urban ambiances: temporal sections. It is about a recontextualization of a graphical tool that represents urban ambiance as a sort of palimpsest.

2 Rereading ambiance as temporal encounter

Approaching the temporal complexity as a potentiality of investigation of the sensory phenomena requires a new tool allowing to integrate the temporal depth in representing urban ambiances. Dealing with the diachronic dimension of the sensory experience necessitates a change in the way we represent time; the ambiance constitutes a space-time experience that recall past experiences into the present time in a voluntary or involuntary manner. Through reactivation, the ambiance brings together different strata of the experience, customs of the past and everyday practices creating therefore a temporal conjugation [6].

Conceiving urban ambiances as a temporal encounter necessitates a new way in representing time. It opposes the vectored timeline as a common representation of time because this linearity does not correspond to the reality our experience of time where the past and the future have a dynamic presence in the present time. This very specific way of reading the past is neither about delving into history nor a study of the urban history in time segments *i.e.* dealing with a specific period of time; it is rather about crossing the past and present experiences in order to establish an evolving and dynamic vision of the urban temporality. This way of decoding former ambiances requires another form of access to urban history that give rise to new modalities of storytelling and new means of reconstructing the historical narrative.

In addition, the researcher in this domain is facing a major challenge related to the large amount of information about the site dealing with the different pasts, the present, and the imagination of the urban future. The conception of space ambiance as a temporal encounter reveals different aspects of the complexity of urban history:

- First, *a temporal complexity*: composed of several superimposed layers of time, where the past is no longer singular but multiple;
- Second, a *complexity of urban history* encompassing spatial history, the history of social practices, the sensory history inseparable from memory;
- Finally, a complexity of the *representation of the city*: multiplicity of sources representing the city such as paintings, photography, documentary films, novels, travel stories, films, which record and describe the urban conditions of the past, in addition to other multimedia documents (sonic and visual)... Allowing to preserve certain moments of cities, these documents constitute, from our point of view, a precious source to access the sensory memory. We cannot ignore the role that these documents play in the redemption of physical and social realities [10]. These materials are testimonies of everyday life in several spatial and temporal scales.

In order to manage the big data challenge, it is important to find a suitable tool that can cope with the data capturing, organizing, storage, analysis, allowing also easy research, representing, sharing, visualizing. In this regard, the proposed temporal sections conceptual tool seeks to attain two objectives: first, the tool should allow a very precise organisation of data in both space and time. It should be dynamic enough to allow a continuous updating – addition, removal or modification of the collected database. The second objective deals with the distinct nature of the ambiance as it examines the layering of the physical characteristics (form), sensory phenomena (former) and social practices (formality) [11], this is why it is important to find a way to contextualize the different sensory phenomena and to embody urban memory by placing it in its spatial and social contexts.

3 Transect and stratigraphy: Two sources of inspiration

The idea of temporal sections is to represent an urban palimpsest. This mode of representation is inspired by two sources: the first is the ***urban transects***, a geographical tool [12], and ***stratigraphy***, a geological one [13]. The urban transect is basically a geographical method that has been introduced in the architecture and urban domain. It aims to develop the capacity of the urban section to become an observation tool and representation technique along a linear path. Being developed in different research works at Cresson, the urban transect hybridizes urban sections with sensory itineraries of the city [12]. Its application allows the articulation between different components of the urban environment that had often separately considered: urban form, sensory phenomena and social practices [12].

We called for "the stratigraphy" as a second source of inspiration. It is a branch of geology that seeks to identify the time of earth by using sedimentary piles. It studies the stratigraphic structures and the different traces they contain [13]. In fact, the simple phenomenon of stratification implicates a discontinuity in time progression (flux); the superposition of different layers represents an assembly of heterogeneous temporal layers. Indeed, the concept of stratigraphy permits to access time that, despite being continuous and homogeneous, is only accessible in a heterogeneous and fragmentary way. There are moments when a deviation in the temporal continuity takes place, constituting therefore points of turmoil in history or temporal marks by which we learn history. The past is thus subdivided by these major events into historical phases.

The conjunction of these two methods produces the concept of temporal sections. It is about a recontextualization of a graphical tool that represents urban territory as a sort of palimpsest. The temporal section takes the shape of a "stratigraphic column" representing a temporal sedimented pile of the territory.

The temporal section is elaborated in three actions, collecting, analyzing, representing, that do not follow a chronological order; they rather occur in a simultaneous manner and progress continually along the research.

Before introducing the temporal section conceptual tool, it is worth referring to the methodological protocol that has allowed for gathering the database that constitutes the main input of the proposed tool.

3.1 Constructing and analyzing a database on sensory memory

Constituting a palimpsest of urban ambiances necessitates a prior comprehensive and long methodological protocol in order to help emerging the sensory features of the present, the future and the different pasts. In this phase, we should first and foremost distinguish two sources of memories: the first is ***living memory***, in which we called for the methods that help gathering information on the *recent* past or the present time in order to reveal how existing inhabitants perceive and remember their district. In terms of time, this temporal phase covers the span of a lifetime of the existing generation inhibiting the territory. In this context, we applied methods like the commented-walks [14] and the method of itineraries [15] that permit to harvest the inhabitants' speech and to access their sensory memory.

On the contrary, the ***archived memory*** deepens our survey to cover the territorial memory. It reveals the former spatial and social composition of territories. It can be found in all sorts of sources representing the city, including maps, plans and all urban documents, videos, sounds, films, documentary films, romans, daily journals, trip journals, photography, etc. They permit to reconceptualize more distant past atmospheres and to access the impressions, feelings, lifestyles and social relationships of the former ambiances.

These documents should be analyzed in order to identify the main sensory features of each era, they should also be classified in order to develop a cross vision of the evolution of the urban ambiances. Each document should be as rigorously as possible localized in space and accurately dated. It should be associated with precise descriptions and, if possible, accompanied by photos or sketches.

3.2 Drawing an urban transect

This first step in the methodological protocol is to identify the main skeleton of the palimpsest via settling both spatial and temporal boundaries that delimit the scope of work of the research. This step, drawing an urban transect, is about defining the spatial framework. The urban transect represents the territory in a linear form and crosses diverse spaces that compose the urban territory understudy. It can be drawn in plan that rather deals with the macro scale for highlighting the different urban forms, or in section dealing on the contrary with micro scale for understanding certain spatial configurations: building height, the relation between the façade and the street, social practices, bodily presence, etc. In our work, we have chosen the urban section to represent the existing sensory experience. This choice is due to the capacity of the urban section to navigate between different scales and show the body in the space, the relation between the public and private realms. It also helps to identify the different spatial filters that define the transitional spaces (Fig. 1).

3.3 Layering the sensory history

The objective of this step is to define the temporal framework of the study. To the above-mentioned transect is added a vertical timeline. The emplacement of the urban transect indicates the present time; the upper part represents the future while the lower part marks the past. On the vectored timeline, the initial moment defines the time span or the temporal depth to which the study is limited. The subsequent step is to specify the stratified temporal structure of the territory. The timeline will be hence subdivided into historical phases based on a detailed historical and memorial study of the territory (Fig. 1). These layers of time are defined by horizontal lines placed at the moments of upheaval where radical changes in the sensory history occur. These ***temporal marks*** require a precise rereading of the urban and social history of the territory. They can be related to urban decisions, for example the introduction of a tramway or a metro, planting a garden, proposing a pedestrian zone, etc.; political events, for example wars, revolutions, or a radical change in the political regime; major crises, like epidemics, or even due to a social change where the territory changes the social class by means of gentrification or a simple degradation of the district that welcomes lower social classes, etc. The timeline is hence subdivided into temporal strata as shown in (Fig. 1).

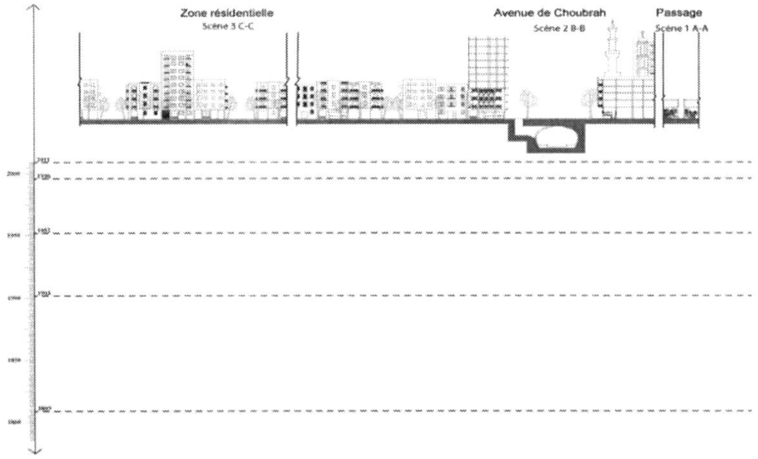

Fig.1. The Spatio-temporal skeleton of the evolution of Choubrah - Cairo. The horizontal lines indicates the moments of major changes in the district's sensory history (Copyright: Noha GAMAL SAID)

3.4 Filling in the spatio-temporal framework

Until then, with the drawing of the urban transect and the subdivision of the timeline into temporal strata, we have constituted the spatio-temporal framework of the study. In the following step, this framework will be progressively filled with information coming from different sources. The chosen data will be placed on the temporal section in the corresponding time (past, present, future). On the different temporal strata representing the past is precisely placed, both vertically in space and horizontally in time, the vast database of varied documents that describe the urban conditions of the past.

The subsequent step is to collect and synthesize the sensory memory in a graphical, iconized and text format and to reorganize it according to the moment of emergence, in the corresponding temporal strata. This step also demands to select the proper representation form for each phenomenon: icons, colors, text, photos, etc. The most relevant comments of inhabitants are selected and presented on the transect. The collected data is gradually positioned on the temporal section by the researcher himself.

4 Choubrah, *a popular district but...*

The methodological protocol and temporal section was applied to Choubrah, a popular district in Cairo. Based on a fieldwork, twenty sensory phenomena have been outlined along the chosen path that unfold a deep anterior temporality of the territory. As an example, we shall detail in the following section the green shadow as a particular sensory phenomenon of the district that unfolds a temporal depth (Fig. 2)

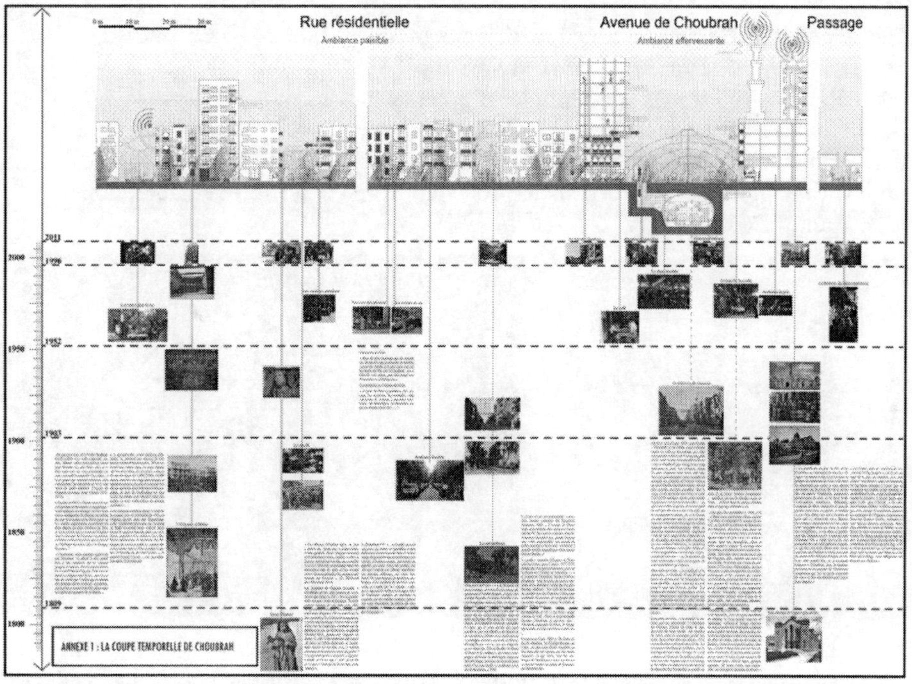

Fig.2. The temporal section of Choubrah - Cairo (Copyright: Noha GAMAL SAID)

4.1 Sequence 1 – Green Shadow

> "In the past, I have always enjoyed strolling in the residential streets to enjoy the rays of the sun filtered by the leaves of the trees forming gold coins scattered on the ground" (Ghada, F., Inhabitant)

Green-shadow is a phenomenon of filtered light caused by the intensive presence of greenery of the aligned trees on both sides of the street. The branches from both-sided trees meet in the sky and form a green vault covering the sky. Found in residential streets and more seldom in the avenue of Choubrah, this phenomenon constitutes a sceno-spatial identical configuration of the district. It unfolds part of the glorious history of greenery in Choubrah that goes back to 1809, the year of constructing Choubrah Boulevard. Since 1809, the avenue of Choubrah has been an important promenade; a linear alley planted with gigantic sycamores and acacias trees aligned on both sides of the avenue. They covered the avenue with intensive impermeable shadows (Fig. 3). This particular configuration had provided an agreeable ambiance for strolling. Despite their minimal presence in the existing urban context when compared to the past, green shadow played a very important role in the constructing a former collective memory of the district.

Moreover, this phenomenon was the main reason to introduce the action of strolling in the urban history of Cairo. The ambiance of strolling in Choubrah Avenue is described in the travel diaries written during the nineteenth century. Green Shadows has also encouraged the presence of women in public spaces, described as 'black angels' in their elegant and well-equipped carriages. Moreover, the intensive greenery in Choubrah constructed a strong image of the site. It is often compared to other famous green spaces in other cities such as Le Prater in Vienna, Le Prado in Spain, Hyde Park in London. The alley of Choubrah has been described in several travel diaries as the Champs-Élysées of Cairo.

> "The avenue of Choubrah, lined with ancient trees, under which one finds always freshness, even in the heaviest heat. Along this promenade of predilection Cairotes rises a multitude of country houses more or less elegant, like the palace of Kasr el Noussa." [16]. "The avenue of Choubrah presents thick and luxurious masses of vegetation that are truly superb. The eye plunges into it with delight and one rests there protected from the terrible reverberations of the sand of Egypt." [17]

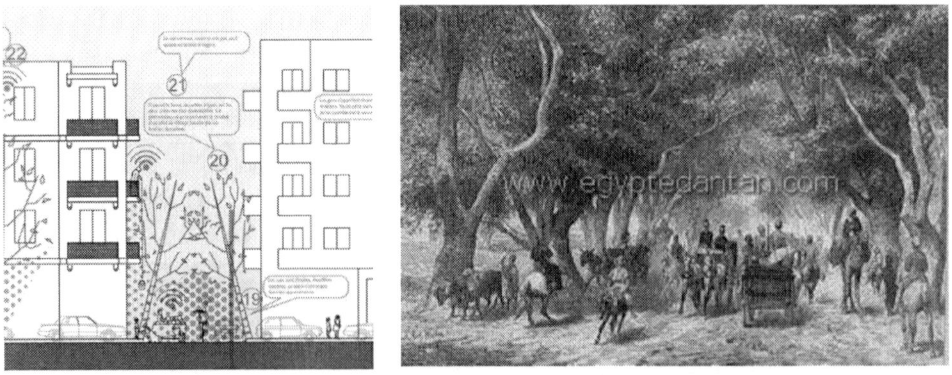

Fig. 3. The Green Shadow phenomenon in the past and the present. The photo to the right shows the green shadow phenomenon as a major component of the former ambiance in the first half of the 19th century. (Copyright @ Noha GAMAL SAID; @ M. Karegl, *L'Egypte d'Antan*.)

5 Reading and writing the ambiance in depth

The temporal section proposes a rereading of the space and the lived experience as a sort of palimpsest. The proposed tool opens a new way of reading and writing the ambiance in depth as they help decoding the urban memory and social heritage by creating links between past and present, comparing temporal strata between them and identifying both traces and oblivion whether invisible, material or intangible.

In this stratigraphic vision, the vertical stacking that crosses time allows to develop different scenarios that help understanding the different behaviors of each urban, social or sensory phenomenon over time: certain are ephemeral and others are persistent. Thanks to this type of representation, it is possible to trace vertically the evolution of the different sensory phenomena; while horizontally it permits to understand the different features of the ambiance at a specific moment of its history. The overall vision unveils the temporal pattern of the sensory phenomena and helps understanding the transformation dynamics of the territory.

Conclusion

The temporal complexity that characterizes contemporary cities necessitates the creation of a new tool that helps rereading the intangible layers composing the urban palimpsest. Urban stratigraphy and temporal section suggest new possibilities for the design of future spaces by considering the evolution of territories in time; they underline a sensory archeology and highlight the memorial dimension of the space by studying the lived experience of the present.

In terms of representation of ambiances, temporal sections mark a deviation from a ***cartography*** identifying a now and here to ***stratigraphy*** by integrating the diachronic dimension; in other words, a deviation from spatial to spatio-temporal representation. It represents a condensed format of the ambiance in both space and time.

Proposing a temporal journey through space, temporal sections provide a very specific way of immersion in different temporalities when experiencing the present time. Linking subjective and objective data, the tool provides a holistic vision of how spatial context is shaped, lived, perceived and memorized by gathering different information concerning the three components of ambiance: spatial configuration, social practices and sensory phenomena.

The keen interest in temporal sections as a conceptual representation tool and a methodological protocol is, on the one hand, to deepen the territory by its ambiance, *épaississement du territoire par son ambiance*; incorporating historical strata brings out the places and phenomena of memory, former uses and symbolic values and the memory traces through highlighting the different components that constitute a social and sensory heritage composing therefore a particular storytelling. On the other hand, it introduces the sensory dimension as part of immaterial heritage in order to sensitize city planners, decision-makers and stakeholders to this dimension in the way they conceive and project cities. Temporal sections help architects and urban designers to develop a specific look at the future while taking into account the sensory values of the past and to recompose them in the forthcoming projects: designing a retrospective future.

References

1. T. Ingold, The temporality of the landscape, World Archaeology. **25**, **2**, 152–174 (2009)
2. A. Corboz, Le territoire comme palimpseste, Diogène. **121**, 14 – 35 (1983)
3. F. Laplantine, A. Nouss, *Le métissage : un exposé pour comprendre, un essai pour réfléchir* (Téraèdre, 2009)
4. M. Roncayolo, *Lectures de villes : formes et temps* (Parenthèses, Eupalinos, 2002)
5. S. Rufat, La rue, palimpseste et vitrine, Tracés, **5**, 111-116 (2004).
6. N. Gamal Said. *Vers une écologie sensible des rues du Caire: le palimpseste des ambiances d'une ville en transition*, Thèse, Université de Grenoble (2014)
7. S. Marot, *Palimpsestuous Ithaca*, Thèse, Paris : EHESS (2008).
8. S. Marot, Sub-urbanisme / sur-urbanisme, Marnes. **1**, 301-351 (2011)
9. J. Chesneaux, Mémoire urbaine et projet urbain In T. Paquot, *Le quotidien urbain : essais sur les temps des villes*. 107-127 (Éd. La Découverte / Institut des villes, 2001)
10. S. Kracauer, *Le voyage et la danse. Figures de ville et vues de films* (Presses de l'Université Laval ; Éd. de la Maison des sciences de l'homme, 2008)
11. G. Chelkoff, Formes, formants, formalités In *L'espace urbain en méthodes*, 101-124 (Parenthèses, 2001)
12. N. Tixier, D. Masson, C. Okamura. *L'ambiance est dans l'air*, Rapport (Cresson ; PIRVE, 2011)
13. P. De Wever, *Temps de la Terre, temps de l'Homme* (Albin Michel, 2012)
14. J.P. Thibaud, La méthode des parcours commentés In *L'espace urbain en méthode*, 79-99 (Parenthèses, 2001)
15. J.Y. Petiteau, E. Pasquier, La méthode des itinéraires In *L'espace urbain en méthode*, p. 99 – 110 (Parenthèses, 2001)
16. H. De Vaujany, *Description de l'Egypte* (E. Plon et Gte, 1883)
17. G. De Nerval, Gérard. *Voyage en orient* (Calmann Lévy, 1884)

Envisioning Ambiances of the Past

Guzden Varinlioglu[1] and *Suheyla Muge* Halici[2]

[1] Izmir University of Economics, Department of Architecture, Izmir, Turkey

[2] Istanbul Technical University, Department of Informatics, Istanbul, Turkey

Abstract. Outdoor museums of archaeological excavations function as sites of both scientific research and public display. Often, long–term archaeological research means postponing the preparation of the site for visitors. This paper focuses on digital tools for the representation of architectural reconstructions, i.e. possibilities for representing a range of hypotheses regarding the past ambiances of the ancient city. It proposes an augmented immersive revisit of the cultural heritage through mobile devices. Based on mobile phones' current technical capacities, which enables rendering of 3D content combined with camera input, we developed the proposed mobile AR application for mobile Android devices. TeosAR offers a real-time, in-situ 3D depiction and visualization of architectural artifacts of the ancient city implementing model-based tracking methods.

1 Introduction

Heritage developed from the concern of a small group of professionals into a key interest of the wider society. With the introduction of "World Heritage List" by UNESCO, there is a heterogeneous piling up of monumental, iconic, intangible, and quotidian heritage entities, representing a "crisis of accumulation" [1]. Due to the heritage boom in global tourism and the memory crisis of late-modernity, doubt has arisen over our ability to create collective memories and cultural identity. At this point, digitization of heritage helped to resolve the issue of the valorization of the heritage boom. Digitization turns the cultural resources into an important pillar of the digital economy. Digital heritage is "created digitally, or converted into digital form, from existing analog resources" [2].

Studies on digital heritage offer an interdisciplinary approach to historical sites, buildings, and artifacts, incorporating archaeology, restoration, and architecture. Recent developments in emerging technologies have led to 3D archaeology, in which digital tools provide a new perspective on the archaeological excavations in regard to the collection, analysis, and visualization of data. Among these technologies, augmented reality (AR) has great potential for investigating archaeological objects as it provides a non-destructive way for archaeologists, art historians, and other scholars to analyze objects through high-resolution images, with no risk of damage, and also allows visitors to envision the 3D reconstructions of the ancient sites.

Digital and computational tools, combined with applications of Augmented Reality (AR), are changing archaeological practices in the areas of presentation and visualization. Through an interdisciplinary approach, this paper and its pilot application, known as TeosAR, applies augmented reality to the archaeological site, the ancient city of Teos. The project is designed to support the ongoing visualization efforts. Teos, within the present-day borders of the Izmir province, was an important city in Ionia in terms of military, politics, and commerce connections, due to its geographical features, particularly its two sheltered harbors. As of 2010, Prof. Kadıoğlu, supported by emerging technologies in archaeology, has led the excavation and restoration work in the ancient city. The Digital Teos Project, conducted under his supervision, is an interdisciplinary research project aiming to investigate and digitally animate the architectural heritage of Teos. The project consisted of a static public setup at the excavation site, a virtual revisit through VR devices, and a mobile game application. The Digital Teos Project, finalized in February 2018, features a repository of 3D virtual reconstructions of the ancient city.

The Digital Teos project was the first step in an interdisciplinary investigation that enquires into, and digitally animates the archaeological site – one of the most important cities of Ionia. The realization of this project involved close coordination between the fields of archaeology and informatics in design and visualization. In this study, the TeosAR is conceptualized as a digital medium that converts archaeological knowledge into interactions, focusing on a case study in the ancient city of Teos. Our focus is on real-time recognition, visualization, and dissemination of archaeological information. TeosAR is a digital platform that allows the interested general public to explore current archeological findings, and to experience the ancient urban setup without disturbing the progress of the archaeological work. The spatial configuration of the immersive reality presents the architectural information with archaeological accuracy.

This study aims to develop a mobile AR application based on model-based AR tracking methods to present findings from previous studies combined with 3D virtual archeological objects and data from the most recent site research through an interactive mobile interface (for visualization and dissemination the information). Augmented reality in open-air museums is mainly achieved by GPS tracking or image-based marking; however, our project is different in that it explores the use of model-based AR tracking technology to superimpose information on the artifacts themselves. Using model-based tracking technology within augmented reality environments, we aim to develop a mobile application that provides information to expert archaeologists, and non-expert end users alike. The information pertains to artifacts found in the field, and is conveyed by means of mobile devices, and thus to provide a new layer of overall archaeological site experience.

2 Literature Review

Augmented Reality (AR) is the process of simultaneous viewing and interacting with 3D virtual objects in the real world [3]. Milgram and Kishino defined the virtuality continuum as extending from the completely real through to the completely virtual environment, with the area between consisting of augmented reality and augmented virtuality. Virtual Reality (VR) environment is the total immersion in a synthetic world; Mixed Reality (MR) environment is the merging of real and virtual worlds to produce a new environment, and AR completes reality without completely replacing it [4]. Since the 1960s, AR technology allowed the development of series of applications in various fields, including culture, archaeology, tourism, sports, entertainment, architecture, art, defense, education, health, commerce, design and task support. Regardless of the research field, the essential aspects of augmented, virtual, mixed reality applications are: tracking and registration of the user's viewpoint, virtual environment modeling, the key devices required for AR systems (computers, display, and devices for input and tracking), and the interaction interfaces between user and the virtual information [5].

In the field of cultural heritage, AR technology has been used for presentation and dissemination through informative or educational apps for indoor and outdoor museum visitors [6]. AR is more commonly applied in an indoor museum context, however, applications to the broader suite of tourism experiences at outdoor destinations are rapidly emerging [7]. According to Dieck and Jung, AR will be increasingly valuable for the presentation of outdoor historic and enlisted sites, as it eliminates the requirement for numerous information boards and other interpretive panels (including photographs, drawings, reconstructions, 3D views, models, maps, audio materials, etc.). The interaction between visitors of a cultural heritage site and AR systems will contribute to a richer experience [8]. Currently, there are two main AR methods to enrich the visitor experience and subsequent engagement with history: these are georeferenced/location-based, and marker-based AR applications.

Currently, in outdoor cultural heritage, georeferenced and/or hybrid AR applications are more common than solely marker-based AR applications. A prominent example of location-based AR application for archaeological sites is Archeoguide for the ancient city and ruins of Olympia, which is based on user's position and orientation in the cultural site, together with real-time rendering [9]. Archeoguide uses three basic subsystems, a site information server, mobile units and the network infrastructure. While carrying the mobile units, the visitors' positions, and other parameters are calculated by GPS. The system corrects the calculations' accuracy using a Differential GPS. As a georeferenced study, AR application on Pisa Monuments by Duguleana et. al [10] co-locates additional information about existing cultural heritage entities of interest located nearby. This study combines a non-optical tracking algorithm through GPS technology with optical tracking using simultaneous localization and mapping (SLAM) based markerless methods. This system allows identifying not only the location points near the visitors but also it displays additional contextual historical information co-located in the real environment. KnossosAR is an outdoor mobile augmented reality (MAR) guide implemented for the archaeological site of Knossos in Crete, which is on the UNESCO World Heritage List. This guide integrates AR projections of interpretive information with a non-linear storytelling [11]. KnossosAR focuses on the occlusion problem, which is the main issue for location-based AR applications. In order to eliminate this problem, KnossosAR employs a method for estimating the field of view (FoV), when physical obstacles (e.g. buildings, monuments) occludes users' own FOV.

Difficulties in enabling controlled outdoor environments are caused by issues such as changing light conditions, moving objects and possible changes in the defined target. To overcome these problems, georeferenced AR applications are widely used for large-scale outdoor environments. However, location-based AR applications have disadvantages of juxtaposition/overlaying of virtual objects over real objects. With small-scale objects, in particular, the juxtaposition of the tracked object and the virtual reconstruction become crucial. In this situation, researchers and developer prefer the more precise registration of marker-based AR applications. However, marker-based outdoor AR applications introduce new challenges due to the uncontrolled access by the public, and environmental conditions. Thus in this study, we focused on the use of archaeological artifacts as targets in the construction of an outdoor AR application. We used the model-based tracking method to reach the AR environment, and develop mobile AR guide for Teos.

3 Methodology

Archaeological research often fails to go beyond the systematic collection of data, site conservation/restoration, and limited/delayed dissemination to the public. In order to define, interpret, and categorize multi-layered data, assistive technologies must be produced for both archaeologists and visitors alike. This study allows interested visitors to experience the archaeological site by using model-based tracking methods for detection and overlaying created 3D reconstructions through an AR environment. Using a model-based augmented reality mobile, actual architectural artifacts become informed objects without disturbing the archaeological site.

3.1 Model-based Augmented Reality

Image-based target and GPS are traditional methods for applying AR environments to archeological sites in digital heritage [5]. However, the mobility of the visitor in an unrestricted and unprepared environment makes the localization of process more difficult. Fortunately, the emerging technology of model-based tracking method makes it possible to track any object in the physical world. The system requires pre-defined 3D models to track real objects in the environment and to superimpose virtual annotations. Recent personal mobile device technology allows model-based tracking methods to be used on site. The potential of this tracking system has been highlighted in several studies (e.g., [12]).

Among various augmented reality software development kits, Vuforia [13] combined with Unity Game Engine, is one of the most popular AR developer portal. The recent release of the Vuforia Object Scanner App (see Figure 1) helps to create Object Data (OD) files to collect feature points of model (object targets). Developers can generate, test and edit OD files by scanning a physical 3D object through the App. The App also provides a visualization of the object's features, as well as their coverage across the object.

Fig. 1. The screen view of Vuforia Object Scanner application.

In the Vuforia Developer Portal, we defined a target database by uploading scanned objects. The target database is then extracted from Vuforia Developer Portal for uploading into Unity, a cross-platform engine developed by Unity Technologies to develop video games for PCs, mobile devices, and websites. Subsequently, the target database is used for developing the proposed app by combining real objects and virtual reconstruction through the mobile device's display. The model-based tracking allows the user to interact with the virtual objects in real-time through a mobile device's screen. The system architecture of TeosAR application consists of five steps: the image capturing, object detection, tracking, overlaying virtual data, and display (see Figure 2).

Fig. 2. The system architecture of TeosAR application.

3.2 Case Study

The pilot study aims to develop an augmented reality mobile application to provide information for archaeologists or visitors to an archaeological area, by using model-based tracking technology. Augmented reality is a platform where three-dimensional virtual entities can be presented interactively in real space images. In the past, archaeological sites were only able to present the actual appearance of historical remains; however, now virtual animation and information screens, experienced with the help of mobile devices, can be overlaid on this real space. Thus, it is possible to utilize technology to produce visually rich, interactive products for users. Previously, the cultural heritage community were restricted to sharing 2D data such as plan, section, elevation, etc.; subsequently, 3D data became possible through interactive digital tools, allowing for the addition of a time dimension, namely 4D [14]. The transformation brought by digital heritage representations has enabled virtual heritage to spread rapidly through this extensive, well-established field, which has generally embraced the new technologies in recognition of their obvious superiority over previous approaches [15].

Fig. 3. Detection of artifact through TeosAR application.

According to Bekele et al. [5], AR in cultural heritage has three main research focuses: enhancing visitors' experience, heritage reconstruction, and heritage data management and exploration. Also, the essential aspects of augmented reality applications are: (1) tracking and registration of the user's viewpoint, (2) virtual environment modeling, (3) the key devices required for AR systems, and (4) the interaction interfaces between user and the virtual information [5]. Model-based targets helped us to track and register the user's viewpoint in an outdoor environment. Model-based tracking, which is a tracking phase of lines, edges, or shapes, uses a model of the features of tracked objects based on distinguishable features. For virtual environment modeling, we utilized the restitution models and juxtaposed them onto the tracked object while integrating updated and related information from other excavated objects on site. This also allowed us to convey archaeological excavation progress information with visitors and archaeologists in terms of the fourth dimension. In the developed application featured two Corinthian capitals of Teos, located near to the archaeological site entrance. The key devices required for AR systems are mobile phones or PDAs. Users direct their mobile device cameras to detect one of these capitals. When detected, archaeological reconstructions, i.e. possibilities for representing a range of hypotheses for the capital are shown. For the Interaction interfaces between users and virtual information, each archaeological artifact has its own text information, interactive 3D contents (see Figure 3), and maps other locations if similar artifacts are on site (see Figure 4). These features allow users to be more engaged and informed about the detected objects. By using this method, we can integrate audio-visual content and interactive games.

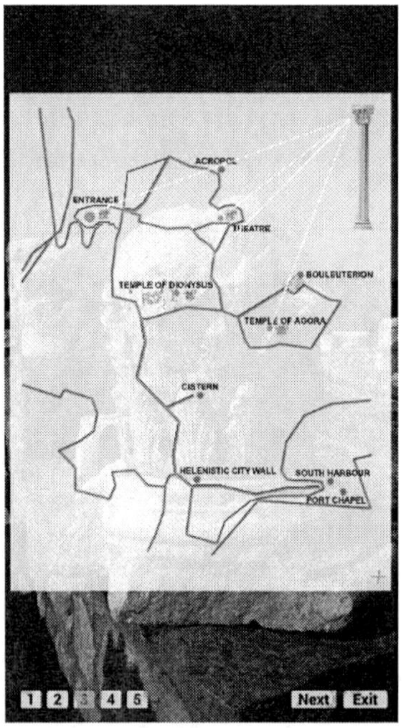

Fig. 4. Pinned locations of similar artifacts on the site map of Teos.

Conclusion and Future Studies

The progress of archaeology is now inextricably linked to applied modern technologies. It is no longer possible to conceive of research projects that do not support informatics; or an excavation that does not extensively use electronic or computerized survey. While all these tools are now accepted as standard, access to data for the site visitor has generally been a low priority. However, computer science and communication networks have become tools for scientists of humanistic science, which offers variety of principles, techniques, systems, and tools for creating effective digital libraries accessible to all end users. New applications of AR have provided opportunities for producing innovative tools for documenting, managing and visiting cultural heritage. As a result of the model-based tracking system developed for TeosAR, both visitors and archaeologists will have access to real-time information on the artifacts they encounter.

In this study, a low-budget augmented reality application is tested and implemented to the archaeological site of Teos. Common computing devices, e.g. Vuforia Scanner App, allow cost-efficient solutions for creating semi-automated guided tours in the outdoor environment. This is an efficient solution for outdoor applications in unrestricted and unprepared environment (with varying light conditions, uncontrolled weather conditions, uncontrolled access of the visitors). The initial results of our pilot application of TeosAR indicated that Vuforia Scanner App was successful in detecting and tracking the targets in various light conditions. In sharp and direct sunlight, as well as incomplete features of the targets, the App successfully detected the target. Comparing with an image-based marker, model-based AR gives an opportunity to look at the artifact in different viewpoints and view-angles which offers users better mobility around the artifacts.

In other words, cultural heritage objects become a smart heritage for a total participation in the promotion of cultural heritage. The study highlights an approach to the smart enhancement of cultural heritage experiences of curious visitors. In this way, archaeologists can use digital technologies to improve the dissemination of information resulting from their research into ancient architecture. Future studies will focus on the visualization of uncertainty in archaeological reconstructions, i.e. possibilities for representing a range of hypotheses regarding the past ambiances of the ancient city. In our virtual system, sliders will allow the user to move from physical to extrapolated data and have instant access to different solutions. As this is an excavation site with limited restitution data, the research will offer a 3D display system proposal, allowing the incorporation of archeologists' observations and interpretations. Thus, this multi-layered information should reflect the past together with a range of hypotheses on the ambiances of the ancient city.

References

1. R. Harrison, *Heritage: critical approaches* (2012)
2. Url 1, http://portal.unesco.org/en/ev.php-URL_ID=17721&URL_DO=DO_TOPIC&URL_SECTION=201.html
3. R. Azuma, PTVE, **6**, 4 (1997)
4. P. Milgram, F. Kishino, IEICE TOIAS, **77,** 12 (1994)
5. M. Bekele, R. Pierdicca, E. Frontoni, E. Malinverni, J. Gain, JOCACH, **11**, 2 (2018)
6. A.-C. Haugstvedt, J. Krogstie, *International Symposium on Mixed and Augmented Reality* (2012)
7. T. Jung, M.C. tom Dieck, H. Lee, N. Chung *Information and Communication Technologies in Tourism* (2016)
8. M.C. tom Dieck, T. Jung, CIIT, **21**, 2 (2018)
9. V. Vlahakis, J. Karigiannis, M. Tsotros, M. Gounaris, L. Almeida, D. Stricker, T. Gleue, I.T. Christou, R. Carlucci, N. Ioannidis, *Virtual Reality, Archeology, and Cultural Heritage* (2001)

10. M. Duguleana, F. Girbacia, C. Postelnicu, R. Brodi, M. Carrozzino, IJOCEACIE, **10**, 11 (2016)
11. P. Galatis, D. Gavalas, V. Kasapakis, G. Pantziou, C. Zaroliagis, *International Conference on Mobile Computing, Applications and Services* (2016)
12. A. Selvam, T. T. V. Yap, H. Ng, H. L. Tong, C. C. Ho, JOEAEAAS, **100**, 3 (2016
13. Url 2, https://developer.vuforia.com/
14. L. Marques, J.A. Tenedório, M. Burns, T. Romão, F. Birra, A. Pires, J. Marques, ACE, **11**, 33 (2017)
15. S. Eve, JOAMT, 19,4 (2012)

The atmosphere of a street: Experimental fieldwork on urban ambiances

Andreas Wesener[1,*]

[1]Lincoln University, Faculty of Environment, Society and Design, School of Landscape Architecture, PO Box 85084, Lincoln 7647, Canterbury, New Zealand

Abstract. The paper discusses phenomenological fieldwork carried out by third- and fourth-year students enrolled in the Bachelor of Landscape Architecture programme at the School of Landscape Architecture at Lincoln University. It focuses on students' sensory experiences while surveying a lively inner-urban street in Wellington, New Zealand, and discusses related (objective) circumstances, sensations and interpretations. Students were asked to describe their experiences while moving through the street and to record them in a field book in the form of notes and sketches. The goal of the paper is to capture, analyse and discuss students' individual experiences of different atmospheric facets of an urban streetscape. Preliminary findings are presented and discussed.

1 Introduction

Atmospheres or ambiances are phenomena that are emotionally-corporally 'felt' rather than cognitively 'understood' [1-4]. They have been conceptualised as "*Halb-Dinge*" (half-things) [5, 6], *"immaterielle Sonderdinge"* (immaterial exceptional things) [7], and mediations between subject and object [8] that may not be directly perceived, however, may affect a person's bodily and/or metal condition. According to Schmitz's [9] concept of *"affektives Betroffensein"* (affective involvement), atmospheres may significantly influence people's moods when spatially accessed or 'entered'.

In an urban context, atmospheres are noticeable but transitory qualities that may capture people's senses. Their particular spatial position is not defined; however, "[t]hey seem to fill the space hazily with an emotional tone"[†] [10]. Although not directly attached to material objects, atmospheres rarely do exist without physical space and have accordingly been considered as a powerful "link between space and sensitivity" [11]. While atmospheres cannot be measured directly, particular spatial attributes related to atmospheres such as narrowness, width, depth and surface have been identified [12]. Böhme suggests that objects are not only determined by their physical characteristics, but also by the way they 'come out'

[*] Corresponding author: andreas.wesener@lincoln.ac.nz

[†] Original German phrases in this paper have been translated into English by the author.

and provide "weight and orientation" to the surrounding space. He concludes that it is possible to conceptualise atmospheres as spaces "'tinted' by the presence of objects, people, or environmental constellations" [10].

Architecture can be experienced visually but also through "tactile reception" [13], e.g. the way the built environment is entered, walked through, and used. Experiencing architecture and the built environment may therefore be a matter of 'feeling' as much as 'seeing' [14]. While the urban built environment supports the production of atmospheres [15], urban landscapes are experienced as holistic combinations of sensory, habitual, and socio-cultural phenomena including class- or milieu-orientated self-representations [1]. Experiencing and remembering vital qualities of a place [cf. 15] might be a matter of holistic atmospheres rather than singular phenomena:

> *"Cities have atmospheres. They are fleeting like the scent of a rose. As this is not part of the blossom, the atmosphere is not attached to the physical space of the city. And yet, without the materiality of the blossom there is not its scent and without the materiality of the city not its atmosphere."* [1]

The study focuses on the sensory experience of streetscapes and related atmospheres. Urban streets are public open spaces that allow people to move around freely and experience the city. They are places where urban life rolls by [16] offering particular sensory experiences that only streets can provide. However, such experiences can differ significantly depending on the geographical, historical and socio-cultural context. The paper analyses and discusses the phenomenological fieldwork of third- and fourth-year undergraduate students enrolled in the Bachelor of Landscape Architecture (BLA) programme at the School of Landscape Architecture (SOLA) at Lincoln University. The students were asked to describe the atmosphere(s) of a section of Cuba Street in central Wellington, the capital of New Zealand, while moving through the space. Students were asked to refer to physical characteristics, smells, sounds, etc., in relation to their own experience and to describe any change of atmosphere while continuing their walk. The paper follows the theoretical and methodological framework used in a similar study [16] carried out with geography students in the historic 'Drosselgasse' in Rüdesheim, Germany. It presents and discusses preliminary findings.

2 Theoretical and methodological framework

Following Hasse's Drosselgasse study [16], the paper combines the philosophical-theoretical framework of Hermann Schmitz's [17] *Neue Phänomenologie* (New Phenomenology) with empirical qualitative social research methods. Schmitz regards the study of human subjectivity and related feelings and bodily experiences as a relevant epistemological complement to the rationalistic-reductive forms of scientific inquiry. At the centre of his investigation are phenomena that show or manifest themselves and their corresponding effect or impact on individual human beings that experience them. While clearly linked to the five senses of human perception, phenomenological experience is symptomatically a "diffuse unity between subject and object" [16]; a constructive process between subjective feeling and interpretative cognition that includes associative and 'learned' socio-cultural norms, constructs, symbols and interpretative patterns. Experiences can holistically combine sensations and interpretations to an extent that interpretations are experienced as feelings [16].

In our study, twenty third- and fourth-year BLA students were asked to describe the atmosphere(s) of a part of Cuba Street in central Wellington (Figure 1) while moving through the space. The students were on a one-week field trip in April 2017. Their programme included several site visits, a design charette with students from other New Zealand universities, and attending a conference.

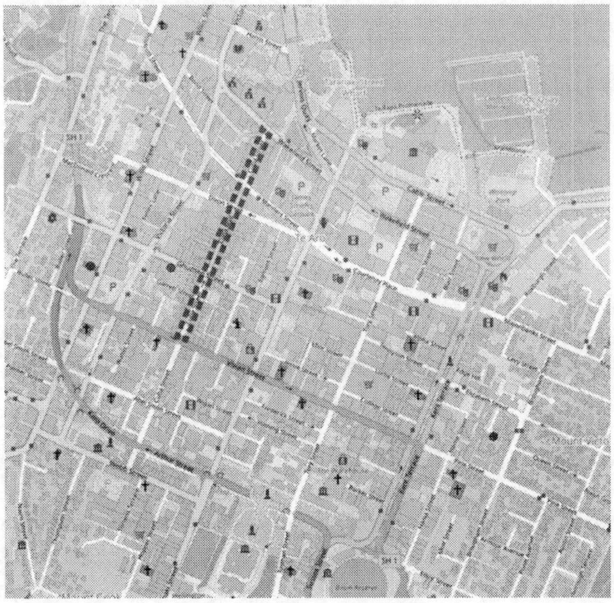

Fig. 1. Wellington city centre (not to scale); case study area highlighted. *Contains information from OpenStreetMap (http://openstreetmap.org/copyright; http://openstreetmap.org), which is made available under the Open Database License (ODbL).*

Students were asked to refer to the visual, physical-material characteristics, smells, sounds, etc. of the streetscape in relation to their own experience that may include interpretations and feelings including different moods, fears, joys, excitements, etc. They were asked not to share and discuss their experiences with other students during the study. The field work took about two hours to be completed. Students had to submit up to two A4 pages including detailed descriptions and sketches to help illustrate their experiences. From 20 submitted descriptions, three were excluded as they contained only irrelevant information.

Descriptions of holistic 'situations' combine objective and subjective circumstances. Following Hasse's approach [16], however using a slightly revised and simplified category system, students' holistic descriptions were systematically coded through deductive qualitative content analysis. Three main deductive categories were used to systematise students' heterogenous descriptions of perceptions, feelings and interpretations: (1) objective circumstances (*objektive Sachverhalte*); (2) sensations (*sinnliche und leibliche Eindrücke*); and (3) interpretations (*Deutungen*). The three main categories were subdivided in first level sub-codes, and – if necessary – second level sub-codes.

Following Schmitz [18], objective circumstances are based on sensory perceptions (mainly visual; auditory; olfactory; tactile). However, they are not necessarily factually 'true' or indisputable. Their main characteristic is that they are communicable – principally by everyone. Objective circumstances are core elements of "inter-subjectively similar perceptions" [16]. They include physical-material characteristics, such as buildings, shops, architectural details, plants and vegetation, construction materials; visual impressions, such as colours, light; people; sounds including music and smells, e.g. from food outlets, cafes and bars.

In contrast, subjective circumstances can only be communicated by one person [18]. They are based on individual experiences and feelings and related to a person's subjective or affective involvement in a situation. Although often related to objective circumstances, subjective circumstances can be clearly distinguished [16]. The paper defines sensations as situative sensory and bodily experiences that are emotionally-corporally 'felt'. For instance, a narrow spatial situation (objective circumstance) can be experienced as 'crammed' and 'uninviting' or 'enticing' and 'welcoming'. Finally, subjective involvement can also be expressed in the form of interpretations. These are descriptions of 'learned' symbolic or socio-cultural relations including subjective programmes, problems or wishes that are cognitively 'understood' [16]. For example, a shop could be experienced as 'unique' as well as 'commercialised'.

Hasse [16] recognises a relationship between sensations and interpretations. Feelings and bodily experiences can be expressed through cultural symbols and triggered by them. Both emotionally-corporally experiences (feeling) and interpretative reflection (thinking) are part of a person's holistic assessment of a situation. Objective circumstances, particularly physical-material characteristics, have a relevant, however, indirect influence on the atmosphere of a place. While an atmosphere is an amalgamation of objective and subjective circumstances, it can only be 'accessed' through situative experience, here operationalised in the form of sensations and interpretations (cf. Figure 9). From a methodological perspective, an unequivocal distinction between sensations and interpretations is not always possible. In consequence, the systematisation and transferability of empirical categories related to atmospheres remains difficult [16].

Hasse [16] utilised a fourth category called objective programmes. Based on Schmitz's *Programme* [18], an objective programme is an interpretative-symbolic category that includes formalised or institutionalised socio-cultural constructs such as norms and rules. Objective programmes and (subjective) interpretations are linked; making an explicit distinction between them can be ambiguous, particularly if the empirical data is based on written descriptions alone. Therefore, it was decided to exclude this category in this study.

3 The case study: Cuba Street, Wellington

Cuba Street is a prominent urban street in Wellington's inner city. It evolved from a former red-light district into a busy, partly pedestrianised shopping street including numerous cafes, food outlets, and boutique shops. Since 1995, it is a registered historic area under the Historic Places Act 1993 due to relevant historic buildings, facades, and architectural details from the nineteenth and early twentieth century. Various key destinations including the Wellington Arts Centre and Victoria University's Faculty of Architecture and Design are nearby. For this study, Cuba Street between Vivian Street to the south and Wakefield Street to the north was used as a case study for phenomenological inquiry.

Fig. 2. Southern part of Cuba Street between Vivian and Ghuznee Street with third-year student making field notes

Fig. 3. Pedestrianised mall with 'Bucket Fountain'

Fig. 4. Shared space with food container between Manners and Wakefield Street

The southern part between Vivian and Ghuznee Street (Figure 2) is a regular (vehicular) one-way street featuring predominantly small-scale street-lining buildings with overhanging canopy structures that cover the pavement. The street is dominated by food outlets on the ground floor. The middle part between Ghuznee and Manners Street is a pedestrianised mall. It features coloured pavement, planters, trees and sculptures including the iconic 1969 'Bucket Fountain' (Figure 3). The northern end between Manners and Wakefield Street is a one-way street with shared space design characteristics (Figure 4). It includes a mix of larger historical and mid- to late twentieth century buildings.

4 Preliminary findings

4.1 The changing atmosphere of Cuba Street

Students noticed explicit and subtle changes of the urban atmosphere while walking through Cuba Street from south to north. The upper (southern) part evoked descriptions of car traffic, narrow roads, and the canopy above the pavement that prevented to see the sky and created a sort of tunnel effect (Figure 5). The place was described as full of people and food smells were noted. Related sensations were often negative; the place felt busy but uninviting, noisy, tight, crammed, dense, messy and enclosed. However, the street also "pulls you in" (student 11) awaking feelings of curiosity. Cognitive interpretations included 'diverse' and 'cultural' – particular in relation to people – and vehicle-dominated. Some students experienced an amalgamation of impressions including objective circumstances and related – sometimes contradicting – sensations and interpretations. For example, student 16 noted with regard to the upper part of Cuba Street: "Enclosure; food smells amazing; [...] a sense of belonging, even though it's a strange environment".

The atmospheric accounts changed after the crossing of Ghuznee Street with the beginning of the pedestrian mall. A feeling of openness and space was expressed; to the extent that it was felt that a weight had been lifted: "[...] then after feel[ing] enclosed, I've been released from the pressure because we suddenly enter an open space" (student 14). Students' descriptions became generally richer and included street furniture, architectural details, trees and urban green, sculptures and playground elements. The presence of people that gather in social places (cafes, outdoor seating) and the openness and permeability between inside and outside areas (Figure 6), buildings and open space ('active frontages', 'open shop fronts', 'lots of windows') were frequently noted including positive atmospheric descriptions such as 'welcoming', 'comfortable' and 'inviting'.

Sensations related to physical characteristics and people made part of amalgamated atmospheric accounts: "Changes after the first street. Bustling, colourful, unique, peaceful, messy – but in a good way" (student 8). Trees, gardens and greenery were generally appreciated and related to a 'calming feeling'. The absence of cars and the larger, coloured paving were related to feelings of safety; the absence of traffic noises was described as being pleasant. A general change of environmental sounds was noted: "less cars, more chatter, music" (student 7) and "pigeons, talking, footsteps, light sound, jackets rustling" (student 8).

Fig. 5. Southern part of Cuba Street – 'tunnel effect' (field sketch; student 16)

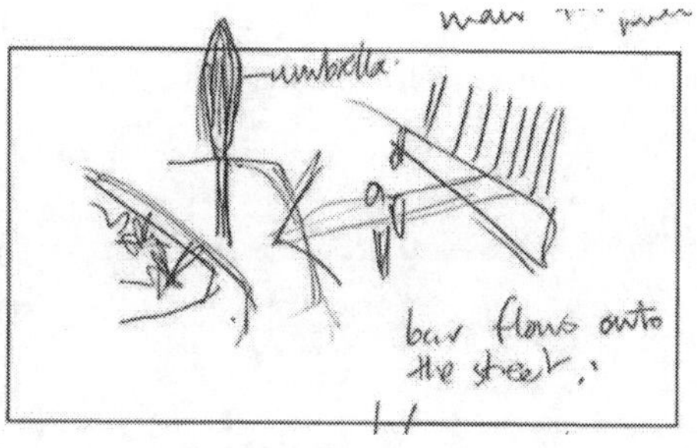

Fig. 6. Inside/outside relationship; socialising area (field sketch; student 11)

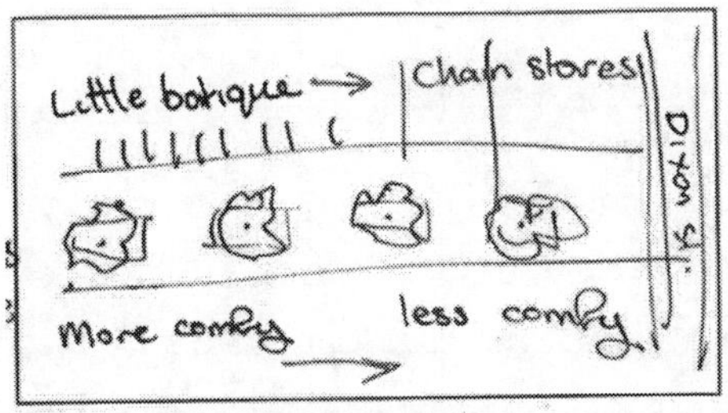

Fig. 7. Small boutique shops vs. big chain stores (field sketch; student 1)

Fig. 8. Small boutique shops vs. big chain stores (field sketch; student 8)

However, students also noticed an atmospheric change within Cuba Street mall when smaller 'boutique' shops made way for larger 'commercial' stores: "Boutique stores are more comfortable to walk past with higher variety of people; once the larger stores start, you feel like you have to be doing something/going somewhere" (Figure 7). Another student noted: "Contrast between small boutique stores and large commercial stores – they ruin the homely feel!" (Figure 8). The same student observed a "divide between unique = inviting to bright and commercial", exemplifying a "diffuse unity" [16] between objective circumstances, sensations and interpretations. Students' activities were occasionally noticed by passers-by who inquired about what students were doing. Such interactions with strangers were experienced as friendly, safe, and welcoming.

At the northern end of Cuba Street, atmospheric accounts changed again. The reappearance of a vehicular road and more cars was perceived as "less inviting" and "out of place". The (larger) buildings were experienced as "overpowering, rugged, unpolished, starch, finished and clean" (student 7). Some experienced the area as "pedestrian-focused" (student 11) and still pleasant but more commercial and "losing its touch" (student 9). The street was also identified as a mixed-use area that "draws people in" (student 5); public transport and temporary uses of space were noticed. Sounds and smells did not play a significant experiential role in this section of Cuba Street.

4.2 Atmospheric components

Students collected almost 200 descriptions of objective circumstances in their field notes and sketches. These were categorised and subdivided into eight sub-codes including (visual) physical-material characteristics, people (moving around & doing things), music and other sounds, silence/quietness, food and other smells, weather/microclimate, tactile experiences and amalgamated impressions (e.g. "open noisy traffic"). Visual physical characteristics were most frequently mentioned as a sub-code within the objective circumstances category, followed by people. Sounds and smells were less frequently mentioned in students' accounts.

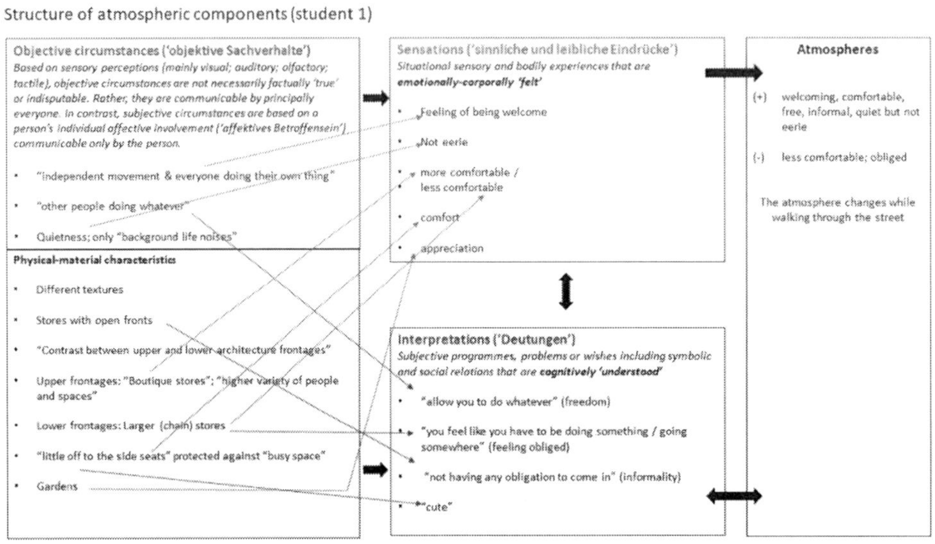

Fig. 9. Structure of atmospheric components (exemplary; based on descriptions of student 1)

Figure 9 depicts an exemplary structure of atmospheric components based on the field notes of student 1. It relates to Hasse's "Emergence of atmospheres" diagrams 4 and 5 [16]; however, without the objective programmes category (cf. section 2 in this paper). The diagram depicts the relationship between the three main categories of analysis (objective circumstances, sensations and interpretations). It shows the indirect influence of objective circumstances on atmospheres; (visual) physical-material characteristics are depicted separately. The green arrows depict the relationships between objective circumstances and sensations; the red arrows show relations between objective circumstances and interpretations. The atmosphere field in the diagram includes positive (+) and (-) negative accounts of atmospheres derived from the sensations and interpretations categories. The relationship between interpretations and atmospheres is potentially leading in both directions; atmospheres may influence interpretations and vice-versa.

Students provided almost 70 accounts of sensations. These were subdivided into positive, negative and neutral sensations. In addition, there were some accounts of amalgamated or 'chain' sensations that combined different sensations that apparently occurred together and sometimes added up to a particular attitude towards a place: "Feel increasing humidity; cooler; feeling strange; doesn't feel lively; no particular place to go; I'd rather avoid this place" (student 14). The total number of interpretations in students' descriptions was similar to the number of sensations. They were subdivided in 10 sub-codes. The majority of interpretations were of positive nature; however often unspecified such as 'cute', 'beautiful', 'nice', 'good', 'hip', 'interesting', 'great', or 'gorgeous'. Frequently described, specified positive interpretations were 'unique' (e.g. with regard to small boutique stores and different people), 'active' and 'lively' (e.g. with regard to people and their activities, but also visual stimuli such as colours), and 'not boring' (e.g. with regard to spatial design details and visual stimuli such as diverse colours). Unspecified negative expressions included 'ugly' and 'tacky' (e.g. with regard to street furniture and buildings). A frequent specified negative impressions was 'commercial', e.g. with regard to (large) shops and the streetscape, particularly towards the lower end of Cuba Street. Neutral interpretations included 'hidden', 'permeable', 'cultural', and 'diverse'.

4.3 The built environment

Students' disciplinary backgrounds (landscape architecture and design) raise questions about the influence of education and pre-existing knowledge on the experience of urban atmospheres. In order to analyse this influence, the objective circumstances' visual physical-material sub-code was further sub-divided to examine elements related to architecture, landscape architecture and spatial design including spatial change. Perhaps not surprisingly, 91 out of 126 descriptions of visual physical-material objective circumstances related to architecture, landscape elements, and spatial characteristics including change. The occurrence of 'disciplinary jargon' (e.g. active frontages; mixed-use; temporary uses; density) was, however, limited.

It seems relevant that students, although explicitly asked to focus on the entire range of sensory experiences and related atmospheres, predominantly communicated visual experiences related to their professional knowledge that is part of landscape architecture and urban design education. Descriptions were influenced by (learned) inventory/analysis techniques and focused predominantly on material-physical characteristics. However, even though students' spatial design background may have caused a sort of sensory imbalance, their accounts provide nevertheless a rich description of atmospheric components. Expressed sensations and interpretations, related to the spatial contrast between narrow/enclosed and wide/open space and between a fine and coarse urban grain, as much as experiences around active frontages and social open space, pedestrian-focused streets and green urban infrastructure provide relevant information about the 'feel' of the street.

Some preliminary conclusions

Atmospheres are influenced by objective and subjective components and can be empirically examined through phenomenological analysis. In this paper, atmospheric analysis has been operationalised in the form of objective circumstances, sensations and interpretations following Hermann Schmitz's *Neue Phänomenologie* framework [17] and Hasse's influential pilot study in Rüdesheim [16]. The preliminary findings of the study show that 'lived urban space' can provide holistic and dynamic experiences including changing urban atmospheres that unfold while moving through the space. Students' descriptions provide a diverse account of changing atmospheres based on their feelings, bodily experiences and cognitive interpretations while walking through different sections of Cuba Street.

While students' descriptions showed a sensory preference for physical-material and spatial objective circumstances – likely based on their disciplinary background – some accounts revealed complex experiential amalgamations that included objective and subjective components. The combined experience of particular spatial configurations, people and sounds such as the combination of a narrow and crowded street, overhead structures that limit views of the sky, cars and noisy traffic created a specific atmosphere, in this case connected to negative feelings of enclosure that, however, disappeared as soon as the space changed again. The study confirms that urban design and architecture have an important influence on atmospheres. However, urban atmospheres cannot be 'designed' in isolation. They depend on the people who contribute to them: "The lived life turns spaces into 'lived places' and historic places" [16].

As with all single case studies, the transferability of the findings in this paper is limited. However, the study contributes an antipodean perspective to an otherwise predominantly Eurocentric discussion. It strengthens the argument that qualitative empirical research with regard to experiential aspects of urban space remains highly relevant for the spatial disciplines including urban design, planning and (landscape) architecture. Urban design research has an increasing focus on public open space as a place for community interaction and social exchange [19-22]. Popular concepts and international movements such as 'place making' have evolved in reaction to a perceived decline of experiential qualities and alienated 'non-places' [23], characterized by a lack of history, identity, social relations and, eventually, authenticity [24-26]. Urban ambiances may be fleeting; however, they are relevant constituents of places. It would be rather senseless to ponder on 'a sense of place' without considering its atmospheric qualities.

References

1. J. Hasse, Die alte Stadt. **35**, 2, 103-116 (2008)
2. R. Kazig, Die alte Stadt. **35**, 2, 147-160 (2008)
3. J.-P. Thibaud, *En quête d'ambiances. Éprouver la ville en passant* (MētisPresses, Genève, 2015)
4. T. Griffero *The atmospheric "skin" of the city*. Ambiances – International Journal of Sensory Environment, Architecture and Urban Space, 2013.
5. G. Böhme, *Aisthetik. Vorlesungen über Ästhetik als allgemeine Wahrnehmungslehre* (Wilhelm Fink Verlag, Munich, 2001)
6. H. Schmitz, *Der Leib, der Raum und die Gefühle* (Aisthesis Verlag, Bielefeld, 2007)
7. J. Hasse, *Zum Verhältnis von Stadt und Atmosphäre. Wo sind die Räume der Urbanität?*, in *Subjektivität in der Stadtforschung (*Institut für Didaktik der Geographie, Frankfurt am Main, 19-40, 2002)
8. R. Kazig, *Atmosphären - Konzept für einen nicht repräsentationellen Zugang zum Raum*, in *Kulturelle Geographien (*Transcript-Verlag, Bielefeld, 167-187, 2007)
9. H. Schmitz, *Leib und Gefühl. Materialien zu einer philosophischen Therapeutik* (Junfermann, Paderborn, 1992)
10. G. Böhme, *Atmosphäre* (Suhrkamp, Frankfurt/Main, 1995)
11. G. Böhme, Oase - Journal for Architecture. **91**, 21-32 (2013)
12. H. Schmitz, *System der Philosophie. Dritter Band: Der Raum. Zweiter Teil: Der Gefühlsraum* (Bouvier Verlag, Bonn, 2005)
13. W. Benjamin, *The Work of Art in the Age of Mechanical Reproduction* (Penguin, London, 2008 [1936])
14. G. Böhme, *Architektur und Atmosphäre* (Fink, Paderborn, 2006)
15. S. Runkel and A. Wesener, *"Rencontre sur les lieux": memory construction in urban ambiances*, in *Ambiances in action - ambiances en actes(s). Proceedings of the 2nd International Congress on Ambiances (*Ambiances International Network, Montreal, 121-126, 2012)
16. J. Hasse, *Die Atmosphäre einer Straße. Die Drosselgasse in Rüdesheim am Rhein*, in *Subjektivität in der Stadtforschung (*Institut für Didaktik der Geographie, Frankfurt am Main, 61-113, 2002)
17. H. Schmitz, *Neue Phänomenologie* (Bouvier, Bonn, 1980)
18. H. Schmitz, *Neue Grundlagen der Erkenntnistheorie* (Bouvier, Bonn, 1994)
19. J. Gehl, *Cities for People* (Island Press, Washington D.C., 2010)
20. J. M. Jacobs, *The Death and Life of Great American Cities* (Random House, New York, 1961)
21. W. H. Whyte, *The Social Life of Small Urban Spaces* (Conservation Foundation, Washington D.C., 1980)
22. M. Carmona and F. M. Wunderlich, *Capital Spaces: The multiple complex public spaces of a global city* (Routledge, Oxon, 2012)
23. M. Augé, *Non-Places: Introduction to an Anthropology of Supermodernity* (Verso, London, 1995)
24. A. Wesener, *The Good, the Bad, the Authentic: An evaluation of concepts, experiences, and considered values of authenticity of place in the context of the post-industrial city* (Königshausen & Neumann, Würzburg, 2014)
25. A. Wesener, Journal of Urban Design. **21**, 1, 67-83 (2016)
26. A. Wesener, International Journal of Heritage Studies. **23**, 2, 141-155 (2017)

Between ambiance and perception: Heritage decoding

Anetta Kepczynska-Walczak[1,*]

[1] Department of Digital Technologies in Architecture and Urban Planning, Institute of Architecture and Urban Planning, Lodz University of Technology, Lodz (Łódź), Poland

Abstract. The aim of the study is to grasp and analyse the intangible process taking place between ambiance and perception. In this regard, ambiance should be understood as a state or aura, as opposed to perception, assumed as a process. The transition of sensorial immersion from ambiance to perception requires the process of decoding. Decoding might be explained as the process running between ambiance and perception and, furthermore, leading to understanding. Therefore, the outcome of the overall process is of cognitive characteristics. Furthermore, it draws research into a relationship between what is projected and actually experienced, between the tangible and the intangible and reveals the process of decoding ambiances where heritage plays a dominant role. The paper discusses the two concepts in the context of the post-industrial heritage in Łódź, Poland. The set of reflections evokes numerous questions: are we into the heritage as it was, or as it is at present, or rather to its projection, tuned to current socio-cultural standards? Does carefully planned and created ambiance give a clear unambiguous sense of a place? Or, it is not ambiance but the processes of decoding and perception that result in various interpretations of the same experience of atmosphere?

1 Introduction

The aim of the study is to grasp and analyse the intangible process taking place between ambiance and perception. The paper discusses two concepts: ambiance and perception, in the context of the post-industrial heritage in Łódź, Poland. It is necessary to stress that the study focuses rather on the phenomena and relation between ambiance and perception as well as on senses than on tools. It draws research into a relationship between what is projected and actually experienced, between the tangible and the intangible and reveals the process of decoding ambiances where heritage plays a dominant role. In order to analyse the process, a brief introduction to background ideas should be presented.

1.1 The concept of ambiance and perception

The theory of ambiance is not new, and moreover, it has been developing dynamically over the past twenty years, focusing in particular on "material atmospheres, perceptive

* Corresponding author: anetta@p.lodz.pl

configurations, physical phenomena, "affective tonality" of places and everyday practices of city dwellers. It lies at the very heart of any sensorial approach of inhabited spaces and reflects the sheer diversity of the senses (light, sound, heat, smell, air flows, kinaesthetic, touch, etc.)" [1]. As Bohme believes, the concept of ambiances occupies the space between subject and object [2]. Referring to Siret, since the past few decades, there has been a growing need "to grasp the architectural and urban aesthetic in all its human sensory dimensions, sonic, tactile, thermal, aeraulic, kinaesthetic or olfactive" [3]. This need is based on the belief that various sensory tonalities affect perception of built environments and their ambiances in which users of a given space are immersed.

It is worth noting here the difference between sensation and perception. Perception is understood as the effect of how the brain processes information from the senses. Hence, perception is multi-sensorial. Furthermore, the notion of ambiances makes us realise that our immediate perceptions of the world have both visible and non-visible components and therefore requires the consideration of all senses and their interplay in the process of perception.

As Thibaud states, "ambiances are both subjective and objective and thus involve people's lived experience as well as their physical environment" [4]. However, according to Pallasmaa, people feel atmospheres immediately and without being conscious of the process, since the ambiance of a given space is the overall feeling that tunes their minds in a specific way [5]. Another perspective has been presented by Morello and Piga [6]. The researchers claim the perception is based on the physical and sensorial experience of places, which includes a multiplicity of aspects, and contributes to the personal construction of the mental image of places. Their statement is based on Kevin Lynch's understanding of the image of a place as a mental construction derived from the analysis of a direct and indirect man-environment relationship [7]. This logic allows to draw a conclusion that perception is a highly subjective notion.

1.2 Theory of direct and indirect perception

In psychology, there are two main theories of perception, direct and indirect, to be distinguished, and two types of processes in perception: bottom-up processing and top-down processing.

In an article titled "Visual Perception Theory", Saul McLeod wrote, "in order to receive information from the environment we are equipped with sense organs e.g. eye, ear, nose. Each sense organ is part of a sensory system which receives sensory inputs and transmits sensory information to the brain" [8]. Thus, the spontaneous activities of looking, listening, and touching, together with the satisfaction of noticing are considered to feature a direct perception [9]. According to Gibson, direct perception is not just a mode of cognition but "the simplest and best kind of knowing" [10]. In other words, Gibson's ultimate sense of direct perception is that of perception based upon information obtained through free exploration [11]. Gibson claims that perceiving is "an automatic response to a stimulus" [10].

On the contrary, some argue that perceptual processes are not direct, but depend on the perceiver's expectations and previous knowledge, as well as the information available in the stimulus itself. Psychologist Richard Gregory has proposed a theory of indirect perception [12]. According to his concept, stimulus information from our environment is frequently ambiguous, so to interpret it, we require higher cognitive information either from past experiences or stored knowledge in order to makes inferences about what we perceive.

As it has been mentioned already, psychologists distinguish between two types of processes in perception: bottom-up processing and top-down processing. Gibson's theory reflects a 'bottom-up' idea while Gregory's vision presents a 'top-down' approach. The first concept is also known as data-driven processing, because perception begins with the stimulus itself. Whereas top-down processing refers to the use of contextual information in pattern recognition since it is assumed a context aids understanding [8].

2 Case studies

In all cases described in this paper, a natural ambiance was replaced by artificially made atmosphere to enhance the sensation of a place. Actually, it was the state of a building or of an urban space that induced an ambiance leading to perceptual experience and, in consequence provoked the action. Therefore, the creators of the ambiances built up a set of multi-sensorial experiences to allow immersion and the feeling of a place. The results contribute to the quest of architecture and environment envisioning and genius loci interpretation.

The research into ambiance and perception focuses on three aspects exemplified by carefully chosen case studies, viz.: a social aesthetic that examines shared sensorial experiences in the first case, on a physical context dealing with the propagation of signals in urban space in the second example, and on design that involves the sensorial aspects of built spaces in the last one.

2.1 WIMA post-industrial interior awaking

In 1877-1879, a large cotton factory was established by two industrialists, Juliusz Heinzel and Juliusz Kunitzer in Widzew – at that time a village at the outskirts of Łódź. Twenty years later the enterprise already consisted of a spinning mill, a weaving mill, a bleach mill, finishing and printing departments. There were own gasworks, a power station and a railway siding. The number of employees was over 3,500. Under the leadership of Oskar Kon – the new president and major shareholder – the company was transformed into a multi-branch combine in the 1920s. A decade later, the firm already employed 10,000 people and was considered as one of the most modern textile factories in Europe. After World War II, the nationalized enterprise was divided into three independent plants producing cotton goods, synthetic yarns, and textile machinery. The cotton production continued until 2009, when it was no longer competitive on the market, dominated by Asian yarn. This was the last cotton spinning mill in Poland [13].

Fig. 1. The WIMA factory complex in Łódź. Source: Google Maps.

Since the collapse of production, a new use of the buildings has to be found. For the company owners the main asset and the burden at the same time is a huge three-story spinning mill. The building is one of the largest built in Łódź with an impressive width of 69 metres (fig. 1.). It is not surprising then that it is challenging to re-use it. The mammoth structure is surrounded by other post-industrial objects successfully let to a number of creative businesses – companies, foundations and associations dealing with culture and art. There are photographers, actors, filmmakers, designers, musicians, DIY enthusiasts and graduates of art schools in Lodz. As a result, recently the place has been re-branded officially from the Cotton Industry Company into the Creative Industry Company.

Fig. 2. The same interior of a disused mill in the WIMA complex – as it is usually (top) and during the described event (bottom). Source: eCAADe 2018 archive.

For the purpose of this study, a perception of an abandoned textile mill interior has been confronted with the immersive experience of the same space during a social event (fig. 2.). The empty unused interior by the means of light and sound effects as well as real objects was transformed into an eclectic environment, balancing between past, present and future (fig. 3.). A multifaceted appearance was matching the nature of the local – strongly eclectic – heritage, reflecting diverse cultural roots of the city dwellers. Ultra-modern magnetic light effects contrasted with historical costumes of dancers who were giving lessons of Polonaise (Polonez), the national Polish traditional dance, to the audience; a delicious smell of food representing four cultures that constituted the heritage of Łódź, viz.: Polish, German, Jewish and Russian; a jazz music band evoking sentiments to the prosperous times of the past, yet to the birth of modern times – all together and interfering with each other built the richness of multisensorial atmosphere and generated unique moments of shared social sensorial experience (fig. 4.).

Fig. 3. The interior transformed by the means of light and sound effects as well as real objects. Source: eCAADe 2018 archive.

Fig. 4. The lesson of of Polonaise (Polonez), the national Polish traditional dance in the ultra-modern magnetic lights. Source: eCAADe 2018 archive.

There is no doubt that, in this case, perception was based on the culture and heritage literacy. By decoding the meanings, the ambiance offered a number of layers unveiling to individuals, depending on their background and comprehension abilities without affecting overall attractiveness. As the event gathered over 200 participants from more than 40 countries from all over the world, for some foreigners it could be just exotic or intriguing experience, while among local residents or those representing the four mentioned cultures, it could stimulate a feeling of deeper reflection or even a patriotic elation. However, what would be their (visitors and locals alike) perception of the same place, if they found themselves in a room full of operating textile machines, producing unbearable noise, dust and smell? Or what would be their perception of this production room as it is nowadays – with countless layers of paint peeling off the wall, piles of unidentifiable rubbish in the corners and a dead pigeon on the floor? The set of reflections brings yet another questions: are we into the heritage as it was, or as it is at present, or rather to its projection, tuned to current socio-cultural standards? Does carefully planned and created ambiance give a clear unambiguous sense of a place? Or, it is not ambiance but the processes of decoding and perception that result in various interpretations of the same experience of atmosphere?

2.2 Light Move Festival

Another issue to be discussed is the ambiance of special open events and its role in promoting local heritage. This case refers to a physical context dealing with the propagation of signals in urban space. It will be illustrated with the immense success of a locally organised Light Move Festival, the most important event of this kind in Poland. It is worth mentioning the idea of organising such festival in Łódź has been strongly inspired by the example of la Fête des Lumières in Lyon [14], Łódź partner city. What is interesting is that kinetic art and 3D mapping on historical facades, offering local heritage-related storytelling, are among the most popular installations during the festival [15]. They are appealing for a broader audience, and each year there are hundreds of thousands of visitors to the city, who for three days admire unique visual effects on facades of historic buildings. According to the local press, this event has already contributed significantly to the change in perception of Łódź from the gloomy post-industrial city into an attractive, and (literally) colourful place. At the same time, and by using the same means the same event encapsulates different message to members of a local community. It offers a specially prepared ambiance, touching heritage-related sentiments, and strengthening the feeling of belonging to the place, of identity, safety and continuity. Raising awareness through events of this kind may contribute significantly to a process of urban regeneration and indicate values of architectural environment, which remain unnoticed or even disobeyed by a local society during their everyday walk of life.

2.3 Revitalisation project of Lürkens factory in Łódź

The third case study presents a master diploma design, successfully submitted and defended by architecture student Katarzyna Romanowicz, entitled: "Lürkens' Spinning Mill – A project of converting the former knitwear factory of Wilhelm Lürkens into a creative co-working space", supervised by the author of this paper.

Wilhelm Lürkens' residential and industrial complex is located in Łódź, in the southern part of a small urban block limited by the following streets: from the north – 6 Sierpnia Street, from the south – Andrzej Struga Street, from the west - Wólczańska Street, and from the east – Kościuszki Avenue. The premises are located in the inner city centre, in the immediate vicinity of Piotrkowska Street, the main historical axis of Łódź (fig. 5.).

The Lürkens factory belonged to medium-sized textile enterprises in Łódź. The employment did not exceed 150 people even when cotton production was booming. The industrial, commercial and residential buildings were built in three stages between 1889 and 1912. From the very beginning, the production activities took place along Wólczańska Street, while the residential and administrative function was concentrated along Aleja Spacerowa (now Kościuszki Avenue). In terms of architecture, the most interesting is the complex of a spinning mill with a water tower, adjoining boiler and engine house, and a high factory chimney, erected at the end of the first decade of the 20th century. These objects have received a more careful architectural treatment as compared to the more severe objects from the earlier period. Facades have the features of late Art Nouveau, with simplified decorative details in the spirit of early modernism.

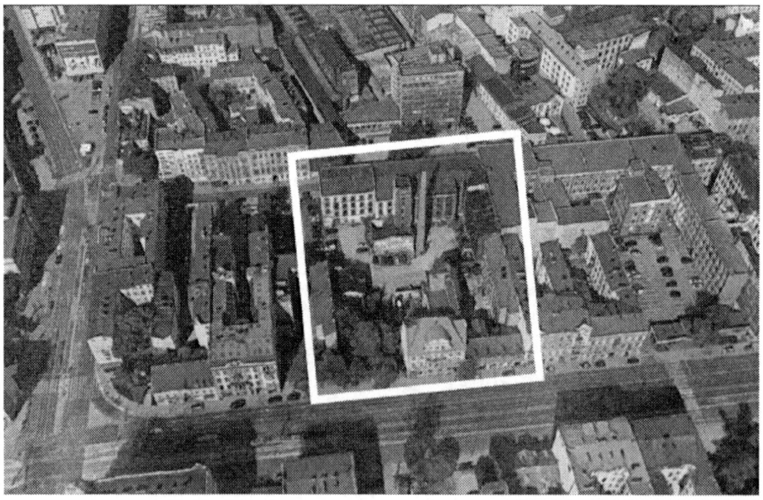

Fig. 5. The Lürkens factory complex in Łódź. Source: Google Maps.

After 1945, the factory was nationalized. Since 1969, Polboy clothing factory operated there, later incorporated into Próchnik, a company specializing in men's wear, mainly produced for export. In 2006, industrial buildings were bought by Spanish investors with an aim to convert them into a hotel. These plans have not been implemented until today, and buildings are deteriorating [16].

Fig. 6. Dilapidated buildings of the Lürkens factory complex in Łódź. Source: images by Katarzyna Romanowicz and Bartosz M. Walczak.

The state of the buildings has been getting worse with each year and some parts of the factory complex have already collapsed. Uncontrolled wild greenery started to grow interfering with existing structures (fig. 6.). Therefore, the project of converting the Lürkens' built legacy into a dynamic, modern and full of life place required deep studies, understanding of a history and above all, grasping the image of a place to create an ambiance that would attract visitors and potential business.

A palette of artistic means was used to (re-)create the ambiance of abandoned factory waiting for successful revitalisation. So, the study covered aspects of searching for reminiscences of the past, the atmosphere of the place, both in urban courtyards and architectural interiors, and original materials that would influence the perception of the physicality of the place. However, the idea was not to reconstruct the original genius loci of the place but to define a set of attractors addressed to contemporary society and taste. Therefore, a proposed design emanates with neat and cosy colours and lights. Furthermore, the refreshed post-industrial space was furnished in a non-invasive modern style what might evoke a feeling of comfort and high quality (fig. 7-8).

Fig. 7. Main building – current state (left) and envisioned future (right). Source: images by Katarzyna Romanowicz.

Recalling the past literally is not what is expected. To expand on it, there is a textile machine in the Central Museum of Textiles in Łódź operating during special shows. The noise a single machine makes is unbearable. Behind the machine, on the wall, there is a visual projection of the mill interior filled with working machines. The scenery informs clearly about the noisy working environment. Moreover, the smell of oil and metal reminds of another negative characteristic that would not be appreciated nowadays. Therefore, the careful selection of sensorial factors in the project led to a decision to enhance urban and architectural space perceived visually, whereas omitting less attractive elements of the past realm. To sum up, the project positioned the past legacy as an anchor and a link to local identity but let imagination to supplement the dominant visual stimulus.

Fig. 8. Modernity inspired by post-industrial decay – vision of re-used interiors of the Lürkens factory. Source: images by Katarzyna Romanowicz.

3 Discussion and concluding remarks

Selected case studies allowed to analyse aspects of sensorial stimulation, such as light, sound, or smell and to investigate the embodied and affective experience of spatial settings.

The relationship between the place, its atmosphere and its perception by various audiences has been observed. It allows to drive the first conclusion that the ambiance acts as a source for various modes of perceiving a given place. However, what ambiance encapsulates is not equal to perception. In this regard, ambiance should be understood as a state or aura as opposed to perception assumed as a process.

The transition of sensorial immersion from ambiance to perception requires the process of decoding. Decoding might be explained as the process running between ambiance and perception and, furthermore, leading to understanding. Therefore, the outcome of the overall process is of cognitive characteristics. However, either we consider direct or indirect perception, to run such process a decoding key is necessary. The case studies, described in previous paragraphs, underline the meaning of the process of decoding ambiances where heritage plays a dominant role.

The ambiance of a place can be natural or enhanced with some additional sensorial effects, or, even more, it can be artificially created. The purpose of the latest one can be to evoke sensation, mood, emotions, reactions, memory and alike.

Considering the ambiance created intentionally, a palette of stimuli it offers might not be the same as what is absorbed. In other words, a creator of the atmosphere might not succeed with communicating intensions. Furthermore, ambiance can be perceived totally or partially or not noticed at all since the perception skills depend on personal abilities, experience and knowledge.

Decoding can be a background action, uncontrolled and free, often happening on emotional level. Even though, the resultant processes of interpretation are knowledge and cognitive-based. They rely mainly on abilities to associate direct and indirect past experiences.

Even if ambiance is composed of a selective sensorial palette, as described in the third case study, the post-industrial legacy is recognised as the heritage to be protected and as the anchor of a local identity. The anchoring in the past gives a sense of security, continuity, identity. It is associated with the issues of sentiment, looking at the past as if it was better than present times and therefore choosing "flavours" and genius loci pre-set ambiance to build an encoded image of a place.

To conclude with final remarks, the representation of atmospheres is a powerful tool to understand the immaterial aspects of lived space. In the respect of cultural heritage, through creation of ambiance, such rich envisioning experience can integrate knowledge about the place, time, society, factors affecting space, as well as other elements that shape the cultural environment.

References

1. https://www.ambiances.net/international-ambiances-journal.html
2. G. Böhme, Atmosphere as the fundamental concept of a new aesthetics, Thesis Eleven 36, 113-126 (1993)
3. D. Siret, Understanding past sensibility to grasp present architecture: the example of solar radiation in: A. Kepczynska-Walczak (ed.), *Envisioning Architecture. Image, Perception and Communication of Heritage* (2015)
4. J-P. Thibaud, The sensory fabric of urban ambiances, Senses & Society, **6 (2)**, 203-215 (2011)
5. M. Amundsen, Q&A with Juhani Pallasmaa on Architecture, Aesthetics of Atmospheres and the Passage of Time, Ambiances (2018)
6. E. Morello, B.E.A. Piga, The Image of Productive Landscapes: A Method for the Perceptual Analysis of Industrial Heritage in: A. Kepczynska-Walczak (ed.), *Envisioning Architecture: Image, Perception and Communication of Heritage* (2015)
7. K. Lynch, *The Image of the City* (1960)
8. S. McLeod, *Visual Perception Theory* (2008)
9. J.J. Gibson, *The senses considered as perceptual systems* (1966)
10. J.J. Gibson, *The ecological approach to visual perception* (1979)
11. A. Costall, A. Still, Gibson's Theory of Direct Perception and the Problem of Cultural Relativism. Journal for the Theory of Social Behaviour, **19:4**, 433-441 (1987)
12. R. Gregory, *The Intelligent Eye* (1970)
13. B.M. Walczak, *Zespoły fabryczno-mieszkalne w europejskim przemyśle włókienniczym w latach 1771-1914* (2010)
14. http://www.fetedeslumieres.lyon.fr/en (access: 20.10.2018)
15. http://www.lmf.com.pl/#/page/14 (access: 20.10.2018)
16. J. Kusinski, R. Bonislawski, M. Janik, B.M. Walczak, *Łódź. Księga fabryk* (2016)

Embraced by swirls of ether. Atmospheric imagination and the representation of assimilation and effect

Juan Elvira[1]

[1] Departamento de Proyectos Arquitectónicos, Escuela Técnica de Arquitectura de Madrid, Universidad Politécnica, Av. Juan de Herrera, 4, 28040 Madrid, Spain

Abstract. Atmospheres have a diffuse, ungraspable nature that make them problematic to represent. Diverse artistic manifestations have relentlessly tried to grasp air infused with properties. Ether is the concept chosen to test the way the history of atmospheric representation was initiated; a cultural artefact whose representation permeates our present imaginary. Such substance has been the image of sensual assimilation, of psychological life and complex ambient effects. Before ambient was subject to technical representation, images of air within can be called 'a state of animation' opened up the ways of seeing and producing ambients. Several case studies are analysed as a useful vehicle for atmospheric representation. Two main strategies have been identified. The first one is the representation of chance, ambiguity and openness, the creation of conditions to let atmospheres leave its accidental traces. The second consists in the precise distribution in space of particles or discrete elements whose initial position is irritated by diverse agents. Both are still recognisable in our days.

1 Introduction

Ether is a cultural artefact whose representation permeates our present imaginary. Diverse artistic manifestations have relentlessly searched for the representation of air infused with properties. Before it was subject to technical representation, images of air in a 'state of animation' opened up new ways of not only seeing but producing ambients. As this article will develop, such substance has been the image of sensual assimilation, of psychological life and complex ambient effects. Therefore, it is a useful vehicle for the exploration of the representation of ambients, of which several examples will be analysed.

The first two will open up two main ways for visualising ether. *Jupiter e Io* (1531-32) by Antonio da Correggio (fig. 1) shows the god, in the shape of a perfumed cloud, embracing Io to possess her. Among the branches of trees, a blurry cloud, translucent and slightly anthropomorphic, cuddles a naked feminine figure. The aqueous body of the cloud-god and that of the woman dissolve in a single medium. This is a representation of the erotic assimilation, where water vapour comes to life and turns into a vivified and active entity.

Likewise, this painting is a literal definition of ambient: this term, stemming from Latin 'ambire,' stands originally for 'both arms', an immersion-producing embrace [1].

Thus, from the conceptual point of view, we are beholding an unequivocal representation of an atmosphere. It is also an ambiguous one. Our gaze stops at the vague shape, recognizing alternatively the shape of a human body and something informal and cloud-like.

Dario Gamboni coined the term 'Potential Images' for those depictions infused with ambiguity which "depend on the onlooker state of mind and come fully into being, in conformity with the artist's intentions, only through the participation of the onlooker" [2]. Jupiter's body is partially hidden in a cloud, or in reverse, the cloud is hidden when mingled with the man's figure. This oscillating condition is also a quality of such images.

Fig. 1, 2. Antonio da Correggio. Jupiter e Io (1531-32), Kunst Historisches Museum Wien. Jan Gossaert, Danae (1527), Alte Pinakothek, Monaco di Baviera.

They are indeterminate, polysemic, and challenge the notion of representation in its usual forms. Thus, the creation of potential images is one of the first pictorial strategies used in the representation of atmospheres.

The next key representational strategy can be found in a painting realised barely in the same years. Again, a Greek mythology character possessed by an atmospherically transfigured god is shown. *Danae* (1527) by Jan Gossaert (fig. 2), depicts a scene where the daughter of king Acrisio receives Zeus, who descends upon her in the shape of golden rain, and the future hero Perseus is conceived. The feminine body sits on the floor and is surrounded by an architectural structure built in blood-red marble. She is only wearing an indigo robe, revealing one of her breasts. The legs, half open, are crossed at her feet. Like a profane annunciation, golden luminous particles are directed towards her in an inverted cone whose vortex points to her lap.

Again, Correggio's painting is the image of assimilation and erotic impregnation. But the scenario is different this time. A classic peristyle surrounds the inverted cone of rain, whose base relates concentrically to the dome. This atomised shape is composed by golden motes acting within space, embodying it in sharp contrast with the solid stone columns. Here the graphic strategy is opposed to that of Correggio's painting. The representation is solved through geometrically precise particles distributed in space. Of the many examples of art history where sensuality and fecundity – in direct relationship with spiritual life – have included an ethereal materiality, the former two are especially relevant in terms of figuration. Both show with precision the mediating nature of atmospheres and the transition between environment and affect which finally expands into a subjective and psychological notion of atmosphere. Both embody the act of erotic conjugation of the body with its medium in a mystical experience. In short, both show ether, vivified air infused with properties. It can be said that both strategies, representing the duality between fields of ambiguity and the distribution of discrete elements, summarise the way atmospheres will be depicted from this moment on.

2 Danae's way. Particles and vectors in the etheric field

The history of ether has found many ups and downs. From essential scientific model, it soon was doomed to the realm of discarded theories. Nonetheless, its dismissal as scientific model gave way to a means of metaphysical speculation and aesthetic production. Although it has known many definitions in time, ether traditionally has been defined as a continuous transmitting medium which fills the void and penetrates matter. This scientific figuration of continuity, which died with the irruption of Einstein's relativity theories, tried to fill the voids which stopped science from giving a satisfactory explanation for certain natural phenomena. Paradoxically, it was Albert Einstein who manifested its coming cultural impact: "[with etheric models] space gave up its passive role as a mere stage for [physical events] The ether was invented, penetrating everything, filling the whole of space, and was admitted as a new kind of matter. Thus, it was overlooked that by this procedure space itself had been brought to life." [3]

The first two paintings presented here are brilliantly intuitive in the way this fuzzy medium is represented artistically. Nonetheless, putting the scientific models of ether on paper was problematic by its own nature. How to capture the invisible, or the fleetingly visible and ephemeral at most? How to grasp the fluid and dynamic? A task like this needs precision and imagination hand in hand.

Fig. 3, 4. Maxwell's lines of magnetic force. Source: J. M. Maxwell, *Treatise on Electricity and Magnetism*, 2 vols. vol. 2, fig. xvii, Oxford, 1873. Faraday's lines of magnetic force. Source: M. Faraday, Experimental Researches in Electricity, 3 vols. vol. 3, plate IV, fig. 3, London, 1839-55.

Scientific representations of models of ether usually revolve around *Danae's* strategy, or the distribution of discrete elements. It does so in two ways. The first one works on the idea of the field, a set of elements ordered in space. In close up, they are only tiny motes. When the whole picture is revealed, they show relational forces entangling them. It is the case of

Faraday's (fig. 3) or Maxwell's (fig. 4) lines of magnetic force, where vectors of energy and activity order space. Magnets and electric currents are the field activators around them.

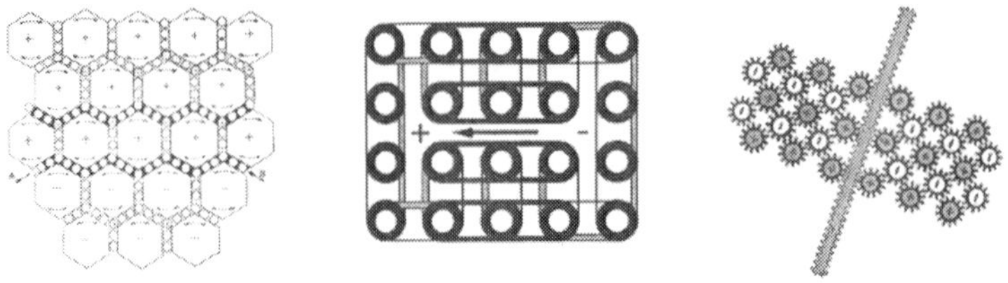

Fig. 5, 6, 7. Maxwell's vortex ether model. (Source: J. C. Maxwell, "On physical lines of force", *Philosophical Magazine*, plate I, 1862). FitzGerald's wheel-and-band ether model. (Source: B. J. Hunt, "Lines of Force, Swirls of Ether"; in Clarke, Bruce and Dalrymple Henderson, Linda; *From Energy to Information,* Stanford University Press, Stanford, California, p. 104-105, 2002). Lodge's cogwheel ether model (Source: O. J. Lodge, *Modern Views of Electricity*, London, p. 187, 1889).

The second graphic paradigm consists in a mechanical infrastructure linking together everything else. A grill of vortexes or 'gears' allow for the transmission of motion and energy in space. It is the case of Maxwell's vortex ether model (fig. 5). When something happens in a certain point of the network, changes will necessarily be transmitted to the rest. Other models, like FitzGerald's, show wheel-and-band images (fig. 6) where the vortexes are connected by 'elastic bands'. Electrical phenomena distort the structure by means of the bands' elasticity, thus representing the effect of opposing charges. Other models replace this elastic bands by cogwheels crossed by a longitudinal gear, like Lodge's cogwheel ether model (fig. 7).

Whether a field or a mechanical infrastructure, ether is shown as a precise distribution of particles in space, activated by diverse generators, a kind of infrastructure that sustains a new spatial materiality [4].

3 Io's way. Archeiropoietic images

Soon at the beginning of the 19th century, the transference from scientific ether models to para-scientific and cultural speculations took place. While dismissed by science, ether quickly entered the realm of the subjective and the irrational, linking scientific theories with psychological ones. When trying to represent such substance, many experiments ambitioned to give a glimpse of the transitions between the aerial and mental worlds. In doing so, the results obtained were fictitious or simply failed, but many of them did succeed in contributing to a fertile atmospheric imaginary. In nineteenth-century bourgeois living rooms, spiritist sessions were often celebrated. All kinds of manifestations took place in the form of what we can call *psycho-aerial phenomena*: misplaced cold drafts, the flickering of a flame, unexpected perfumes diffused in the air, ectoplasms or spontaneous spatial densifications emerging from the medium's body… Etheric theories were embraced by theosophists and occultists alike. With them, another strategy of atmospheric representation emerged.

When grasping the invisible and the unknown, the strategy of the archeiropoietic – from medieval Greek *a·cheiro·poietos*, 'made without hands' – is recurrent. The classic example is a privileged or miraculous image in the Christian tradition, an instrument of heavenly intervention imprinted in a cloth or in a wall without human intervention. The origin of the unpainted image can be celestial or due to the direct contact of the body, like in the cloth of Veronica – *vera icon* – or the Christ's shroud [5]. This image self-generates by means of the physical laws governing the material's behaviour, laws that usually remain hidden and invisible. It is proto photographic in the sense of the 'authenticity' of the image itself, because the image and what it represents converge.

August Strindberg (1849-1912) was a Swedish playwright, novelist, poet and essayist, but for the purpose of this article we will focus in his important activity as painter. After having an active artistic and social life in Paris and Berlin, Strindberg settled in Dornach, Austria. He soon devised an innovative pictorial process based on chance and not in the permanent intervention of the artist. In 1894, he summarised his experiences on automatic art in the text "On chance in artistic creation," in the French magazine *Revue des Reves* [6]. Here he briefly described different material processes and personal experiences in which artistic creation is co-produced by the randomness of invisible natural forces:

"I've been told that the Malays drill holes in the bamboo stalks that grow in their forests. When the wind blows, they lie on the ground to listen to the symphonies produced by these gigantic Aeolian harps. The strange thing is that each listener hears a unique tune and a unique harmony, all according to the whim of the wind. It is a fact that weavers use a kaleidoscope to discover new patterns, leaving it to chance to arrange the bits of coloured glass" [7].

Strindberg recognises the perceptive value of modern art as a search for intelligibility, where initially a chaos of shapes and color opens up a subjective recognition of the pictorial matter, where the image finally presents itself to the viewer. The aim of the work of art would be to inhabit this pareidolic recurrence.

The text closes with the following conclusion: "The art of the future: [...] Imitate nature in an approximate way; imitate in particular nature's way of creating!" [8] The artistic practice understood as the creation of material conditions in order to reveal the intervention of nature is also reminiscent of photography pioneer William Fox Talbot (1800-1877). Talbot was the inventor of the *Photogenic Drawings*, a process where photographic images where obtained by direct contact of the photographic plaque over the represented object. Photogenic images were produced, in his words, by the "pencil of nature" acting upon the sensitive photographic surface [9]. Fox Talbot describes such images in archeiropoietic terms, done "without any aid whatever from the artist's pencil", only by "nature's hand; and what they want as yet of delicacy and finish of execution arises chiefly from our want of sufficient knowledge of her laws" [10]. Thus, here the role of chance is the foundation of this artistic practice in order to let nature reveal itself.

Fig. 8. August Strindberg. *Celestograph* n. XII, 1893-94. Source: National Library of Sweden, Collection of Manuscripts, Strindbergsrummet.

Strindberg experimented with photography in analogous terms, with the purpose of capturing ethereal images of the firmament, drawn by scientific and poetic drives. "Celestographies" were realised in Dornach between 1983 and 1894 with the only aim of letting nature depict itself, capturing for example the true nature of the celestial dome (fig. 8) or the presence of the sun. For him, photography was a means of capturing what the eye alone could not see [11]. He rejected the use of lens and camera in favour of a more archaic and unmediated technique: he simply soaked the photographic plate in developing solution and exposed it to the environment. For him, the accidental images obtained were the true imprint of heavens, a *non manufactum* of the ungraspable forces of nature. The result is also a *Potential Image* as described before. In its modest dimensions of 9x12cm, although the result of dirt, dust and random chemical reactions, it shows with pareidolic latency the indeterminacy of the celestial atmospheres.

Strindberg's interest in occultism and alchemy grew in the following years, later described in his autobiographical novel *Inferno* [12]. He read the work of theosophists like fellow countryman Swedenborg and other authors who experimented with photography.

Two years after the last Celestography was made, French physician and spiritualist Hippolite Baraduc combined Strindberg's and Fox Talbots' techniques in order to grasp images of his own states of mind. In 1896, he published *L'âme humaine: ses mouvements, ses lumières et l'iconographie de l'invisible fluidique* where he included several pictures of images of the spirit [13] or 'psychicones'.

He affirmed that he could grasp the fluid of conscience by simply applying the sensitive photographic plaque onto his subject's foreheads or other parts of their body. The images again are merely accidental, but effectively join the iconography of the swirls of etheric *humours* within a dark empty space. Psychions, then, were interpreted at the time as *vera icons* of the photographer's mind. The cloth applied onto the saint's face is replaced by the sensitive and revelatory photographic plaque (fig. 9).

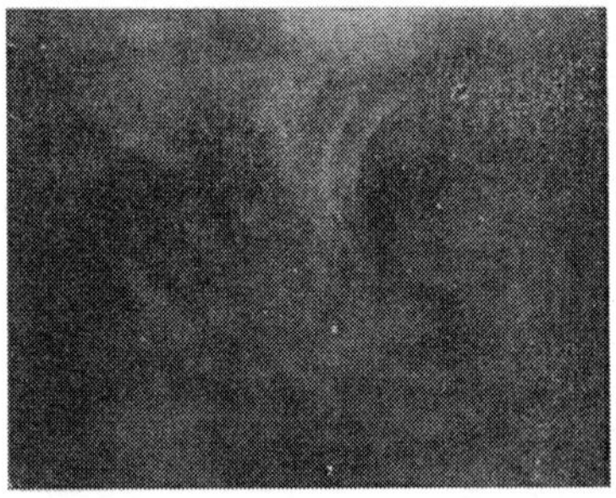

Fig. 9. Hippolite Baraduc. "XXIII. Projection of special state of soul". Source: *L'âme humaine: ses mouvements, ses lumières et l'iconographie de l'invisible fluidique,* 1896.

Other psycho-aereal 'states of the soul' were interpreted and carefully designed in 1901 by British theosophists Annie Besant and Charles Webster Leadbeater in their work *Thought Forms* [14]. The book depicted a series of graphic correspondences of auras according to the personality or the mood of the subject ("love," "devotion," "intellect," "wrath," etc.) and experiences of various kinds ("meeting a friend," "in a funeral," etc.). They gave graphic body (an obviously subjective one) to the emotions related by specific situations, thus literally representing atmospheres.

For German philosopher Gernot Böhme (1937-), the production of atmospheres consists in the creation of conditions for a phenomenon to happen, not the fabrication of a particular object. To define such assemblages, Böhme uses two expressions, 'generators' and 'condensators', which seem very appropriate in light of the former experiences described so far. Such producers of atmospheres emit everything we experience without us initially being aware and ignoring where they come from. Böhme uses the term *ecstasy* to designate such presence, the series of sensitive traits, or in a word its atmospheric qualities [15]. Ecstasies are able to 'dye' space and announce the presence of something independent from the conditions they proceed [16]. This concept stresses the interdependence among a series of basic components of an atmospheric architecture: the cosmetic emission, the situation of the receptor and finally the relationship between the ecstatic assemblage and the former two.

American critic Jeffrey Kipnis approximated the 'ecstasies of things' when he used the term 'cosmetic explosion' when describing the capacity of architecture to infuse space with qualities beyond superficial ornamentation [17]. This "erotic camouflage" is a transubstantiation of the skin's surface where the superficial and the purely ornamental is sublimated. It transcends the structure it adheres and gain a separate identity, an atmosphere that is able to destabilise the structure it comes from. He writes: "where ornaments retain its identity as entities, cosmetics work as fields, like a blush, a shadow, the air or auras" [18]. It is a field of effects which dissolves into pure affect. Like Odyssey's sirens, although they do not explicitly say or promise anything, they produce an irresistible attraction.

The main innovation in 'Thought Forms' consists in the inclusion of plastic representations of musical works in the context of the architectural containers where they take place. Music provokes temporal cosmetic diffusions, or in the author's words, structures "radiating in all directions its characteristic vibrations" [19]. Two of these documents summarise the graphic strategies analysed so far. In Mendelssohn's *Songs without words, Book 2, n. 3,* (fig. 10) large concentric halos and semicircular traces in vertical succession emerge upwards from the architectural container. *Faust*'s "Soldier Chorus" by Richard Wagner is devised as a set of formidable multicolour auras in vertical convulsions, almost protruding the terrestrial atmosphere. Hovering over a stable object, built with stone, fleeting luminous structures hover. The tiny silhouette of a cathedral is the synesthetic witness who stands under some ambiguous shape that can be apprehended in different ways: maybe similar to a gigantic atomic mushroom cloud, maybe to the concentric waves in water. Such architectonic representation is subjective as it is original and unheard of at the moment. Historically, architecture has been represented as something stable and permanent, walls and roofs immutable forever. Now for the first time the staging of an activated space is depicted, with ephemeral and aerial qualities engaging experience and the content of space itself, full of complex effects.

The duality between object and atmosphere is emphasised. To do so, some scientific graphic resources are used: some reminiscent of weather map isobars, concentric waves as in magnetic fields, sets of particles in different colours… Above the old church stone walls the trees and the green meadows, fleeting etheric light structures swirl.

Different means of atmospheric representation, which begun with the pictorial representation of natural forces and the substance of space, begin to integrate the two main strategies defined so far, Io's strategy or the ambiguity of the abstract accidental form and Danae's strategy or the precise representation of particles in an activated field.

Fig. 10. Annie Besant and C.W. Leadbeater, "Songs without words, book 2, nº 3" by Felix Mendelssohn. Source: Thought Forms, Percy Lund, Humphries & Co Ltd, The Country Press, Bradford, 1901.

Conclusions

Atmospheres have a diffuse, ungraspable nature that makes them particularly difficult to represent and manipulate. Ether is the concept chosen in this article to test the way the history of atmospheric representation was initiated. It represents the problematisation of the will to grasp the invisible in different ways. The representation of ether has shifted from scientific to cultural artefact and finally to a vehicle to bring forward our desire for spatial emotion. Historic etheric representations have permeated the way atmospheres are represented and conceived in our days. Although subject to different interpretations in time, the representation of ether is always that of activated atmospheric space, of its aerial qualities related to personal experience and composed by all kinds of ambient effects. The aforementioned case studies of 'animated air,' are Potential Images embodying the transition between the rational to the irrational, from a conceptual model of continuity to a tool for precise ambience creation.

Two images of etheric immersion, close in time and content, have been chosen to bring forward two main graphic strategies. Images which open two main directions when visualising atmospheres. The first one is the representation of chance, ambiguity and openness, the creation of conditions to let atmospheres leave its accidental traces, such in archeiropoietic images. The second consists in the precise distribution in space of particles or discrete elements whose initial position is irritated by certain objects or generators of atmospheres.

Both strategies are not mutually exclusive, but encompass two opposing modalities that are recognisable today in the representation of atmospheric architecture. Frequently both are combined to keep the two conditions present in the eyes of the beholder, in a balance that depends on every singular atmosphere.

References

1. J.P. Thibaud, "Petit Archeologie de la notion d'ambiance". CRESSON / UMR 1563 *Ambiances Architecturales et Urbaines* http:// cressound.grenoble.archi.fr/fichier_pdf/librairie_ambiance/ Thibaud _2012_petite_ archeologie _notion_ambiance.pdf (2012)
2. D. Gamboni, *Ambiguity and Indeterminacy in Modern Art.* Reaktion Books, London, p. 9, (2002)
3. A. Einstein, Albert, "The History of Field Theory: Olds and News of Field Theory". *New York Times*, (February 3rd 1929)
4. For a detailed account of the representation of ether, see B. J. Hunt, "Lines of Force, Swirls of Ether"; Clarke, Bruce and Dalrymple Henderson, Linda (eds); *From Energy to Information,* Stanford University Press, Stanford, California, p. 104-105, (2002)
5. H. Belting, *Imagen y Culto. Una historia de la imagen anterior a la edad del arte.* Akal, Madrid, p. 69, (2009). Originally published as *Bild und Kult. Eine Geschichte des Bildes vor dem Zeitalter der Kunst*, Beck'sche Verlagsbuchhandlung, Munich, (1990)
6. A. Strindberg, "On chance in artistic creation," *Revue des Reves*, Nov. 15, Paris, (1894). Consulted edition: *Cabinet magazine, Issue 3, Weather*, New York, (2001)
7. Ibid
8. Ibid
9. W. H. Fox Talbot, *The pencil of nature*, Longman, Brown, Greens and Longmans ed. London, no pagination, (1844). http://www.gutenberg.org/files/33447/33447-h/33447-h.html
10. ibid
11. R. Rosemblum, *The painting of August Strindberg: The structure of chaos*, Blondal, (1995)
12. A. Strindberg, *Inferno*, Albert Bonniers förlag, Stockholm, (1897)
13. H. Baraduc, *L'ame humaine : ses mouvements, ses lumières et l'iconographie de l'invisible fluidique,* G. Carré, Paris, (1896)
14. A. Besant, C. W. Leadbeater, *Thought Forms,* Percy Lund, Humphries & Co Ltd, The Country Press, Bradford, (1901)
15. G. Böhme, *Aisthetik. Vorlesungen über Ästhetik als allgemeine Wahrnehmungslehre.* Wilhem Fink Verlag, München, (2001). Consulted edition: *Atmosfere, estasi, messe in scena. L'estetica come teoria generale della percezione.* Christian Marinotti Edizioni, Milan, p. 242.
16. G. Böhme, "Atmosphere as the fundamental concept for a new aesthetics". *Thesis Eleven,* nr. 36. Sage Publications, p. 120, (1993)
17. J. Kipnis, Jeffrey. "The cunning of cosmetics: a personal reflection on the architecture of Herzog and de Meuron". *El Croquis* 84, Madrid, (1997)
18. Ibid
19. A. Besant, C. W. Leadbeater, "Forms built by music", en *Thought Forms*, Percy Lund, Humphries & Co Ltd, The Country Press, Bradford, (1901)

Branding Space: Deliberate Formations in Beginning Design

Margaret McManus[1,*]

[1]Marywood University, School of Architecture, Scranton, Pennsylvania, USA

Abstract. Arguably, ambiance in our culture is created in large by architecture and media. The former consumes us as we inhabit spaces both interior and exterior, allowing physical and material constraints — and the programmatic functions they permit — to create ambiance. The latter, media, inhabits *us* as we consume *it*. It creates or perhaps manipulates an ambiance within our minds. Through signage, television, and the internet we have graphic design flooding our brains via advertising and marketing: emotionally directing our decisions, telling us to buy this or invest in that. Analogous to material and physical choices that create the ambiance of architecture, we can blame branding strategies and the practices thereof for the strength and power that they give media to create ambiance. While students use branding strategies to assist themselves in creating ambiance through architecture, they are becoming grounded in a process that compels them to carry out objectives and promises made. Ultimately, this undertone of the pedagogical approach is what they can take forward as they embark on more complex projects, where timely decisions must coincide with a singular message and clear vision.

1 Introduction

Architecture and branding can physically and psychologically enforce ambiance upon its receiver. They can both be large and small, inanimate and animate, quiet and loud, and with a similar capacity to persuade, dictate, and influence. While both branding and architecture fall under creative disciplines, they are traditionally different in their outcomes or output medium, and perhaps their means to an end. Architecture imposes an ambiance through external factors, where branding attacks from within and aims to reach internally to impose or suggest ambiance. While the merging of the two is growing ever closer as media and materiality become integrated and as we navigate space in digital, virtual, and actual realms, this paper focuses on fundamental principles of branding that infiltrate architecture design studios.

Admittedly, as a professor of architecture, when speaking of branding in this paper it is in more of a general nature—closer to how the general public may understand it, yet, with the help of research as noted within, but ultimately, architecture is the primary source of study. These architecture studio topics are not about creating architecture that supports a

* Corresponding author: mcmanus@marywood.edu

brand, yet it looks into the process of branding and aims to parallel it in design studios in an attempt to harness the influential ambiance that branding inherently strategically creates.

Branding strategies that tend to be successful are so because they adhere strictly to a singular message—a promise—to the public. Bob Ulrich, CEO of Target states that "At the heart of the strategy is our commitment to delight our guests…This is our 'expect more. Pay less' brand promise."[1] In general, this brand promise becomes something internal that the user accepts as truth, and subsequently the user forms a bond that self-creates an associated ambiance. Branding through media has incredible power to dictate and influence. Successful architecture does, too, hence the exercise to parallel these approaches in architecture education. Beginning architectural design students often struggle with finding their own process as they embark on semester-long studio projects. Why not leverage the processes in which companies express and carry-out their commitments to the public through successful branding strategies? After all, university students may be much more aware of the influence—though not the processes behind them—that media and branding have on their everyday lives versus that of the influence of architecture. If a tried-and-true branding process can focus students on a common goal, message, and promise of their design project, then they can begin with a platform that can assist as a checks-and-balances system for which to create deliberate formations, and in the meantime create remarkable spaces.

At a superficial level, the creative processes related to problem-solving for clients in both architecture and branding industries are quite similar: they involve gathering a vast amount of information, sorting it per priorities as necessary, defining an objective, iterating resolutions, and releasing solutions. A pivotal point is in "defining an objective" and this point of departure related to the processes is closely related in the emphasis of language. Words are paramount in carefully crafting a descriptive experience that defines a brand. They become the cerebral link from message to ambiance—both media-defined ambiance and self-defined ambiance. Often times these words are not stated or printed, yet they are operating psychologically to govern the output message. A brief quote from "Designing Brand Identity" by Alina Wheeler: "Brand is not what you say it is. It's what *they* say it is." (credited to Marty Neumeier, The Brand Gap) [1]. Here the power is given to the consumer and their words (relative to an experience) equal truth.

To understand that words can work to quarterback a result is where an architectural pedagogy steps in. It is through the generation of carefully chosen words—more accurately, verbs and adjectives—that students are tasked with defining a deliberate experience through architectural intervention. Void of a product to brand, they must generate words that can begin to brand a space: to create a defined, intentional ambiance. These adjectives are translated into formal models that merge with physical site constructs, where emphasis is more on experiential engagement than programmatic constraints (see Fig. 1). These words can then assist in creating a coherent 'language' that the architecture speaks fluently.

Fig. 1. A student's study models depicting various words inspired by a project site visit; still devoid of context and scale. Left to right the words indicate *romantic, wild, boundless,* and *hidden* (B. Dolgas).

Decision-making for students becomes manageable as they consistently seek to obey their project's defined language on all scales. This "branded" space is defined by decisions that are submissive to an overall experience: the orientation of form, the shape of a roof, the placement of windows, the details of a railing—all seek to adhere to the "brand". Figure 2 showcases a beginning design student's final project, where even the representation (the process of "release") is accountable for upholding the defined promise.

Fig. 2. A student's final rendering upholds the branding strategy used in defining its spatial construct through form, composition, vantage point, surrounding environment and context (J. Leone).

1.1 Basic Branding Principles followed

Branding strategies and ideals have been observed, albeit on a perhaps superficial level and through the lens of a proposed architectural pedagogy. These have come from various sources, and more pointedly a tried and true reference in its fifth edition: "Designing Brand Identity," by Alina Wheeler. [1]

Some brief parallels included in the resource to the process used in the architectural studio can be seen in the brand's clarifying strategy: *Overview and Market Research* (parallel "gather"), *Narrowing and Market Auditing* (parallel "sorting"), *Positioning, and Naming* (parallel Define). Next, there is *Design Identity,* which encompasses subtitles such as *look and feel, color, more color, typography*, etc., and this parallels the more concise phase of "Shape" within the architecture studio's process.

The last of the categories in Wheeler's resource called *Creating Touchpoints* parallels the concise "Release" portion of the architecture studio. This phase, however, admittedly, deserves the breadth of attention that Wheeler proposes in her book that moves beyond that of the pedagogical architecture studio. It supports the idea that "presentation" or "release" of your creation can take on many forms and methods, yet in this paper, many of those other topics are still under-explored as they relate to the pedagogy of an architecture studio.

2 The Studio

2.1 Process

The branding process used to outline the studio semester consisted of a cycle that is not unlike a typical architectural process; though this paper looks at it under the lens of a branding strategy revealed by the principal of an internationally-recognized design firm that focuses on brand campaigns and promotions. The first part of the process is to *gather*. This involves heavy research into client history, previous branding if applicable, the client's target audience, avenues and mediums of communication, external influences that may impact current and future trends, etc. Step two is to *sort* everything that has been gathered, and to consider what is the single greatest challenge, opportunity or message for the client/brand [community/user]? Step three is to *define* the brand platform—this is not a slogan or saying, but the statement from which all aspects of your brand must answer back to, and yes, this is where the descriptive term lies. In a single statement, the brand is to be simplified into one phrase: "Brand X is [insert verb or adjective here]." That term deserves careful consideration along with word association exercises and emotional investment before being finalized. See Figure 3 for an example where the descriptive word "romantic" helped to steer a student's desired output in conveying an experiential site visit. Author of "A Branded World," Michael Levine, also reinforces this step as paramount in conceiving a clear, realistic brand identity: "These promises, which should be written down in the simplest language possible and distributed on a regular basis to every employee of the company, are covenant made to the public." [2] Once this pivotal endeavor has been realized, then *shape* can take place in Step four. This is where creativity and design take place in an effort to support and promote the brand platform. Step five is to *release*. The branding process is more than graphic design relating to a project. It is a holistic strategy that comprises everything including, but not limited to font type, commercials, and timing in which to engage and to *release* to the public. The promise or defined message should emerge through the presentation and communication of the student project.

Fig. 3. A student's early representation of the site's atmosphere conveying the adjective: "romantic". (B. Dolgas)

2.2 The Studio Assignments

Experimenting with the notion of following the branding process outlined above takes place in second-year design studios dedicated to both landscape and urban conditions (perhaps further study can focus on insights found in these distinct contexts). Rather than promote the word "branding" for fear of direct relation to advertisement and product associations, the pedagogical strategy was to task students with creating intentional and deliberate experiences that were rooted *in* and rooting *for* specific word descriptors. And similar to the phases of a branding process—and as mentioned above, the studio followed the process related to creating architectural outcomes: *Gather, Sort, Define, Shape, Release.*

Concentrating on language and words and word-associations is a central focus for marketing and branding strategies. Therefore, assigning such words in the context of second-year design studios became the main catalyst in attempting to reinvent the architectural process for this studio assignment. Focusing on both verbs and adjectives, the words were translated into study models that were, at first, devoid of context and scale, yet gradually became introduced and rooted in the projects' physical site. The introduction of such words, however, does not happen until Step 3: Define. Those words have to be inspired by relevant research (Step 1: Gather) that has been sorted into various layers that expose various (and sometimes personal) hierarchies (Step 2: Sort). Once information becomes organized, the designer can begin to make those important design decisions that put forth a proposal and defined project goal (again, Step 3: Define). When putting forth such provocative and branded "atmospheres", language can help communicate a desired outcome.

Words, however, cannot do it alone. Step 4: Shape must come in and give form to spaces that uphold such atmospheric desires. While second-year studios often miss the opportunity to dive into details of their project for various reasons (including a heavy emphasis on schematic design), this strategy remains in hopes that design decisions made-and not yet made—in the *shape* phase of Step 4 will continue to be in response to the project identity (adjective) as defined in Step 3. For example, if the sole identity/word that had been narrowed down from the *Gathering, Sorting* and *Defining* stages resulted in *romantic,* ask: can your project--form, materials, handrail details, etc.—support that identity? Can they strive to?

Even as the project approaches Step 5: Release, ask "how can even the presentation of the project reinforce the identity of *romantic?"* See Figure 4 for an example of a student's final architectural section drawings (often merely technical) meant to convey the term *romantic.* Here the student strives to use graphics alone to evoke a feeling of romanticism as it relates to her provocative intervention in a landscape. Just as brands perpetually communicate a message, so does architecture and the make-up of architecture. It is a challenging predicament both to communicate architecture and to communicate *through* architecture; yet to have a singular identity platform can prove to be a reliable basis or starting point from which to aim to communicate.

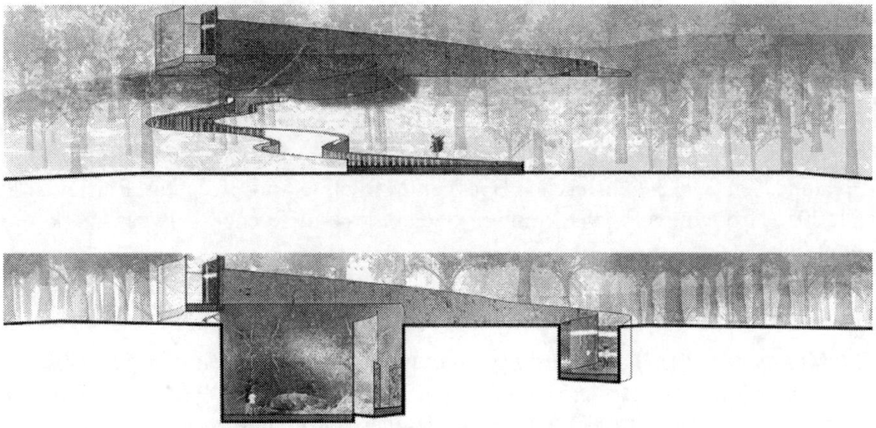

Fig. 4. A student's final sectional representation of their design proposal upholding the notion of "romantic" in both presentation and form.

3 After-thoughts

There may still exist a question as to whether the students are creating "Branded Space" as the title suggests, but it is clear that by following (or paralleling) a rigorous, creative, user-centered process such as that of a branding process, the students are making deliberate decisions that inform creative spatial constructs rooted in meaning. Project decisions involving superficial and spatial complexities, formal volumes, entry, thresholds, furniture and even presentation drawings can all be supported by a clear, defined [brand] identity while also performing their unique functions.

The comparison between branding and architecture is increasingly intriguing, as much of 'space' is become increasingly virtual, artificial, and digital—a prime home for all that is graphic and branding. In Wheeler's book "Designing Brand Identity", there is a side quote from Paul Pearson, Managing Partner for a prominent branding agency, in which the term *brands* can be seamlessly replaced for the word *architecture* (having to do with the physical, built environment as discussed in this paper) and carry the same, coherent meaning and definition [1]:

> *Brands* [Architecture] *can amplify engagement and solidify positive impressions by carefully considering the customers' journey, and by seeing the interplay between physical sensory experiences and compelling digital experiences.*

This quote begins to immerse itself into a virtual yet physical and spatial world, all the more relating the two hierarchical topics together. This begs the question as to whether or not, in fact, the output medium of branding and architecture will be vastly different in the future.

References

1. Wheeler, A. (2018). *Designing brand identity: An essential guide for the entire branding team*. Hoboken, NJ: Wiley.
2. Levine, M. (2003). *A branded world: Adventures in public relations and the creation of superbrands*. Hoboken, NJ: J. Wiley.
3. Heaton, J. (n.d.). The Difference Between Marketing and Branding. Retrieved January 21, 2017, from https://www.tronviggroup.com/the-difference-between-marketing-and-branding/
4. Pine, B. J., & Gilmore, J. H. (2011). *The experience economy*. Boston, Mass: Harvard Business.
5. McManus, M. (2017). *Addressing Adjectives: Mimicking Branding Strategies to Promote Meaningful Architecture* in *NCBDS 33 | Begin w/ Why: Ethics and values in beginning design.* Salt Lake City, Utah: University of Utah.

Architecture as Landscape

Yasmine Abbas[1]

[1]The Pennsylvania State University, University Park, PA 16801, USA

Abstract. This study constitutes the latest reflection on pedagogical research and experimental pedagogical projects involving the representation, design, and computation of ambiances. Led by the author at various architecture schools in France, Japan, and the United States, these creative explorations involving drawings and models offer ways to realize, feel, and fabricate architecture. The projects described were conducted in 2018 in courses offered by the Department of Architecture, Stuckeman School, College of Arts and Architecture at the Pennsylvania State University. They show that architectural productions are not static objects, but instead render a dynamic landscape itself nested within a changing milieu. Through these projects, by looking closely at the parameters of spatial effects, students engaged in processes of design taking movement into account in meaningful ways.

1 Milieu in motion

Considering architecture as a landscape implies appreciating motion, in its many incarnations, as essential to the making of a moving (fluctuating and touching) atmosphere that enchants us and connects us to the world. Individuals experience space through movement, get in touch with their surroundings through the means of sensory perception. Further, most of the elements that constitute the atmosphere of a place are dynamic. A strong light washing across a vivid pink wall [1: p. 149] creates, for example, a particular effect that contributes to the prevailing ambiance at a given time. Buildings themselves undergo transformations that participate in changing perceptions of the architectural object: a building's function (and form) changes over time and marks caused by wear and tear give surfaces a certain patina. The materials in use, the building orientation, the locations of openings, and the articulation of interior spaces each contribute to the creation of microclimates (e.g., warm and humid as in greenhouses), sensations (e.g., shivering brought on by a draught) and atmospheres. A building is a fluid milieu nested in a larger one. Each to one degree or another adjusts to the other.

Architects have investigated the sensory and experiential relationships between architecture and surrounding landscapes by exploring phenomena that occur in the interplay of material and light, material and sound, and color and light and by using "prospect and refuge" strategies to stage buildings as shelters [2]. What other ingredients, easily perceptible and less perceptible, could architectural designers experiment with to create atmospheric spaces? The movement of humans and non-humans, animals, insects, etc., in a milifeu; the temperature; the relative humidity; the direction and velocity of air flow; the presence, strength, and nature of fragrances; the qualities of materials including their ability to absorb or reflect light, which, in turn, impacts perceptions of color; and many more invisible constituents—ultraviolet or infrared light for example—contribute to the making of fluid spaces that are experienced dynamically.

In this study, we examine the performative potentials of architectural drawings and models that bring architecture closer to landscape architecture. Overall, we demonstrate ways to design and teach for spatial delight based on a range of sensory and atmospheric parameters. Described are several experimentations that took place in 2018 in courses offered by the Department of Architecture at the Stuckeman School of Architecture and Landscape Architecture, College of Arts and Architecture, the Pennsylvania State University.

2 Atmospheric grammar

Since 2014, in multiple pedagogical frameworks, i.e. research seminars, workshops, intensive studios, and single assignments over the timeline of a semester-long architecture studio, conducted in different cultural contexts (France, Japan, and the United States), undergraduate or graduate students, sometimes working together, have explored the potential of drawings and models to serve as both representational and projective atmospheric tools.

On the one hand, the cartography of ambiance [3] not only charts the landscape or "map[s] the intangible," as Stan Allen argues in his exploration of notations and diagrams [4: pp. 40–69] but also seeks a "pattern language" [5], not of architectural forms, but of atmospheric conditions. By manipulating notational strategies, the cartography of ambiance offers a way to script "sensoryscapes" that pulls inspiration from the drawings of Bernard Tschumi [6], the collective Archizoom [7], and polymath Iannis Xenakis [8].

On the other hand, atmospheric devices or models explore given anticipated phenomena in a more immediate and tactile way than drawings do. They demonstrate, for example, the combined effect of light, materiality, and reflection as in Peter Zumthor's large-scale model for the Therme Vals [9]. Conceiving of a given model as a series or ensemble of models is meant to acknowledge the many and various effects that can be achieved using just one set of parameters.

Historian Mark Wigley argues that architectural representation is inherently atmospheric [10]. If we agree, then the cartographies of ambiances and atmospheric devices have a particular feel. Yet, they are also "operative" [11] or "generative" of atmospheric grammars that relate sensory and atmospheric parameters. They contribute to a deeper understanding of the relationship between architecture and landscape.

2.1 The cartography of ambiance

2.1.1 The field

"Practicing the landscape," understood as going to the field or experiencing a milieu in movement, is, for the architect, an important immersive and relational experience that s/he embodies and translates into an architectural projection. The "field condition," as Allen calls it, "implies the acceptance of the real in all its messiness and unpredictability" [4: p. 217]. In the realm of contemporary "visual practices," which rely on computer programming capabilities, the field condition, as commonly illustrated by mesmerizing bird formations, is also preoccupied with localized types of relations between single elements [4]. So, as Allen writes, "form matters, but not so much the forms of things as the forms between things" [4: p. 218]. The field condition is inherently dynamic and literally "stages" an ambiance, contributing to what philosopher Gernot Böhme calls the "paradigm for an aesthetic of atmospheres" [12]. As an example, Jean Dubuffet's Coucou Bazar "practicables" construct on stage a dynamic field condition of lines and shapes [13] that relate in a variable manner.

In the context of the Atmospheres: Perception, Design and Fabrication spring 2018 course, the author challenged students to individually explore an atmospheric field condition, to stage an atmosphere based on the vivid memory of a place. The author asked students to devise an "atmospheric formula" incorporating a consideration of environmental aspects such as smell intensity, color, wind direction, and decibel level, and to create a two-dimensional representation of the ambiance inspired by the popular representation of the No-Stop City project by architecture collective Archizoom.

Andrea Branzi of Archizoom uses the grid as an "open field" [4] to represent "a continuous space without clear limits" [7: p. 130]. Clusters of "calligraphic signs, forming clouds" populate some areas of the grid [7]. The grid does not function as a traditional architecture plan. Instead, architectural ambiance or atmosphere is suggested (i.e., architecture in its gaseous form), represented as field of things within the structure of the grid. Archizoom's drawings, the group's political message aside, therefore, offered a good starting point for imagining how to visualize ambiances in two dimensions.

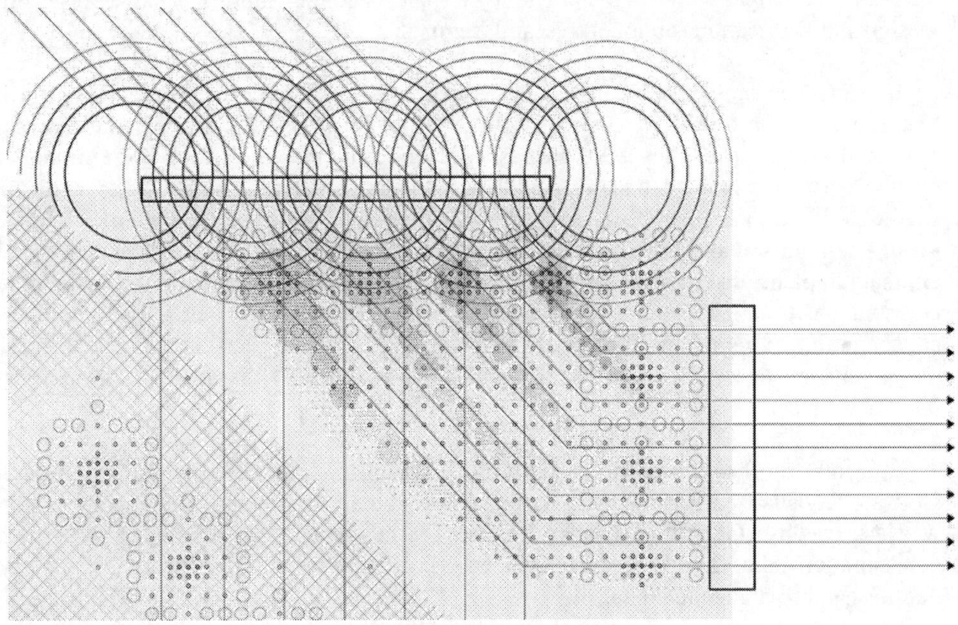

Fig. 1. Atmospheric map (excerpt) by In Pun. Image © In Pun.

The dynamic nature of atmospheres prompted everyone in the class to discuss ways to incorporate notational elements and other graphical techniques to indicate movement. In Pun, for example, decoded the atmosphere of a temple that she had once visited, which had greatly impressed her. To show the soft and fluid feel of the ambiance, the student used arrows to indicate the direction of the air flow, patterns that became lighter and smaller in scale as the scents they represented faded away, and concentric circles to signal the propagation of sound (Fig. 1). Atmospheric maps *à la Archizoom* called for more dynamic visualizations of atmospheres, for more "operative" than "representative" drawings [11], useful to guide the design of an ambiance, if not to replicate it. Atmospheres that "emanate" [10] from either architectural or landscape architecture arrangements are similar in nature, i.e., fluid (in their gaseous state) and temporal. Finding a process to embed mobility into a design to build meaningful atmospheres has driven this pedagogical research since 2014.

2.1.2 From sensory mapping to architecture

On a sunny Thursday afternoon in the fall of 2018, teams of students enrolled in their fourth-year architectural studio went to the field, to visit their project site. As a way to closely engage them with the context, they were asked to sample data related to each of the senses (each team selected a sense, and all senses needed to be addressed) using tools of their own choosing, from pencils and sticky notes to record imprints of textures to mobile phone applications for recoding decibel levels. Given the expanse of the project site and the time to be spent in the field (an afternoon), the students were directed to collect data within a given arbitrary grid of twelve feet by twelve feet with no architectural indication. This assignment was given at the beginning of the fall 2018 semester and served both the Architectural Design VI studio (co-taught with Nathaniel Quincy Belcher) and the Atmospheres: Representation, Design and Fabrication (optional) research seminar[1].

In a manner similar to but not quite as codified as the scientific soil sampling in the Amazon forest described by Bruno Latour [14], the students took notes pertaining to the sensory data collected. They were then asked to code and map their data. For example, as a response to the assignment, a team of students created a soundscape map. Circles on the map, smaller to larger, indicated sound intensity, whereas color indicated the kind of sound. As a result, the loud sound that pedestrians generated by walking on hard surfaces rendered the circulation paths discernible throughout the site without the need to draw them in a conventional way. The soundscape map also revealed different patterns: for example, the concentration of sound in an area of the map, which raised questions about the reasons underlying the occurrence of such phenomena and increased the students' awareness of intangible effects.

Students who had elected to enroll in the atmospheres class used parametric modeling software to visualize the same dataset: they reported their data in an Excel document, which was later read by a Grasshopper script designed for the seminar,[2] and through which patterns, concentrations, dips, and rises were revealed. The advantage of the software is its ability to manifest and spatialize the data.

---

[1] Department of Architecture, Stuckeman School of Architecture and Landscape Architecture, College of Arts and Architecture, The Pennsylvania State University.
[2] Many thanks to Grasshopper wizard Eduardo Raposoda Silva De Castro for his assistance

 https://doi.org/10.1051/shsconf/20196402002

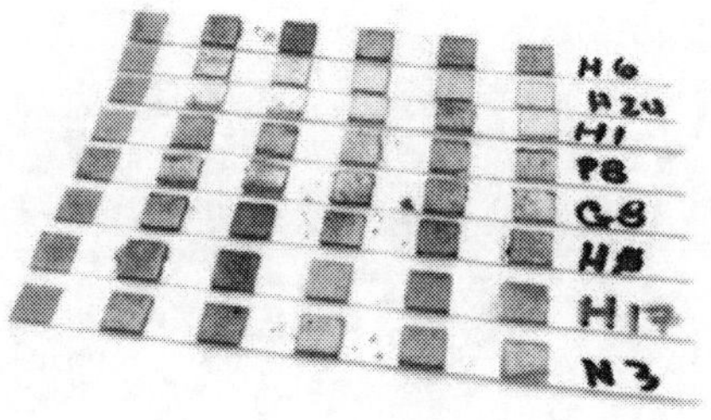

Fig. 2. Tray of pH samples collected by students Audrey Buck and Teagan Marron, fall 2018. Image © Author.

Sensory data collection has proven to be the easiest entry point to exploring the parameters of architectural ambiances. Students sample decibels and sound quality (mechanical, human, or natural sounds), textures (hard to soft), variations in canopy density (in connection to the sense of vision, the percentage of sky and foliage determined when taking a photograph looking up), or types of fragrances (plant-based, chemical, or else). However, collecting data related to the sense of taste is difficult. Given that this is the case, a team of students used the following approach: they collected soil samples, added water to them, and using off-the-shelf pH strips, determined the acidity or alkalinity of each (Fig. 2).

Through this investigation at the architectural studio, a new avenue of design inquiry opened up related to the impact of architecture on the landscape and the potential of spatial strategies for mitigating soil pollution. Based on the pH map (Fig. 3), the team decided to explore the flow of water on site, to design a roof capable of collecting rainwater and distributing it to the soil via specific landscape and architectural features at the ground level, and to select building materials (for example, copper or wood) to modify the pH of the water on contact.

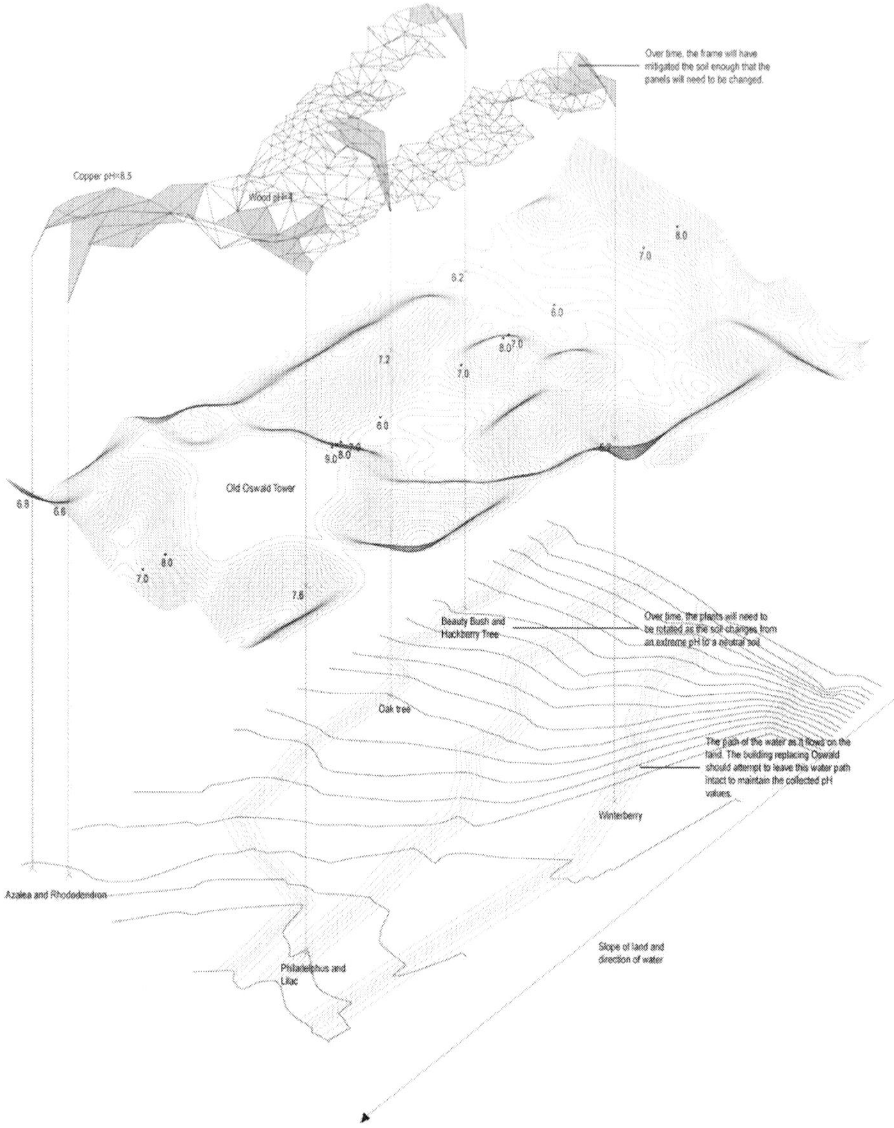

Fig. 3. Map of sampled pH and the roof imagined by Audrey Buck and Teagan Marron, fall 2018. Image © Audrey Buck and Teagan Marron.

The team, whose members were interested in minimizing the impact of architectural production on the natural environment, noted that when the pH of the soil had reached a "neutral" level, key elements of the building and the site—i.e., the building material of the roof and the locations of plants—would have to be changed to avoid a reverse effect. Thus, the students joined a broader conversation about architecture and the landscape through creating engaging spaces.

The meticulous (and strenuous) process of sensory data collection, however extensive and varied the data, will never be sufficient to capture the atmosphere of a place. Yet, the processes of collecting and analyzing data proved to be a worthy pedagogical exercise in terms of connecting students more closely to the context and helping them to see the scope and potential impact of architectural (and atmospheric) intervention.

2.2 Atmospheric devices

To feel the atmosphere they imagined, students translated the atmospheric grammar into architectural form, thereby engaging in a process of thinking through making [15]. The scale model is not only a miniature copy or representation of what ought to be or what is imagined, or an instrument of communication between an architect and its client. The model also constitutes a "machine" through which an architect can define and present his/her own worldview [16: pp. 61–68] "as a thinking mechanism used in making the invisible visible" [16: p. xxviii]. In class, models were also considered as "atmospherical devices" akin to the kinetic machines designed by Lásló Moholy-Nagy at the Bauhaus [17], which created spatial effects by using light in conjunction with perforated and transparent materials to cast changing shadows on surrounding surfaces. If the atmospherical prototypes were used to estimate and calculate atmospherical effects, they could come close to "qualitative" or "engineering models" [16: p. xx].

Instruction was given to create one or a series of atmospheric models using the same set of parameters, each of which suggests a variation of a general ambiance. One example (Fig. 3) is an interpretation of a fire camp set in the middle of a forest. The locations of the origin of heat radiation, color, and spacing of the surrounding trees (translated as columns) suggests the potential for differences in translation, and for varied effects. In regard to the latter, the vertical elements of the architectural model would, for example, grow in number and decrease in dimension, thus changing the shadows cast and filtering in a different way color, light, and even the temperature radiating from the center.

Fig. 4. Translation of the atmosphere of a fire camp surrounded by trees by Joshua Huang, Stuckeman School, spring 2018 course on atmospheres, Penn State. Image © Author.

The series of bathing experiences of the Therme Vals [9] designed by architect Peter Zumthor stands as a clear precedent, a spatial symphony around the different states and experiences of water, warm, cold, still, or in movement, infused with flower petals for aromatherapy[3], to be tasted at time, resulting in an assemblage of atmospheres in connected parallelepipedal boxes, sharing similarities, yet different in regard to subtle variations in the atmospheric parameters, for example, temperature, sound, color, and light.

Parametric modeling tools enable the rapid production of a series of models, emphasizing the possible variation of form based on (sensory) data entry (Fig. 4). The exquisite 3D printed models that Thomas Leonard generated by transforming sound data into form are atmospheric because of their materiality and sensuous forms.

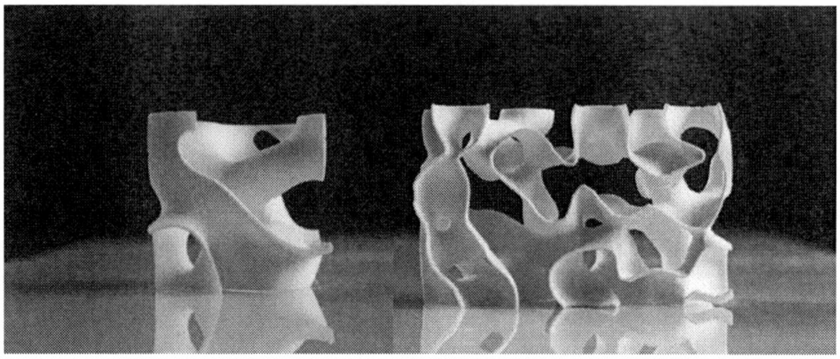

Fig. 5. Thomas Leonard, *Turn Up The Volume*, fall 2018. Atmospheres: Representation, Design and Fabrication research seminar, Penn State. Images © Thomas Leonard.

Through making atmospheric models, designers think of incorporating other parameters of ambiances into their atmospheric devices (for example, which absorbing or reflecting materials to use) to create an architecture that performs, akin to, for example, an orchestra of musical instruments or a fire camp in a forest to produce enchanting building atmospheres.

3 Translation

To this point, the discussion has focused on experimental ways to design meaningful atmospheres starting with choosing, sampling, and translating atmospheric and sensory data. Yet, many questions remain: What are we to make of parameters that are fleeting and intimately related to each other? Sound and smell, for example, travel with wind, which flows in particular directions in response to a variety of obstacles and surfaces. The students converted the data they collected, thereby iteratively transforming "matter" into "form," and with and between each operation, a "gap that no resemblance could fill" [14: p. 125] became apparent. For example, from soil to pH, from pH to a roof with a rainwater collecting system. Unlike the scientific process, the design process is not so concerned with "reversibility" [14]. When the scientific process requires "circulating references," i.e., that one must be able to track original data entry [14], the "designerly ways of knowing" [18] welcome leaps from "matter" to "form" that creatively incorporate additional knowledge learned from generating the form itself. Thus, the data at the origin of the process becomes less distinct.

Because an atmosphere is dynamic, parametric modeling software and methods offer special opportunities to explore the computation of sensuous and immersive atmospheres.

[3] As seen in the video by Richard Compans, *Les Thermes de Pierre* (Arte + Centre Pompidou, 2001): https://www.youtube.com/watch?v=V1UVmNevN5s (Last accessed, February, 4, 2019).

The experimentations described herein translate dynamic sensory data into atmospheric parameters and, more importantly, feelings into forms for informed spatial effects. Thus, in this work, we describe and propose an enhanced architectural curriculum that invites students to realize the importance of designing for and with the senses to create quality spaces. By exploring atmospheric variability that incorporates sensory data (related not only to vision, hearing, and touch, but also to smell and taste), changing environmental factors (air flow, light, and temperature) or situational "porosity," experienced as we move through space [19], designers explore hybridizing architecture with landscape. To "free architecture" from function [20], as architect Junya Ishigami puts it, to let architecture blend in with, become a part of, its milieu is integral to the purpose of the pedagogical practice described herein.

Through the various assignments focused on decoding and encoding a collection of sensory data, and inspired by lessons learned from, for instance, the drawings of architectural collective Archizoom, or the work of architect Peter Zumthor [9, 21], it became increasingly liberating to explore the performative potentials of drawing and modeling. These are the very processes used by architects to project—imagine and create— spaces of effect and affect to which we can memorably connect.

References

1. Pauly, D., Barragán, L. & Habersetzer, J., *Barragán: Space and Shadow, Walls and Colour* (Birkhäuser, 2002)
2. Hildebrand, G. The Wright Space: Pattern and Meaning in Frank Lloyd Wright's Houses (The University of Washington Press, 1991)
3. Abbas, Y. "The Cartography of Ambiance," in *Ambiances, Tomorrow–Ambiances 2016 Third International Congress*, September 21–24, Proceedings, **1**, pp. 253–258
4. Allen, S., Practice—Architecture, Technique and Representation (Routledge, 2009)
5. Alexander, C. A Pattern Language: Towns, Buildings, Construction (Oxford University Press, 1977)
6. Tschumi, B., Migayrou, F. & Centre Georges Pompidou, Bernard *Tschumi: Architecture: Concept & Notation* (Editions Du Centre Pompidou, 2014)
7. Capdevila, P. M., "The Interior City: Infinity and Concavity in No-stop City," *Cuadernos de Proyectos Arquitectónicos*, 2013, no. **4**, pp. 130–132
8. Kanach, S. E. & Lovelace, C. (Eds.), *Iannis Xenakis: Composer, Architect, Visionary* (Drawing Center, 2010)
9. Hauser, S. & Zumthor, P., *Peter Zumthor Therme Vals* (Scheidegger & Spiess, 2007)
10. Wigley, M., "Architecture of Atmospheres," *Daidalos,* 1998, no. **68**, Constructing Atmospheres, pp. 18–27
11. Tschumi, B., "Operative Drawings" in Constant et al., The Activist Drawing: Retracing Situationist Architectures from Constant's New Babylon to Beyond (Drawing Center, 2001), pp. 134–137
12. Böhme, G., *The Aesthetics of Atmospheres* (Routledge, 2017)
13. Duffy, J. H. 2016, "Dubuffet plays hide-and-seek: Lineage, reflexivity, and perception in "Coucou bazar," *Art Bulletin*, vol. **98**, no. 2, pp. 237–260
14. Latour, B., *Pandora's Hope: Essays on the Reality of Science Studies* (Harvard University Press, 1999) in Mello, L. and Byvanck, V. (Eds.) *Dreaming Awake*, Automatica, Marres, 2018)

15. Ingold, T., Making: Anthropology, Archeology, Art and Architecture (Routledge, 2013)
16. Smith, A. Architectural Model as Machine: A New View of Models from Antiquity to the Present Day (Routledge, 2004)
17. Blume, T., Hiller, C., Stiftung Bauhaus Dessau, Henie-Onstad kunstsenter & Kungnip Hyŏndae Misulgwan, Human-Space-Machine: Stage Experiments at the Bauhaus (Spector Books, 2014)
18. Cross, N., "Designerly Ways of Knowing," *Design Studies*, vol. **3**, no. 4, October 1982, pp. 221–227
19. Janson, A., "Porosity: Ambiguous Figure and Cloud." In Feldhusen, S., Poerschke, U
20. Weidinger, J. (Eds.): "Mixings in Architecture and Landscape Architecture." *CloudCuckoo-Land, International Journal of Architectural Theory*, vol. **21**, no. 35, 2016, http://cloud-cuckoo.net/fileadmin/issues_en/issue_35/article_janson.pdf (last accessed, February 9, 2019): pp. 35–46
21. Ishigami, J., *Freeing Architecture* (Fondation Cartier pour l'art contemporain/ LIXIL, 2018)
22. Zumthor, P., Atmospheres: Architectural Environments: Surrounding Objects (Birkhäuser, 2006)

Perceptions of the nocturnal public space and technological innovations – citizen evaluation of a smart public lighting installation in Nantes

Nicolas Houel[1,*], *Dany* Joly[2], *Laurent* Lescop[3]

[1]AAU Laboratory, CRENAU team, ENSA Nantes 6 quai François Mitterrand 44000 Nantes, France
[2]EPICE department, Nantes Métropole, 5 rue de Saverne 44000 Nantes,
[3] AAU Laboratory, CRENAU team, ENSA Nantes 6 quai François Mitterrand 44000 Nantes, France

Abstract. The recent introduction of digital tools in public lighting represents a milestone in the evolution of street lighting. Simultaneously to this innovation phase, urban lighting is currently being studied in terms of energy and environmental sobriety, to reduce its financial and ecological impact. This study presents the observations and evaluations carried out by nocturnal urban dwellers of an experimental public lighting system, set up in real conditions. The evaluations question the relevance of sensors integrated to the lighting material in real time via a series of survey managed autonomously by residents of the experimental environment.

1 Introduction

The leading position of public lighting in urban spaces, determined by progressive and innovative approaches that are still topical [1][2] has since long been supported by initiatives dealing with the safety of people and property, traffic, and the balance between the electricity production and consumption in France [3]. Beyond deeply contrasted nocturnal landscapes between rural and urban environments [4], this attention has consolidated and generated some beliefs in terms of nocturnal urban practices and related artificial lighting. Among those, while the connection between lighting and feelings of safety is definitely rooted [5][6], and the association between visual comfort and the amount of light is present in several urban lighting projects, new lighting and communication technologies, i.e. LED and wireless networks, introduce a new type of beliefs: one in technology for communities and citizens. With technological development comes a wider range of choices, criteria, accessories and ambiances than before, i.e. high-pressure sodium lights and metal halide lamps. In that sense, how can communities, with a public lighting stock in constant renovation, face the choices and arguments of the manufacturers? More specifically, at a time when lights can provide additional services, adjust their lighting levels in real time and change their light temperature to fit the needs, the following question arises: how to illuminate? Led by the leitmotiv by the *Association Française de l'Éclairage,* some communities are currently exploring the necessary tools to determine the needs and uses for their nocturnal territories, as to provide

* Corresponding author: nicolas.houel@crenau.archi.fr

adapted lighting configurations. Thereon, Nantes Métropole, with its advisory experiences (Major debates on energy transition[2], permanent conference on the Loire[3], and, at the municipal level, the *Conseil Nantais de la Nuit*[4]), is interested in the ways to encourage citizen involvement in the development of their Public Lighting Equipment Scheme (*Schéma de Cohérence d'Aménagement Lumière – SCAL*), designed to support the public lighting renovation in the metropolis' 24 municipalities. In that respect, the perception studies on artificial lighting carried out in the 2000s in France [7][8][9], and the current initiatives on nocturnal wandering to observe public space at night (Sara Castagné[5] and Isabelle Corten[6], lighting designers in France, and Léni Schendinger[7] in the United States) strengthen the baseline for qualitative studies for the observation of the nocturnal public space, and allow to explore tools to collect physical (urban fabric configurations, lamps) and sensory (perceptions of the nocturnal public space, of the lighting) indicators. The citizens are thus invested and contribute to the knowledge input that is wanted nowadays. As part of SCAL, the first stage – evaluation of tools and methods to collect indicators – was carried out on the Nantes Métropole territory in 2018. The chosen protocol questioned several beliefs supported by the citizens and the communities, namely: Are the citizens' views relevant in terms of environmental considerations about artificial lighting? Is technological innovation, in particular the solution of embarked sensors, of interest for the communities? Are lighting level and comfort (vision, feeling of safety) connected? Lastly, the experiment led to the evaluation of acceptability regarding a dynamic lighting system, in which the light intensity changed depending on the movements detected in the street. For this study, the experiment relied on the implementation of ten experimental lamps equipped with various sensors, including a motion sensor, set up in Nantes Métropole's public space. Configurable in terms of intensity level and dimming speed, they were assessed by citizens twice. Carried out independently, the citizen evaluations led to the creation of two grading systems: one was physical (paper) and the other digital (online survey). Based on the results, it aims to question the relevance of a lighting system with varying intensity and its acceptability according to different parameters (intensity level, dimming speed). The first results of the investigation, taking place at the level of a street, aim to contribute to the elaboration of a collection tool for physical and sensory indicators on a bigger scale, to build a quantitative and qualitative database about nocturnal public space and its perceptions. Therefore, the study questions the current intuitions and beliefs regarding public space at night, lighting material and their evolutions. Indeed, if the first results appear to highlight contradictions with the established beliefs, the limited number of the panel calls for reproducing this evaluation on a bigger scale, in order to confront the results and assess the state of beliefs and potential objections in a quantitative way. Lastly, reproducing this system on a bigger scale would allow to participate in the implementation of a form of education about lighting sobriety for the communities and their users, thus allowing to talk about light extinction and dimming policies in public space, or the concept of dark scheme in a collective way, to contribute to the current research on fair lighting.

[2] https://www.nantestransitionenergetique.fr/
[3] https://www.nanteslaloireetnous.fr/
[4] https://www.nantes.fr/conseil-nuit
[5]https://www.lightzoomlumiere.fr/interview/sara-castagne-mettre-en-lumiere-en-co-conception/
[6] http://www.radiance35.eu/fr/agence
[7] http://www.nightseeing.net/

2 Spatial and material environment

The study took place in the rue La Noue Bras de Fer in Nantes, on a 250m linear segment, located between the Boulevard Léon Bureau/rue La Noue Bras de Fer junction and the Manny building. Over this area, ten 150W high-pressure sodium lamps were substituted with ten LED dimmable lamps, which are 15W in standby mode and 45W in activity. The location for the experiment was chosen according to the following indicators: presence of a bi-directional traffic lane, of pedestrian accesses on each side of the road, of parking spots on one side, commuting traffic at the beginning and end of the day, presence of vegetation (medium-height trees), diversity in the uses of the site (residential, professional, commercial, vehicular and soft traffic). The road and the pavements are lined with medium-height contemporary buildings with different façade treatments, two junctions are identified, as well as vegetal space. Lastly, one of the determining elements was the presence of a Linky metre connected to the control cabinet for the selected light points. It allows to quantify the expected energy savings. The ten smart lamps that were deployed are the results of the Interactive Data Light project[8], award-winner of the 2017 *Challenge pour l'Accélération Digitale de l'Ouest*. Designed by a consortium of Nantes-based companies, they are equipped with the following: air quality sensor, decibel level sensor, temperature and hygrometry sensor, energy consumption, motion sensor. Each lamp independently captures the data related to the sensors, then sends it via a wireless communication protocol to the master-lamp (n°1), located at the beginning of the street and equipped with a 3G communication system to send it to a server. The communication protocol enables a precise activation of the lamps: when a n lamp senses movement, it changes its light intensity (10% of the flux to 30%) and communicates the same order to its $n-1$ and $n+1$. Therefore, regardless of the direction of the detected object, its path is illuminated. On each lighting column, a kakemono banner is set up to indicate the experimentation object and invite the pedestrians to fill out an online questionnaire. It aims to collect the sensory indicators from the users of the experimentation space, in terms of their perceptions of dimmable lighting.

3 First study: voluntary and autonomous assessment of lighting

To this day, lighting with varying intensity is little developed in public space. While its values regarding energy savings and reduction of light pollution can be verified – during the evaluation –, knowledge of the users' sensory perceptions of the nocturnal public space is currently researched [10][11], in particular regarding lighting levels, illuminated areas and related feeling of safety. The observation of the current evaluation protocols for public lighting demonstrates that their spatial and temporal conditions are determined for the assessment. Therefore, the site or itinerary and the beginning and end of the evaluation are established a priori, and often realized in a controlled environment (street without cars, enclosed university campus, etc.). Moreover, a panel of evaluators is constituted in advance, and they sometimes receive financial compensation. For the first study, the environment to assess is not adapted to the evaluation: it maintains its usual traffic flows and uses. Additionally, no panel of evaluators is determined in advance. These two elements allow to consider the collection of sensory indicators in real conditions, and to observe a pedestrian's proper initiative, to be completed with the online questionnaire.

The ten-question survey aims to identify the evaluation conditions (time when they crossed the street, means of transport), profile of the evaluator (age, gender) and the perceptions regarding lighting (identification of the variation in intensity, sensation of comfort, sensation of discomfort, feeling of safety). Fifteen evaluators participated, on their

[8] https://www.nicolashouel.fr/?p=493

own accord and autonomously, in the survey. Parity is almost total, with seven women and eight men. Among them, two are between 18 and 24 years old, ten between 25 and 45 years old, one is between 46 and 59 years old and one is over 60 years old.

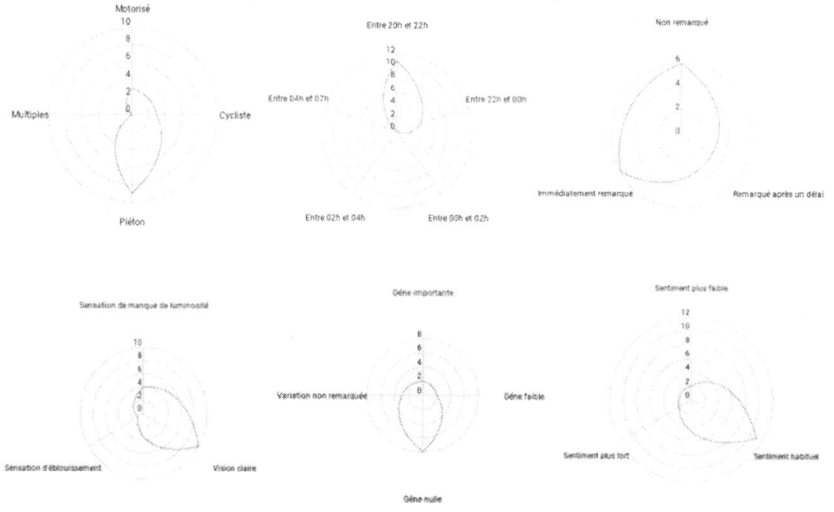

Fig. 1. Radar profiles of the collected data – 1st line (a), (b), (c) ; 2nd line (d), (e), (f).

The collected data allows us to see that most evaluators are pedestrians (Figure 1a, val. 9), and cross the street between 8pm and 10pm (Figure 1b, val. 11). The variation of intensity by motion sensor is balanced between a *complete absence of notification* (Figure 1c, 6), and a *direct notification* of the variation (Figure 1c, val. 6). The luminosity proposed in *active* mode (30% of intensity, around 18 lux) is largely considered as *correct* (Figure 1d, val. 10). Increasing the light level, from 10% (around 8 lux) to 30% (around 18 lux) causes a *slight discomfort* in 13% of the cases. The same number corresponds to *strong discomfort.* However, the discomfort is *null* in 73% of the evaluations (Figure 1e). Lastly, the feeling of safety is assessed as *usual* in 73% of the cases (Figure 1f, val. 11). Two evaluators consider that they feel *less safe*, and two people think that they are *safer*. These first observations allow us to hypothesise that a public lighting system with variation of intensity, respectful of the normative values of lighting at high level and inferior to the same criteria in reduced flux, could be considered as acceptable by the pedestrians, at least in the temporal and spatial conditions of the experiment site. Set up between the end of April 2018 and the middle of September 2018, the promotion of the evaluation of this experimental system was disseminated in the media and on the social media pages of some of the project's institutional partners. The relatively small number of evaluators lastly questions the importance given by individuals to public lighting. Several hypotheses can explain this phenomenon: poor legibility of the communication, set up above eye level, reduced typographical characters; the length and complexity of the URL to access the questionnaire (http://www.jeparticipe.idl-smartcity.com/), and the evaluation period, given that night falls later during the spring and summer. The number of pedestrians at this time might have been too small to result in a quantitative study. One last element remains to be studied: the comments, left to the discretion of the evaluators. Out of the fifteen participants, four of them have left a comment:

1/ Very useful against light and atmospheric pollution.

2/ Try this on the Pont de Chèvre *[Editor's note: Sèvre] that has not been illuminated for years. Unacceptable in rainy times in the winter, we can't see anything, not even the exits. Spend tax money efficiently.*
3/ This is a nice project, to be multiplied in other streets in Nantes and around.
4/ Light out of service on May 1st, at 9.45pm.

These comments seem to be coherent with the sensory evaluations collected before the questionnaire. Moreover, we can note an awareness to environmental concerns (1), a possible way of identifying breakdowns (4), and the possibility to think about the design of public lighting in collaboration with the citizens, comments (2) and (3) stating the desire to multiply the system and the identification of a potential intervention site.

4 Study n°2: evaluation of lighting by residents

Integrated within a broader research project about education on lighting sobriety, the second study aims to observe the ability of a panel of evaluators to contribute to the establishment of a knowledge base on nocturnal uses of a territory. This study also aims to evaluate the awareness level regarding current considerations on artificial lighting (energy savings, night sky pollution, damage to the fauna and flora). In that sense, an evaluation protocol on nocturnal perceptions in the rue La Noue Bras de Fer was established as follows: four successive weeks for the evaluation with six questionnaires, three meetings between evaluators and project leaders (presentation, collaboration, conclusion). During these four weeks, three changes in the light intensity of the lamps (Table 1) were implemented, then three changes in the dimming parameters (Table 1).

Table 1. Configuration of the parameters for the variation in intensity and dimming, by stage.

Étape	Intensité *(veille > activité, en % d'intensité)*	vitesse *(veille > activité > veille, en ms)*
1	10 > 30	500 > 2000 > 500
2	10 > 30	500 > 2000 > 500
3	10 > 20	500 > 2000 > 500
4	10 > 20	100 > 2000 > 500
5	10 > 20	500 > 1000 > 500
6	10 > 20	500 > 3000 > 500

Following the first study, the evaluations were carried out autonomously by the residents. Only a reminder via email was sent the day before a change in the lighting parameters. No information was given about the modifications themselves, as the aim was to observe the general perceptions of the street ambiance, independent from the sole observation of artificial lighting. In order to help the procedure, the evaluation questionnaire came in two ways: a paper and a digital format. The tester's log book included, for each step, a transcript of the physical conditions of the evaluation (time where they crossed the street, weather conditions, site traffic, traffic direction of the evaluator) and five questions according to the protocol established by Boomsma & Steg (2014). Within the study, the questions suggested by Boomsma & Steg, on the feeling of insecurity, were rewritten to focus on the feeling of tranquillity. To this end, we used IATE[9] and the CNRTL[10] thesaurus to study the potential positivity in the perceptions, with the hypothesis that a series of affirmative questions could perhaps guide the evaluator's perceptions. Figure 4 presents the questions that were submitted to the participants. Each one was subject to a notation system, like Lickert's scale, from 1 (complete disagreement) to 5 (complete agreement).

A. *I found the ambiance pleasant.*
B. *I felt safe.*
C. *If I didn't have to go through this street for the evaluation today, I would have liked to go through another street.*
D. *I changed my walking pace in the street.*
E. *I would have liked to see more people in the street.*

Fig. 2. series of questions asked to the evaluators, for each stage of the evaluation.

The data collected during the second study seems to present positive results in terms of modification of public lighting parameters. As shown in Figure 3, the perceptions of a nocturnal public space in real conditions (automobile traffic, construction works in the street, varying weather conditions, varying crowd, potential glare by car lights, etc.) seem little affected by the transformations in artificial lighting. More specifically, only the dimming speed parameters, in particular during the transition between standby and active mode are perceived as blinding when they are too quick (100ms). Otherwise, the decrease in lighting level is not perceived, and thus does not seem to affect the feeling of comfort and safety.

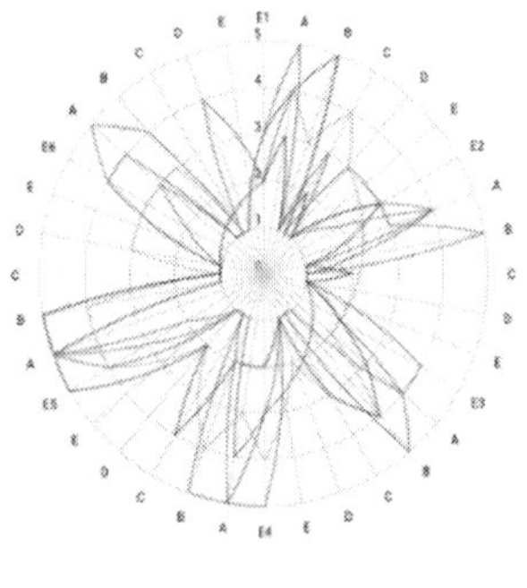

Fig. 3. Overview of the evaluations. The E*n*s represent the evaluation stages, the A to E letters represent the questions (cf. Fig. 2) for each stage.

The decision to select a panel of residents from the experiment site aimed to facilitate the autonomous evaluation over a certain period of time and the implementation of meetings. The first one was a presentation of the evaluation project on nocturnal perceptions of the rue La Noue Bras de Fer, and the collection tools of these indicators. The second meeting was a

[9] https://iate.europa.eu/home
[10] http://www.cnrtl.fr/

restitution stage by the evaluators (three filled-out questionnaires per person) and allowed to test an evaluation tool on current public lighting, i.e. an interactive digital map that displays colour gradients regarding the power of the 95,000 lamps in Nantes Métropole. Projected horizontally on a glass table, the map allowed to learn about Nantes Métropole's entire public lighting stock, as well as zoom in and out of specific areas. With this tool, the residents were able to give a series of sensory indicators on nocturnal perceptions of spaces they are used to visit. The first evaluator told us that a road along the Loire could benefit from varying lighting to reduce the light flux outside the commuting times, and thus sustaining biodiversity. Another evaluator then shared his difficulties to ride his bike on a specific route, as the path was not properly illuminated. Lastly, a third evaluator mentioned that she did not feel safe on a section of her everyday walk, in an area where the lighting was lacking, in her opinion. While this last contribution suggests the current importance of lighting in the feeling of safety [5], the first two comments on artificial lighting, i.e. nuisance for the ecosystems and the need to accompany the rise of urban soft mobilities with lighting equipment that is adapted at nightfall. Moreover, this last point is one of the major entries in the revision of the Nantes Métropole's Public Lighting Equipment Scheme and is in continuation with the renewed use of soft mobilities (cycles, gyropods, scooters, etc.) that recently appeared in the city. The simultaneity of the SCAL development and of these new transportation modes could be studied in more details, for a way to illuminate soft modes of transport and understand nocturnal landscapes with darkness [12][13][14].

Conclusion

The two studies aimed to explore three current questions: Are the citizens' views relevant in terms of environmental considerations about artificial lighting? Is technological innovation, in particular the solution of mobile sensors, of interest for the communities? Are lighting level and comfort (vision, feeling of safety) connected? For each question, the results collected in the studies are encouraging. With their comments during the second study, and in particular during the manipulation of the cartographic tool, the evaluators seemed aware of environmental questions regarding artificial lighting, and willing to reduce lighting to sustain the ecosystems. Moreover, these comments are in line with Boomsma & Steg's research (2014), on the abilities of individuals who are aware of environmental considerations to accept reduced lighting levels. Within the two studies, the solution of mobile sensors is appreciated for their ability to automatize the variation of intensity and, according to one comment, to give "the impression of being welcomed by the light!". Beyond the detection of moving objects, the other integrated sensors did not succeed in verifying the interest of this innovation to the community. Indeed, the data on air quality, sound, temperature and hygrometry levels have not yet found a way to be used by public services. One element, currently under study, could be of interest: the measure of the reduction in energy consumption of one lamp with intensity variation. As it is factual, the measure cannot be a precise estimate on a big scale, as the variety of lighting material and associated electricity consumption require deeper and more targeted studies, in particular in terms of nocturnal uses of one space – a space that is little visited at night could claim more significant energy savings than a frequently visited site, the increase in light flux is different depending on the case. Lastly, the sensory indicators collected during the studies provide results in terms of the relation between light level and feeling of comfort: the evaluators mainly consider that the variation in intensity was not bothering them – except when it was too brutal – and that the proposed light levels, sometimes lower than normative indications, ensured a pleasant ambiance in the street.

Discussion

The results from these field studies allowed to collect factual elements on the perception of public lighting with a variation in intensity, in real conditions. The implementation of physical and digital survey tools, to be completed autonomously by the evaluators, was appreciated, and the different lighting speed parameters allowed to identify sources of discomfort for the evaluators (glare from the harsh light flux increase). The results support the citizens' views as relevant and the important of technological innovation for the communities – in particular about energy savings and light nuisance reduction – and the study of the connections between comfort and lighting level. However, they must be modulated by two elements: the scale of the studied site, i.e. rue La Noue Bras de Fer in Nantes, and the evaluators' familiarity with their environment, especially those of the second study, recruited as residents of the site. Indeed, the potential reproduction of this type of evaluation on a bigger scale (neighbourhood, city, metropolis) can raise questions on the varieties of public spaces. Could a lighting device with intensity variations and the same dimming parameters create the same perceptions in various environments? The scope of the evaluators' knowledge of the place should also be taken into consideration: would a casual walker give the same answers as a resident? Lastly, the results from these citizen consultations could become elements to support renovation policies in public lighting, which are led by an increasing number of French and international municipalities and metropolises since the arrival of LED systems on the lighting market. In that sense, we can question the desire of cities and metropoles to rely on citizen consultation tools to accompany their actions for the renewal of the lighting stock.

References

1. W. Schivelbusch, *La nuit désenchantée*, Éditions Le Promeneur (1993)
2. A. Bovet-Pavy, *Lumières sur la ville : Une histoire de l'éclairage urbain*, Coéditions François Bourin (2018)
3. S. Challéat, *« Sauver la nuit » : empreinte lumineuse, urbanisme et gouvernance des territoires*, sous la direction de A. Larceneux, 281-295 (2010)
4. S. Bertin, S. Paquette, *Apprendre à regarder dans l'obscurité : les « entre-deux » du paysage urbain nocturne*, Environnement urbain /Urban Environment, **9**, 1-20 (2017)
5. S. Mosser, *Éclairage et sécurité en ville : l'état des savoirs*, Déviance et Société, **31**, 77-100 (2007)
6. A. Peña-Garcia, A. Hurtado, M. Aguilar-Luzón, *Considerations about the impact of public lighting on pedestrians' perception of safety and well-being*, Safety Science, **89**, 315-318 (2016)
7. J.M. Deleuil, *Evalum : de l'évaluation de différentes sources lumineuses par les usagers*, in J.M. Deleuil (dir.), Éclairer la ville autrement, Innovations et expérimentations en éclairage public, Presses polytechniques et universitaires romandes, 37-52 (2009)
8. J.M. Deleuil, *De l'acceptabilité sociale des diminutions d'éclairement : une expérimentation lyonnaise*, in J.M. Deleuil (dir.), Éclairer la ville autrement, Innovations et expérimentations en éclairage public, Presses polytechniques et universitaires romandes, 55-67 (2009)

9. M. Fontoynont, J.M. Deleuil, *Sodium haute pression vs iodures métalliques comparaison des performances visuelles sur un site urbain*, in J.M. Deleuil (dir.), Éclairer la ville autrement, Innovations et expérimentations en éclairage public, Presses polytechniques et universitaires romandes, 70-83 (2009)
10. C. Boomsma, L. Steg, *The effect of information and values on acceptability of reduced street lighting*, Journal of Environnemental Psychology, **39**, 22-31 (2014)
11. C. Boomsma, L. Steg, *Feeling safe in the dark : examining the effect of entrapment, lighting levels, and gender on feelings of safety and lighting policy acceptability*, Environment and Behavior, **46**, 193-212 (2014)
12. T. Edensor, *Reconnecting with darkness : gloomy landscapes, lightless places*, Social & Cultural Geography, **14**, 446-465 (2013)
13. T. Edensor, *Rethninking the relationship between light and dark*, Urban Studies, **52**, 422-438 (2015)
14. T. Edensor, M. Cook, *Cycling through dark space : apprehending landscape otherwise*, Mobilities, **12**, 1-19 (2017),C. Boomsma, L. Steg, *The effect of information and values on acceptability of reduced street lighting*, Journal of Environmental Psychology, **39**, 22-31 (2014)

Studio_ L28: From a Socially Engaged Sound Art Practice to an Open Training Ground for Sonic Design Experimentation

Caroline Claus[1] and *Burak* Pak[2]

[1] Ph.D. Student, Altering Practices for Urban Inclusion Research Group, Department of Architecture, KU Leuven, Brussels
[2] Professor, Altering Practices for Urban Inclusion Research Group, Faculty and Department of Architecture, KU Leuven, Brussels

Abstract. This paper reflects the preliminary findings of a PhD research on the spatial politics and potentials of noise and vibration, and the affective or attractive and repulsive power of sonic force. We focus on the public space of a railway area in transformation in Brussels, where sonic conflicts are prevalent. To explore the affordances of a sonic urbanism as critical spatial practice and thus to break free from prevailing modes of urbanism which focus on sonic risk and vibrational nuisance - we constitute a working practice exploiting and nurturing the productive encounters between disciplines such as sound art, urbanism and urban architecture. By setting up an experimental design studio at the KU Leuven Faculty of Architecture, embedded in local auditory culture and in connection to ongoing planning processes, we aimed to facilitate an open learning ground for sonic design experimentation in the development of innovative sonic spatial tools and approaches. The studio was oriented to students of the International Master in Architecture summoned to research the multiple (sonic) vibrations of the L28 railway area, to exploit and contrast these vibrational forces, transforming them to into actions and opportunities. From a critical sonic understanding of urban space, students played and explored a contradictory role compared to the widespread noise control practices, reformulated environments, perimeters and relations of urban phenomena and searched for interactivity with vibrational dynamics that already exist in the territory.

1 Introduction

Under the guidance of Prof. Dr. Burak Pak and Caroline Claus, eleven students of the KU Leuven International Master in Architecture enrolled in a fourteen-week urban sound design research project. In this paper we introduce the experimental design studio operating as a reflexive inquiry, an engagement with students in architecture and as to its own purpose. In what follows we will explore the following questions:

- What are the affordances of a sonic urbanism as critical spatial practice situated in a design studio?
- How can an urban-architecture design studio be an open testing ground for learning and exploring novel forms of sonic urbanism?
- How can practitioners and students learn from different disciplines, and reformulate it into a new way of operation?
- What are the alternative strategies for engagement, critical and spatial design?

2 Studio_L28 as critical learning experiment

The Masterstudio_L28 was a fourteen-week design studio we organized in autumn 2018 at the KU Leuven Faculty of Architecture, Brussels. Together with sound artists and curators, and students of the KU Leuven International Master's in Architecture, the potential of urban sound design for public urban space has been explored from different critical spatial perspectives. Rather than an educational program the KU Leuven Masterstudio_L28 was conceived as a platform for learning. In this context learning is understood as means of developing ideas in collaboration with students and sound artists, outreach workers and architects. We have aimed to facilitate the production of critical ideas and motivated the students to create exploratory projects as a means for presenting and testing these ideas.

2.1. Research Context

KU Leuven Masterstudio_L28 is part of the ongoing PhD project: The Vibrational Nexus with Prof. Dr. Burak Pak as supervisor and Peter Cusack as co-supervisor. The research has its roots in an interdisciplinary project Caroline Claus realized in collaboration with the Brussels workspace for sound art, Q-O2 and public school for architecture Brussels [1] and her long-term involvement in the planning processes for public railway space along the railway line L28 [2]. The PhD research project explores the position and role of (sonic) vibrations in public space development along Brussels Western ring railway line L28.

Open space along the Western ring railway L28 has long time been marginalized in Brussels planning processes [3]. The development of public space on former railway land is relatively new here. Because of its natural, historical and ecological richness and the recent plans for development and transition [4, 5], the railway area of the L28 line can be considered as an excellent object for research into sound and urban public space development. In the official planning processes (sonic) vibrations are dominantly discussed as a nuisance to suppress. Our interest, by contrast, is focused on the quality of sound and vibration in relation to urban transformation, mobilization and transition.

For his ecological theory on sonic warfare, philosopher, musician and label owner, Steve Goodman [6] points to the importance of understanding the agency of vibrational matter and the shortcoming of anthropocentric conceptions. Accounts of the sonic in terms of conscious hearing or listening are troubled by Goodman in favour of an unconscious, affective account of sound as material impact, hereby opening up to the inaudible frequencies. Goodman describes the vibrational nexus as each actual occasion of experience that populates the discontinuum, drawing in an array of elements into its collective shiver. The conception of a differential ecology of vibrational effects leads him towards a non- anthropocentric ontology of ubiquitous media, a topology in which every resonant surface is potentially a host for contagious concepts, percepts, and affects.

Understanding the open space of an urban railway area in transition as a discontinuum of sonic and vibrational possibilities determined by more actants than human sounds, encourages a revision of dominant sonic approaches, method and tools used in the urban design process as we know it. A decentering of the human sense and perspective seems to have implications for the urban design it constrains. Building on Goodman's theoretical work, we adopt Augoyard and Henri Torgue's definition of a sonic ecology [7] in a proposal for a sonic urbanism as critical spatial practice.

From a questioning of the ontological turn in sound studies the Ph.D. research project elaborates upon the following positions and propositions: 1) human and non-human actants co-constitute a discontinuum of sonic and vibrational possibilities, 2) we need an a-disciplinary rethinking of sonic forms and methods for urban sound design and, 3) (sonic) vibrations as design material opens up to a re-negotiation of urban transition.

2.2 Critical learning

The design studio is the heart of architecture and urban design education [8]. It is a place for "reflection-in-action" through which students learn experientially by designing their own projects. Yet, this reflectivity is mostly limited to the individual interactions in a traditional studio setting: design studio in this limited form is predominantly mono-disciplinary, predominantly teacher and student-centered [9]. In this context, rethinking the design studio as a medium for learning critical spatial practice required the use of particular strategies and methods.

Critical spatial design aims to go beyond architecture as a physical construction and explore the *construction of alternate realities, criticizing existing protocols, and generating new protocols for this venture* [10]. In the studio we introduced sonic spatial perspectives for understanding and intervening in public space as a means to enable a critical version of the 'reflective practicum in designing' [11]. This involved developing the students' capacity of reflecting-in-action and reflecting-on-action as a *crossbencher* and *going off of autopilot,* in other words, instead of blindly following and repeating what the students learned in the past, questioning the existing protocols within the field of architecture with the will to develop new critical practices. This required students and studio coordinators to get out of the comfortable boundaries of traditional expertise in architecture, towards the unknown, the intentional and skilful mastering of incompetence in the ocean of practices [12].

An important challenge in this task was to extend the understanding of architecture towards the non-physical. Architectural design as a discipline has long been suffering from the long-time isolation of *the spatial* from the social and the sonic which assumes space an independent construct. However, it is impossible to understand the space without taking into consideration the social conditions as well as its sonic qualities. Rethinking architecture as a sonic and social composite opened up possibilities to develop a *polyphonic approach* to the *re-articulation of the design practice* [13] and enabled us to go beyond the physical and explore new realities, critically questioning the existing protocols and rewriting them when necessary.

For Schön each design move of the student is a l*ocal experiment which contributed to the global experiment of* reframing [14, 15]. Motivating making experimental actions-design moves generated effects in the *situation*. These effects formed a circular feedback loop, a *conversation* between the student, studio coordinator and the *situation* to articulate, modify or reframe the existing design approach, method, tools and protocols [16]. While the situation was changing, the design ideas and protocols shifted, and at the end of the process the student in conversation with the studio coordinators invented a new way of thinking about a particular situation. Examples of these were sonic sections and sonic-spatial cartographies discussed in the following section.

2.3 Program, methods and tools

The KU Leuven Masterstudio_L28 happened in fourteen weeks and was built around six strategies for facilitating sonic urbanism as a critical spatial practice:

1. A series of class sessions
2. Soundwalks
3. Individual field sessions
4. Exercises in sonic representation

5. Interdisciplinary conversation
6. Synthetic assignment

In a series of eighteen *class sessions*, we have introduced and discussed different sonic perspectives for a re-reading of existing planning and design tools. This has facilitated an iterative design process. During the process we have urged students to equip themselves with the needed skills to redeploy techniques and invent novel cognitive and material tools.

Building on Guy Debord's [17] theoretical work on drifting and Hildegard Westerkamp's [18] definition, we have curated a *soundwalk* for an explorative observation of the open railway space in relation to the subjective experience of place, and the idea of moving narratives while walking and listening. As a warm-up before the walk, we also did an exercise in voice spatialization. The soundwalk was introduced as a tool for the training of a heightened listening awareness of urban context and place, to explore the idea of a non-human experience of (sonic) vibrations, and to learn how to employ that into design research.

The program has encouraged *individual fieldwork* on site. Field recording was explored as a step towards critical cartography. Students were asked to search individually for the sonic qualities of an open railway space they were interested in. We have introduced them to different techniques for listening, recording and representing to grasp and map urban sonic quality. Students were asked to develop a graphical notation system which integrates the sonic, spatial and social characteristics of the railway space of their choice.

For a training in the *representation of urban sound* we have worked with exercises on how to make a sonic image of an architectural project. We have discussed the idea of a sonic section as a tool for the systematic study of the sonic and spatial qualities of an urban space and the option to work with the sonification of design ideas. Students were introduced to various composition techniques coming from sound art, music and/or choreography and a series of computer-aided simulation and sonification tools as possible resources for an advanced elaboration of their design project.

Through network practice we have facilitated an *interdisciplinary conversation* on methods and tools applied in a context of sound art, urbanism, and urban activism. Students were asked to exploit a contradictory role compared to the widespread noise control practices, reformulated environments, perimeters and relations of urban facts and to search for interactivity with (sonic) vibrational dynamics on the territory.

For the final assignment we asked students to develop an urban architectural project consisting of experimental sonic spatial forms with reference to sonic theory, histories of sound art and experimental music. Each student was asked to represent an urban sound strategy by use of graphical images, written text, and a sonic image or sound file. One of the things we as atelier coordinators have stressed along the process, was that we were less interested in formalistic urban architecture and more in the possibilities of intervening in the agency, the experience of formal-mechanic dimensions of urban sonic vibrations as a source for a re-negotiating of urban transformation.

3 Studio_L28 as facilitator for sonic spatial design

Having outlined the context of the design exercise, an a-disciplinary research project in which urban sound design forms a methodically bridging between the concerns of sound art, urbanism and urban architecture, in what follows we examine student's work as experimentations in urban sonic form and architectural design. In our dispositive of critical spatial practice-based learning: we endeavour to identify and assess critical positions, the search for new methods and tools for the analysis and definition of urban sonic forms taking into account dispositives coming from sound arts, architecture and the urban context, that is the transforming railway area of the line L28. In the next section, we will reflect on students' projects in relation to the PhD research project positions and propositions. We will discuss sonic approaches, methods and tools employed, whether or not a transdisciplinary shift appeared, and to what extent role expectations and requirements led to friction. In the following part we therefore discuss the final design proposals, the introduction to new sonic spatial forms for urban re-negotiation.

3.1 Positions

The interdisciplinary notion of *sonic effect* as developed by Augoyard and Henri Torgue's [19] has been adopted in half of the projects. It was used as a building block for new tools such as a new typology of design interventions. Some students have directly applied one or more of the defined effects in their project proposal. Although originally conceived as a human centered approach, it was also used for a connection with non-human agency. One student, Alex Auris Gonzales has used it as a starting point for his experimental approach to sonic vibrations of an underground railway track. Sarah Thorisaen researched by design how non-human agency can play a role in the improvement of urban quality. She has explored possible sonic effects of different types of plants. The focus on *sonic experience* of urban space has prompted most of the students to experiment with techniques such as the interview or survey, narrative explorations, field recording of sounds in relation to the built environment, contrasting qualitative research with the decibel approach used by authorities and planners. From this position, different other disciplines were approached, without really leaving the discipline of architecture.

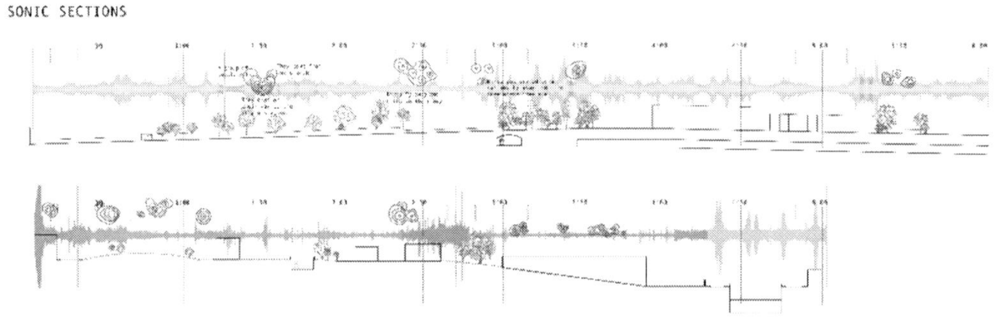

Fig. 1. Sonic sections by Alex Auris Gonzale

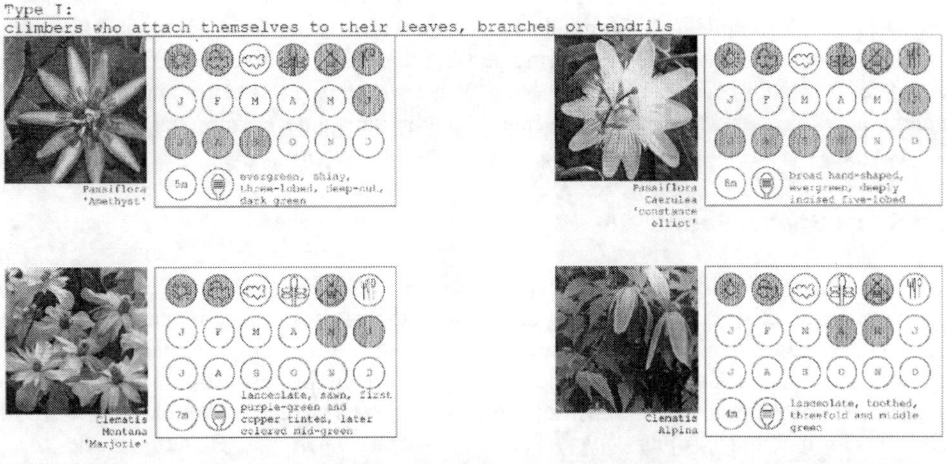

Fig. 2. Fragment of the study of sonic effects of different plants by Sarah Thorisaen

Two students have searched for a positive approach to urban noise in relation to the urban context of the railway space. Both positioned themselves, as crossbencher, in opposite to the dominant approach followed by Brussels authorities and urban planners and architects involved in the on-going development of a socio- recreational green infrastructure space along the line L28. From these radical positions the students have developed a series of punctual solutions for railways space bordering the railway line. Their design approach we consider as transdisciplinary in nature. Vilius Balčiūnas, has integrated techniques such as soundwalk, field recording, literature studying a critical cartography. Melissa Ince went for a study of the experience of noise in the existing socio- recreational green infrastructure space by focussing on different types of users. Both students have integrated methods and techniques coming from different disciplines for combining spatial qualities, urban culture and sonic aesthetics in their sonic design approach. Two other students, Daniel Dent Murgui and Eman Afanen have used a transdisciplinary approach as the starting point for the design of a performative installation they. Finally, we found these crossbenchers adopting an artistic position.

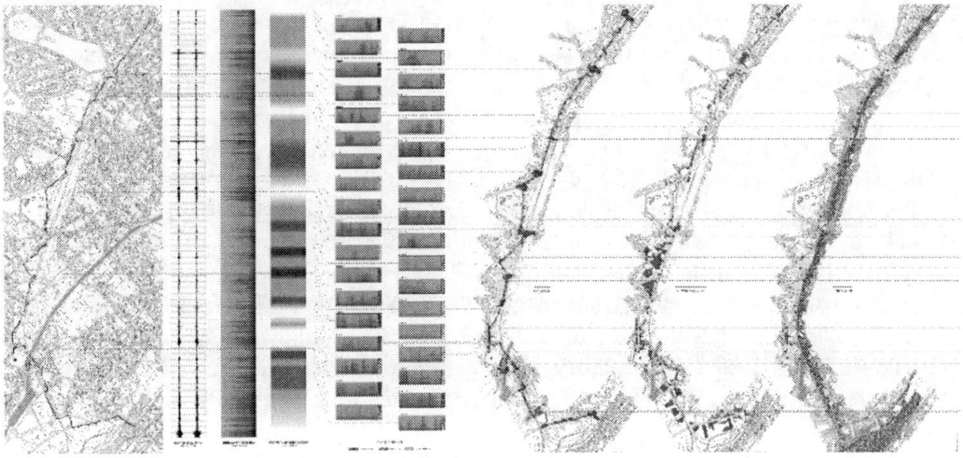

Fig. 3. Fragment of the research into sound out of place by Vilius Balčiūnas

Three students have operated from within the discipline of architecture without much confrontation with other disciplines. Using the theoretical work on *aural architecture* as a starting point, Leonie Van Bellingen has developed a personal approach that allowed her to go beyond the physical, hereby exploring new connections for urban inclusion. Two students opted for a system-based design approach. Where Mateusz Modzelewski, inspired by the *open structure* of Oskar Hansen, ended up with a human centered approach, Asa Pelin opted for an approach based on *parametric and algorithmic design*.

3.2 Sonic Spatial Forms

Dealing with sonic spatial qualities not present in other analyses of the area, we have touched upon the limits of dominant planning and design approaches. What appeared in the design process are spatial qualities, urban atmospheres contributing to a particular identity for the urban space in transformation. The final projects highlight different possible sonic approaches to a variation of transitioning aspects. Each project illuminates or introduces elements for listening or other ways of sonic vibrational experience. Different modes of sonic experience are articulated through design: from a focus on hearing a sound source, to aural attention to formal or structural details and their arrangement in space and time. Students relate differently to transitioning space and facilitate new ways of negotiation, inclusion, autonomy - who gets it in/out? Not all students have designed for a re-negotiation of urban transition by virtue of being made aware of the contingent aspects of a sonic experience, and capable of re-engineering the control-panel. Some design projects facilitated rule governed selection, the installation of a protocol, an exercise in autonomy: a subjective experience or relation to urban transition as negotiable.

In several projects the adoption of a phenomenological approach led to a punctual urban sound design proposal. Two students proposed a series of sonic spatial design interventions based on different sonic effects listed in the catalogue of Augoyard and Torgue [20]. The design projects introduced a collection of new aesthetic qualities to strengthen the social and cultural appropriation of a future railway park/place. Punctual interventions are added to stimulate the participation of an audience. Another project consists of a spatial structure that offers space for non-human agency in support of a new sonic identity for the Weststation housing complex area.

Following a positive noise approach two students have designed for a controlled aesthetic acceptance of urban noise. The experience of a series of types of urban noise is included in the design for a network of public railway park/spaces. The idea of a sonic sequence is also present in a landscape project that starts from the vibrational experience of an underground railway. These design solutions are transdisciplinary in nature and are oriented towards a future development of new sonic spatial forms for socio - recreational green infrastructure space in Brussels Capital Region. The rather experimental forms open up to a multiple appropriation of future public railway space.

Performative installations designed by Daniel Dent Murgui and Eman Afaneh are conceived as an infrastructure to augment peoples experience of railway space in transformation. The installations reinforce the interplay between human and non-human relations on one side and between sound and vision on the other. The temporary infrastructure project by Daniel Dent Murgui collects various passive audio techniques and converts them into active tools for a participatory intervening in railway rhythms. Both installations demonstrate how a critical sonic approach can play a role in a (re) negotiation of urban transition.

Fig. 4. Project for empowerment by Daniel Dent Murgui

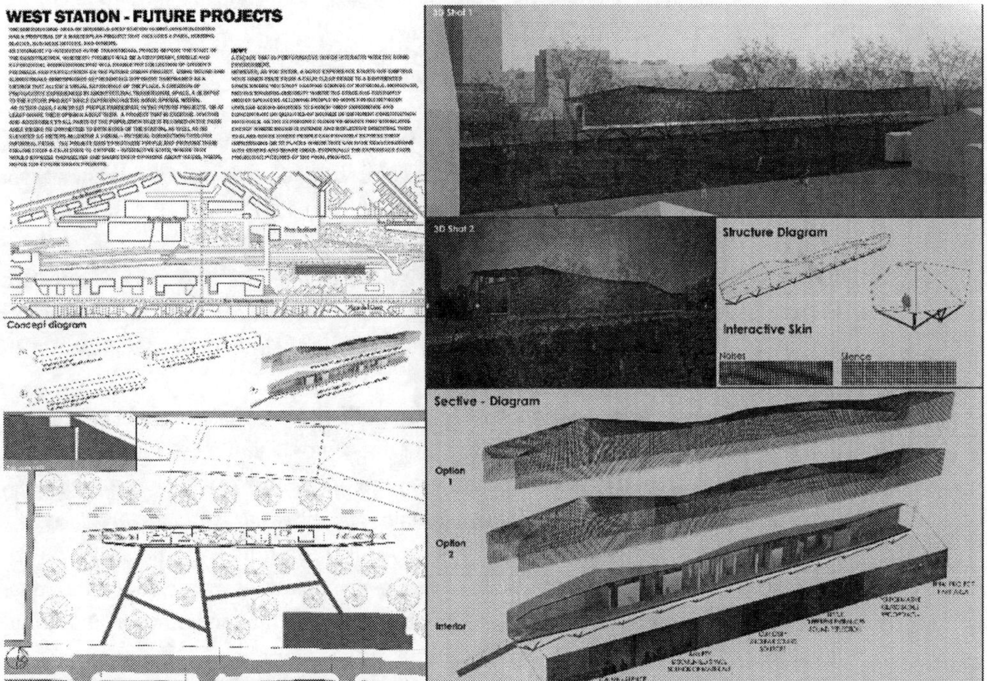

Fig. 5. Transitional Architecture by Eman Afaneh

Three projects evolved from a more practical perspective and concrete architectural approaches. Via an interweaving of acoustic elements in the design for tactile interventions the students have explored and illustrated the negotiability of sound in the architectural scale of urban space. Inspired by the idea of an *acoustic arena* as defined by Blesser en Salter [21], Leonie Van Bellingen has designed a series of three arena's that act as an interactive and socially, as well as physically, connecting narrative for the Weststation area in transition. Basic acoustic principles form the foundation of the system-based design projects by Pelin Asa and Matheusz Modzelewski. Easy technical methods for filtering and amplifying sound waves have been used to facilitate a multiple use of public and private railway spaces.

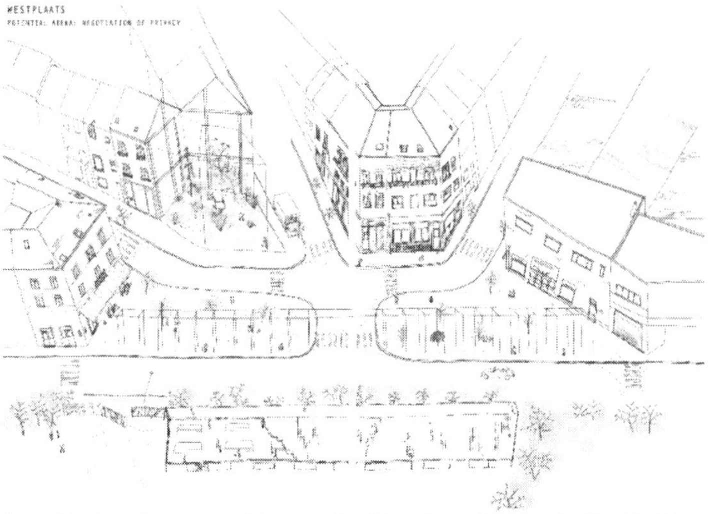

Fig. 6. Design for a potential arena for Westplaats by Leonie Van Bellingen

3.3 Urban Sonic Design Typology

On the basis of the design approach introduced and the final design proposals and propositions produced during the design studio, we identify four types of design approaches in relation to the disciplinary position and urban sonic forms applied in the projects:

- *An interdisciplinary urban sound design approach* in line with Augoyard and Torgue's suggestion of playing the city via its sonic Instrumentarium by modulating its vibrational effects. Within this we can distinguish a human versus non-human design approach.
- *The noise landscape as opportunity for urban design.* A transdisciplinary noise perspective allows for a gradually embracing of noise and an attitude of experiment. Noise oriented features are integrated in a green framework for urban design.
- *The design of transitional architectural projects in which (sonic) vibrations are used as material for a re-negotiation.* Ephemeral installations by means of artistic sonic strategies and techniques facilitate a disclosure of urban experience in a context of urban transformation. The experimental nature of the installation encourages an exploration of non-human agency in transitional planning.
- *Concrete architectural approaches and projects* that work with sound as material, and plan a design idea from an explicit acoustic perspective. The implementation of basic acoustic principles in relation to place-making allow for a controlled (re-) negotiation.

4 Reflections and future directions

In 2018, eleven students of the KU Leuven International Master in Architecture enrolled in a fourteen-week urban sound design research project under the guidance of Prof. Dr. Burak Pak and Drs. Caroline Claus. From an a-disciplinary questioning of human centered sonic conceptions and existing noise policies studies, students were challenged to move beyond the human centred sonic conceptions and to explore in practice, an a-disciplinary selection of sonic methods and forms, to exploit (sonic) vibrations as design material for a possible re-negotiation of urban transition. From a mapping of the design practices, we have looked to the disciplinary positions the students departed from, and the urban sonic forms they designed for. From here we have set out the first lines of a typology that summarizes the practices and from where we proceed to the creation of concepts and/or a set of work arrangements that are at once material and processual.

Completing the evaluations of the Masterstudio_L28, at present, several challenges related to the a-disciplinary approach become clear. Due to the short duration of the project, the combination of different disciplines was difficult for teachers and students. Breaking the autopilot was not possible for all students. Increased complexity at all levels demands more participation, more knowledge and theoretical perspectives, introduction and experimentation with new tools and methods coming from multiple disciplines. The same applies to the reaching out to the super-diversity and complexity of Brussels. We observed how students can get lost in the ocean of practices. A form of guidance is necessary but more towards enabling the students to push through the frontiers they chose to challenge. Incompetence is bliss but also reduces the quality of the output in short time. At the level of the Faculty of Architecture and the research team, a shift from a-disciplinarity to transdisciplinarity demands more collaboration, connection with research groups active in other disciplines such as sound art and engineering. From a practical point of view, we also see the value in more cooperation for the supply of infrastructure, tools and space for this type of research and education.

Acknowledgements

KU Leuven Masterstudio_L28 is part of the ongoing PhD project: The Vibrational Nexus with Prof. Dr. Burak Pak as supervisor and Peter Cusack as co-supervisor. Participating students were Vilius Balčiūnas, Afanen Eman, Thorisaen Sarah, Ince Melisa, Modzelewski Mateusz, Vasudeo Doyel, Asa Pelin, Yin QI, Auris Alexander, Van Bellingen Leonie, and Daniel Dent Murgui.

References

1. Claus, C. (2018). Studio_L28: Sonic Perspectives on Urbanism. Brussels. Q-O2.
2. Kandjee, T., Pferdmenges, P. and Persijn, F. (2019) 'Project van Richtplan van Aanleg Weststation'. Perspective.
3. Mabilde, J. and Loeckx, A. (2016). Metropolitan landscapes. Brussel: Vlaams Bouwmeester.
4. "Het groene netwerk | Leefmilieu Brussel." 22 Sep. (2017), https://leefmilieu.brussels/groen-netwerk. Accessed 24 Sep. 2018.
5. "Gare de l'Ouest | Perspective Brussels." http://perspective.brussels/en/node/602. Accessed 24 Sep. 2018.
6. Goodman, Steve. (2010). *Sonic Warfare*. Cambridge, Mass: MIT Press.
7. Augoyard, Jean François, and Henry Torgue. (2005). *Sonic Experience*. Montréal: McGill-Queen's University Press.

8. Schön, D. A. (1987). Educating the reflective practitioner. San Francisco: Jossey-Bass.
9. Newton C., Pak B. (2015) Virtuality and Fostering Critical Design Thinking: An Exploration of the Possibilities Through Critical Theory, Design Practices and Networked Learning. In: Jandrić P., Boras D. (eds) Critical Learning in Digital Networks. Research in Networked Learning. Springer, Cham.
10. Miessen, M. (2017). Crossbenching : towards a proactive mode of participation as a Critical Spatial Practice. Doctoral thesis, Goldsmiths, University of London
11. Schön, D. A. (1987).
12. Schön, D. A. (1983). The reflective practitioner: How professionals think in action. New York: Basic Books.
13. Miessen, M. (2017).
14. Schön, D. A. (1987).
15. Schön, D. A. (1983).
16. Schön, D. A. (1983).
17. Debord, G. (1958). Theory of the dérive. In K. Knabb (Ed. & Trans.), (2006), *Situationist International anthology* (pp. 62–66). Berkeley, CA: Bureau of Public Secrets.
18. Westerkamp, H. (2007). Soundwalking. In A. Carlyle (ed.) *Autumn Leaves: Sound and the Environment in Artistic Practice*. Paris: Association Double-Entendre.
19. Augoyard, Jean François, and Henry Torgue. (2005).
20. Augoyard, Jean François, and Henry Torgue. (2005).
21. Blesser, B. (2007). Spaces Speak, Are You Listening? Experiencing Aural Architecture. The Journal of the Acoustical Society of America., 121(4), 1820-1821.

SHS Web of Conferences **64**, 02005 (2019) https://doi.org/10.1051/shsconf/20196402005
EAEA14 2019

Traversing and sculpting scape through pattern development: Blending the natural and the built to promote ambiance in early architectural education

Kristin Barry[1,*] and *Sean* Burns[1]

[1]Ball State University, College of Architecture and Planning, Department of Architecture, 2000 W University Ave, Muncie, Indiana, USA

Abstract. In the formative years of their education, beginning design students are often equipped with several predetermined biases and stereotypical notions about architecture, with specific design intentions centered around how architecture should look, rather than how it should perform. The intent of this paper is to dispel these preconceptions by addressing the iterative design process through conceptual parameters of pattern and module creation, population, and manipulation in an effort to provide students with an alternative perspective for how architecture can create a dialogue with its environment and facilitate the transition from "thing seeing" to "pattern seeing." In two projects examined in this paper, students were asked to initiate the design process by exploring and testing the inherent strengths and constraints of pattern formations through the lens of a scape and constructed assembly, each of which would focus on principles of ambiance and spatial qualities, rather than relying on traditional architectural typologies. By developing these conceptual projects, students were able to break away from their preconceptions of what architecture is, and instead focus on what architecture could be with a reinvention of the design process.

1 Introduction

Acknowledging, employing, and manipulating patterns allow beginner design students the opportunity to give order to a system and test its constraints and capabilities, as influenced by the behaviors of constituent parts, to produce a performance within a composite assembly. Patterns have a measure of logic and rhythmic predictably due to the use of similar or modified elements arranged in an anticipated manner. There is also a degree of pliability within patterns, as inherent strengths and flexibilities allow them to differentiate, respond, and adapt to introduced criteria, demands, and desires, while remaining compositionally and systematically cohesive. Discovering all limitable bound conditions within pattern

* Corresponding author: kmbarry@bsu.edu

configurations and alterations can encourage beginning design students to focus their attention on the realized parts-to-whole relationships towards appreciating how things behave and respond to one another internally, rather than championing the external appearance of their creations as they begin to create various ambiance settings within environmental design.

2 Pattern as order, function, and design

As designer and theorist Gyrogy Kepes introduces his essay Thing, Structure, Pattern, Process: "By deepest instinct, we reject chaos and try to order our experience. We seize upon means to bring coherence into the welter of signals furnished by nature as though our life depended on it--as it does" [1]. Kepes' ideas propose that inherently, beginner designers understand the world through known signs, symbols, and typologies, and it is only when they break from this preconception that they can fully understand and effectively communicate design. Paul Anderson and David Salomon expand upon this principle by introducing pattern as a way to deconstruct preconceptions and transition to the examination and construction of an unknown. They suggest that patterns can be multi-functional in their inherent capabilities and break from a canonical trope of architectural design, asserting that "Instead of form following function, patterns produce performances" [2]. Anderson and Salomon emphasize the performance or use of the object as that which matters most, as patterns do not simply relay identity, but become an active agent in the design process. The dichotomy of an object's use and its perception therefore is an instigator in the design, and consequently aesthetic qualities become the result rather than the driver.

Using the theories presented by Kepes, Anderson, and Salomon, the examination of understood or preconceived context can result in the deconstruction of known entities into fragments or patterns, and can encourage the reconstruction of these elements into a built environment within a new and innovative context. Deconstructing this environment into two elements, "constructed assembly" and "scape," helps to break down known typologies for beginning design students and focus on working within patterns, which can then purposefully interact to develop a contextual ambiance. As Kepes suggests, this allows students to move beyond "thing seeing" and toward "pattern seeing" [3]. Analyzing patterns, through the means by which the constituent elements are populated, deployed, and encouraged to advantageously interact, helps students identify inherent fundamentals of design such as connection and order, which can then be used in operational developments.

3 Pattern seeing as ambiance

To facilitate the initiation of new design iteration through more conceptual strategies, two projects given to beginner architecture students were developed to help students deconstruct their preconceived "thing seeing" and conversely transition to "pattern seeing" as a way to conceptualize ambient space. Each project required the design of a constructed assembly and accompanying scape through modules or patterns that would sensitively interact for a total user experience. Through an analysis of constraints and opportunities, students were encouraged to give the pattern, or module, agency in how it behaves, reacts, or responds to given criteria or the designer's intentions. Ambiance, therefore, is found in the understanding and purposeful sculpting of the scapes through the implementation of lighting, passage or movement, and exposure or enclosure strategies. As part of the design process, ambiance questions the interactions of these paired contexts and is driven by the way that the module or pattern responds, and not by an imposed or preconceived aesthetic ideal.

Addressing design from the perspective of a module is one method of providing constraints and breaking down preconceptions for students to work within when initializing architectural design. In a project titled 'Part to Whole,' students were asked first to develop two disparate three-dimensional modules, or tessellations, and analyze them for constraints and possibilities with regard to *flexibility*, *connection*, and *strength*. Developing the modules as an individual action outside of a greater architectural design process encouraged students to think about the module as an individual actionable element with its own design criteria before thinking about its application as part of a total design. Students were not given any instruction on what the modules would later be used for, but instead were encouraged to think about the development of the small, repeatable object for its own outlined qualities, resulting in modules that responded to the characteristics given. As this project occurred early in the semester, students struggled with abstracting their ideas about the module from greater architectural thinking, and had difficulty focusing on the modules as individual elements and not as part of a greater scheme. Despite the hesitation felt, this ensured that students were not influenced by how the module would eventually be used and instead forced an individual design process centered on the element.

Following a critique of the modules with regard to flexibility, connection, and strength, students were asked to adapt the modules and then multiply them to be utilized in an ambient design where one module became an architectural constructed assembly and the other became an accompanying scape. While the modules were able to be adapted to better react based on the critique, they had to maintain the "spirit" and general aesthetic of the original modules to ensure that the students did not develop totally new designs based on the new requirements given, which would have negated the secrecy in the first part of the project. While difficult for students, blindly developing the module first was integral, to teach the students to design within restrictions and adapt based on critique. This ensured that the approach was on the part-to-whole development based on the reaction of the module and not on aesthetic intentions from the outset.

The second portion of the project required students to create a structure/enclosure/building and accompany scape using the adapted modules as the new constraint, and to think about the inherent qualities of the modules and their impact on space, with regard to ambiance. The new design process asked students to conceptualize ambiance through the purposeful application of lighting strategies, scale, and, enclosure, all of which should be facilitated by the design of the adapted modules. If the modules themselves were too enclosed and did not allow for light permeation individually, then spacing them out in the design to allow natural light through the offset of multiple modules was required. This presented a way of encouraging pattern seeing within the new building as a whole, suggesting to students that by creating multiple patterns with the modules that they could achieve an ambient space through the characteristics given of light, scale, and enclosure in different ways.

As a strategy to avoid common replication of patterns and encourage the adaptation of ambiance in both the structure and scape, students were asked to establish their own scale and program, and ensure that the building and scape had a sympathetic relationship. Rather than fall back on the common beginning design student preconception that buildings and landscapes exist independently and minimally react to each other, the project required that the structure and scape influence each other in multiple ways and to be designed to accommodate each other, suggesting that the ambient qualities of one could be carried into the design of the other. To reinforce this, both the scape and the architecture were viewed as permeable, emphasizing a three-dimensional relationship that would enhance experience, ambiance, and program.

The resulting designs exhibited a variety of strategies in the creation of ambiance through the modules, with some students choosing to identify a recognizable architectural program and work toward the creation of ambiance based on precedent, and others developing the building for ambiance unrelated to an established typology and instead focused specifically on the opportunities provided by the developed modules. Ultimately, the projects that used the modules as the design inspiration were more successful in creating ambiance qualities in the building or scape. The projects that relied on an existing building typology struggled overall to break away from preconceived notions about architectural design, as emphasized in the two parts of the project being assigned independently.

Student Jesse Lindenfeld's project [Fig. 1] embraced the idea that the landscape and building could both be structural and the design facilitated a difference in lighting and spatial qualities by embedding portions of the building within the permeable scape. The resulting design was strategic in its programmatic development of a library and maker-space, where the ambient qualities of the library required natural light, but the maker-space was embedded into a faceted landscape and purposefully fractured into different interacting programs. The design encouraged patrons to move between enclosed and open spaces, light and dark spaces, and into and out of the landscape as the program dictated, providing multiple ambient opportunities for exploration. The spaces were created through a three-dimensional module that was repeated and scaled to produce the building element and a folded module to create the landscape, which was flexible. Jesse's impression of the constraints was similar to many of the other students. He noted: "The constraint that provided the most challenge within this project was that of flexibility. The modules created had to be flexible enough to orient in different ways but also be manufacturable and sturdy. The way that I resolved this issue was starting from basic geometry, duplicating and folding from a base equilateral triangle. This repetition at a non-orthogonal angle allowed for easy reproducibility and interesting formal qualities" [4].

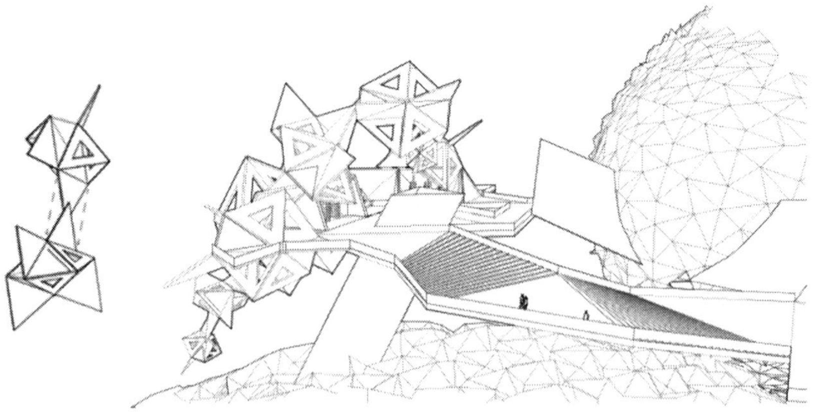

Fig. 1. One developed module and the corresponding building, incorporating a folded scape and constructed modular assembly as landscape and building for project 'Part to Whole'. Project and images by student Jesse Lindenfeld.

Student Darrin Shedrow developed one of the more complex but adaptable modules for his building, which allowed for substantial flexibility in the repeated patterning that eventually made up the structure. By manipulating the connections between the modules, Darrin was able to open up the building to light in different ways and design a curvilinear structure that was impacted by site wind conditions. The landscape module, however, was much simpler and did not allow for the same flexibility in the design, resulting in less interaction between the landscape and building than the other two projects [Fig. 2]. All used ambiance and feeling within the space as inspiration behind the design instead of relying on building typology to initiate design, resulting in projects that were unique and creative in their approach for the first architectural project completed in their educational career.

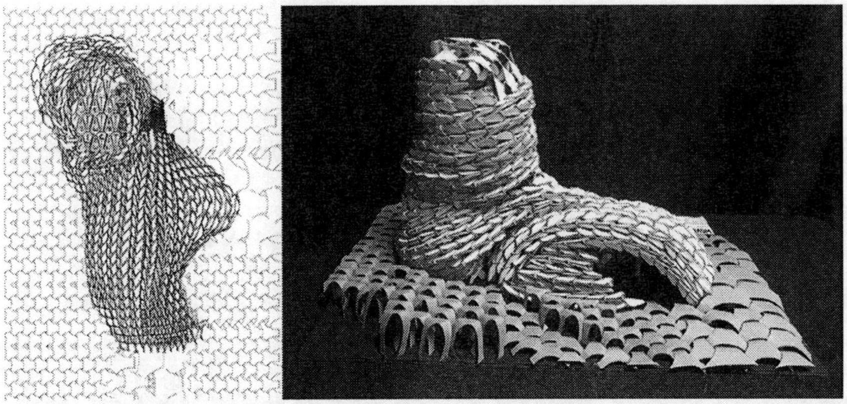

Fig. 2. Project module and diagram showing the flexibility of the constructed assembly module and scaling of the scape module to adapt to spatial requirements for project 'Part to Whole'. Project and images by student Darrin Shedrow.

4 Performing patterns towards creating a contextual ambiance

As with developing an adaptable module, folding is a procedure that demonstrates intention and consequence within the manipulation of an uncompromised membrane. The act of folding is indicative of creating or interrupting established pattern formations to achieve a desired response for a surface. A project entitled 'Pliant, [Dis]Engaged, Captured' asked first-year environmental design students to blur the concepts of scape and constructed assembly by studying the performance capacities and compositional organizations of pattern formations as a means to create an ambiance within an existing underappreciated site. By considering folding, with its inherent consequences of seams, triangulated creases, pleats, and divisions, students created or manipulated pattern configurations as a procedure to capture and sculpt spaces with coveted programmatic intentions at a 1:1 scale. A separate structural framework system, which was responsive but disengaged from the folded skin at various moments, was implemented for the project. The structural framework and folded skin were required to sympathetically acknowledge and join one another in a manner that achieved a symbiotic relationship between the *skin* and *bones* within their project. As a result, folding became a vehicle to entice students to acknowledge the responsive behaviors within pattern formations, allowing the students to discover ways in which manipulated pattern formations might mediate light infiltration, capture and sculpt spaces, and suggest a new purpose for a site through creating ambient settings within their projects.

Working in teams, students began by studying the work of geometrist Ron Resch [5] to gain a better understanding of how folded tessellation systems allow paper to become both structural and flexible, as dependent upon the arrangement of various pattern formations and repetitive seams. Student teams translated their findings through iterations of folding operations to create a series of surfaces that explored the conditions of seams, triangulated creases, pleats, and divisions of a paper as informative elements to the overall surface membrane of a single sheet of paper. Additional sheets of paper were able to be included, only if attached along the full length of a creased edge, to examine how various strategies of combining multiple pattern formations might offer potentials for enclosure if imagined at a larger scale. As a final step, the students were asked to develop a framework system of *bones*, that would respond to their folded paper *skin* model by carrying forth its inherent geometric logic beyond the created surface as a means to provide additional structural support strategies for consideration within their designs [Fig. 3].

Upon completion of their preliminary studies, the student teams selected a site that was currently deemed uninformative in its programmed intentions and underutilized as a defined space within the interior confines of the College of Architecture and Planning at Ball State University. After observing the sites, the teams were tasked with activating the spaces by creating a 1:1 scale installation comprised of *skin* and *bones* to create a thematic ambiance using the folding operations and tectonic framework strategies learned in the previous explorations. So not to damage the existing conditions of the site, the students were not allowed to use any secondary means of connecting, bracing, or adhering their installation to the site beyond the direct contact of the framework or folded scape to the existing surroundings.

Fig. 3. Initial folded paper and tectonic structure studies by first-year environmental design students for 'Pliant, [Dis]Engaged, Captured' project.

First-year students Hao Xu and Nate Brower chose an interior site at a corner condition within a collaborate space of the building. The student team's initial analysis of the site determined that the space prohibited light intervention, was spatially undefined, and ineffective in promoting students' abilities to communicate and/or study. To address these concerns, Xu and Brower utilized the folded surface within their project to define and sculpt a space that encouraged peer interaction and discussions. The gesture of the skin was such that it followed the intended movement of the users as they traversed among the repurposed space. The formal gesture of the skin commenced by creating a portal overhead to invite users within the sculpted space. Upon entering the site, the folded membrane then descended to a establish a more private setting with an accompanying built-in seating area. Upon this moment of pause, the skin then ascended to lead the users out of the project and back through the created entry portal. The student team utilized a singular folded strategy for the geometric pattern formations of the skin and sympathetic bones, yet modulated and adapted the size of this pattern to best achieve the desired formal gesture along the path of travel for the project. Further, as the absence of light was noted as a flaw of the original environment of the site, the students integrated artificial lighting within the interior envelope of their folded skin and removed panels of their skin's undulating pattern at specific moments to reveal the interplay between the tectonic bones and the folded membrane [Fig. 4].

Fig. 4. Final installation design and fabrication and detail of symbiotic connection between skin and bones by first-year environmental design students Hao Xu and Nate Brower for 'Pliant, [Dis]Engaged, Captured' project.

The student team of Mia Baker and Kirsten Toole adopted a portion of the building's interior four-story tall atrium that was contiguous to the architectural library as their site for the project. Upon studying how people circulated and interacted within and around their designated space in its existing setting, the student team decided to focus on redefining the boundary conditions, in the x-y-z axes, to establish a more intimate scale as their overarching strategy and repurpose their site as a place to read and contemplate, given its proximity to the library. To achieve this charge, the students studied several iterations of scaled study-models to strategize various means to soften the height of their site, as well as capture and define a personal space that was to be shielded from other activities within the atrium. In its final configuration, the pattern formation of the folded skin was predicated on its ability to pivot at specific creases to achieve the form of the newly-established sculpted demarcation. Ultimately, the students' design of the skin was such that it emerged from the ground, utilized the tectonic framework bones to brace itself to a low-height wall near an open staircase, and then reposition itself to create a canopy overheard of the implemented seating area. The formal gesture of the project was further influenced by the project team's desire to mitigate the noise of the atrium, protect the inhabitant of their site form the wind emanating from the front exterior doors of the atrium space, and deflect the abundance of outdoor light at the newly created seating area [Fig. 5].

Fig. 5. Folded paper process study model and final installation design and fabrication by first-year environmental design students Mia Baker and Kirsten Toole for 'Pliant, [Dis]Engaged, Captured' project.As commented by student Mia Baker, "The project demonstrated that design does not always commence with trace paper, but instead, it can start with observing, listening and developing a design language. Our team used a process of making and collaborating, all new skills to us at the time, to help cultivate a specific ambiance based on our personal shared experience of our given site. I learned how to abstract and reinterpret structure and surface as both interconnected and interdependent systems within a compositional design for an intended environment. Further, as the project was at full scale, it allowed us to directly and physically create and refine our solution to encourage users to interact with the created scape" [6].

Conclusions

Each of these projects introduced principles of pattern manipulation towards conceptualizing architectural design as a marriage of elements associated with the ground and among the sky. Having students conceptualize ambiance through these principles is integral to their development as beginner architectural students, as it introduces inherent qualities or principles of design and their influence on the final product, focusing on the iterative process and emphasizing concept and reaction as developmental tools. Establishing parameters ensures that students apply rigorous analysis of behaviors in the creation of the scapes, rather than allow aesthetics to drive the process when designing ambient space.

References

1. G. Kepes, *The New Landscape in Art and Science*, 204 (1956)
2. P. Anderson, D. Salomon, *The Architecture of Patterns*, 33 (2010)
3. G. Kepes, *The New Landscape in Art and Science*, 204-207 (1956)
4. J. Lindenfeld, "Part to Whole." E-mail message to primary author, (December 3, 2018)
5. R. Resch, E. Armstrong, Paper and Stick Film, (1970)
6. M. Baker, "Pliant, Dis[Engaged], Captured Project." E-mail message to secondary author, (November 19, 2018)

Teaching a thorough, repeatable design process for ambiances in architecture

Thomas M. Lesko[1]

[1]College of Architecture, Design and Management, Wentworth Institute of Technology

Abstract. Le Corbusier, it has been said, defined architecture as "the poetry of construction" and as such, the term "ambiances" is reminiscent of those great words. In order to create that special quality, that poetry, that ambiance in architecture, a design concept that is the result of a thorough, repeatable design process must be employed. The purpose of this paper/presentation is to explain the design thinking/design tools we need to teach our students, i.e. the future Le Corbusiers of the world, how to create those special qualities that make ambiances/poetry in architecture possible.

A design process for architecture, or other design fields for that matter, begins with seeking a full understanding of the problem, which can be broken down into several identifiable and teachable aspects, and ends with the generation of viable and defendable design alternatives. A study of the processes utilized by great architects shows that each goes through a similar process, one that begins with steps that define all facets of the problem, including context, program and client issues.

Several cases in point will be used in this paper that illustrate, by professional examples, what great architects do regarding process to create the beautiful ambiances inherent in their artistic creations. For this paper, selected works by Herzog and de Meuron and by Renzo Piano will be used; these are architects that students normally have a familiarity with. I chose these selected projects because of the "perceived" difficulties that each problem presented. Those difficulties can become poetry of course… problems are most often opportunities for design expression. How does one design poetry into a simple warehouse on the site of a former rock quarry?

Herzog and de Meuron, when they created the wonderful buildings for Ricola, in Laufen, Switzerland, found the special "design generating" data that helped them design buildings having that special quality. The design process that they utilize seeks out those important elements of ambiance-related, ambiance generating knowledge.

Fig. 1. Herzog and de Meuron, Ricola Warehouse, Laufen, Switzerland

Fig. 2. Herzog and de Meuron, Ricola Marketing Building, Laufen, Switzerland

Renzo Piano finds the special ambiance-related elements of his projects by utilizing a design process that also seeks out those special pieces of information. How does one design poetry in an urban context, when it is necessary to extend the work of another great architect? In the expansion of the High Museum of Art in Atlanta, Piano had to deal with an existing context that was not only the work of Richard Meier, who designed the original museum, but also surrounding buildings of performance art and art studios.

These are both situations that students can relate to because they struggle to understand the problems given to them just as these examples are difficult to see where the poetry can come from.

Fig. 3. Renzo Piano, High Museum Expansion, Atlanta, GA

Each of these examples presented an unusually difficult challenge for the designers. To be most valuable to students, one must show 'how' they were able to be successful and 'where' the design inspirations came from. Doing this provides the invaluable tools that form a solid process for design and serve to illustrate just what great architecture in the twenty first century.

1 Definition of Architecture and Abstraction

It is important to define architecture in words that are not arbitrary opinion statements. To that end, I state that architecture is an abstract visual representation of elements of context and elements of program. Moreover, it is also necessary to state in a non-arbitrary opinion statement what abstraction is. Abstraction is the removal of blatant, obvious visual information so that more information and meaning can be included. One does not need an entire tree to communicate what the essence of the tree is.

This presentation gives students the tools to do that work, presented in a definable, teachable, eight-step design process that yields poetry. . . ambiances in architecture.

1.1 Step one: Background and Context Data Collection

Problem-"seeking" is a valuable term in the studio context because it not only defines a necessary part of a design process, but it also raises awareness to the fact that when a problem is presented to students and architects, it is not the complete problem. There are multiple facets to each and every problem, and they all need to be defined and explored. Each and every piece of a problem holds design ambiance generating potential.

Along with an initial familiarization of the problem, an initial familiarization with the context provides an overall picture of where the design will be located. This is an invaluable "stepping into" the problem step, without putting pressure on a student to come up with initial evaluations, students can allow their senses and intuitions regarding the context to come into play. Every project must begin at some point; typically, students will begin by reading the project statement given by the professor, visiting the site, taking pictures and gathering site/contextual information that will be put onto a site plan in no organized fashion, but perhaps with some good graphics to make it look the part. This is a start, but hardly an analysis or good background. A project background statement should explain the origins of the problem in detail, the desires of the client and any information that is available to the problem at the beginning of the process, drawing on data collected and organized well.

1.1.1 Data Collection

Once familiar with the context and once the problem has been defined in full, more detail is sought. Providing seventeen different categories for data collection, with some prodding descriptions to get students to think through each, provides a great frame for this segment. Not to be confused with what some call "site analysis", a thorough collection of context data provides the opportunity for students to fully understand the context and subsequent idea generating potential of each category of data.

Context analysis is a two-part process: the collection and organization of data into distinct categories and the subsequent analysis of the data. The data analysis can take the form of a design idea generated by those data or can be an answer to the question "so what?". So what could the data mean to the problem? How could it be utilized? Is it a good thing or a bad thing?

The site and its context should be analyzed by first collecting data, and then, in a separate step, analyzing those data, an often misunderstood or omitted step. Data should be collected in at least the following categories: history, site location, site size, site legalities, physical neighborhood, sun direction, movement and effect on the site, existing vegetation, topography and drainage, built elements found on the existing site, pedestrian circulation, vehicular circulation, noise, views from, to and through the site, and human, cultural and social elements relating to the site. [1] The information found in those categories should be organized, and then, in a separate step, responded to with words and sketches of analysis.

1.2 Step Two: Context Data Analysis

Typically, students are asked to use one sheet of paper for each category above. This acts as an organizing element. The paper is divided into two parts, the top has a small picture or sketch of the site and the bottom also has the same picture or site sketch. The top is where the data is placed, relating to the site with arrows or other graphics. The bottom is where the "responses" go. This is the analysis. Without some response to the data, there is no analysis. The results from a pool of ideas from which the concept may be drawn. Because this produces many pages of well-organized information and analysis, an overview is written for the actual concept statement. The following is the example sheet I give to students:

 https://doi.org/10.1051/shsconf/20196402006

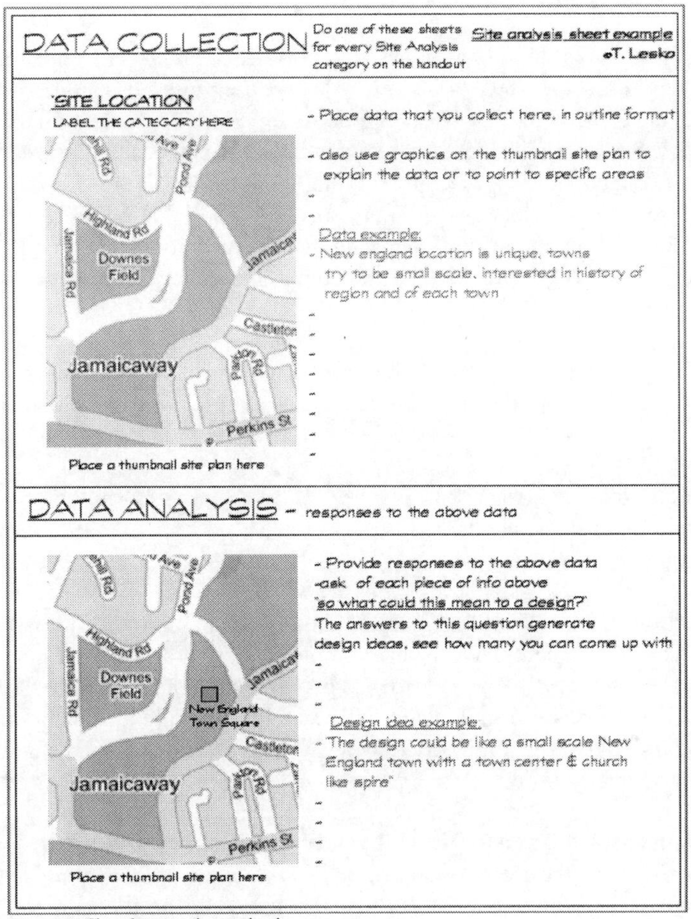

Fig. 4. Context Data Collection and Analysis

As mentioned, some professors advise students to do a "site analysis" that only consists of information gathering and perhaps a graphic site plan with information on that plan. Unfortunately, with that approach, there is little analysis done. It is with meaningful analysis that idea generation in its full glory can be accomplished – searching for ideas with the potential for the creation of the poetry of construction, as Corbu would say. With pages of data collected, there is one key to idea generation and creating design alternatives from the context. That key is to ask "So what" of the data. So what… does it mean? So what… 'could' it mean? With words and necessary thumbnail sketches, literally dozens of ideas will be generated and the search for the ambiances that make great architecture is easily accomplished.

1.2.1 Design ideas from the context

With an abundance of design ideas, some method of judgement needs to be employed. What is most important? What is yielding the most excitement? What satisfies the designer's artistic imagination? Qualitative and quantitative questioning the ideas by the designer, from both contextual analysis and program/problem analysis, one can simply state design goals. The stating of design goals clarifies the design direction and moves toward the clear statement of a solid, meaningful concept statement.

1.3 Step Three: Idea generation – creating design ideas from the problem:

With the belief that great architecture in the twenty-first century is an abstract expression of context and program, there needs to be a phase where the program is "mined" for architectural expression. In each program, there exist important elements or qualities that can be discerned from the others. Searching for these "profound qualities" is the step required to move toward a series of working concept ideas. For example, the Ricola warehouse, was to be… a place to stack and store the products of the company. By itself, this does not sound profound, however, if seen through the lens of poetry and ambiance, a profound quality was stated and explored. How can storage and stacking of product be artistic? A patient search yielded the result.

With each design problem, we must be able to find and state some 'profound' or otherwise 'special' quality to build a concept around. These qualities will be found in the following categories:

- site/background analysis,
- the particular user or client,
- the project.

These qualities should be stated clearly, elaborated on, and will be part of the decision process to develop a concept. Here is an example of stating a profound quality by Terrence Harkness. It is in reference to a project to design a garden in the relatively flat land of Illinois. This example is used because students initially see little to build a concept upon – the land is flat farmland, and little is seen initially to work with. It provides a good example of what can be done with apparently little.

"Living in the Middle West (of the U.S.) and experiencing it's particular light, the expanse of sky and horizon and the effect of sunset on the flat plane of farm lands – these have been particularly powerful components of my visual sense. During the summer, impressions of changing field patterns have provided experience in spaciousness, incised pattern and exploration…"important qualities present in the landscape of east central Illinois: the richness and detail of winter trees, for example, seen against the sky. Ground fog of late fall shrouds the intermittent lines of hedgerows."[2]

Fig. 5. Site Image

Sketches were made to explore the qualities presented and to translate the written qualities into graphic images which can be used to form design ideas. The translation of written words into visual images is a most important step in design. And it is addressed after the concept pyramid is created. These sketches take the first steps into that translation.

Fig. 6. Site Sketch A (Harkness)

Fig. 7. Site Image B (Harkness)

Fig. 8. Site Ariel Photo (Harkness)

Fig. 9. Site Interpretation Sketches (Harkness)

1.4 Step Four: Expressing Personal Vision:

Each designer possesses a background and personal vision that is unique. It is important that students understand this and are able to make some statement about what they love… what captivates them regarding the design that they are about to create. If they have trouble stating this, it serves as a reminder that this is something missing from their development and must be addressed. This uniqueness is what makes a design wonderful and personal. An example of this kind of statement for the above garden design:

"One's vision of the world is part of the feeling behind – and subject of – much of one's work. This garden expresses my deep affection for the Illinois landscape. As expressed in this garden, the vision originates in the rural Middle West. The garden reflects my preoccupation with the visual qualities of these places, and my attempt to confirm and reveal them to others. Its creation has required a conscious acknowledgement of the importance and power of both past and current experience." [2]

1.5 Step Five: Choosing a Design Direction:

The information in steps 1, 2 and 3 provide the basis for a design concept. Once that is achieved, a design "direction" can be chosen. This is not a concept necessarily, but is identifying an area of interest. Students choose a "design direction" based on the important elements, and then explain how that choice comes from something in categories 1 and/or 2 above. They especially build on one of the profound or special qualities found above and describe it in a paragraph.

Words or themes must be identified that can be translated into some physical form. For example, if we were to design a "Fire Station", we could analyze the context and analyze the project, the building type and the clients… but we are still wondering: what will the fire station look like? What is its architectural statement, its concept?

Using the example of the Illinois garden, a design direction would involve the use of the unique landscape of that region, specifically the flatness of the fields, the looming horizon, the straightness of lines forming roads, fields and fences as explored in the sketches. These elements are periodically punctuated by intersections where trees, houses, barns and silos exist. The choice of dealing with these elements defines a direction, and also defines elements that can be transformed into design elements – the verbal to written translation. A concept statement can now be formed.

1.6 Step Six: Forming a Concept Title

The concept title provides the beginning of and the ultimate "focus" for a concept, it sums up the concept in precious few words. By defining a title, a statement is made: "my design will be this". It may change, however: a stand must be taken! A decision must be made and subsequently explored to find its image. Here are two ways to write a concept title. Students define a very short title for their project using one of these methods:

#1) The Simile/Metaphor method

- An example of a concept title using the "simile/metaphor" method using "As or Like" to start a concept title:

Concept Title Example:
"The Illinois Garden *as* a Regional Landscape" or
"The Illinois Garden *like* a Series of Regional Elements"
(I provide definitions for students who should but may not know what these are)

Simile: a figure of speech comparing two unlike things that is often introduced by like or as (as in: "cheeks like roses")
Metaphor: a figure of speech in which a word or phrase literally denoting one kind of object or idea is used in place of another to suggest a likeness or analogy between them (e.g.: drowning in money). Moreover, metaphors are comparisons that show how two things that are not alike in most ways are similar in one important way. Metaphors are a way to describe something: "Cindy was such a mule. We couldn't get her to change her mind."

#2) The Title: subtitle method

- An example of a concept title using the "Title: subtitle" method:

Concept Title Example:
"An East Central Illinois Garden: A 'Regional' Garden"

1.7 Step Seven: Defining Project/Design Goals

The definition of project goals is a difficult one for students, however, providing them with a good example that they can compare their statement to is quite helpful. In reality, the actual concept statement is imbedded in this section. It is an elaboration of the concept title and is about the student's design direction, it is not about the program at this point. Again, using in part the Harkness project statement as a good example of a goals statement:

"The goal of the East Central Illinois garden is to recreate the extreme flatness of the landscape – the strong horizon, the essentially open and treeless terrain, the tension between the crispness of edge and the warping of the land plane... The patterns are seldom symmetrical but are all contained in the Jeffersonian square-mile grid." [2]

1.8 Step Eight: Convey the Project's Purpose

At this point, the purpose of the design should be stated. There are subtle differences between the goals above and the purpose. The purpose of the project is further elaboration of the concept title, but it also should include programmatic information. This combination of concept and program is an important junction, one that will be explored with design sketches and helps make that important leap from verbal information to graphic imaging.

"This garden's purpose is to reflect these characteristics (the profound qualities, referenced above, ed.) as a visible, built expression, to be experienced through time, seasons and changing light. The garden focuses on and celebrates the landscape; it should demonstrate the power and qualities of this unique place. The design sources for this garden are in the landscape itself. The garden and the landscape are inseparable." [2]

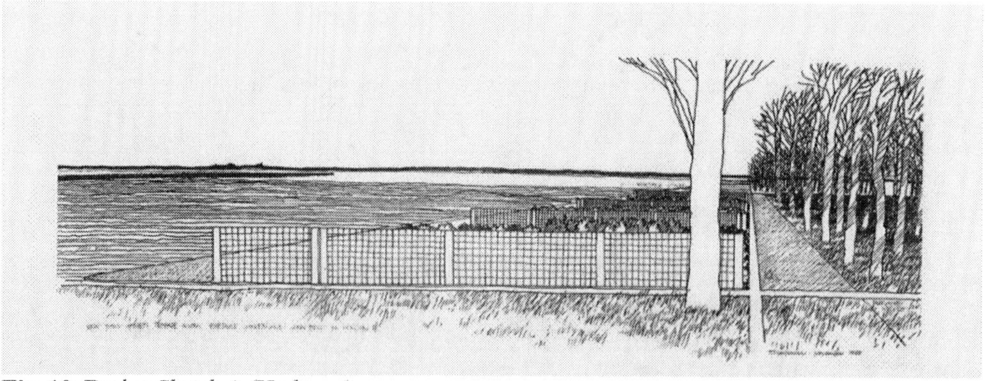

Fig. 10. Design Sketch A (Harkness)

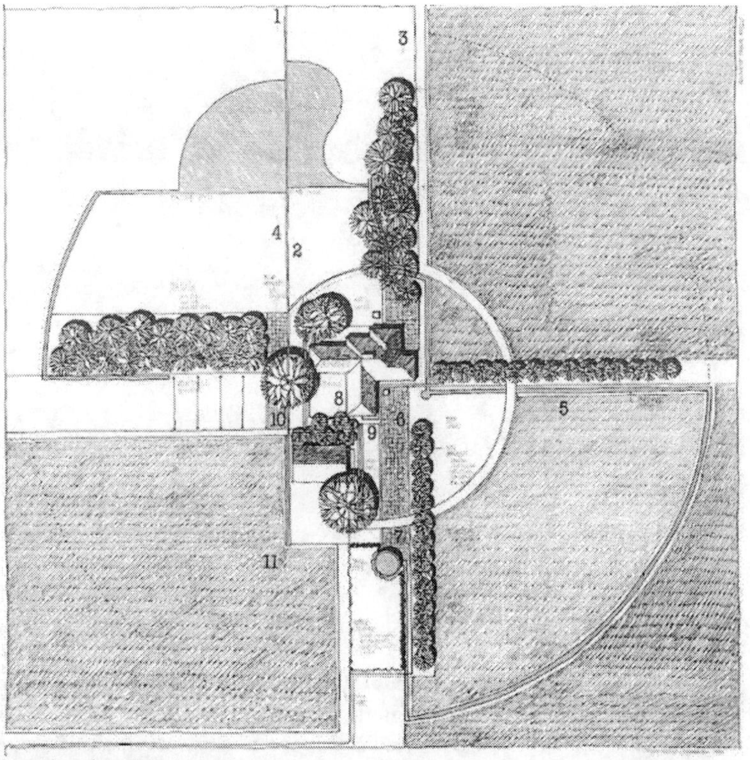

Fig. 11. Design Sketch B (Harkness)

The above steps are a definition of what steps designers go through to design. Based on the belief that architecture is not an arbitrary thing, that architecture is an abstract visual representation of elements of context and elements of program woven together with the art and skill of the individual pursuing personal vision and/or ideology. Moreover, that abstraction is the removal of blatant, obvious visual information so that more information and meaning can be included. It has been said that design is a patient search, so, now that the problem has been elaborated and ideas, goals and concepts have been identified, the search begins (continues?). The generation of alternatives – of possibilities that bring to life the ideas, goals and concepts defined is now the task at hand. If one believes that the work done to this point is valid, that it is correct and exciting, sketching possibilities of those things are the final part of the design process. The design process is identifiable and deliberate and illustrates that good design does not fall from the sky in a burst of light, it takes work. That patient search will yield countless possibilities for the creation of architectural poetry.

References

1. White, Edward T. *Site Analysis*. Architectural Media Ltd., 1983, Tucson, 1983, pg. 26. Print

2. Harkness, T.: 1986, "An East Central Illinois Garden," Transforming the American Garden: 12 New Landscape Designs, Cambridge, MA: *Harvard University Graduate School of Design, Exhibit Brochure* pg. 40

Envisioning the atmospheric effect through (im)materiality

Nilay Ünsal Gülmez[1], *Dürnev* Atılgan Yağan[2], *Murat* Şahin[3], *Efsun* Ekenyazıcı Güney[1], and *Hande* Tulum[1]
[1]Bahçeşehir University, Department of Interior Architecture and Environmental Design, Istanbul, Turkey
[2]Bahçeşehir University, Department of Architecture, Istanbul, Turkey
[3]Özyeğin University, Department of Architecture, Istanbul, Turkey

Abstract. In an attempt to bridge the gap between architectural/interior design practice and education, 'atmosphere' as a prolific contemporary architectural debate in practice and theory is covered by the experiment of 'Staging Poe' carried out as a first year Design Studio through the study of Edgar Allen Poe's selected poems. Poe's 1846 text of 'The Philosophy of Composition', unfolding his analytical method of writing and emphasis on "effect" in poetry, provides a ground for experimenting with facets of materiality and structuring the studio. Aiming to cultivate intuitive design experiments of students into informed processes in hybridizing conceptual/textual and material/sensual aspects, studio is structured in two phases. In the first phase, "materialization", idiosyncratic interpretations of students from words to materials with a focus on tectonic experiments and haptic experiences are sought in between materializing and dematerializing processes. In the second phase, the "atmospheric", emphasis on dematerialization of the perception of materials through tools, such as light, color and sound is exercised to transform the object into a performance stage. Outcomes of the studio on aspects pertaining to material and materialities in creation of the immaterial that is the atmosphere is followed and evaluated through responses of students' weekly reports.

1 Introduction

The concept of the atmosphere is very much on the agenda of contemporary architectural debates revolving around materiality and immateriality through projective theory [1, 2]. Even though all attempts at re-sensualizing architecture, there are multiple approaches to presence and production of atmosphere in design disciplines. In professional practice, one can mention the literal inclusion of infrastructures making the immaterial air an atmospheric design element as in the case of Blur Building by Diller Scofidio + Renfro [2] and projects of Philippe Rahm [3] that can be traced back to the campfire phenomenon of Reyner Banham [4]. On the other hand, there is the condensation of sensorial aspects of space and materiality which gives room to haptic experience (tangibles) alongside all the other senses as in the works of Peter Zumthor, or immaterialization efforts through light, illusion, transparency and dissolving of the boundaries that can be observed in the works of Olafur Eliasson, Toyo Ito and SANAA with a special focus on vision and auditory senses (intangibles) [2].

However, within pedagogical contexts, it is possible to claim that such re-sensualizing attempts have not yet been incorporated enough into design education. Departing from the necessity of bridging the gap between architectural (more specifically interior) design practice, where atmospheric thinking is already influential, and design education, this paper covers the presentation of the process and outcomes of a project-based experimental study carried out in a first year Design Studio aimed at materializing/immaterializing the atmosphere of Edgar Allen Poe's selected poems. Poems; their verses and words are thought as conceptual appetizers for the making of atmospheres.

Two main questions will be debated in relation to the outcomes of this first-year studio experiment: First, how to think on the relationship between the conceptual/textual and the material/sensual through the concept of atmosphere and second, how to allow possible tools and opportunities to hybridize those seemingly opposing tenets while structuring the studio.

2 Conceptual vs. Sensual or the Search for a Hybrid Becoming

Relationship between the conceptual/textual and the material/sensual has always been and is still the main point of debate in architectural theory and in its epistemological origins. Immanuel Kant, as a key figure to differentiate the physical world as 'noumenal' and the world experienced as 'phenomenal', offers the possibility of understanding the object by a subject through experiences, unfolding how our bodily contact with the world and the conceptual thinking complement one another in explanation of knowledge [5]. Stretching from Kant to two "Peters" in 'Peter versus Peter: Eisenman and Zumthor's Theoretical Throwndown', A.J. Artemer (2016) presents the split in contemporary architectural scene, diagnosing a battle between Zumthor, and Eisenman as '*... between phenomenology and autonomy, emotion and rationality, between materiality and concept, and being in the world and transcending it*', respectively [6]. Meanwhile, Antoine Picon points out to a materiality as a hybrid between seemingly opposing categories of abstract and concrete in the age of computer [7]. At present, a paradigm shift to reinstate the authority of making (praxis) in design and to find alternative ways to heal the split between thinking and making like integration of architecture/design with visual arts and performing arts is creating an transdisciplinary platform that can be observed in a multitude of current debates on the concept of atmosphere and its various understandings in relation to aesthetics [8, 9], embodied spatial experience [10], politics [11] and philosophy [12].

3 Unfolding Materiality

The education of architecture and design also took its share from the split between the conceptual and the sensual, the mind and matter. The importance given to unveil the already given information in the world waiting to be decoded with extensive analysis on one hand, versus the belief in subjective creative talent of the designer, the idea of genius on the other have been influential in structuring the design schools and their curriculums. Moreover, from the day education is liberated from the material world – when the act of thinking/ideas considered superior to act making – onwards the education has long been dependent of immaterial presentation of thoughts, the drawings [13]. The Bauhaus school has reminded the significance of 'making' with its emphasis on ateliers and craftsmanship and acted as a melting pot for opposing tenets in its ambitious intend to bring art and crafts, the subject and object, the mind and world together [14]. The technology now is hybridizing thinking with producing, reorienting them as simultaneous acts and facilitating the way thoughts are materialized in virtual or actual environments. Due to the zeitgeist of the era, design schools are welcoming technologically apt students and curriculums are to transform accordingly.

We argue that all the process of atmospheric creation could be regarded as a constant dialogue between facets of materiality (material, immaterial and non-material) and this dialogue shall be incorporated into the education of design disciplines from the very beginning. Yet, this requires unfolding the term 'materiality' to reveal it potentials in cultivating design education.

Pointing out to a change in understanding of materiality from the past to present, Gernot Böhme [14], expresses the change from an idea of integrity and inconspicuousness of materials; into one where 'materiality is to show itself, to come forward, to help shape the atmospheres'. Referring to the valeurs d'ambiance of Baudriallard, Böhme proposes the term 'theatrical value' in describing the new understanding of materiality and its role in formation of character of space that oriented towards 'staging materiality'[15]. He defines three types of relationship with the materials: work-related, perceptual and the medial. The work-related relationship is about the way we deal and experiment with the raw material, the way we search for its nature even force or change it to an extent. In the perceptual relationship, we are involved with the manifestations of materiality, 'the pure form of its appearance' that includes dimensions of physiognomy, its synesthetic and social character. The third one, medial relationship is experiential. Referring to Aristo, Böhme defines it as sensing the environment we exist with our flesh physically that is about to 'remain bodies among other bodies and live within the media'. This is more about the memory of our flesh, rather than touching [16].

The new understanding of materiality embraces both the physical substance of material itself and all the immaterial aspects like (light, sound, water vapor) that add theatrical value to the materials themselves by concealing, augmenting or changing/transforming physical and perceivable features of materials. This broader conception of materiality also helps in rethinking space as a performative domain, a container (void) setting the stage for ephemeral becoming. At the end, all contribute to a (creation of) non-material issue, the spirit of place (the genius loci) described as 'ephemeral and unfocused' related to 'its unique perceptual and memorable character' by Juhani Pallasmaa [17] that approximates to the definition of atmosphere. Such a definition, the feeling emanating from the character of place/space highlights its 'effect' on the viewer/perceiver and refers to the third definition of materiality by Böhme, the medial and experiential one.

Unfolding materiality, the emergence of new tropes in current architectural discourse in tandem with the pace of technology like 'material responsiveness', 'trans-materiality' and 'dematerialization', also helps interacting and approximating the scientific/objective facts with artistic/subjective imagination [18]. Sandra Karina Löschke uses the term 'immaterial materiality' to express the immediate effect of such an understanding of materiality on the process of design of objects to be conceived as "not constructed to satisfy aspects of monumentality, formal style or functionality, but to generate a multitude of temporal perfections and effects' [19]. Dematerialization in its reference to the process, is about an addition of ineffable elements that dissolves the appearance or perception of physical substance of the materials, their character. Toyo Ito's use of daylight and artificial light adding lightness to his buildings pushing them to the edge of dematerialization [20], or SANAA's play with light and translucency creating layers of cloudy surfaces and shadowy silhouettes in space epitomize such practices in architecture.

4 Bridging the Gap in the Beginnings of Design Education

Recently, there has been an increasing tendency towards bringing sensuous aspects of design and related tools forward in the first-year studios. However, the atmospheric issues that we suggest as appropriate in first year design studios to bridge the gap between the mind and the matter; and to unfold the concept of materiality embracing all its components be it material, quasi-material and immaterial have not been widely incorporated into design education. The efforts encountered in the literature so far are mostly addressing senior students such as John Stanislaw Sadar's [21] redesign of an environmental technology coursework to work with atmospheric energies; Gerlinde Verhaeghe's [22] 'mixed media atlas' that challenge' the split between idea and the material' and Sandra Löschke and Michael Day's [23] *Coproductions* graduate studio that focused on the 'interplay between material, light and space.

Cultivating the intuitive elements pertaining to materials and materialities embedded in the already existing experience of anybody, a group of first-year interior design students, in our case, is mediated through the study of atmospheres, that is another embedded experience. Böhme uncovers this embedded experience as a dominant relationship with materials starting from early childhood that operates in the "background memory" and calls it 'the medial' [24]. In studying an immaterial phenomenon that is the atmosphere, the medial experience of students is to be unfolded through to bring out the role of material and materiality into a conscious and informed design process and experimentation for the students.

We argue that, for novice design students, beginning with narratives, some literary texts and/or cult authors/poets they are familiar with might provide a potential for the creation of (im)material constructs that speak to the senses and could be taken as a starting point to bridge the gap between the conceptual and material. For our case, Edgar Allen Poe has been introduced as a key figure to contemplate on the mystery of the beginnings. Anthony C. Antoniades [25], who underlines the fertility of the involvement of "obscure angles" utilized by many arts, such as poetry, literature, music and film, also points to Edgar Allen Poe's techniques of atmosphere and mood creation in prose and poetry to stimulate experiments in design education. Poe and his emphasis on 'effect' in poetry emanating from the 'atmosphere' provided a ground for experimenting with facets of materiality in multiple varieties.

After a thorough analysis of his anthology, five poems having strong spatial atmospheric tones, namely The Valley of Unrest, The City in the Sea, Dream-land, Fairyland, and Silence Sonnet are selected to be envisioned and staged by the students. Poe's seminal text 'The Philosophy of Composition' [26], where he unfolds how he wrote – his famous poem – 'The Raven', has been influential in choosing his poems and structuring the experiment 'Staging Poe': Firstly, through the manifestation of his analytical method of writing with a mathematical precision proceeding step by step to the final goal of transferring the intended 'effect' which contradicts with the so-called assumption that he wrote these romantic poems intuitively; Secondly, through the exploration behind transformation of the raven, main character of his famous poem, the narration of which is established upon its reality, into the metaphor of pain and remembrance that made us contemplate on transitions from the abstract to material and immaterial respectively and their cyclical presence.

5 Staging Poe

The proposed approach for translation of literary effect into a spatial atmosphere/ambiance in the studio is composed of sequential yet cyclical steps that could be mainly discussed in two phases: First, the materialization phase in the search for idiosyncratic translations of students from words to materials with a specific focus on tectonic experiments and haptic experiences oscillating between materializing and dematerializing (forcing the limits of materials or transforming their nature by hybridization processes); second, the atmospheric phase, making use of light and sound to dematerialize the appearance of materials used that accompanies the ongoing tectonic experiments (figure 1).

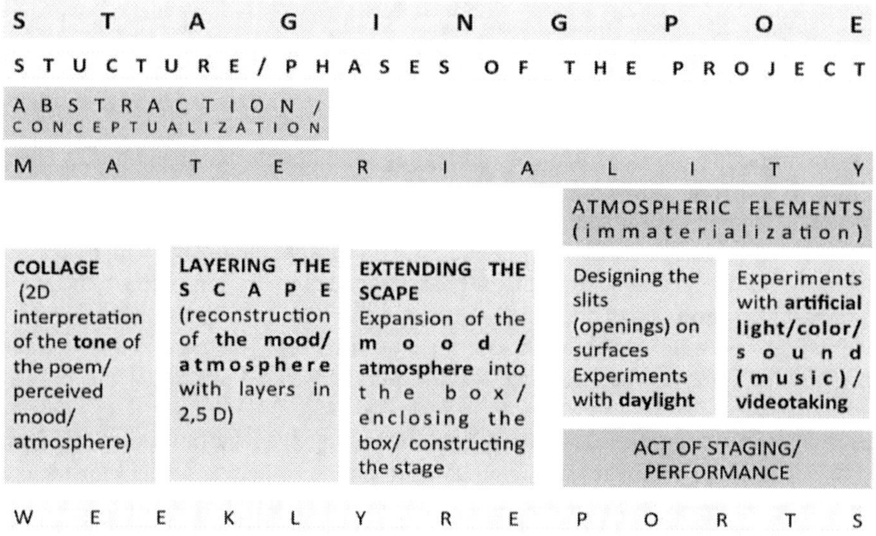

Fig. 1. The Flow of the Project 'Staging Poe'.

The materialization phase starts with (a) collage making and (b) layering the scape that foster students to contemplate on the tone of the poem and their personal response while thinking on abstraction in translating the perceived mood/atmosphere into materials, beginning from two dimensional to gradually develop into the third in the following one, that is (c) extending the scape into a boxy frame. This step paves the way for tectonic experiments that make students get acquainted with the nature of materials, compel their possibilities to the limits and recognize the character of materials. Formal searches for articulating and sequencing elements appear as integral part of material and structural experiments. The material phase concludes with (d) enclosing the frame of the box and opening a peek-hole to be able to have visual access to the interior. Once the frame is enclosed, it leaves the body of the student out limiting the experience of proximate senses, specifically haptic experience of the atmospheric field within.

Overlapping with the materialization phase to some extent, the atmospheric phase begins with (e) opening slits on the surfaces of the box (f) to experiment with daylight, artificial colored-light and with multiple sources of light consecutively (figure 2). The position and amount of slits are repeatedly articulated with respect to the light experiments in search for capturing the imagined atmosphere. This step is followed by (g) the selection of a music piece by the students that would enhance the intended atmosphere inside the box. Light, color and sound act as ephemeral dematerializing agents transforming the whole box-space into a 'theatre stage'.

In order to understand the mutual and sometimes tense relationship between materialization and dematerialization processes and to conceive the ephemeral quality of immaterial atmospheric agents, unconventional tools have also been utilized in the studio. Considering the box as atmosphere/ambiance generator, the students have recorded multiple performances under changing conditions of light and sound/music. Selections of music have been observed to be very much related to the rhythm of their camera moves and envisioned atmosphere of the stage. Affirming Pallasmaa [27], music pieces helped creating 'lived and existential atmospheric interior spaces, ephemeral and dynamic experiential fields'. In its resemblance to the experience of dioramas, students experimented and assessed the atmospheres/ambiances generated in viewing inside of the boxes through the peephole under constantly changing conditions. At the final stage, short videos are taken by the students and shared via a social networking app (#stagingpoe2018) to open up the atmospheric performance of the closed mysterious box to the viewing and comments of all studio participants (figure 3).

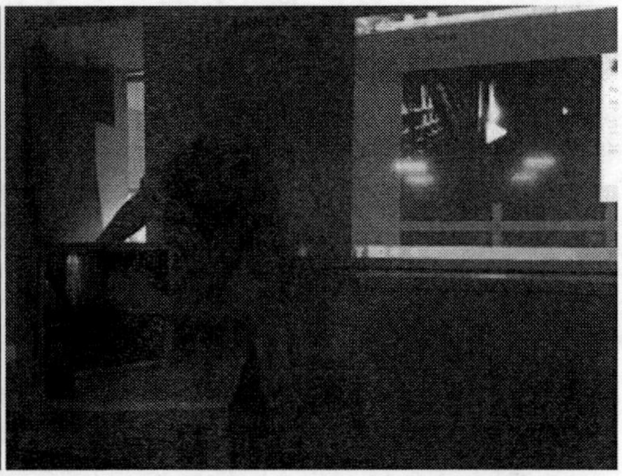

Fig. 2. (left). Students experiencing the atmospheres inside the boxes under changing light conditions
Fig. 3. (right). Final presentation of project 'Staging Poe', the box and the video.

Discussion

In this project, referring to Böhme's types of relationship with the materials, all three (making, perceptual and medial) have been experienced by students in sequential phases: First, in recognizing the nature of materials by making, then by observing the change in appearance of materials after enclosing the box and structuring the light experiments, and lastly by uncovering their embedded relationship with materials and associated memories that have been operational in their material choices and making decisions.

During the process, students were asked to write online weekly-reports where they replied to questions regarding the process. Writing is instrumentalized as a pedagogical tool that aimed at initiating a dialogue between how students think and make. The act of writing through the process also made a subtle reference to the literary pieces that have been used as appetizers at the beginning. We observed that, starting with the effect of Poe's poems, experimenting with facets of materiality in search for staging the atmosphere of the poems by intuitive use of material-immaterial agents and incorporation of dematerializing processes, culminating with videos that present the ephemeral, transient nature of the atmospheres have provided a holistic introduction and understanding on both the nature and processes of design.

The reports revealed that most of the students associated the mood of the poems with the character of the materials. The perceptions regarding the character of materials were idiosyncratic that can be related with students' medial relationship with materials. Many students started considering and working with the texture and surface quality of materials (transparency, translucency, porosity) according their performance under atmospheric agents. They achieved awareness about the facets of materiality (material, immaterial and non-material) and their role in creation of the ephemeral, performative atmospheres.

References

1. R. Somol, S. Whiting. 'Notes around the Doppler Effect and Other Moods of Modernism', in *Constructing a New Agenda: Architectural Theory 1993-2009*. A. K. Sykes (ed.), New York: Princeton Architectural Press (2010)
2. O. W. Fisher. Atmospheres-Architectural Spaces between Critical Reading and Immersive Presence, *Field*, **1**, (1), pp. 24-41 (2007)
3. http://www.philipperahm.com/data/index.html (accessed: 01.02.2019)
4. R. Banham. 'A Home is not a House' in *Housing and Dwelling. Perspectives on Modern Domestic Architecture*, B. Miller Lane (ed.), London and New York: Routledge, pp. 54-60 (2007)
5. M. Gelertner. *Sources of Architectural Form. A Critical History of Western Design Theory*. Manchester and New York: Manchester University Press, p.177 (1995)
6. AJ. Artemer. https://architizer.com/blog/practice/tools/peter-versus-peter/ (accessed: 29.01.2018)
7. A. Picon. "Architecture and the Virtual: Towards a New Materiality" in *Constructing a New Agenda. Architectural Theory 1993-2009*. A. K. Skyes (ed.), New York: Princeton Architectural Press. p. 282 (2010)
8. G. Böhme. *The Aesthetics of Atmospheres. Ambiances, Atmosphrese and Sensory Experiences of Spaces*, J. P. Thibaud (ed.), London and New York: Routledge, pp.144-145. (2017)
9. T. Grifero. *Quasi-things: The paradigm of atmospheres*. New York: State University of New York. (2017)
10. P. Zumthor. *Atmospheres*. Basel: Birkhäuser. (2006)
11. C. Borch. *Architectural Atmospheres. On the Experience and Politics of Architecture*. Basel: Birkhauser. (2014)
12. P. Sloterdijk. *Foams: Spheres Volume III, Plural Spherology,* Los Angeles: Semiotext(e). (2016)
13. J. Hill. 'Hunting the shadow-immaterial architecture', The Journal of Architecture, **8** (2), pp. 165-179. (2003)
14. Gelertner. *Sources of Architectural Form*. pp.238-250
15. Böhme. *The Aesthetics of Atmospheres,* p.143
16. Ibid, p.147
17. J. Pallasmaa. 'Space, Place and Atmosphere: Peripheral Perception in Existential Experience' (pp. 18-41) in *Architectural Atmospheres on the Experience and Politics of Architecture*, C. Borch (ed.), Basel: Birkhauser, p.20. (2014)
18. S. K. Löschke. 'Immaterial materialities: Aspects of materiality and interactivity in art and architecture' in *Interstices 14, Journal of Architecture and Related Arts*, (pp. 5-11) p. 5

19. https://www.interstices.ac.nz/published-journals/interstices-14-immaterial-materialities/ (accessed 01.02.2019). (2013)
20. Löschke. Immaterial materialities, p. 6
21. S. Allen. 'Toyo Ito's Patient Search' In *Toyo Ito: Forces of Nature* (ed. Jessie Turnbull), New York: Princeton Architectural Press. p.22. (2012)
22. J. S. Sadar. Quasi-Materials and the Making of Interior Atmospheres. Interiority, 2018, **1**, (1), pp. 49-63. (2018)
23. G. Verhaeghe. 'Learning Architectural Design. Material and Immaterial Aspects' *Design for Next, 12th EAD Conference, Sapienza University of Rome,* pp. 12-14 (2017)
24. S. Löschke, S. and M. Day. *Co-Productions: Material Light Architecture Studio (https://www.uts.edu.au/sites/default/files/UTS-dab-ug-2014.pdf)*(accessed 01.02.2019)
25. G. Böhme. *The Aesthetics of Atmospheres*, p.147
26. A. C. Antoniades. *Poetics of Architecture: Theory of Design*. New York: Van Nostrand Reinhold. p. 99. (1990)
27. E. A. Poe. *The Philosophy of Composition*, CreateSpace Independent Publishing Platform (first published in 1846). (2015)
28. J. Pallasmaa. 'Space, Place and Atmosphere: Peripheral Perception in Existential Experience' (pp. 18-41) in *Architectural Atmospheres on the Experience and Politics of Architecture*, C. Borch (ed.), Basel: Birkhauser. (2014)

Experiments with visible and invisible materials: designing and building an ambiance

Céline Drozd[1,*], *Virginie* Meunier[1], and *Antoine* Mabire[2]

[1]Graduate School of Architecture of Nantes, AAU Laboratory, 44262 Nantes Cedex 2, France
[2]Mabire&Reich Agency, Graduate School of Architecture of Nantes, 44262 Nantes Cedex 2, France

Abstract. This article relies on a workshop called "materials of ambiances" that is taking place at the Graduate School of Architecture of Nantes. It aims for the students to question the qualities of ambiances within spaces with an approach through the concept of material, making them aware of invisible materials (ambiances) from the manipulation of visible materials (building materials). The experimentation with materials holds a significant place: the students are asked to build an ambiance device to create a sensory experience in order to highlight every sensory quality of a material capable of producing an ambiance. The materials used for this experiment are mostly from fields beyond architecture as to think of potential misappropriations. It is about making sure that the student tests by himself the materials and recreates the connection between data from the experiment, and the physical characteristics given by the manufacturers and industrialists. The pedagogical experience that we present aims at creating within students an interest for the built material, to develop ambiance intentions to qualify projected spaces, which are not always measurable but always noticeable. This article proposes to turn back to the way the qualities of ambiances are felt, the misappropriations of materials are perceived, as well as the perspectives on the evolution of the architectural design process. This contribution is illustrated by an ambiance device whose the name is "Cosmic dream" built in 2018/2019 by Marilou Bach, Hugo Falaise, Carole Lyssandre and Charlotte Say.

While using digital tools within architectural design can give the impression of distance with concerns about body experience in architecture [1], environmental questions seem to have resulted in a growing interest for the relation between humans and their environment, which translates in particular into an aesthetical approach of ambiances [2]. Therefore, in order to connect the technical and sensory dimensions of architectural design, a lecture given at the Graduate School of Architecture of Nantes aims for students to question the qualities of ambiances within spaces with an approach through the concept of material. This workshop, called "materials of ambiances", grants a large part to the experimentation with materials to make them work with invisible materials (ambiances) from the manipulation of visible ones (building materials). Indeed, as Jean-Claude Prinz [3] suggests, materials "take part in the general ambiance of a place. […] By creating an exchange with our senses, they send us back

* Corresponding author: celine.drozd@crenau.archi.fr

to sensations, emotions, values that elicit reactions of appropriation, acceptance or rejection" (p.8). Moreover, as Jacques Ferrier* explains, in the current context of reduction of buildings' energy consumption, we should not forget that architecture, beyond the precise control of ambiances, is also a "question of atmosphere. Nowadays, it is possible to create environments that are more comfortable and more economical in terms of energy without sacrificing the architect's primary intention, which guides him to decide on an atmosphere." Following this objective, the pedagogical experience that we present aims at creating within students an interest in the built material, to develop ambiance intentions to qualify projected spaces, which are not always measurable but noticeable nonetheless.

1 Experimenting with materials as a pedagogical way to design and build an ambiance

As Grégoire Chelkoff [4] states, "(...) talking about architecture in terms of ambiance means conceiving it as a sensory culture of the inhabited environment. It requires a modal, rather than causal, approach regarding physical and sensory dimensions (...)." (p.19). In that sense, we offer at ENSAN an incremental approach on the pedagogy of architectural and urban ambiances throughout the training in architecture, from awareness in first year (Licence 1) to the design and building of an ambiance at Master's level, which allows to experience and live the proposed atmosphere. Over the course of the training years, the common objective of these different approaches is for the students to realise the importance of crossing the physical and sensory dimensions in the design and construction of an ambiance, so that they can, as Louis Kahn [5] explains, "(...) start from what is non-measurable, go through measurable means during the project and, ultimately, be non-measurable." (p.50-51). Therefore, in the Master's degree, the deliberate point is to make the students work on ambiance with their sensory component as an entry point into the design process. In fact, we agree with Grégoire Chelkoff [4] when he says that it would be beneficial to "refine the knowledge and the processes of training (of practices and ordinary experiences of spaces) (...) facing design modes that favour formalism, which tends to impose its vision or effects, or functionalism, which is often too reductive in terms of human variations of inhabiting." (p.19).

We made the choice of addressing the design of ambiance from the work on materials in order to connect material preoccupations (implementation of materials) with immaterial ones (ambiances). The joint work on visible and invisible materials is different from usual lectures on that subject in architecture school, which usually distinguish the lessons on the structural and physical characteristics on the one hand, and the poetic and symbolic elements on the other hand. To this end, we establish a pedagogy on the experimentation of materials that allows the students to place themselves within a "deliberately active and transformative dimension within the research process" [6] (p.2). By going beyond the stage of spatial outline, they are confronted to the reality of materials, to the consideration of ambiances and the complexity of uses.

Experimenting ambiances via materials is an approach that is still little developed, as Grégoire Chelkoff [6] stresses: "If we know about experimentations that are essentially constructive, we know less about those that are directly based on qualities or criteria regarding ambiance." (p.5). Indeed, the concept of teaching by experimenting traditionally takes a constructive approach that aims at reproducing elements from reference buildings to

* Jacques Ferrier, on the sensory city, online: http://www.sensual-city.com/ consulted on 28/09/2018

understand spatial forms and structural compositions. Without questioning their pedagogical interest, this is a partial approach to material, which we tried to overcome. We think that, as Tadao Ando [7] explains, "architecture is made of two elements. One is intellectual, since one needs to create a logical and clear space, with a logical or intellectual order. Furthermore, we need to use our senses to give life to space. Those are the two fundamental aspects of the creation of architectural space. One is theoretical and practical, the other is sensory and intuitive." (p.53). With the manipulation of materials, the sensory experience becomes a way to apprehend space and design it, for students who are future architects.

2 "Materials of ambiances": objectives of and expectations from the workshop

The students are expected to design and build an inhabited ambiance device, allowing to live a sensory experience that includes the body, or part of it. The design of that device relies essentially on one material, selected for its potential of ambiances. The point is for the students themselves to test the materials, and to re-establish the connection between the data from the experimentation, i.e. their own observations and analyses, and the physical characteristics provided by the manufacturers and suppliers, while going beyond this and reveal the material in all its qualities and potentialities. According to us, it seems indeed necessary to bridge theoretical knowledge and a physical, manual approach of material, as Peter Zumthor [8] does in his work: "the idea (is) that the task of creating architectural atmosphere also comes down to craft and graft." (p. 21). Therefore, some materials are analysed and studied to reveal an aesthetics of ambiances through an experience to be lived, which implies a particular position given the selected material. Moreover, the ambiance device includes climatic and ambiance conditions that are already within the studied site (solar radiation, natural light, air flow, rain, noise…). Those parameters are created and/or emphasised by artificial devices (radiator, speakers, fan, hairdryer, spotlight, smoke machine…), and it allows to reveal the ambiance qualities of the materials selected by the students, and to highlight senses besides vision.

The design of the device is composed of back and forth motions between different stages to reach its construction, then its perception within the built space: knowledge of the material via tests and manipulations to evaluate the possible transformations and transfigurations, narrative that leads to the formulation of an intention in terms of ambiance that stimulates creativity, definition of a scenario of sensitive experiences implying several senses and one or several body positions, outline of the device via the formalization of models on different scales to better envision the 1:1 scale construction, carry out survey of the ambiances in the implementation space for the device, in order to determine the adequate place for the projected ambiance intentions. The design process continues in the construction phase of the device, since adaptations for the system are not unusual at this stage. Lastly, the presentation of the ambiance device is accompanied by displays of the material, to introduce it to the school users in general. The device is also tested by asking a public who is not integrated to the class to experience the ambiance, so as to determine some feedback on the sensory experience of the device.

The materials used by the students are provided by suppliers who have been contacted, and who are interested in the experimental approach of this class. For the most part, they are coming from other fields than architecture, in order to consider possible other uses, and they were selected a-priori, depending on the ambiance potential they seemed to hold. Beyond the visual aspect, some of them have obvious sound and/or thermal insulation qualities, have materials with strong olfactory power, produce unprecedented sounds under certain conditions, reveal their characteristics upon touching, as opposed to what they offer to see, etc. This evocative potential, regarding sensations and emotions, is not always perceptible at first glance. This is why manipulation tests and the development of imagination are necessary stages to quickly overcome the emerging prejudices and open the scope of possibilities: "Architecture does not only implement spaces and compositions, but also materials that offer an array of perceptions. The implementation of materials that have been little transformed unfolds a range of evocations. Through their seeds, their surfaces, their textures, they provide perceptions that go beyond the strict architectural framework to include the perception of the material, its power, its rest." [9] (p.120).

To ensure that they will go beyond the first contact with the material, and to stimulate their creativity to fuel the design process, we ask the students to put together a narrative, a story around the material, which refers to the different senses, to a poetics, to an imagination… Indeed, as Nathalie Tornay [10] recalls, materials hold an imaginary dimension "that relies on metaphors and connotations to make sense" (p.93) and allow thus to develop within the design process "sensory aspects by games of surface, material, etc., making the sensory perceptions even more intense (vision, touch)." (p.92). Other authors [3, 11, 12], such as Antoine Picon [13], also insist on the symbolic and cultural dimensions of the perception of materials: "(…) our body, similarly to the perceptions and sensations that represent its base, is marked by the society in which we are living. Indeed, we do not perceive our surroundings without the context in which we have grown up. This context, in a way, filters out the diversity in outside stimuli and gives them a shape. (…) This learning process happens in close relation to the materials, the tools, the instruments and the machines that are at our disposal." (p.61-62).

For example, we will explicit the work of a group of students that was produced with a piece of fabric, whose surface was made of glass beads and thus reflected the light (retro-reflective fabric from the Super Black line by JRC Reflex)*. Usually, it is used for safety clothing, in particular in the road sector. Based on their manipulations, the students have developed an imaginary around a "cosmic" vocabulary to translate their sensations into a narrative: "Put down your foot… it sinks, then bounces back. Step by step, the material bends under your weight, then regains its original appearance. Surprisingly elastic, incredibly resistant (…)! You feel like you are balancing, as if in the air, a moon landscape appears before your eyes (…). The light travels, dances with the fabric, reverberates. The ground blends with the walls, the walls with the ceiling. You lose the sensation of depth… You decide to change your viewpoint, and suddenly… A rainbow. A psychedelic spectrum crashes down onto your field of vision!"

The resulting device highlights the temporary deformation characteristics in the fabric, as well as its resistance, the silky touch, the thin rustle we can hear during the deformation, the lighting effects that are produced (fig.1).

The students introduce it: "The fabric that composes it, with its flexibility and fluidity, wraps around steel arms. Minimal, almost shy, it is still a long way from unveiling its hidden world. Without obvious invitation, the visitor has to come close and, like an explorer, to find an entry point. But this new universe is not reached easily. One needs skills, force and flexibility to expand the space and create a new dimension. That space is no longer a plan, but a moving volume in constant metamorphosis. Drawn in by this unique universe, disconnected from the outside world, the experiment has started. We are now starting an

* http://www.jrc-reflex.com/Products/SuperBlack.aspx

immersive and sensory ride. The journey is strange, difficult but pleasant, led by a weakened gravity and a flexible, expandable medium. The material that makes the walls of this universe represents an incredible, almost improbable, paradox. Even though it is made of glass, it is soft to the touch in a soporific way, and has a mesmerizing flexibility. The body ultimately sits or lies down to enjoy the visual show. Similar to a psychedelic daydreaming, the glass material decomposes the light. Reflected and turned into a multi-coloured spectrum, it moves and accompanies its disturbed universe." The efforts made by the body to enter this space are proportional to the efforts the fabric makes to welcome its guest.

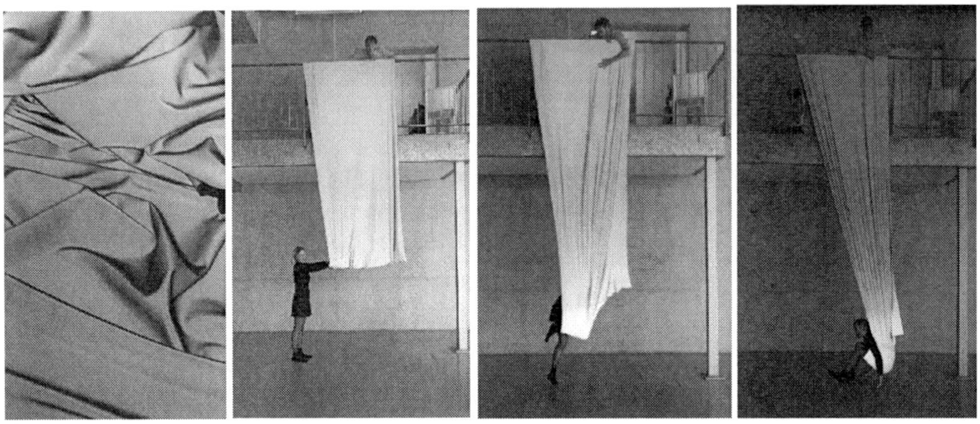

Fig. 1. Tests and manipulation by students with the retro-reflective fabric: silky aspect, deformation and resistance, lighting reflections.

The "cosmic" imagination that was developed by the students is recalled by the users of the device: "With this light, we feel like we're somewhere else… The little rainbow reflections are great!", "It's like a parallel world! It was almost stressful to be inside! Because you want to try every position, but it's not easy physically.", "To see the seam closing up again is almost like seeing the universe turning grey, it is very enveloping", "I feel like I'm in a very comfortable pouch… As if I was hovering above the ground, we're not touching anything, it's like flying!", "The material is elastic, it's like a hammock, you are falling towards the ground and the material… it's soft!" (fig. 2).

Fig. 2. Device from the outside to the inside.

3 Pedagogical feedback on the evolution of architectural practice

The subsequent session of the workshop allowed to identify some pedagogical contributions benefitting architectural design. Among those, we can mention the constitution of a reference corpus on singular materials that generate ambiances, but in particular a reference corpus on experienced ambiances, which will support architectural design and research. As Xavier Bonnaud [9] says, "in education, (…) a portion of the learning processes on design are accomplished in the domestication of the subjective reception of places, the ability to feel and perceive non-rational qualities a-priori, and to decipher the secret manufacturing conditions and to try mastering them." (p.34).

One of the purposes of this workshop is to encourage students to look differently at materials, to reveal their potentialities, and thus to go beyond the usual implementation of this material, to enrich the architectural design and construction. Some architects implement this approach, such as Antoine Mabire in the case of a house project in La Rochelle, in which a widespread constructive mode (concrete formwork blocks ÉCLAIR) was used in an unusual way: "The idea was to say: the walls are made like this, we insulate from the outside and our walls, they're like that. In the result, we will have an entry"*. Moreover, this is used for the aggregate of materials used throughout the different architectural production stages of the agency: "(…) We have constants. For example, we like using the material for what it's good at, and then we keep it visible, to establish ambiances"*. To this end, models are used during the design process, and they allow to project ambiances. The difficulty to translate the material into the different scales arises, as Antoine Mabire points out when mentioning the design of the vegetable processing area at the Lycée Rieffel in Saint Herblain, France (fig. 3&4). Three of the walls are made out of weaved wicker: "We made (…) a big part of the model where we actually wove wicker, but we faced something unexpected: finding the right scale for the material. We tried different things, we wanted a rather big scale to see well. But the problem was that wicker can be worked on only when it is wet…"* (fig. 5).

Fig. 3. Lycée Rieffel, Mabire&Reich Agency, Saint Herblain, France.

* Antoine Mabire, architect, interview conducted on 20 December 2018.

Fig. 4. Tests for weaved wicker, Lycée Rieffel, Mabire&Reich Agency, Saint Herblain, France.

Fig. 5. Ambiances in the building Lycée Rieffel, Mabire&Reich Agency, Saint Herblain, France.

However, as architecture is defined as the passage from a conceptual idea to its materialisation, working via the material seems to ensure a deeper reflection on the sensory qualities of space, to reach "a harmony between the project and the materiality, until we find a perfect agreement between the programme, the form and the material." [11] (p.85), which goes through "creativity, which invites the comparison of materials, to take them out of their context or to think them under new forms and structures" [11] (p.85).

The other essential dimension in such an approach is the consideration of the lived dimension in the design process. It is the reason why, throughout the different workshop sessions, we emphasised the sensory feedback surrounding the built device. This addition in the pedagogical development aims to help the students understand more easily their experience in the process of architectural design. Indeed, it appears to us necessary to give the students the dimension of the body experience of space to "humanise architecture" [14] (p.104). This concern is connected to different fields of activity around architectural design, whether via materials, as it is presented here, or via the physical reading of a place, like for example at ENSA Strasbourg. The professors explain: "Training in sensory experiences of space could without a doubt fill the gap created by the abandonment of this physical dimension from the project. It should raise awareness of the magic about reality, and enable practices of project that integrate the imagined experience of architecture in a better way." [15] (p.306).

The workshop sessions presented here have been taking place at the Graduate School of Architecture of Nantes since 2014. However, as an optional course, it is catered for a limited number of students. For this reason, given the interest it has for all the students of the school and the professors, we decided to implement built devices in a location of the school that is frequently visited, as an exhibition over a few days, to give time to discover the visible material in a more dynamic way than in a material library, but also to test the built devices and experience the proposed ambiances to uncover the invisible material. This workshop has led to fruitful collaborations between professor-architects and manufacturers, between manufacturers and students, and even collaborations between designers, manufacturers and students that went beyond the frame of the course, to become a professional setting of architectural production. This allowed to start an evolution of the design process around the involvement of stakeholders during different stages of the project. The multiplicity of experiences represents a wealth of knowledge about visible materials, and in terms of corpus constitution about sensory experiences, which will support architectural design in its sensory, aesthetic and poetic dimension.

References

1. V. Lebois, D. Laburte, Ambiances, **4** (2018)
2. S. Malfroy, Faces, **67**, 4 (2010)
3. J.C. Prinz, O. Gerval, *Matières et matériaux : architecture, design et mode* (Eyrolles, Paris, 2012)
4. G. Chelkoff, Faces, **67**, 18 (2010)
5. L. I. Kahn, *Silence et lumière* (Editions du linteau, Paris, 1996)
6. G. Chelkoff, Ambiances, **4** (2018)
7. T. Ando, *Du béton et d'autres secrets de l'architecture* (L'Arche, Paris, 2007)
8. P. Zumthor, *Atmospheres: architectural environments – surrounding objects* (Birkhäuser, Bâle, 2006)
9. X. Bonnaud, *L'expérience architecturale : réflexion sur une notion, points de vue sur une discipline* (Université Paris 8 Vincennes - Saint Denis, 2013)
10. N. Tornay, *Vers des outils d'aide à la conception pour intégrer les dimensions techniques, écologiques et sensibles des matériaux de construction* (LRA, Toulouse, 2011)
11. H. Manfred, D. Hans, Z. Martin, *Matérialité* (Birkhäuser, Bâle, 2007)
12. P. Von Meiss, *De la forme au lieu + de la tectonique, Une introduction à l'étude de l'architecture* (PPUR, Paris, 2012)
13. A. Picon, *La matérialité de l'architecture* (Parenthèses, Marseille, 2018)
14. A. Aalto, *La table blanche et autres textes* (Parenthèses, Marseille, 2012)
15. V. Lebois, D. Laburte, *Ambiances, tomorrow*, 301 (2016)

Community-based grassroot communication strategies, process and product: The drawings that saved our hospital

Michel Mounayar*, RA; Professor of Architecture[1]

[1]Ball State University, Muncie Indiana. USA

Abstract. A case-study analysis of an urban design communication strategy employed by our university-based design team entrusted with re-envisioning the uncertain future of a local small-town community hospital in Indiana. The design process is carefully constructed from structured public input, and community participation, whereby students, faculty, physicians, nurses, as well as ordinary citizens combine their efforts to strategically develop their 'plan for planning'. Finding a strategy to define the scope of their future needs and the definition of important priorities to organize the project scope prior to engaging professional consultants. In this scenario, the design team is only the guide and translator, working closely with stakeholders to help them visualize and clarify the aspirations of their town. This paper will present our community-based design methods and most importantly our graphic communication techniques, specifically formulated to envision and facilitate consensus for a new unified public health system in a small Midwestern American city.

1 Community-based design

Over the last five decades, our university program used grassroot methodologies to demystify the design process and democratize the town planning decision-making. To that end, we developed and refined methods to engage the public through specific graphical techniques especially designed for public participation and input. In the following, we will explore one such case-study for Dekalb Health System in Dekalb county Indiana, looking specifically to present our method and outcomes.

As urban designers, how do we converse with the lay public? What graphic tools must we use to help our communities invent and envision a prosperous and designed future? Employing community-based design processes are we able to assist an endangered small county hospital in northern Indiana plan for its public health strategy? What graphic tools do we deploy to change public preconception, while also resolving ongoing community conflicts? These communication questions are the principal focus for this analysis.

*Corresponding author: Professor Michel Mounayar, RA mmounaya@bsu.edu

The process succeeded, and this small county overcame the challenges confronting it. To this date, the hospital continues to provide high quality, locally-sourced, healthcare for its citizen. In this age of mega corporate mergers, where bigger is better, we propose to illustrate a unique story where the large corporate takeover threat ceded control to health managed at the local level through a modest 50-bed not-for-profit county-owned and managed hospital.

Fig. 1. Team discussion, Dekalb Charrette team, students, faculty, and hospital physicians and administrators working to map community input.

Such a small hospital is predictably under continual hostile attacks from larger, profit-driven, health systems in the region. Health is big business in our country. The United States is the last industrialized country with no national healthcare system that guarantees health coverage to all its citizens. Moderate to low income individuals and most families struggle to meet unexpected rising costs of healthcare, and medicine. "On a per capita basis, health spending has increased over 30-fold in the last four decades, from $355 per person in 1970 to $10,739 in 2017. In constant 2017 Dollars, the increase was almost 6-fold (600% percent) from $1,797 in 1970 to $10,739 in 2017" based on data from the Kaiser Family Foundation Analysis (12/2018). The challenge seems to grow exponentially every year, resulting in double digit price increases. What might be perceived as a right in some countries, is surprisingly considered as a privilege in ours.

As our national healthcare debate matures, designers must actively participate in the conversation. The health care problem is fundamentally a "design project" requiring creativity and the integration of disparate and competing interests into synthesized practical and workable solutions. The challenge is complex and requires tools and techniques that will clarify needs and constraints, artfully weaving both into a comprehensive complex working system. Design graphics and communication strategy are central for both success and innovation. Our techniques include onsite photography, digital modeling, and digital and analog sketching. However, the public nature of our visioning process is perhaps the secret for our success. Bringing the planning process into the open requires engaging the different stakeholders in public prompting all involved to use a common language for these discussions. Preferably, one that is equally new and fresh to all participants, so as preconceptions and restrictive and convoluted history does not get in the way of progress. Design language and graphics provide such a common channel for all participants to communicate with one another. Napoleon Bonaparte said "a good sketch is better than a long speech", expressing the power of graphics to bridge differences and create vision for action. So, our sketches and plans form the foundation for a new open conversation based on candidness and clarity, helping us overcome the public's bias and leading the group to invest in imagination. Our process is rooted in our belief that success hinges on open data driven decision making. Consequently, graphically interpreting and visualizing data is the secret to our success.

2 The Charrette Process

At the invitation of the community, our university-based student/faculty design team, joined efforts with local health professionals, physicians, politicians, business managers, and citizens to safeguard a fragile but successful hospital, preventing its demise due to the threat of a looming corporate takeover, helping it and the community avoid entering their county into health desert status.

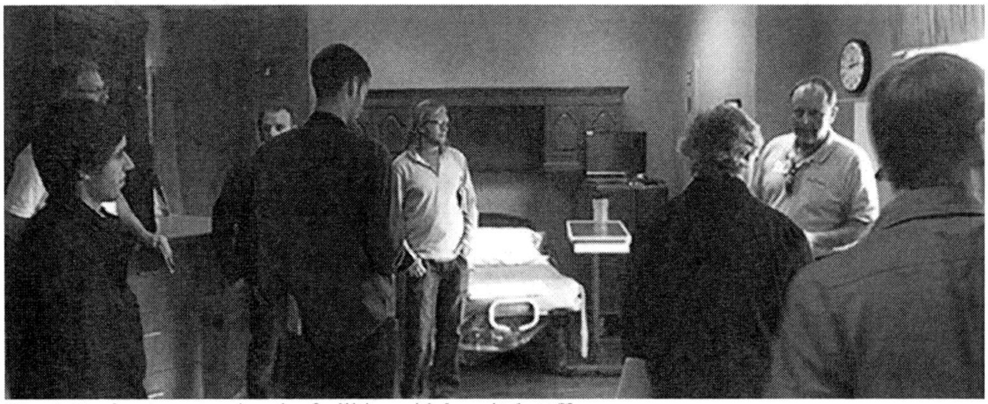

Fig. 2. Design team touring the facilities with hospital staff.

This case-study also examines our public participatory process, graphics, and techniques used to safeguard the hospital. Our design concept builds on the importance of adopting total community wellness through design, where the hospital joins local partners to provide a synergistic comprehensive health and wellness program for the entire county. Community organizations such as the YMCA, City Hall, organic food growers, school districts, as well as cultural and artistic not-for-profit organizations join forces to develop what we termed "Complete Community Wellness" (CCW). Architecture students' involvement in the project was invaluable and benefited both the community and the academy. Students offer a level of neutrality facilitating stakeholders' collaboration allowing them to find common ground, so they may set aside their political rivalries, encouraging them to collaborate. In return, the students receive a unique and special educational experience not available in the confines of a studio or classroom. Realtime progress and community feedback show our young designers their social responsibility and their immeasurable impact on the community. Our university is committed to immersive learning, where students and faculty are imbedded in towns and cities working on real problems affecting real people.

Fig. 3. Public presentation of analysis data and findings.

Fig. 4. The studio is public open to all participants. The drawings develop under the preview of the community.

2.1 The Storefront Studio

The studio must be open and accessible to all citizens. Often improvised by the host town using donated vacant storefront retail spaces, city hall meeting rooms, church basements, library multipurpose rooms, or any available and open real estate in town. However, the studio must be visible and central to the community. Participants must feel invited and free to visit at their convenience. Keeping the studio public helps the team gain credibility. The studio must facilitate involving the community in both the definition and resolution of the design challenge.

2.2 The Charrette Team

For the duration of the charrette, the team lives with volunteer host families. Taking residence in the community demonstrates the team's commitment towards its citizens. The interaction and stories exchanged in residence are an essential part of the process.

Pedagogically, removing students from the confines of the academy and allowing them to enter our communities is greatly impactful to their professional growth. The students find purpose in engaging real design problems. Charrettes often start with strategic meetings engaging special focus groups seldom in contact with the academy. In the meetings, students conduct interviews with diverse stakeholders and community leaders. Providing the students direct contact with mayors, council members, city crews, transportation specialist, police chiefs, educators, professionals, and technicians. These focus groups assist the team with clear formulation of the project's problem definition. For the students, learning in the field is memorable and dynamic. Student feedback shows that they garner high level of value from their workshop participation. Such concentrated field experiences are typically inspirational and life changing for them. The essence of the experience is the blurring of boundaries between learners and teachers. The obscuring of these boundaries allows for open and clear communication. Students forget their obsession with grades and focus on outcomes. In this context, feedback is blunt, clear, and immediate. Success is measured by the level of public understanding of the designers' proposals. Grades are circumstantial and meaningless. Effectiveness and success are ubiquitously on display, all the time, as the project unfolds. This format makes learning a byproduct of doing. However, success is guaranteed by the simple act of participation. The value rests in experiencing all the community forces, in real time, as they struggle to shape their decision-making, while actively searching for, and finding their illusive community leaders. On display for the students is also the role of the architects as agents of positive change. The facilitators and problem solvers who synthesize complex demands and incongruent information into meaningful and wholistic design solutions.

Fig. 5. The team relies on both analogue and digital graphics.

2.3 The communication strategies

The public and the open nature of the process is what is critical here. The fact that the drawings are forming in real time and in public for all to see, gives them deeper meaning for the community. Consequently, the magic is not really about the superiority or any uniqueness of our drawings, as much as where we draw them! Taking the design process to the public is the secret!

Employing a myriad of graphic tools, developed during our university's long history (50+ years) of community-based projects, we can change the conversation for all concerned citizens and town elders in support of their community hospital. The strategy consisted of five phases. First, the development of an expanded steering committee (Growing leadership). Second, undertake a factfinding search through interviews with focus groups. Third, the design team prepares a clear presentation showing a visitor's perspective of the context.

Forth, an open public presentation of finding, showing several potential strategies for future development. Fifth, a final project steering committee meeting dedicated to analyzing the charrette drawings and developing a detailed prioritized project list, as well as, an implementation schedule. As the process matures, the community concludes that the hospital is more than a place to visit when one is sick. Through the years, the facility and programs became an integral civic part of the town. Stakeholders graphicly discover illustrated strategies that could assist the group secure state and federal funding.

3 The drawings

Public conversations must be free of technical jargon. Two-dimensional graphics such as plans, sections, and elevations are too abstract for most community members. Our team mostly relies on three-dimensional graphics such as eye-level perspectives and axonometric drawings. We find that atmospheric drawings are more accessible to residents. Drawings must also remain in progress, unfinished, as to keep them focused on ideas and not become finished pieces of detailed architecture. If the drawings are too realistic the team risks the conversation shifting from general concepts to specific opinion or objections about style and taste. Ambiguity in this case is powerful! It allows the viewers' imagination to inhabit and complete the images in their mind. Furthermore, the designers should always present multiple options of the plan. Developing a singular direction for the design increases the potential for binary positive or negative acceptance. While having multiple proposals facilitates the conversation about potential alternate designs and concepts. Keeping the dialogue alive.

Fig. 6. Emphasizing the need to improve community walkability. The team uses digital sketching to illustrate the potential specific street improvement to achieve this goal.

Fig. 7. Community entries may be improved with strategic interventions. Digital sketches showing proposed median, bikeways, and landscape.

4 A vision of Complete Community Wellness (CCW)

The wellbeing of a community is beyond just the quality of hospitals and doctors. Health and wellness rely on a complex system with multiple moving parts, pressure points, and components. Our analysis identified both assets and liabilities for the project area. On the adverse side significant air and water pollution challenges were mapped and recorded graphically. Also, access to medical professionals and the ability to afford care were equally identified as major challenges for residents. Conversely, the intimate scale of the hospital, and the personal familiarity of the community with the medical staff are complemented among the most important assets to the group. Doctors, nurses, and hospital administrators are neighbors and friends. Patients feel a precious connection to the hospital staff. Perhaps, this is the most significant factor that motivate the community to fend off the takeover of their facility and safeguard this institution. Dekalb Health truly existed as a community hospital since the early 1920s.

Our vision imagines the hospital transforming into a community hub focused on the concept of "Complete Community Wellness". The hospital through this concept provides the community with meeting space, art galleries, a YMCA like exercise gym, pools, and programs in additions to the best medical facilities in the area. The grounds will promote health through walking trails and open gardens that might even provide community gardening and other public and civic amenities. In this manner, the community hospital connects closely with its public. Schools, parks, playgrounds, and gyms are part of the system that will keep us active and healthy. After several days of focus group meetings with community stakeholders, the hospital staff identified the following important health indicators the community must monitor and manage:

- Obesity rates
- Chronic illness associated with being overweight such as diabetes, High blood pressure, and cardiac disease
- Residents' Physical activity levels
- Healthy eating options
- Smoking rates
- Community aesthetics
- Educational and employment achievement

These important markers exceed strictly medical data, suggesting the importance of context and location. Where you live and what you eat are as much a factor to one's wellbeing as biology and genetics. The physical environmental design must take into consideration the level of satisfaction, fulfillment, and activity of the inhabitants. We need to achieve health and wellness through the implementation of designed environments that prevent destructive ailments such as obesity, stress, and depression. The intent here is rooted in the idea that the county must establish an ongoing partnership with the hospital for one's life-long wellness. A preventive system that cares equally for both healthy, as well as, ill individuals. Complete Community Wellness (CCW) is a holistic and system wide concern for: young and old, physical and emotional wellbeing, short and long-term health. CCW is more inclusive in its consideration of healthy living. Exercise, fun, and socializing are as important as food, medicine, technology, and science. The goal is to design an arrangement of infrastructure, programs, facilities, support technical staff, physicians, nurses that make up a total system that promote health through wellness. Consequently, access to gyms, sidewalks, bikeways, open fields, sports facilities, parks, healthy and safe food supply, clean water, places to meet and socialize are equally important to the availability of top-notch physicians, surgical suites, MRI units, or any other contemporary high-tech medical technology.

The concept is built on a total and complete approach to body, mind, and spirit, tended to and managed by the community hospital. All aspects of human life, spiritual, mental, and physical is orchestrated in partnership between the hospital, the county, educational, and social organizations, food growers, and the business community. Keeping our citizen healthy is dependent on how active and lively our cities and towns become. DeKalb Community Hospital is already engaged with its partners to seek such an innovative and exciting future.

5 Three design concepts

The team develops three different concepts specifically designed for the community's and the hospital's consideration. All three concepts recommend the transformation of the hospital into a community cultural and activities center, where medicine, arts, culture, exercise, and social gatherings are open to all residents. Beyond the center, the hospital also is recommended to establish smaller neighborhood-based outreach clinics that might be located although the county in both urban and rural settings. The future hospital master plan is imagined using three different approaches:

5.1 Filling-in the gap

Just like most hospitals, Dekalb health grew organically through the years without the guidance of a clear master plan. A maze of corridors and hallways confuse patients and visitors alike. Floor plate heights and existing mechanical installations limit the ability to update the required medical systems. Consequently, the intervention uses direct connection between important nodes, in the spirit of "a straight line is the shortest distance between two points." The principal goal for this approach is to work just outside the existing infrastructure on the perimeter of the existing buildings to provide strategic infill pieces designed specially to improve overall patient orientation, enhance staff circulation, while also expand the site with new quiet exterior gardens, as well as, new community-wide active public trails. The proposal adds new sections just outside the existing building façades providing for new nurses' stations, clearly connected corridors, and state of the arts patient rooms. Intervening to the front of the existing building also allows the hospital to showcase the organization's new progressive, and environmentally responsive image. Additionally, the new structure may be constructed without substantial day-to-day operational disruption. As requested by the project steering committee, new informal meeting rooms are located at the extreme tips of the new structure taking advantage of views and sunlight. These rooms are designed to serve as multipurpose rooms for the main patient floors.

Fig. 8. Sketch showing how "filling in the gap" may provide the hospital a new image while updating and modernizing the existing infrastructure.

Fig. 9. Digital sketch showing the proposed courtyard campus.

5.2 Courtyard wellness

This concept emphasizes the importance of exterior natural spaces for the speedy recovery of patients. New construction provides a series of connected courtyards. Each designed to fulfill different wellness activities. The courtyards have the potential to bring young and old together for special activities including gardening and food production. These spaces also become an integral part of wayfinding landmarks for patients and visitors. If each takes on a unique character, then visitors may use them as orientation markers. Although, such open space arrangement surrounded by programmed buildings, may increase daily walking distances for occupants, but the advantages of daylighting and the availability of natural views outweigh this inconvenience. The addition of a new YMCA sports facility, outpatient doctor offices, plus new eateries and art galleries provide the themes for the new courtyard masterplan.

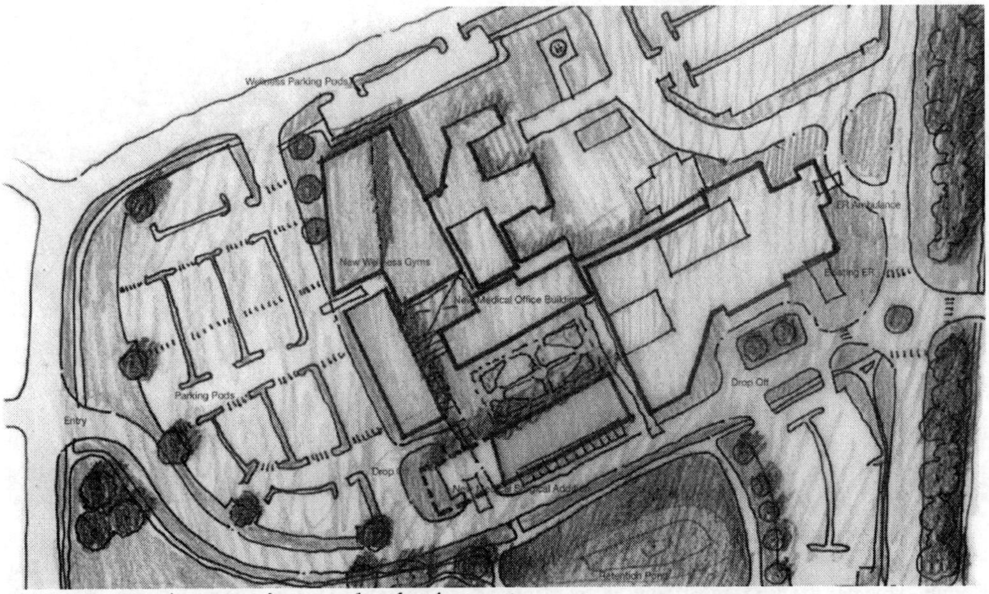

Fig. 10. Proposed courtyard masterplan development.

5.3 New campus

Fig. 11. Team's illustration of potential all new wellness/hospital campus near the interstate highway.

Influential members of the steering committee were intrigued by the possibility of moving the entire hospital out of the downtown and developing a clean-slate totally new campus in the hinterlands of the county, on farmland, near the main interstate highway. They believed that the old structure limits the hospital options and would increase the ultimate cost for the project. Most corporate hospitals are building new operations around Dekalb county and this newly establish campus, in their mind, has a better chance to fend off the takeover attempts and would safeguard the hospital.

Fig. 12. View from the interstate showing the new hospital neighborhood.

Although our design team disagreed with this assessment, it was important for us to seriously explore this alternative. To ignore it, and disregard this faction's interest, would have cost us our neutrality as project facilitators, and would have potentially led to the overall failure of our team. So, we assigned one of our most senior medical facilities expert colleague for the development of this design option. The group put together an extensive site selection process, as well as a fully conceived regional hospital/office/housing neighborhood master plan. This development option may be now readily evaluated more effectively compared with the two other options presented. The issue was no longer abstract, nor emotional for all concerned. Having these drawings is very important for the project steering committee.

Conclusion

Architects and designers are important to the healthcare debate in the United States. Health should be considered as an integral part of town design. Once patients enter the complex and costly system of hospitals, physicians, and medication the opportunity to prevent and protect is missed. Our physical environment must promote healthy living. We must provide our citizens active, engaging, inspiring, and esthetically pleasing neighborhoods, towns and cities. Walking and physical activity must be designed as part of daily life. Even if one never visits the gym, the act of living in a certain location should results in an active lifestyle. Activity, clean air, and food are all improved through design. The designers' ability to illustrate and organize information gives them the capacity to help with such complex problems. But unfortunately, often, they are not consulted. The Dekalb Health case-study showcases how the work of a university community-based projects design team of students and faculty, using participatory illustrations, successfully conceived a new and exciting healthcare future for a small Indiana town.

References

1. H Leyland, Ø Naess, *The effect of area of residence over the life course on subsequent mortality*, J R Stat Soc Ser A Stat Soc. 2009 Jun ; 172(3) : 555–578

2. R Kamal, C Cox, Kaiser Family foundation, *How has U.S. spending on healthcare changed over time? (2016)* https://www.healthsystemtracker.org/chart-collection

Out of "touch"? – An experiential pedagogical approach to daylighting in architecture and interior design education

Gillian Treacy[1]

[1]School of Design, Edinburgh College of Art, The University of Edinburgh, UK

Abstract. A new challenge is emerging. Contemporary built environment pedagogy demands engagement with both analogue and digital tools for simulation and verification of lit architectural environments. The use of analogue tools within architectural design education grasps onto the historically valued craftsmanship of drawing and physical models to measure, represent and understand our lit environment ambiance. Digital tools can provide efficient, simultaneous and precise verification of lit architectural interior space through 3D computer modelling and calculation software. However, the understanding and representation of daylit scenarios is becoming more numerically complex as lighting metrics and software gain in accuracy and dynamic range. With the majority of easily accessible software tools focussing on numerical verification, the ephemeral ambience that daylight in particular creates in interior architectural spaces is becoming ever more difficult to grasp for the architectural design student and practitioner. This paper seeks to challenge the exclusive use of digital tools for the understanding and representation of lit interiors by proposing that this methodology cultivates design epistemologies that are out of "touch". Questionnaire findings and workshop studies are presented as pedagogical constructs are proposed inviting physical, experiential learning of lighting principles *in collaboration with* numerical and digital modes of learning to provide connections and translations to develop through 'touch'ing light.

1 The new challenge emerging

Contemporary built environment pedagogy invites engagement with both analogue and digital tools for simulation and verification of lit architectural environments. An attunement to the atmosphere created by light in an architectural space is necessary for every architectural designer (architect or interior designer). This understanding of the character of light allows some certainty in the prediction of the lit ambiance of a space. However light, with its intangible and contextually dependent characteristics can be difficult to grasp and challenging for architectural designers to work with, particularly in the realm of lit ambiance. Now more than ever, a deeper understanding of the phenomenon of daylight is necessary to engage fully with contemporary holistic approaches within architectural design; light touches upon many inter-connected elements, such as building location climate, orientation, adjacent contexts, spatial volumes, materiality (such as texture, transparency and reflection), and its physiological and psychological effects on users of the space. The use of analogue tools within architectural design education grasps on to the historically valued craftsmanship of drawing using physical models and tools to measure, represent and understand our lit environment ambiance. However, in the field of architectural and environmental building design, engagement with digital tools is now commonplace and any architectural designer would be viewed as 'out of touch' if they avoided engagement with these technologies. Digital technologies are providing progressively more efficient and precise verification of lit architectural interior space through 3D computer modelling of virtual environments and calculation software. It is apparent that daylighting design as a generator of interior ambiance is quickly becoming a lost art in the face of numerical verification. It has become too simple to calculate light as 'energy'.

> *As designers, we are trained to place value in the concept of spatial experience; however, we are increasingly asked to quantify our design intentions in terms of net energy balance. (1)*

This ontological understanding of light provides an alternative approach to designing with light exposed and expressed through the use of the analogue draughting tools; significant value is now given to accuracy in numbers and large data sets. Energy efficiency, lighting efficacy and solar gain can all be designed with digital tools to easily provide highly predicable data defining energy usage and compliance. It is apparent that the understanding and representation of daylit scenarios is becoming more numerically complex as lighting metrics and software gain in speed, accuracy and dynamic range. Although these data can provide some guidance on thermal comfort or even task lighting levels this provides little indication of the lit ambiance of the architectural daylit space. With the majority of easily accessible software tools for environmental analysis focussing on numerical verification, the ephemeral and embodied ambience that daylight in particular creates in interior architectural spaces is becoming ever more difficult to grasp for the design student and new architectural design professional (2). Further, quality and "uniformity of experience" (3) in interior spaces is becoming codified to ensure regulatory compliance where practical. This diminished necessity to value our personal experiences through our "implicit, naturalistic, ecological cognition of everyday existence" (4) denies powerful design motivations and limited ontologies for light.

Therefore, with energy compliance top of the agenda it might be argued that it is no longer necessary for designers to explore this phenomenological embodied relationship with the characteristics of daylight. The subtle and inviting revealing of the "thingness of light" (5) to occupants of the space is no longer a necessary attribute and skill of the architectural designer. In spite of this view, I would argue that architectural design demands an emotional understanding of spatial environments and this cannot be solely described using the daylighting metrics and tools currently available. Dean Hawkes (6), in his book "The Environmental Imagination", poses a question that invites discussion as to the place of science and technological devices in architectural design.

> *The ability to envision the outcome of the conjunction of form and material, **set within** the physical facts of the climate and locale, in ways that inform and enhance the purpose and meaning of a building...lies at the very heart of the architectural project.(6)*

He proposes that the nature of architecture requires an understanding of both realms, the quantitative and qualitative, the physical and virtual, the real challenge being to understand the unquantifiable aspects of architectural design.

> *I only wish that the first really worthwhile discovery of science would be that it is recognised that the unmeasurable is what they're really fighting to understand. (7)*

Pallasmaa argues that "The quality of a space or place is not merely a visual perceptual quality as it is usually assumed. The judgement of environmental character is a complex multisensory fusion of countless factors, which are immediately and synthetically grasped as an overall atmosphere, feeling, mood or ambiance" (3). Plummer (8) advocates through photographic and written narrative research that daylight heightens the "immaterial aspects of architecture" and the human experience of being in a space touches "perceptual", psychological", "emotional" and "spiritual" realms. He suggests that this emotional experience of light in a space is a necessary personal exploration for anyone aspiring to gain insights into daylighting. With this in mind, it can therefore be suggested that the architectural designer needs an ever expanding and flexible skillset benefitting from the ontologies of both the quantitative 'science' and qualitative 'art' of light. The increasing variety of tools, outputs and competencies and their understanding in use demands an "eclectic and heterogeneous skills set" (9) with an underlying knowledge of the characteristics of daylight to use the tools available effectively.

This paper therefore suggests that the educational methods to invite authentic understanding of the particular phenomenon of daylight are critical, valuable and necessary. It is proposed that an understanding of and the ability to predict lit ambiance cannot be disregarded in architectural communities and educational establishments in favour of tools that allow for numerical building design energy output data sets or 'optimised' solutions with little engagement or skill in their creation and complexities in their interpretation. The design tools used for the analysis of daylighting in architectural contexts currently generate unrecognisable data for architectural designers (see questionnaire results below), as outputs give little indication of the architectural space and the contributing factors directly affecting the results. The understanding of this data requires translations into 'designerly ways of knowing' (10) to ensure intentions are understood and the ability of architectural designers in the understanding and prediction of lighting ambiance does not become a lost skill or undervalued architectural design element.

> *As long as educational frameworks relegate the emotional and experiential to the place of a supplement, then our design processes will continue to unconsciously promote environments of thinness and superficiality. (11)*

It is therefore proposed that tools, providing primarily quantifiable data are in themselves 'out of touch' with design methodologies. They do little to assist designers with the prediction of important insights into lit environment ambiances that relate to increasing understanding in the field of physiology and psychology of the visual and non-visual effects of light. This paper seeks to challenge the exclusive use of either method, whether analogue or digital, for the understanding and representation of lit interiors. Either methodology used in isolation encourages the cultivation of design epistemologies and ontologies for light that are out of "touch". Instead, this paper aims to rise to the challenge of these contemporary complex understandings of daylight, the science and the art of the light and questions: If the demand for metrics to predict the energy value of daylight continues to grow how might this ontology of daylight and the lighting data generated help architectural designers to improve their understanding of lit ambience of a space? Findings from a questionnaire are presented to support the case for understanding and interpretation of quantifiable measures of daylighting in architectural communities. Further, pedagogical constructs based on the problems identified in the questionnaire are evaluated through findings from a series of educational workshops that invite physical, experiential learning of daylight's characteristics *in collaboration with* numerical data and digital modes of learning to provide connections and translations (12) to develop.

> *The very character and purpose of light is dependent on a set of design principles which are revealed to the observer through experience, and not through a planar map of illumination levels. (1)*

2 Translational Explorations – Design Practitioner's Questionnaire

This paper first reports on the findings from a questionnaire of over one hundred architectural professionals in the UK. The questionnaire was initially created to explore the broader conversation of how architectural designers perceive architectural lighting information and to determine a common knowledge base or ontology for light across varying scales of architectural practice and communities. The questions were created by introducing a series of visuals to view on an individual's computer screen. Each visual example provided a different format and level of lighting information to ensure a fairly broad range of contemporary lighting representations/information outputs were tested. Responses to these visuals were recorded by each individual designer as a series of qualitative descriptive written answers to branched, open-ended questions. It was hoped that through analysing these descriptions of the visuals some understanding of the common or diverse interpretations of architectural lighting from analogue and digital representations of lit spaces would emerge.

The questions were created as paired visuals; representations of the lighting of interior spaces. Each question displayed a simple visual showing one representation of lighting information. The following question of the pair added an additional layer of lighting information. The participants were asked to respond to each question in sequence and not to return to a previous question. This guidance was to ensure parity in the sequencing of responses and to guarantee responses were not affected by information gained in later or the second of the pair of questions. This paper discusses the findings from the first four example questions presented.

The selected examples utilise plan views of iconic architectural spaces, one of those selected is discussed in this paper; le Corbusier's Notre Dame du Haut, in Ronchamp, completed in 1955. This building was selected for its notoriety and familiarity in architectural communities in Western Europe (the participant group were UK based) and, most importantly, its large variation in daylighting conditions and ambiance. It was expected that any architectural designers familiar with the work of Le Corbusier would be able to identify the building by its unique plan form and elevational profile. The visual shown in the questionnaire therefore takes the form of an isolux diagram as an orthographic plan projection where contours provide points of equal illuminance. This format of a plan with contour lines is representational of the varying daylight levels throughout the space as calculated on the working plane[1]. Rather than Illumination levels (Lux) being shown, the plan indicated Daylight Factor (DF)[2] numerical values, expressed as a percentage, a visual format used by many daylighting and environmental design software applications. The Daylight Factor is a metric used in the UK and defined by UK building regulations to predict expected daylight

[1] The 'working plane' as defined in https://www.cibse.org
[2] Daylight factor as defined within https://www.cibse.org

levels. Although this metric is now under scrutiny for its accuracy in use[3] it was assumed that DF values often demanded as part of local authority planning and building control guidance would be a familiar lighting metric to many architectural designers in the UK. Of the 110 respondents to the question, only 10% were able to correctly describe the metric used as 'Daylight Factor' and/or most commonly, incorrectly used the term 'luminance' in their response. When asked to discuss the predicted lighting contrast in the space only two of the respondents identified with the numerical characteristics of the isolux diagram and were able to note ratios of 1:5 etc.

The second example used the same plan showing the same Daylight factor values but added a graded white to black monochromatic shading render between each contour relating to the DF values: high (white) to low (black). In analysing the responses, it quickly became apparent that with this additional layer of information it was possible for all those examining the visual to arrive at similar conclusions for the descriptive writing task. Significantly, using the combined rendered and numerical visual, 90% of the participants felt confident enough in their understanding to answer this question and 60% of participants discussed the contrast using the numerical information given on the plan. These data highlight clearly that the visual format, including the shaded patterning helped significantly *in the translation of* the numerical DF information.

The second pair of questions was selected to show a virtual view of the same built space (Notre dame du Haut) created using Sketch-up. For continuity, the same building was chosen and the same main space visualised but in this scenario the format was a virtual representation of the space with shadows turned on and materials as defined by the limitations of the software's materials palette library. The same questions were asked and the results were considerably different. It was assumed that participants would at this point realise that the space they were considering was the same one as that in the previous questions, but it is clear this was not the case. Participants described the space in terms of volume and only 5% of respondents wrote more than a couple of words describing the lighting. The second visual of the pair had luminance information added as measured on site at the time of the photo being taken. Luminance levels were measured with the use of Photolux app[4]. Many more participants responded to this visual and notably 30% discussed the lighting contrast. It was particularly interesting to note that the Sketch-up visual did not show as significant contrast values (to see by the eye) as the numerical values displayed and yet all the descriptions suggested that there was 'little contrast', or the space was 'evenly lit', even although the measured contrast was 1: 5 and were noted as such on the visual. It was apparent that although this numerical information had been added to the visual the actual numbers shown either weren't understood or the participants had an overriding bias to the visual scene shown rather than translation of the numerical results.

The final pair of questions aimed to define a descriptor of the lit ambiance of the space using a photographic visual of the interior. Again, for continuity, the same building was chosen and the same main space visualised but in this scenario the format was a simple colour photograph of the space. All participants responded to this visual and responses used many verbs and adverbs to convey the lighting. When asked to describe the contrast in the space all attempted to discuss the contrast *in relation to* the architectural characteristics such as; "sharp light through the small piercing windows against the gloomy wall', 'tiny windows provide strong contrast', 'the dark areas are contrasted with the very bright clerestory'. The context of the lighting and its relationship to the built space was clearly helpful in providing

[3] Sketch-up - https://www.sketchup.com
[4] https://www.photolux-luminance.com

an ambiance that could be visualised and described easily by the participant group. The second image of the pair used the same base photograph but added in luminance levels as measured on site at the time of the photo being taken. Luminance levels were measured with the use of Photolux app. With this added information 70% of participants chose to discuss the numerical contrast values alongside the text descriptors. Again, for this visual image, the added layer of luminance values provided further information to discuss but was clearly understood as a metric in relation to the view of the interior space.

These findings, although limited due to the participant groups locality (UK based) and number of participants, highlighted points worth investigating further. Respondents showed little awareness of lighting metrics, their numerical relationships and the technical terminology associated with daylighting. It must be presumed that either these architectural designers do not have a need for this knowledge or they are passing these aspects of the design on to another. Were the numeric values simply not understood? Research has shown that our understanding of lit ambiance cannot easily be equated to photometric values (13). As daylighting metrics are derived from schema that are fundamentally mathematical constructions, not physical luminous quantities, (14) it can be challenging to predict the temporal, context specific phenomenon of daylighting ambiance in an interior space.

The results may imply that the rationale behind lighting metrics is not clear and, significantly all the more confusing as we lose our direct engagement with daylight's intrinsic characteristics concealed behind a digital user interface. Digital methods allow data to be generated seamlessly and few architectural designers have shown this to be problematic but clearly the translation of this information into useful project information is less successful.

Many software developers are researching solutions to create improved speeds and accuracy to encourage designers to engage better. However, few architectural designers have reported that the speed of calculation is problematic. Rather, it seems that this research and development may be misaligned. The results of the questionnaire suggest that the output data from the software would be more frequently used if it could be understood and therefore assigned more value by designers as an important part of their design process. The hidden processes that generate the lighting data in digital software may be better understood if exposed to the user in an intermediate form, rather than concealed behind the user interface. As soon as the metric formula with its relatively straightforward relationships, angles of the sun to sill height and areas of window to wall are removed from sight we seem to lose our understanding of how the numerical outputs are generated and the rationale behind the metric itself. In order to understand daylighting software output data fully it is crucial that the designer is given the full story, for example; what parameters were used for this calculation and what values were assumed? This is important information, for without the designer understanding these input values it is impossible to analyse results and revise the design appropriately to suit the architectural rationale and design priorities.

3 A need for experiential explorations

In Pallasmaa's essay on 'Multisensory experience: The significance of Touch', he suggests that genuine architectural works stimulate our ideated sensations of touch, and this stimulation is in itself life-enhancing as it enhances our experience of ourselves (3). This idea of a physical encounter through ideated sensations when we interact with an architectural space has support in contemporary scientific findings. Discoveries reveal a more embodied understanding of emotion, through our sensorimotor circuits providing mirror mechanisms resulting in "embodied simulation" (Gallese 2005 in (15), the physiological rationale behind our enjoyment of watching dance, examining art or listening to music being played. This connection with actions and the environment within which they are taking place can be aligned with our appreciation of an architectural space, its character and the emotions it evokes as we experience it. It suggests too that as we perceive daylight patterning in an architectural space; the ephemeral, transient nature of daylight interacting with the architectural materiality and spatial composition we attune sensorialy and emotionally with these fluctuating surroundings as Edmund Husserl describes "animate organisms". This paper proposes that design educators can learn from Pallasmaa's call to defy this "sensory impoverishment" and Mallgrave's Phenomenological Model for Architectural Research by re-evaluating daylighting pedagogy through valuing experiential learning with regard to lit ambiance.

The following outlines an investigation into a contemporary pedagogical approach to daylighting design that seeks to make transparent the ephemeral and particular ambiance that daylight can create in architectural interior spaces through strategies that encourage "sensory invitation and discovery, mystery and shadow" (16). Using a Constructivist pedagogical approach phenomenological constructs are proposed that align with teaching methods to re-engage with daylighting in experiential ways, allowing designers to "work directly with the object of their thought" (12). Designed workshops aimed to encourage sensory invitations to discover daylighting by re-engaging physically with context specific scenarios. Strelitz (17), anthropologist and interior designer, when discussing her approach to creating 'Buildings that feel good' notes that "Learning from buildings that feel good instructs us in creating buildings of lasting value…as elements with which we engage in the built fabric around us"((17) p3). As we are often unaware of lighting in a room unless it is particularly poor, it was considered important to discuss the rationale of each workshop with participants beforehand to ensure that the processes they became involved with were explicitly and consciously considered. Theodorson (18) has highlighted the difficulties in teaching lighting in this regard:

> *Humans have untold experiences with light; it is the primary means we use to "know" our world. Yet this immense range of exposure is primarily passive and ironically, does not often result in retrievable knowledge about light's physical behaviors, impacts, qualities, or intensities. (18)*

The workshops outlined in this paper challenge participants to "touch" light, implying a physical and emotional connection with daylighting that the study questionnaire suggests we intrinsically value. They sought to provide practical strategies for the unravelling of daylighting design through overlapping digital and analogue methods and how this might enhance understanding of the lit ambiance of a space in current contexts.

4 Translational explorations – 'touch'ing light workshop

A series of workshops was created to explore and evaluate physical, haptic engagement with light. The research explorations utilised design workshops, focus group discussions and questionnaires with the aim of identifying insights into design processes and knowledge acquisition through the application of varying tools in daylighting design scenarios and the subsequent participants' responses. Design studio was chosen to test these methods through controlled scenarios and observational techniques. Studio workshops were designed to fit within a current architectural educational setting to ensure continuity of results through weekly analysis within the same environmental setting and participant group, producing results from real studio contexts without the added complexity of unfamiliar, simulated situations.

Over a period of a month, a group of approximately 30 students attended a two-hour workshop in lighting each week. The student group was a mix of 1st year Interior design students to Masters level Interior design students ranging in nationality and design educational background. Participants were placed into groups of three to encourage discussion and provoke realisation of others values and experiences of the lit environment. The workshops were held in the general studio space, providing access to laptops, model making materials, drawing boards and layout space. Attendance was optional but encouraged. The work created was not assessed but it was assumed that any drawing, model or design work generated would be beneficial for each student's portfolio. The first of these series of workshops is outlined here:

Workshop 1 – Describing lit environments -photographic visuals of an architectural space
Describing lit environments -photometric visuals of an architectural space

Workshop 2 - Creating matching lit environments (2D visual to 2D drawing) through hand drawing
Creating matching lit environments (2D visual to 3D drawing) through physical modelling

Workshop 3 - Measurement of Illuminance and DF using software and 3D VE model
Adapting 3D VE models to comply with metric requirements

Workshop 4 - Measurement of Illuminance and DF using physical modelling
Adapting physical models to comply with metric requirements

Workshop 5 – Focus group meetings

The workshops provided an opportunity for students to work in an experimental way using a broad range of approaches to describe a lit environment. The pedagogical approach of weekly group tasks commenced with shared oral and written descriptions of daylit interior environments followed by a re-creation of one of the spaces as imagined in a three-dimensional physical model to allow open discussions around the subject of daylight, to awaken understanding of others' unique understanding of lit ambiance and common qualitative aspects of daylighting design.

> *A transformation in the architectural design process should be stimulated redirecting attention to the context and the cultural spirit of*

> *place, as well as encouraging a sound understanding of the quality of natural light. (19)*

Through three-dimensional heuristic learning methods and associated formats for expression it was imagined that participants would indeed learn to 'touch' light tempted by the invitation to explore and investigate lit space. The subsequent workshops required engagement with daylighting metrics, using quantitative methods of analysis. Participants were provided with a three-dimensional virtual model of their space to work with in IES VE[5] software and further to a 15-minute demonstration tutorial, each small group followed through calculating the daylit environment using basic sunlight and daylight metrics. This quantitative analysis was then repeated using each group's physical model measuring DF, illuminance and luminance with Edenapp and Photolux apps on mobile phones placed within or against the physical model. These workshops demanded physical model making to encourage haptic engagement with lighting metrics.

The final workshops asked for revisions to both proposals, the three-dimensional virtual model and the three-dimensional physical model. Participants were asked to revise the models physically and virtually to comply with a 4% daylight factor and an average illuminance of 200 Lux over the working plane as an example level requested by local building regulations.

Interior lit environments were selected to be broad ranging yet architectural projects that had a strong interior and daylit aesthetic. Student participants were able to describe the photographic visuals well. Very few were able to discuss the photometric visuals successfully and no participants commented that the same space had been shown in qualitative and quantitative terms. It was interesting to note that the students used a similar amount and type of descriptive language as the practitioners who had been given the same lit interior scenario in their questionnaire.

The participants created physical models of an interior space based on the 2D visual they had been given. Participants worked quickly and confidently to develop spatial voids which they filled with texture, colour and light. Some participants used transient sunlight to explore ephemeral atmospheres. Others used static artificial light and layered the spatial armature with tones and textures to correspond with their understanding of the space. Most models created had no determined scale but were successful nonetheless in realisation as it quickly became apparent to the participants that light has no scale. Many models had a clear observation angle. Conversation analysis during group work made this phenomenon explicit as participants declared "You have to stand here to see it properly!" Most physical models allowed for participation *within* the spatial void created. Other participants were invited to "come, look inside and see how it feels!"

Participants found it quick and straightforward to create photometric data for their 3D virtual environment model. The output data as Isolux diagrams, numeric DF values and contrast ratios could only be explained by a few participants however. The lack of understanding of these numbers and the relational values to the interior space and external context was further demonstrated in the subsequent workshop.

---

5 IES ve - https://www.iesve.com/

 https://doi.org/10.1051/shsconf/20196402010

The following week, participants were asked to adjust their VE model to suit the metrics given for compliance. Participants were unclear as to which input variable to select but very quickly this approach, allowing multiple trial and error scenarios to be evaluated, showed some success in the numerical results. Significantly it was noted by most that the reflectance of ceiling material became crucial to the metric values achieved. When participants were asked to discuss how this change in reflectance might be created in a real project with real materials most assumed the use of mirrored surfaces rather than changing the colour (light absorption properties of the material). It became clear that the participants were making changes to their design without really reflecting or understanding the nature of their revisions and the impact on the architectural aesthetic or change to spatial perception in achieving the required results. Nicholas Carr, in writing an article about on-line learning, discusses controversially that in using digital software we are frequently just 'decoders'(20), considering unrelated parts of information out with known contexts. The findings of this study imply that this may be happening when we use digital software without a basic knowledge of the processes involved. It is proposed that perhaps the lack of haptic engagement presents itself in this specific situation through observing this virtual scenario created using digital lighting software devoid of experiential context. Pallasmaa (11) proposes that "touch is the unconsciousness of vision, and this hidden tactile experience determines the sensuous qualities of the perceived object".

In the second part of the workshop, students sought to measure the lit ambiance of the physical model photometrically (Fig.1.). Students used Edenapp and Photolux to calculate illuminance levels, luminance and DF values. Participants were then asked to adjust these measurements to suit a given lighting metric requirement through manipulation of the model. It became apparent very quickly that this was possible and connections were strong linking the metrics with the physical model materiality, opening sizes etc. Each causal architectural relationship within the architectural space needed consideration and this was easily tested using physical materiality and light source.

Fig. 1. Example drawing to model image to Photolux layer. Source for exploration Bruder Klaus Chapel – Zumthor

Each workshop was discussed in the subsequent focus group sessions and a key element of daylighting knowledge and design epistemology emerged. The exploration of these relationships *between* light and architectural space served to help visualise this embodiment explicitly. It was necessary for these design explorations to be interactive and encourage iterative variables in order to unravel the varying relationships between materials, light and photometric values.

5 Future explorations

In conclusion, this study offers insights into methodologies for design pedagogy that can provide transparent relationships between experiential learning, numeric reasoning and representational techniques seeking to encourage the creation *and translation* of daylit ambiances through overlapping of each mode with numerical digital interactions. Through the demonstration of multi-modal tools in use, this paper proposes future collaborations to reveal understanding of ambiance through experiential design. This application of qualitative and quantitative measures and insights may provide improved understanding of daylight's characteristics and potential in built environment contexts. In so doing this provides a better understanding of lighting metrics and their use in application. This methodology of practice is wide ranging and may provide a basis for further research addressing collaborative pedagogies within related fields of architecture with acoustic, sound design and other intangible sensory applications.

> *Becoming more fully aware of the extent of our biological complication, whose underpinnings reach deeply into the sensory-emotive world that we daily inhabit, is simply a first step in this process". (21)*

This research seeks to challenge the idea that we may be out of 'touch' experientially *when* we use digital tools exclusively rather than, as might be more commonly understood, as being out of 'touch' by *not* using digital methods. The current tools available for daylighting design can meet the expectations of contemporary design students and educators if used in parallel and in haptic and collaborative ways. Using a combination of tools, the architectural designer might predict the lit ambiance of an interior space but only through first understanding how the relationships of light and its interactions with the interior space work – characteristics as experienced. Only with this understanding can a sensitivity to daylighting ambiance be revealed and intentionally designed into our built environment.

> *The computer is an important tool- no one could do without it – but for me it's only a tool and it doesn't replace thinking. It can make you disconnected...bring it out of the computer...use physicalities and models to understand and anticipate what this thing will be in the end: something physical, something real, something for people. (22)*

References

1. Rockcastle S. *Annual Dynamics of Daylight Variability and Contrast A Simulation-Based Approach to Quantifying Visual Effects in Architecture*. Andersen M, editor: London : Springer London : Imprint: Springer; (2013)

2. Treacy G, *Innovative learning technologies for daylighting education*. Professional Lighting Design Convention; Rome: VIA Verlag (2015)

3. Pallasmaa J. *Encounters : architectural essays*. MacKeith PB, editor. Helsinki: Helsinki : Rakennustieto Oy, (200)

4. Storkerson P. *Naturalistic Cognition: A Research Paradigm for Human-Centered Design. Journal of Research Practice*. 6(2):M12.(2010)

5. Holl S. *Parallax*. Basel: Basel : Birkhäuser; 2000.

6. Hawkes D. *Environmental imagination.* London: London : Routledge; (2008)

7. Lobell J. *Between silence and light : spirit in the architecture of Louis I. Kahn*. Kahn LI, editor: Boston ; London : Shambhala; (2008)

8. Plummer H. *Poetics of Light.* A + U (Architecture + Urbanism) Extra Edition.(December no.12). (1987)

9. Mardaljevic J. *Book review: Daylighting handbook I.* (2016)

10. Cross N. *Design thinking : understanding how designers think and work.* Oxford: Berg; (2011)

11. Teal R. *Dismantling the Built Drawing: Working with Mood in Architectural Design.* International Journal of Art & Design Education. 29(1):8-16 (2010)

12. Evans R. *Translations from drawing to building.* AA files. (1986)

13. Boyce PR, Smet KAG. LRT symposium *'Better metrics for better lighting' – a summary.* Lighting Research & Technology;46(6):619-36. (2014)

14. Tregenza P, Mardaljevic J. *Daylighting buildings: Standards and the needs of the designer.* Lighting Research & Technology.50(1):63-7 (2018).

15. Harry Francis M. *Embodiment and Enculturation: The Future of Architectural Design. Frontiers in Psychology.* (2015)

16. Pallasmaa J. *Hapticity and time: Notes on fragile architecture.* The Architectural Review. 207(1239):78-84.(2000)

17. Strelitz Z. *Buildings that feel good.* London: London : RIBA; (2008)

18. Theodorson J. *Teaching light - Constructing knowledge across multiple dimensions.* (2012)

19. Traverso G, editor. *Modelling Daylight.* Germany: VIA Verlag; (201).

20. Carr N, Norvig P. *Is Google Making Us Stupid?* New York Times Upfront.143(3):22. (2010)

21. Mallgrave HF. *The architect's brain : neuroscience, creativity, and architecture.* Wiley I, editor. Chichester, West Sussex, U.K. ; Malden, MA; Wiley-Blackwell; (2010)

22. Gordon A. *"Credit Swiss," Interview with Pierre de Meuron and Jacques Herzog.* The Magazine from the Wall Street Journal. (2008)

A network as a source of knowledge for the education of an architect

Melis Baloglu Asut[1], and *Yüksel* Demir[1]

[1] Istanbul Technical University, Department of Architecture, 34367 Taşkışla, Şişli Istanbul, Turkey

Abstract. In a learning space, whether it is surrounded by digital tools or not, the way a tutor represents the information is usually in a linear fashion. However, e-learning platforms have already started to support the learner and a new paradigm for learning. In light of these thoughts and concerns, this paper is focused on the education of an architect and examines potential e-learning platforms including BlackBoard, Facebook in the case of an architectural course, EdX, and Khan Academy Knowledge Map based on the types of visual narrative, the aim of the platform, the information visualization quality, and data types. As an output, this study creates a comparative visualization to see and discuss the potentials of the tools for the education of an architect. After the analysis of e-learning platforms, this study concludes that (1) an education model based on a network can help architects to see the big picture, however it also causes data reduction and loss, (2) an education model based on a network allows architects to create their own learning paths and share it with their colleagues, (3) an education model based on a network may be a sustainable platform for architects from different experience levels and enhance the communication between them to support the youngers' education, 4) an education model based on network decentralizes the process of learning and enriches it through the interaction between other learners, concepts, thoughts, and objects, (5) an education model based on network bridges the gap between what one already knows and what one needs to know to understand the whole picture, (6) an education model based on network gives the opportunity to compare previous years' feedbacks, concepts, and thought in one medium.

1 Introduction

In a learning space, whether it is surrounded by digital tools or not, the way a tutor represents the information is usually in a linear fashion. For example, a book is covered from the introduction to the conclusion, or a PowerPoint presentation can be shown from the first slide until the end. However, the instruction is becoming more learner-centered and nonlinear as a network where:

(a) the learner is aware of knowledge resources believed to be relevant for her/him,
(b) the learner builds his/her knowledge together with the other members based on the shared interests in the community,

(c) the learner creates her/his personalized plan in the discussion groups.[1]

In such a network, the learner is the center of learning experience, not the teacher nor the institution. The environment is described as a networked learning environment. Networked learning refers to both network and learning. And also, networked learning tools refer to e-learning platforms, social media, forums, blogs, games and so on.

E-learning platforms and social media have already started to support the learner and a new paradigm for learning, working, studying and problem-solving is already defined by them. This new paradigm reflects the connectivity of people and learning sources [2] However, this new paradigm had also brought a new problem called "know-where". Know-where basically means the understanding of where to find data which data may be used as a knowledge source [3]. Although e-learning platforms support;

(a1) the awareness of knowledge sources like (a) Khan Academy Knowledge Map [4], but it can also create the situation called map shock since the map is big and if learners cannot filter or zoom the sources related [5] to their interest and relate one to another.

(b1) learners to build their knowledge with the interaction of other members like they do in EdX [6], but they cannot manipulate their research history with commands like undo, replay etc. [5]

(c1) learners to create their own personalized learning plan (c), like they do with the help of social media as Facebook. Social media allows users to learn, share and communicate with anytime and anywhere learning but learners cannot see the big picture or overview. [5] Social media platforms like Facebook, and an open source learning management software like Moodle, or a learning management software as BlackBoard may be stronger and more sustainable than e-learning platforms developed by formal education institutions. Having said that, social media was not designed to be an e-learning platform, it is used like a communication facilitator between learners. Each system has its own positive and negative effects for its users. But how do we know which kind of platform, what kind of visualization and narrative help an architect to learn continuously in a fast-changing world is the question of the digital world.

In light of these thoughts and concerns, this paper is focused on the education of an architect. How data should be structured and presented in e-learning platforms for the education of an architect is the question of this research. To be able to answer the research question, this study is examined potential e-learning platforms including social media for the education of an architect based on *the types of visual narrative* [7]; *the aim*; *the information visualization mantra for interactivity* [5]; *data types* as 1D, 2D, 3D, network, etc. [5]; and *Edward Tufte's writing about the visual communication of information* under the topics of scale, color etc. [8] and finally it creates comparative visualizations. As a conclusion for this study, a network can be seen as an environment to support know-where. In a network, there is no center; there are just nodes and relations between them. And if the tools to produce information are readily available to any member of society, the question of "who is an expert" may yield entirely different responses than in more hierarchical, structured ones [9]. In such a society, an architect can tease out the data from the big picture/data and can find the data s/he needs where the network visualization assists her/him.

As an outcome, this study proposes a model for the education of an architect. After analysis of e-learning platforms and social media, this study concludes that (1) an education model based on a network can help architects to see the big picture however it also causes data reduction and loss, (2) an education model based on the network allows architects to create their own learning paths and share it with their colleagues, (3) an education model based on the network may be a sustainable platform for architects form different experience levels and enhance the communication between them to support the youngers' education. Since knowledge construction is a unique process and everyone has their own way to interpret information in relation to their personal experience, desires, and hopes [10], in a network anyone can see the one's unique learning process, (4) an education model based on network decentralizes the process of learning and enriches it through the interaction between other learners, concepts, thoughts, and objects, (5) an education model based on network bridges the gap between what one already knows and what one needs to know to understand the whole picture, (6) an education model based on network gives the opportunity to compare previous years' feedbacks, concepts, and thought in one medium.

2 Network, Network Society, and Networked Learning

When we refer to networked learning, we refer to both network and learning. For this reason, this section introduces the network society and the features of its network. Modes of production bring new tools that bring new ways of communication that bring new way of daily life. Nowadays, people can take their personal computers and go wherever they want to study or work. In a such a daily life, there is a network between people and people, tools and tools, tools and people and people and tools. This network is a set of interconnected nodes. A node is a point where the curve intersects itself. A network has no center, just nodes. All nodes are necessary for the network performance. When nodes become redundant or useless, networks tend to reconfigure themselves, deleting some nodes, and adding new ones. [9] Between the nodes in the network and those outside the network, distance is infinite, since there is no access unless the program of the network is changed. Networks became more efficient in terms of their organization as a result of three major features. They are flexibility, scalability, and survivability.

- Flexibility: they can reconfigure according to changing environments, keeping their goals while changing their components. They go around blocking points of communication channels to find new connections.
- Scalability: they can expand or shrink in size with little disruption.
- Survivability: because they have no center, and can operate in a wide range of configurations, they can resist attacks to their nodes and codes, because the codes of the network are contained in multiple nodes, that can reproduce the instructions and find new ways to perform. So, only the material ability to destroy the connecting points can eliminate the network [9].

Network society is a society whose social structure is made of networks powered by microelectronics-based information and communication technologies [9]. The culture of the network society is a culture of protocols of communication between all cultures in the world, developed on the basis of the common belief in the power of networking and the of synergy obtained by giving to the others and receiving from the others [11]. In other words, the culture of network society is more open to share and manipulate like in the network of hackers. The hacker ethics is conceptualized by Pekka Himanen and supports the free sharing of knowledge and the joy of creative making. In a sense, the hacker ethics could open a discussion for the architect ethics since architects and students of architecture use tools to design and during the design process, they use online libraries, they share their representations online. That is why, the hacker ethics is a way to discuss the rules of online communities, licenses and sharing however this topic is not the main issue of this paper.

Networked learning integrates both the new movements and is based on existent learning theories as it consists of building blocks from (1) connectivism, highlighting the relevance of connections for learning, (2) social network analysis, pointing out the different types of connections and their implications for learning, and (3) Collaborative Learning Theories, based on constructivism [12]. Networked learning is learning in which information and communications technology (ICT) is used to promote connections: between one learner and other learners, between learners and tutors; between a learning community and its learning resources [13]. Networked learning has been described as involving learning situations in which collaboration and participation are mediated via technologies [14]. In this study, while it is referred to networked knowledge, it is not only about connecting textual content but connecting different media like video, and audio content, images, and documents. The term "networked knowledge" exemplifies several important facets of knowledge. First, in a "knowledge society", knowledge needs to be connected in order to generate new knowledge or innovation, which can be realized by Semantic Web technologies. Second, knowledge also needs to be shared among people in order to be used effectively, and much of this sharing is based on collaboration, social software, and social networks. And third, knowledge is never isolated but always embedded in a context, connected with other Information [15]. It is important to examine the features of tools used or to be used in networked learning since network society and its demands have been changing according to fabrication and communication technologies and the culture that they are affecting.

2.1 Networked Learning and Tools and Technologies

Mobile, personal, voice-enabled multifunctional devices such as laptops, tablets, and smartphones have made it possible to participate in networked learning 24/7 from almost any location, including in workplaces, home, the bus and the street [14]. According to the Networked learning 2018 report, it is visible that: (a) Learning management systems (i.e. Blackboard, Firstclass, WebCT, Moodle, Fronter) decline, (b) MOOCs seem to follow a wider cultural trend of becoming excessively popular after 2012, following the rise of platforms, such as Coursera, and EdX, (c) YouTube protects its place, but research needs to be done in the field of architecture since students usually watch tutorials from YouTube to learn how to use software, (d) The popular use of Twitter can be understandable in two ways. Firstly, most people use their smartphones with them during the day and Twitter is a simple tool to follow from a small screen, (e) Regarding the use of digital devices by learners, the term laptop is declining, however, the terms phone and mobile are increasing (See Figure 1.)

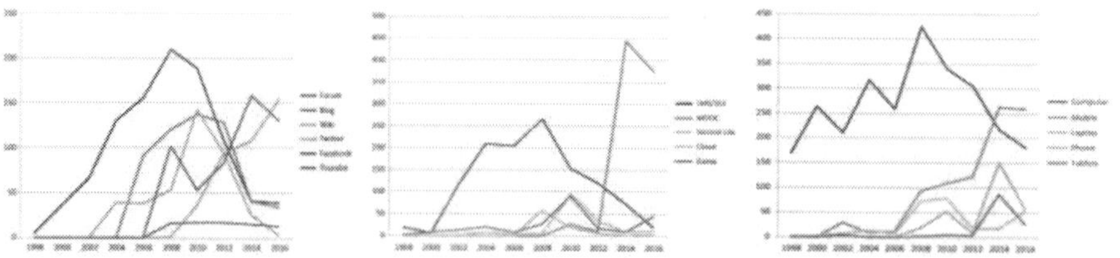

Figure 1. Based on the Networked Learning Report 2018

After a quick analysis of the Networked Learning Report 2018, this study also highlights that the way to represent data is also crucial. In light of these thoughts and concerns, this study examines potential e-learning platforms including social media for the education of an architect based on:

- *the types of visual narrative* as a static visual narrative like comics, a dynamic visual narrative like animation and an interactive visual narrative like interactive games [7] can give an idea on users' learning preferences;
- *the aim,* such as providing access to learning material or visualizing concepts can give an idea of the possible combinations of functions in one tool;
- the information visualization mantra for interactivity, such as the ability to overview, zoom, filter, relate, extract history [5] can give an idea on how one needs what kind of features to reflect content;
- *data types* as 1D, 2D, 3D, network etc. [5] can give an idea of how one can pick the relevant tool in relation to data type.
- *Edward Tufte's writing about the visual communication of information* under the topics of scale, color etc. [8] can give ideas on design concerns of a tool. The following sections introduce the mentioned concepts.

2.1.1 Networked Learning – Tools and Technologies – The Types of Visual Narrative

Human beings have been telling stories in a written and oral way. One of the first examples is cave paintings in Lascaux. A visual is something that can be seen to a human eye. In a simple definition, a narrative is the representation of an event or a series of events [16]. It can be said that a narrative is the representation of a story while a story is a sequence of events. And visual narrative is synonymous with visual stories, narrative images, picture stories, narrative pictures [7]. A visual narrative could be represented in any medium. This study is focused on four case studies that are online platforms by their nature. These case studies are examined through whether they contain static visual narratives, dynamic visual narratives or interactive visual narratives. Simply put, in the case of the static visual narratives, the visual is frozen, but the viewer's eyes and mind move on [17] like cave paintings; in the case of dynamic narrative, the viewer is in a fixed spot, but visuals are replaced by visuals at the same display space like animation and puppet shows; in the case of interactive visual narrative, the viewer and the visual can be both mobile or static as interactive games [7]. In the field of architecture, the static visual narrative is used to show technical drawings and photographs of buildings to students. The dynamic narrative is used both by students to represent their projects via digital tools and also by their tutors to show the spatial quality of a building virtually.

2.1.2 Networked Learning – Tools and Technologies – Aim of the tools

Learning content is often complex, ill-structured and not pre-designed. That is why there are tools for knowledge management, which make use of the human cognitive processing system in order to create and convey content more efficiently. Recently, there has been an increasing interest in visualization for knowledge management systems in academia like Khan Academy Knowledge Map [4]. It is understandable since the conventional knowledge management or content management tools do not allow academics freedom and flexibility in the course design to develop pedagogically effective learning activities, nor do they allow easy resourcing, reusing or re-sharing. The existing body of research on the knowledge and content management systems for the educational use suggests that combining every feature like sharing, content archiving, organizing different types of materials in one tool is a complex task however the visualization can help. There is a need for tools aiming at supporting cognitive processing in generating, representing, structuring and restructuring, retrieving, sharing, and using knowledge.

Helping students to organize their knowledge is as important as the knowledge itself since knowledge organization is likely to affect student's intellectual performance too. There are studies supporting knowledge organization both as linear learning plans which cannot be traced by learners such as Khan Academy Knowledge Map [4]; and as platforms which offer online courses like EdX [6]. Despite all, social media help both instructors and students to increase their motivation, using mobile devices in conjunction with social media, free web tools that allow for communication and enhance learning [18]. Although each tool has its own potential, it is still not known how to combine the most effective features like sharing, content archiving, showing links between concepts in one tool to foster the meaningful learning. To sum up, the aim of the tools can be stated as follows: (a) visualizing concept map, (b) searching for concepts, (c) providing access to learning materials, (d) creating and exploring learning plans, (e) creating concept maps, and (f) adding course materials.

2.1.3 Networked Learning – Tools and Technologies – The Information Visualization mantra and Data Types

"Visualization" is an activity in which human beings are engaged as an internal construction in the mind [19]. It is something that cannot be printed on paper or displayed on a computer screen. This paper refers to the term "visualization" as visually representing data that can be generated and calculated. According to Mazza, Information Visualization lays between data and information [20]. We visualize data to produce information where that information is transformed into knowledge with the help of the user's experience. Visual representations help us to ease that process. However, a human can understand a visual representation when it is well designed. Shneiderman [5] enumerated seven tasks users could perform on the data that can also help to design an efficient tool which are: (a) Overview: Gain an overview of the entire collection, (b) Zoom: Enlarge items of interest, (c) Filter: Filter out uninteresting items, (d) Details on demand: Select an item or group and get details when needed, (e) Relate: View relationships among items, (f) History: Keep a history of actions to support undo, replay, and progressive refinement, (g) Extract: Take out sub-collections of data or history to save and communication.

To be able to select the correct technique of data visualization, one should define the data type first. Shneiderman [5] distinguished seven data types and assumes that all data in information space are collections of items, where items have multiple attributes. These data types are; a one-dimensional like text files and alphanumeric list of names, a two-dimensional like geographic map or book layout, a three-dimensional like real-world objects and chemical molecules, a temporal like time-series and scientific measurement rows, a multidimensional like relational database content, a tree-like structured data collections with hierarchy constraints, and a network which is a structured object sets which do not apply to tree constraints [5].

During the architectural education, a student of an architect come across a 1D data in most of the theory courses, a 2D data in most of the theory courses and in a studio like an urban map, a 3D data such as physical models, a temporal data can be continuous courses in the curriculum, a tree can be seen as the whole education system like each course is designed by a coordinator, taught by a tutor, and the coordinator is assigned by the vice dean, and a networked data is visually not "visible" for everyone since the learners cannot link the ideas easily during a course and create a network for them.

2.1.4 Networked Learning – Tools and Technologies – Tufte's perspective

Edward Tufte's writing about the visual communication of information helps anyone who wants to structure the information they want to present and identify suitable visualization strategies. Edward Tufte points out some criteria to ensure that a visual representation is effective in his books named Envisioning Information, and The Visual Display of Quantitative [8]. There are five chapters from the book to analyze as Small Multiples, Narrative of Space and Time, Macro/Micro Readings (as seen in number 1 below), Layering and Separation, and Escaping Flatland (as seen in number 2 below) [8]. Using scale and sequence can illustrate the concept of time. And also, combining maps (2D) and perspective drawings help people to understand their way if they do not think in 3D.

Figure 2. Examples of good visualization [8]

As Tufte puts in words, we can discuss whether a representation is good or bad under four captions:

- graphical excellence: A good representation aims to provide the viewer with "the greatest number of ideas, in the shortest time, using the least amount of ink, in the smallest space".
- graphical integrity: The representation of numerical data, just as they are physically designed on the surface of the graphic, should be directly proportional to the numerical quantity represented.
- data-ink ratio: It is important not to overload the reader with too many elements, which could end up being unnecessary, if not positively damaging, to the final learning.
- aesthetics: It is good to avoid decorations [5]

In the coming sections, case studies are compared to these features listed and discussed during the body of this study.

3 Case Studies

There are four case studies illustrated, which are BlackBoard, Facebook in the case of Theories of Architectural Design Course in Istanbul Technical University in Turkey, Khan Academy Knowledge Map, and EdX.

□ Learning Management System: Blackboard

According to the criteria of this study, it is visible that Blackboard meets half of the requirements (as seen in Figure 3).

□ Social Media: Facebook in the case of ADT

The case study took place in the Spring semester of 2017-18 in Istanbul Technical University. The course named Theories of Architectural Design was developed and instructed by Assoc. Prof. Dr. Yüksel Demir. The course aims to examine all the dimensions of architecture, through essential "concepts", "questions", "problems", "texts", "works", "individuals" in unity. In this context, the interdisciplinary nature of architecture, the relations of architecture with "nature – culture", "art – science" will have the highest priority. All these studies will be realized considering the inseparable relation of theory and practice. The students are expected to propose their personal agendas and compose the collective agenda within this frame. Since the beginning – the term year 2003-2004, the course enables students to be aware of the various theories, which direct the architectural design. Each term student discussions form an interactive platform, where specific topics come up. Every session the volunteer reporters took notes, prepared a report, and these reports have been shared online after the moderation of the course leader. Although the course is a theoretical course, the assignments are presented usually in a visual format like posters, flyers, GIFs, short videos, and mind maps. The course has different types of communication medium between students and instructors such as blogs, e-mail groups, and Facebook groups.

In the mentioned semester, the course had a closed Facebook group. I am a virtual assistant of the course and I made four Skype meetings with the class, and each week I shared projects, assignments and made comments. The content of the course became richer with the participation of other faculty members. The course content is recreated each semester according to students' agenda. Students started the semester with a proposal consisting of their interest such as concepts, individuals, texts, works and questions. Each student must submit five concepts, five individuals, five texts, five works, and five questions. Then, they need to form a group according to their shared interest. According to a short questionnaire, findings are; *Problem 01:* Interface (Facebook) has a poor resolution and limited bandwidth. *The solution would be*: Putting stress to a verbal channel, presenting words as narration and moving some essential from the visual channel [21].

Problem 02: Reduction of information. *The solution would be*: zoom in & out and also click the text to see the explanation.

Problem 03: Using readymade templates and layouts (FB) creates standardization. *The solution would be:* A tool to create a room for personalization.

Problem 04: showing the content is easy but to be able to understand the relations between concepts, it becomes complicated. *The solution would be*: A network visualization as a concept map.

The feedback was mainly positive according to the survey done by students however it is also visible that Facebook does not meet a high percentage of the criteria. It is obvious that the aim plays a vital role while choosing the tool.

 Learning Platform: EdX

EdX founded by Harvard University and MIT in 2012, edX is an online learning destination and MOOC provider, offering high-quality courses from universities and institutions to learners everywhere [6]. It is also visible that EdX does not meet a high percentage of the criteria (as seen in Figure 3).

 Learning Platform: Khan Academy Knowledge Map is like a big picture of what a person wants to learn. It shows the concepts and their relations to other concepts so that if a learner has an interest in topic A, s(he) can follow topic C which has a relation with topic A. Since it also reflects a network, the platform meets the criteria at a high percent (as seen in Figure 3).

4 Conclusion and Discussion

The visuals can be read from left to right, having in mind that "x" means "the mentioned feature is available" while "-" means "the mentioned feature is not available".

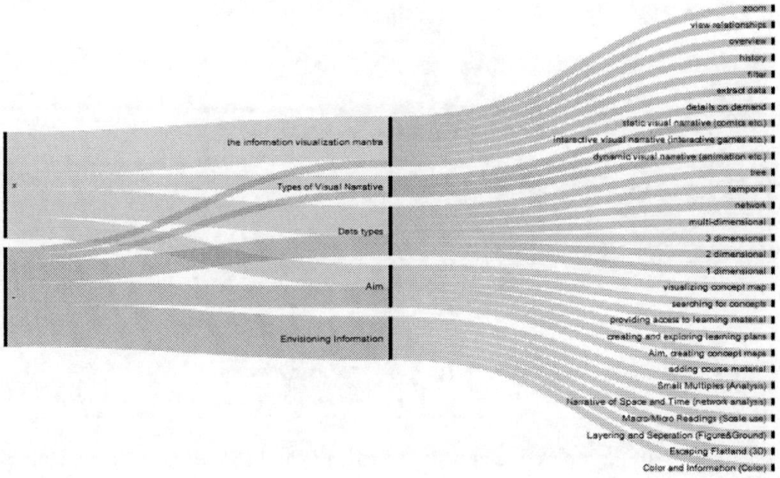

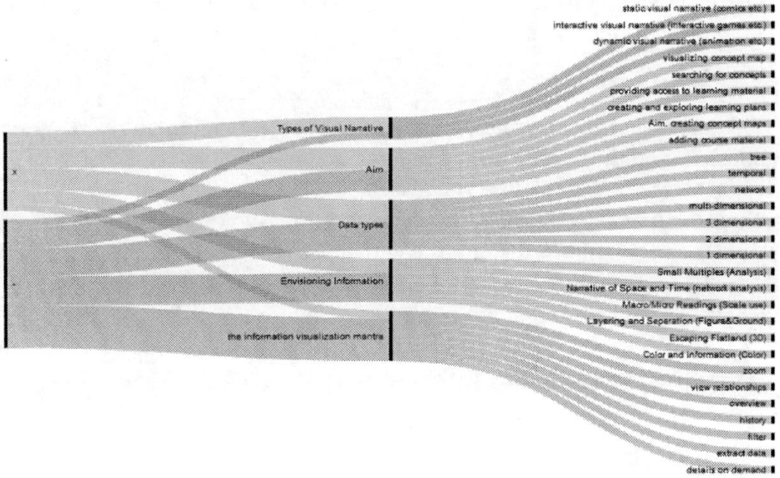

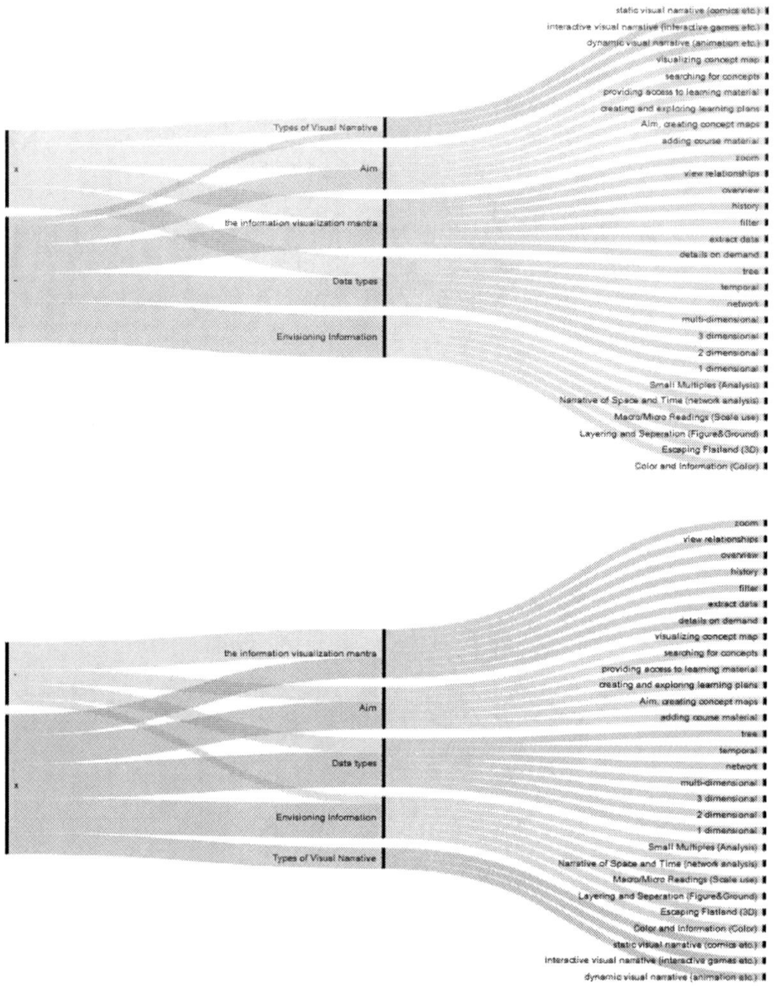

Figure 3. All the case studies are compared according to [5], [7] and [8], illustrated by Baloglu Asut, M. with RawGraphs

All serious projects should start with a rigorous specification of the requirements, collected from potential users of the system through interviews, questionnaires, etc. [20]. To answer :

Functionality: Does the visual representation provide all of the functionalities requested by the instructors and identified during the requirements elicitation?

The answer: All the cases have positive and negative sides on functionality that is the problem to solve. For some of the platforms, it is easy to embed 3D data and videos as EdX, but for some of them there is capacity as Blackboard and the speed of downloading is changing per country.

Effectiveness: Does the visual representation provide the instructors with a better knowledge of the number of messages read and written in a discussion forum than the traditional interfaces provided by the tool? In particular, does the use of visual representations allow the instructors to have information on the number of messages sent and read with better accuracy and precision than other tools? Or, is there additional information that is made available exclusively by the visual representations?

The answer: Rather than Blackboard, Facebook has more advantages on effectiveness since it has a feature that shows who and how many people sees the comments.

Efficiency: Can the visual representation provide the instructors with information more rapidly than the tools provided by the system?

The answer: Efficiency also depends on the speed and quota of the internet in different countries.

Usability: Is the interaction with the graphical interface simple and intuitive enough for the instructors?

The answer: Khan Academy Knowledge Map may create a map shock. On the other hand, Facebook is relatively easy since people use it in their daily lives.

Usefulness: In what way, and in what context, is the information provided by the graphical representation useful to the instructors?

The answer: The visual narrative effects the usefulness of the tools. Each people has a way to learn so that if a tool or a platform serves different kinds of visual narratives, then it has a high chance to reach learners.

As an outcome, this study proposes a model for the education of an architect. After analysis of e-learning platforms, this study concludes that (1) an education model based on a network can help architects to see the big picture however it also causes data reduction and loss, (2) an education model based on network allows architects to create their own learning paths and share it with their colleagues, (3) an education model based on a network may be a sustainable platform for architects form different experience levels and enhance the communication between them to support the youngers' education. Since knowledge construction is a unique process, in a network anyone can see the one's unique learning process, (4) an education model based on a network decentralizes the process of learning and enriches it through the interaction between other learners, concepts, thoughts and objects, (5) an education model based on a network bridges the gap between what one already knows and what one needs to know to understand the whole picture, (6) an education model based on network gives the opportunity to compare previous years' feedbacks, concepts, and thought in one medium.

References

1. G. Ayala, Intelligent Agents Supporting the Social Construction of Knowledge in a Lifelong Learning Environment, Proceedings of the International Workshop on New Technologies for Collaborative Learning NTCL' 2000, November 2000, Hyogo, Japan, 79-88 (2000)
2. J. Mason, Qualitative Researching, Sage, (2002)

3. G. Siemens, Connectivism: A learning theory for the digital, International Journal of Instructional Technology Distance Learning, 2,1, 3-10 (2005)

4. https://tr.khanacademy.org/welcome>; accessed on 20.10.2018

5. B. Shneiderman, The Eyes Have It: A Task by Data Type Taxonomy for Information Visualizations." Proc. IEEE Visual Languages, 336-343. (1996)

6. <https://www.edx.org/&gt>; accessed on 20.10.2018

7. R. Pimenta, R. Poovaiah, On Defining Visual Narratives. Design Thoughts, 25-46, (2010)

8. E. Tufte, Envisioning Information, Graphics Press, Cheshire, 58 (1990)

9. M. Castell, The rise of the network society, Blackwell, Oxford, 3,6 (1996)

10. E. K. Ackermann, Experiences of artefacts: people's appropriations / objects' affordances, in Larochelle M. (ed.) Ernst von Glasersfeld, Key Works on Radical Constructivism, Sense Publishers, (2007)

11. P. Himanen, et al., The Hacker Ethic: and the Spirit of the Information Age. (2002)

12. U-D. Ehlers, A. Seinert, Networked Learning in a Networked World, (2010)

13. P. Goodyear, Pedagogical frameworks and action research in open and distance learning, European Journal of Open and Distance Learning, http://www.eurodl.org/materials/contrib/1999/goodyear/index.html (1999)

14. P. Goodyear, L. Carvalho, Framing the analysis of learning network architectures, in: L.Carvalho & P. Goodyear (Eds) The architecture of productive learning networks, New York, Routledge. (2014)

15. S. Schaffert et al. (Eds.): Networked Knowledge - Networked Media, SCI 221, 1–6, (2009)

16. P. Abbott, The Cambridge Introduction to Narrative,13,(2000)

17. E. Souriau, Time in the Plastic Arts, The Journal of Aesthetics and Art Criticism, 294307, (1949)

18. Rodriguez, (2011)

19. B. Espiau, et al., A new approach to visual serving in robotics, in Geometric Reasoning for Perception and Action, 106–136, (1991)

20. R. Mazza, Introduction to Information Visualization, Springer, 126, (2009)

21. R. Mayer, R. Moreno, Nine Ways to Reduce Cognitive Load in Multimedia Learning. Journal of Educational Psychology, 38, 43-52, (2010).

"Public Participatory Graphic Communications"

Harry A. Eggink[1,*]

[1]College of Architecture and Planning, Ball State University, Muncie, Indiana USA

Abstract. Ball State University is a public university in the state of Indiana which has a College of Architecture and Planning; its Community Based Program (CBP) was developed and created in 1969, and is now one of the three oldest continuous community education and service programs in the US. The program's main objectives are to provide an educational design service to the public sector, to immerse our students in a public participatory urban design environment, and to educate the public sector to become active in the design and planning process of their communities. After my Urban Design Graduate studies at Harvard, I published two urban design booklets (the Urban Design Primer and the Urban Design Dictionary) for public distribution, to be utilized prior to our small-town charrettes. These illustrated booklets were designed to bridge the language and design process gap between the design professional and the public citizen, and to create a more active immersive participatory urban design engagement. Since the introduction and public use of these booklets, I have been involved with over a hundred CBP charrettes. In this paper, I will introduce and present the urban design public booklets, and demonstrate how the urban design graphics and visual communications were utilized effectively through several small-town charrette case studies. The paper will also blend the transition between the analog graphics and the digital imagery.

* Corresponding author: heggink@bsu.edu

These Charrettes are community-based design exercises that are driven by interviews, conversations, and rapid design visualization. In these public engagements, quick graphic communications are critical in order to demonstrate the communities' visions and future aspirations. This presentation draws on methods used historically, while illustrating present day approaches, and collectively, imagining the future of charrette graphics. The author compares the way designers have visually told stories, chronicling the charrette design process over the last four decades, through a selection of traditional and emerging graphic techniques. The author will demonstrate how different practices, which have been developed over time, are shaping the contemporary charrette process in unique ways. The presentation will illustrate how designers have evolved a series of unique traditional and emerging techniques to not only show people what their visions are, but also show them *how* designers are making those images. Charrettes are one of the only times where people actually see the process of making designs. As participants see designs created in real-time, they become part of the thinking and visioning process.

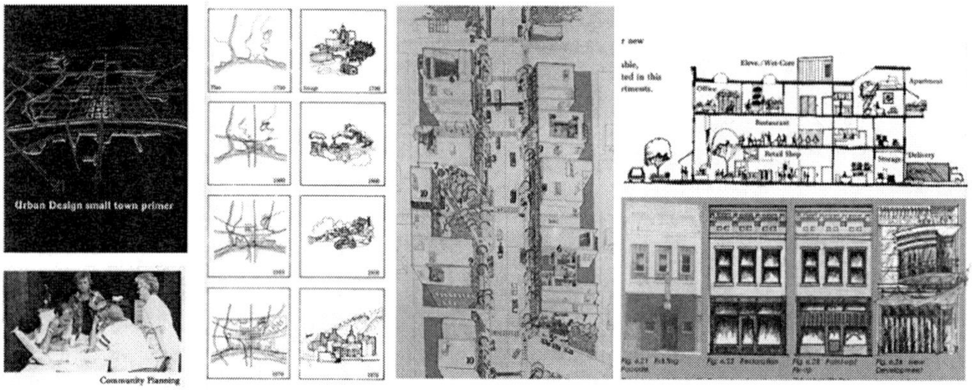

Fig. 1. Samples of illustrations in the Primer, historical character, overall downtown environment to façade composition

1 Introduction: Urban Design small town primer

Urban Design is a discipline, the goal of which is "to improve the quality of the physical environment of cities by understanding the interactions among the social, economic, political/legal, and aesthetic forces that shape cities". Traditionally, in this country, urban design studies have focused on the problems of large cites. However, most of the ideas or approaches of urban design in large cities have also been found useful when applied to small towns. The last decade has seen numerous Indiana cities and towns involved in revitalization or renewal efforts. Over 30 of these towns have been substantially aided through the urban design studies conducted by the Community Based Projects Program of the College of Architecture and Planning at Ball State University at Muncie Indiana. A combined educational/public service activity, the program involves students and faculties from the college's three departments: Architecture, Landscape Architecture, and Urban and Regional planning. These community-based projects have continually confirmed the vital role of community citizens in applying urban design to realizing the opportunities for improvement of small cities. Once these people have an understanding of the processes of urban design and the benefits that can result, they become the main force for positive change in their communities. The impact of average citizens upon the future of their own towns as a result of urban design projects has promoted the following sharing of principles and experience in the urban design of small cities.

1.1 Intent

This booklet is an urban design primer for people living and working in small towns who are concerned about the physical improvement of their community. This is not intended to be a "how to do it" book but rather the introduction to the concept of small towns. Although the principal focus is on common action by citizens, it is hoped that an understanding of cities and their shared environment will help individual members to be more effective contributors to the quality of their city.

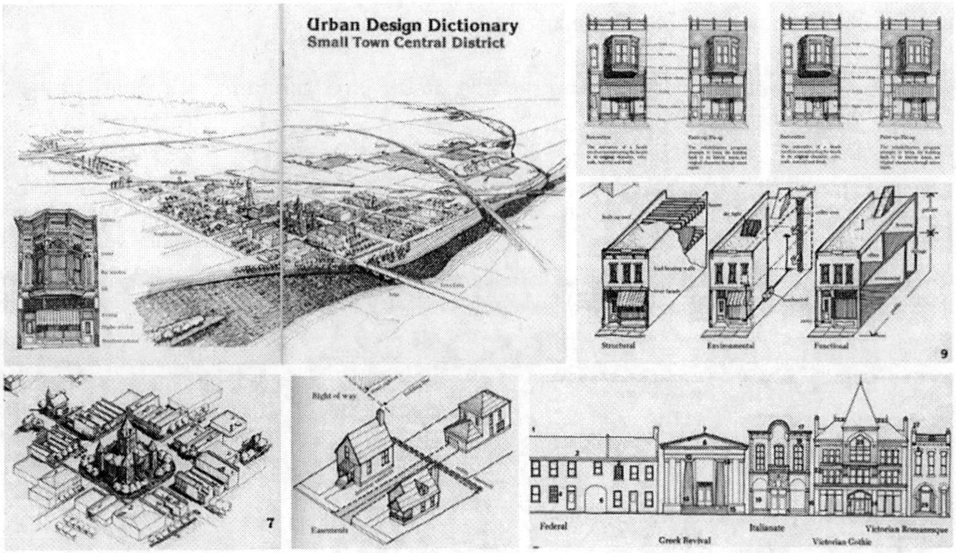

Fig. 2. Samples of the Dictionary illustrations, regional town environment to the façade historical patterns and building elements

2 Urban Design Dictionary, Small Town Central District

Communication between citizens and the professionals is critical to the success of urban design and communities. In the past decade, over 30 towns have been studied in a combined educational/public service activity through the Community Based Projects Program of the College of Architecture and Planning, Ball State University, Muncie, Indiana. These projects have continually confirmed the vital importance of the role of community citizens in applying urban design to improvement and development of small cities. In order to effectively help these communities, citizens, public officials, and professionals must share a common understanding of the community environment and a language which describes that environment.

In this dictionary, terms are compiled that are most frequently used during typical urban design activities in small cities. The terms are defined both verbally and visually in order to clearly relate language and physical environment. The booklet is divided into several parts that deal with the process of design and the physical components of the city.

Design process: a method or methods used to solve the physical/environmental issues.
Townscape: a system of similar or dissimilar parts that combine and connect to project an image often distinctive.
Central District: the city center, downtown, where the main core of the business and public functions are located.
Streetscape: the corridor of space bounded by the facades of the buildings on either side of the street.
Façade: face of the wall of a building, storefront.
Planning and Legal Terms: terms commonly used by the legal and planning professions that are relevant to urban design.

At the present time, 50% of all Americans live in small towns and cities, and have increased in population. These small communities are into a new stage of development in which a clear understanding of the physical assets and needs is required to guide future growth. It is the intent of this dictionary to:

1. Develop a citizen understanding of the physical parts and pieces of the city center.
2. Bridge the technical language gap between the professionals and the community.
3. Create a working urban design tool to aid the community and the professional.

Fig. 3. Elkhart Charrettes: 1 The Downtown, 2 The Civic Plaza, 3 The East Bank, and 4 The River Projects and Connections

3 Case Study: The Elkhart Urban Design Charrettes

The Elkhart case study is unique because the first charrette was followed by the development of the city's adopted Phoenix Plan and three additional charrettes that resulted in major changes and had dramatic impact on the city's physical and cultural environment. In one of the follow-up charrettes, we developed an experimental method, where we tested a more three-dimensional approach and encouraged the public to build physical models of their design ideas. Utilizing a provided Styrofoam site base, nine public teams worked several hours and presented their concepts to each other in a public form. The process created a diverse and creative pool of ideas and generated public awareness of issues and potential urban solutions.

Introduction of the Charrette Process:
"Let's Talk about it" titled The Elkhart Truth editorial of April 10, 1985 after I presented the small-town charrette process at a 7:30 pm public participatory meeting in the City Council Chamber. The meeting outlined the Ball State Community Based Program design workshops, schedule events, and the public participatory planning process in aiding Indiana's towns and cities. The process introduces a series of on-site pre-charrette investigative public organization meetings, physical site documentation through sketches and photography, historic research, newspaper as a communication design tool, community leaders' organization, downtown studio location, local team housing…

"We like the idea of getting the citizens of Elkhart to plan their own destiny, participation in drawing up a plan for the future."

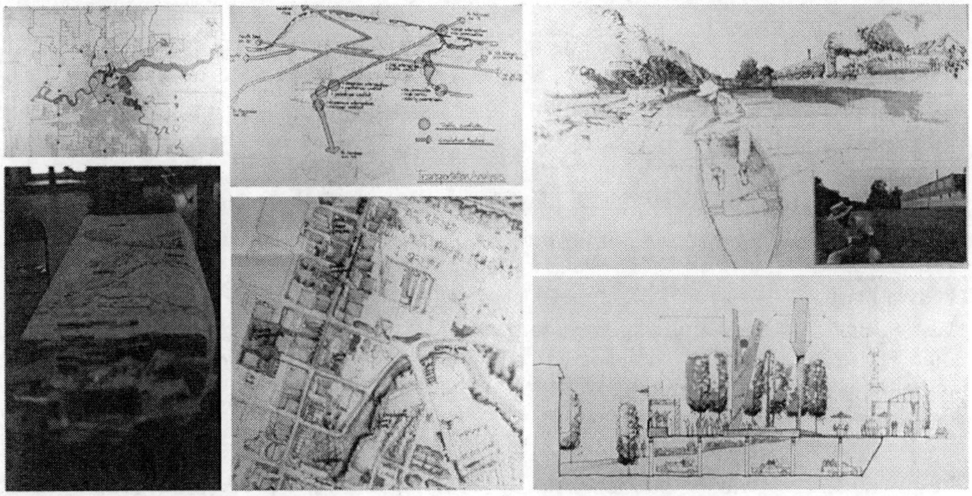

Fig. 4. Series of graphics illustrating regional map, circulation diagram, downtown plan, river perspective, and plaza section

3.1 Charrette 1: The Downtown

"Dream what could be and make it reality" Joseph Rueff president of the Elkhart Redevelopment Commission 5/21/1986

The Charrette concentrated on the downtown area and the immediate surrounding neighborhoods. Elkhart is a city of about 45,000 in population, located in the northern region of Indiana, at the junction of two major rivers (the St. Joseph and the Elkhart River), a potential linear green system that can link the neighborhoods together and a housing stock that was constructed during the rich entrepreneurial history of the community.

The two-day on-site public participatory charrette, located in an empty downtown department store, developed several urban design concepts including a civic plaza along Main Street, a pedestrian river walk following the Elkhart River, an amphitheater on the river banks, parking garages and a building façade restoration program. The conceptual ideas were presented to a well-attended public meeting and documented and assimilated through a special eight-page edition of the local newspaper of the Elkhart Truth. 8/7/1985

These sketches "talk pieces" are the result of the investigation of the analysis drawings, citizen's input, and design opportunities.

Illustration Topics:
1 River edges
With the wealth of the river environment, section/perspective sketches promoted the idea of river housing integrated with a public river landscape and boardwalk.

2 Cityscape
A river belvedere was visualized through a public sculpture to celebrate "Speedy Alka Seltzer", a local icon, a center piece along the Elkhart River with an enhanced river landscape and a historical urban building reuse.

3 River redevelopment
A before photograph and after sketch illustrated the reuse of the old river industrial complex into a living entertainment development.

4 Civic Plaza
An isometric and a section sketch of a future civic gathering plaza was proposed as an important urban component in the overall comprehensive plan, a public space, and an outdoor living room for the community.

The urban architecture
Preserving and revitalizing the rich historical urban storefront architecture was an important component of the future plan for the downtown fabric.

1 Parking system
With the elevation change between Main Street and the river, a section illustrated how the proposed parking garage takes advantage of the elevation change.

2 Opera House study
Three conceptual design proposals illustrated an attempt to save the historic Bucklin Opera House, which has been slowly deteriorating through neglect.

 https://doi.org/10.1051/shsconf/20196402012

3 Façade Program
A before photograph and an after illustration showed the revitalization of the rich historic "Queen Ann Style" storefront architecture prevalent on Main Street and proposed an active urban streetscape.

4 Building Reuse
An urban historic storefront reuse proposal presented a mixed living/working building integrated with sustainable concepts, day-lighting and green roofs.

5 River Architecture
River living and activities were proposed in multiple sketches.

6 Historic Building Reuse
This conceptual elevation diagram proposes the reuse of the old post office and the linkage between two historic structures.

Civic Plaza
"A Public Plaza must be multifunctional and have the comfort able to accommodate the solitary person drinking a cup of coffee 7:30 in the morning to hear the birds sing, to the festive large groups in an active public function".

Fig. 5. 1) The on-site Plaza Concept, 2) Perspective of Concept, 3) The plaza fountain "the four season", 4) The Clock Tower

3.2 Charrette 2: The Civic Plaza

The first charrette expressed the need for a city center, a public place, and a space to celebrate civic events.

The second Charrette was conducted in a two-day public marathon workshop in an old downtown drugstore and utilizing the public design discourse that resulted in a series of drawings and a three-dimensional model of the future plaza. Several concepts were developed featuring an interactive fountain, a shaded urban tree grid, and a geographical illustrated floor, surrounded with a "wall of fame", and dominated by a clock tower. The design team also addressed the calendar of civic events that would activate and give color to the urban culture. "The goal is a new opportunity to complete and give Elkhart a real city center", "a jewel".

1 The clock tower
The Elkhart Tower illustration poster designed by Eggink was used for a public funding and marketing program. The tower designed as an urban landmark, dominate the public plaza, and provide a pavilion for season activities.
2 The fountain
The water fountain "the four seasons" designed by Professor Tony Costello shown here in a model and the construction donated by the local Martin Foundation.

Conceptual studies
Three proposals:
Concept 1: based around and dominated by an interactive water fountain and flanked by an urban amphitheater and green garden.
Concept 2: is dominated by a tower/pavilion, surrounded by a water feature, and a floor plane geo/sculpture of the Elkhart Rivers and the urban grids.
Concept 3: a contemporary urban space graced by an urban pond and sculptured treescape with a suspended public pavilion.

The Clock Tower
The clock tower became the new landmark of the downtown environment and the civic celebration and festival space. The illustrations show an isometric onsite concept drawing, a plaza perspective, the model of the proposed water feature and the tower and pavilion. The plaza and the tower became both a public symbol as well as an environment of memory; illustrated by a wedding on the plaza of a young couple that met at the public "New Year's Eve party" on the plaza, exchanging their vows later in the summer.

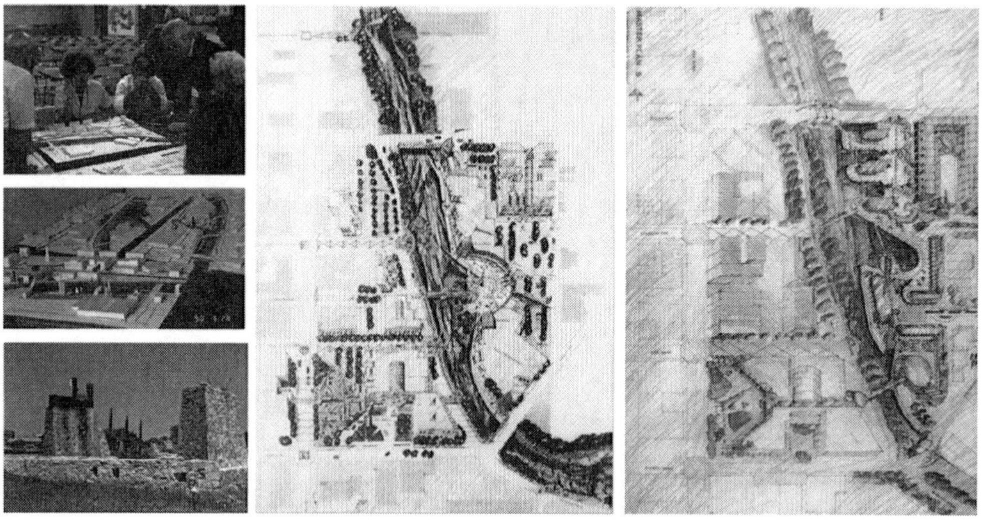

Fig. 6. Public Participatory Concepts, Master Plan A (by Eggink) Master Plan B (by Mounayar)

3.3 Charrette 3: Elkhart River "East Bank"

"We happen to believe that the people from the community are the experts…the people who live here and work here know our community best", Mayor James Perron, June 1990 statement kick off **Charrette 3.**

The Charrette explored the potential and opportunity of activating the Elkhart River that parallels Main Street east by a block, revitalizing an abandoned industrial complex with a direct view and access to Main Street and creating a public plaza on the East Bank. The study was to use the public as the design team to explore new ideas for the east bank.

Nine large preliminary conceptual models were constructed as bases for the public to engage in and develop future proposals for the river edges. The process modeled nine concepts of the river, its edges, new infills, public spaces, and links to Main Street. These nine publicly developed physical models were interpreted later into two master plans and vision sketches for final public review.

The method used in Charrette 3 was to attempt to have the public build physical models of their design ideas. By providing a Styrofoam base model and a collection of objects and materials to represent the media components in creating the final model. Nine public teams were assembled, worked several hours, and presented their concepts to each other in a public format. The process created a diverse and creative pool of ideas as well as a public awareness of the issues and potential of urban solutions.

"Let's make the river dance" let's reinvent the historic raceway and power a carousel" quote from a participant.

East Bank Concepts
These are a series of conceptual diagrams that responded to the public participatory models of the river corridor.

Concept 1 showed the reuse and rehab of the existing industrial architecture on the river bank, creates a public plaza at the historic race way entry to the river, develops a river canoe docking facility, and proposes a river walking/jogging path along the west bank.

Concept 2 created a public pedestrian bridge across the river from Main Street into a new amphitheater/public plaza celebrating the historic turbines of the raceway. The industrial architecture would be revitalized and activated by a mixed-use program.

Concept 3 expanded the public plaza east towards the river and through new architecture creates a public activated river edge.

Conceptual Diagram
Conceptual diagrams were developed to explain the complexity of the master plan. The concept diagram showed the new east river wall/edge, pedestrian bridge, public river plazas, series of river belvederes, and the link from the city plaza on Main Street to the river.

The activity diagram illustrated the newly discovered public uses along the river edge from a river walk, public spaces, river docks, and new building uses.

The circulation diagram highlighted the pedestrian movement patterns, locations of public places, and river walks and links to Main Street and future neighborhoods.

Studies
Master Plan A
1 An illustrative drawing by Eggink of the vision for the east bank of the Elkhart River and its connections to Main Street, this master plan links the existing Main Street with a pedestrian bridge that becomes an entry/gateway to the river plaza on the east bank. The east bank rehabilitates and brings life to the existing industrial architecture, proposes a river restaurant, new urban housing and reinforces the Main Street Plaza by a new urban facility.

2 An isometric of new proposed urban housing by reusing an old industrial facility and inserting contemporary architectural housing.
A before photograph and an after sketch visualized the reuse of an old industrial artifact as a new entry sculpture to the proposed urban housing complex.

Master Plan B
Master Plan 2 executed by Prof. Mounayar creates a strong link from the existing Main Street Plaza through a pedestrian bridge to the east bank. The river is explored by carving out a series of water inlets and docks that are then lined with new river architecture. The mixed-use development is integrated with the proposed urban residential pattern with framework of public river walks and spaces along river.

Fig. 7. Bubble Diagram West Bank Site (by Eggink) Elco & Civic Plaza Site (by Mounayar)

3.4 Charrette 4: A series of River Projects and Connections

The fourth Charrette was to create a series of "connections and visions" for Elkhart.
The urban design study would connect a decade of urban studies and link four critical urban sites together by allowing each site to dominate. The four sites needed to be analyzed, programed, and integrated in to an overall vision.

The first urban site, "Elco Theater", was to connect the historic theater, an Elkhart Landmark, to its surrounding environment.
The second site "Plaza Site" was to redesign, reform, and reinforce the urban architecture around the plaza and link it to the river.

The third site "West Bank" was to develop an active river edge through rehabilitation of the backs of the main street commercial buildings, new river architecture, and vital river pedestrian corridor.

The fourth site "East Bank" was to create contemporary urban river edge and architecture, and integrate a new river urban housing pattern into the context.

1 The bubble diagram shows the individual sites, Elco Theater Site, Plaza Site, West Bank Site, and the East Bank Site, within the overall context of the downtown commercial district of Elkhart.

2 These are a series of isometric diagrams that explain the urban space definers, movement patterns, and location within the Central Business District.

The Elco Site shows the Elco Theater linking with the Elkhart high-rise residential facility to the south and connecting to the Main Street Plaza to the north.
The Plaza Site plans to reinforce the Plaza as an urban space and link the development to the river and the east bank.
The West Bank Site creates an active edge of new community facilities and a landform bridge that crosses the river and connects to the east bank.

The East Bank Site envisions a new architecture that will activate the river edge and link to the proposed river residential neighborhood to the east.

The Elco Site
This plan, as the Elco Theater as a central historic icon, will rotate and create a new cultural center of the future. The vision is to link, across Main Street, and connect to a proposed new museum to celebrate the creative entrepreneurial history of Elkhart. To the south the Elco connects to the renovated senior housing facility and to the north a strong diaangular interior pedestrian walkway system is developed to link to the civic plaza, which is to be reinforced with new urban buildings. The plan also calls for a pedestrian civic boulevard to a sculptured belvedere on the river.

Beneath the master plan are a series of sketch diagrams that explain the urban concept, the vehicular/pedestrian circulation, and the building strategies.

Civic Plaza Site
This vision revisits the urban plaza by proposing a mix-use urban building with a new city hall embracing and surrounding the public plaza. A pedestrian mall would bisect the facility and terminate as a sky walk overlooking the Elkhart River. The proposal also connects the Elco development via a second level walk to the south and activates the river bank with cafes, restaurants, and small shops.

The West Bank Site
This proposal creates a community/convention center on the west bank of the Elkhart River and a large public plaza along the river edge. A land bridge would come out of the plaza, span over the river, and curve to form a land amphitheater on the east bank. The west bank is also edged with a canoe port, a plaza music pavilion, and a festival trellis frame structure for river activities and celebrations.

SHS Web of Conferences **64**, 02012 (2019) https://doi.org/10.1051/shsconf/20196402012
EAEA14 2019

Beneath the master plan are a series of sketch diagrams that explain the urban concept, the urban plan, and the vehicular/pedestrian circulation.

4 Case Study 2: The Warsaw Urban Design Charrette

In this case study, I will present the comparison and bridging between the analog and the digital charrette graphics.

The computer and the digital language has been utilized and embedded into the charrette process for decades. First as a tool for mass distribution of the charrette findings and the engagement of public participatory online discourse complimenting their local newspaper taps.

Today, the digital visual design methods are partners in the onsite charrette visualization and communication environment. For decades, visualization has flattened design ideas into pictures through markers and color pencil drawings. Now designers and members of the community can quickly overlay photos and familiar entourage to create collaborative, lively and dynamic visualizations that are drawn directly from existing imagery. Around the tables in contemporary charrette environment, design thinking is increasingly more inclusive as technology becomes more intuitive, visceral, and collaborative. This accessibility and speed is making the process faster, sometimes bigger (in terms of scale and impact), and arguably more user friendly for everyone involved.

The Warsaw charrette was composed of a constellation of district community redevelopment projects spread throughout the already-beautiful downtown area. Two of the areas included in the site were the "Little Crow Market", a once vibrant but shuttered food market at the east edge of the inner urban core, as well as the "Gatke" site, an old abandoned factory building in the city's periphery on the old Pennsylvania rail line that once connected Washington, DC and Chicago.

Fig. 8. Site Photograph R/R Station analog (by Eggink) R/R Station Plaza digital (by Bussiere)

At the Gatke site, a new high-speed rail station was imagined, complete with a large civic plaza and contemporary sculpture park using work from local artist. Because the old factory shells on the site were intact and for adaptive reuse, a host of new economic uses were projected to be housed within and in the surrounding areas of the new rail station. Programming included shared transportation strategies, mixed-use developments, in-fill neighborhood housing and a variety of business incubators. The designers visualized a contemporary rail station and clock tower with glass enclosures, an art pavilion, and galleries. The graphics express a blended conceptual image and graphic process.

Fig. 9. Little Crow Photo. Market, analog sketch (by Eggink) Market, digital sketch (by Bussiere)

At the Little Crow Market site, contemporary food systems were illustrated that would capitalize on the city's newly forged relationships with local farms, restaurants, and other pop-up food vendors along with a future boutique hotel. Hearing from local business owners who wanted the designers visualized exterior scenes with food trucks, café seating, colorful murals, and gathering spaces at entryways. The graphics illustrate two conceptual ideas utilizing a fast sketch and a fast-digital overlays process using on-site photograph.

References

1. Lasuau, Paul and Crowe, Norman. *Visual Notes for Architects and Designers*, John Wiley sons, 2011
2. Mounayar, Michel. *Design as a Social Act*, Ball State University, Community Based Projects, 2013
3. Bussiere, Simon and Eggink Harry, Charrette Graphics: Old and New School of Visual Thinking, 2014

Experiential Simulation and Psychological assessment as a learning process in architectural higher education

Barbara E. A. Piga[1,*], *Nicola* Rainiso[2], and *Marco* Boffi[2]

[1]Laboratorio di Simulazione Urbana fausto Curti, Dipartimento di Architettura e Studi Urbani, Politecnico di Milano, Milan, Italy

[2]Dipartimento di Beni Culturali e Ambientali, Università degli Studi di Milano, Milan, Italy

Abstract. This paper presents the advancements of an applied methodology developed by the authors, namely an architect and two environmental psychologists, based on Virtual Reality coupled with psychological surveys – developed ad hoc by the authors – for urban design; in particular, the paper discusses its application in the field of architectural higher education. The proposed method is based on the pre-evaluation of people's urban experience for testing the effectiveness of design solutions, discovering the unforeseen, and reducing the risk of failure. The method connects architectural to psychological approaches and investigations. The procedure can be schematized as follows: (1) final users navigate the simulation of the design scheme and feedback is collected; (2) the analysis of these collected data is run by professionals and discussed with the designers that then develop the final design solution.
In this contribution, we illustrate the application to the urban design development carried out by 100 students of the last year of the Master of Science in Architecture and Urban Planning working in groups. The paper will firstly synthetically present the overall research; secondly, it will argue about its application for higher education; finally, it will conclude by describing the pros and cons of the experimentation, especially in relation to the whole learning and maieutic process and its impact on the design solutions adopted by students, with a focus on the experiential multisensory design.

1 Research framework, methodology and case study application

This contribution presents the educational effects of a design methodology developed by the authors that links experiential simulation, i.e. Virtual Reality navigated with Head Mounted Displays (HMD), with psychological investigations, i.e. a questionnaire for assessing the experience in a given space (physical or virtual). The approach and the methodology are part of a wider research under development [1, 2, 3, 4]. The broad goal of such approach is to

* Corresponding author: barbara.piga@polimi.it

increase the awareness of designers about the impact of their design projects on people's experience, relying on multidisciplinary tools and contextual data collection to inform the designers. Using concepts and methodologies derived from social sciences in general, with a specific focus on psychology in this case, and tools such as Virtual Reality, it is possible to investigate and get a picture of the social, cultural, symbolic context induced by design projects before construction. To reach this goal, the authors developed a tool apt to collect quantitative and qualitative data about the experience of people in the environment, that can be used both in currently accessible spaces (physical condition) and not accessible/future spaces (virtual condition). The expectation is to provide the designers with more sources of inspiration for their design process, and to support their creativity in answering people's needs. In this paper, the focus is on the application of the above-mentioned methodology in higher architectural and urban planning education, namely in the course of Architectural and Urban Simulation lead by prof. B. Piga & R. Salerno, that is addressed to students of the Master of Science. The presented learning methodology was applied for two years, involving a total number of 100 students divided in 14 groups.

The learning methodology is based on a combination of architectural and psychological approaches for teaching directed to the same goal, namely, the anticipation of inhabitants' urban experience leads by design projects. The method is a combination of theoretical lectures (with two main subjects: simulations for urban design and environmental psychology) and practical exercises. Most of theoretical lectures regard the possible application of different kinds of simulations for urban design; the main goal is to build a critical capability to evaluate the proper tools in order develop and test design proposals paying attention to people's experience [5, 6, 7]. Specific lectures are instead related to some key concepts in the field of environmental psychology, with the goal to discuss some basic disciplinary concepts with students who, in most of the cases, are facing the subject for the first time. Among the issues presented to explore the features of the interaction between people and environment, a key role is devoted to the Preference Matrix, explaining environmental preferences through four main concepts [8]; a second main topic is instead the Flow Theory [9], which explains how people build their own wellbeing when involved in engaging activities and how this can have an impact on the relationship with the environment [2]. While presenting the specific topics, teachers refer to the other discipline in order to clarify the strong link between the two: the theoretical background is in fact structured in order to deliver a comprehensive educational message regarding the human/environment relationship. The practical application of the method consists in two main phases; the first one is related to the development of the design project and its representation and is articulated in a series of interrelated exercises [1]; the second phase consists in showing inhabitants the produced simulation, and subsequently administering an ad hoc questionnaire. The first phase must necessarily start before the second one and finish after that, it is indeed a continuous and long process.

The entire process aims at developing and testing the effectiveness of the design project. To do that, the procedure makes use of experiential simulation visualized immersively with simple and low-cost Head Mounted Displays (CardBoards). This navigation of the transformed environment was firstly used for training students to envision the subjective experience generated by their design schemes; the recursive procedure of passing from bi-dimensional to tri-dimensional representation along with an immersive simulation of the environment supported participants to connect the geometrical representations to the perceptual dynamic experience, i.e. to develop the ability to read abstract depictions of the urban layout and interpret it as subjective navigation of the environment. This is a crucial learning tool that facilitates the development of the typical architects' ability of encoding and correctly understanding technical representation [10]. Secondly, students are asked to fill in the questionnaire that investigates some psychological constructs in order to collect their design desiderata, i.e. their expectations in terms of people reactions to their design solution. After this, they are trained to administer the same survey to inhabitants that have virtually navigated their proposal. They are supposed to get around fifty respondents in order to complete the procedure. Later, the environmental psychologists analyze the collected data, without having seen the projects, and they present the outcomes of the study to students. The discussion of results is done in a collegial way, firstly by presenting the overall outcomes of all groups in a comparative way, and secondly by commenting group by group. In this phase, the interaction with the professors of the architectural field is of course very high in order to facilitate students in reading the relationship between the psychological reaction of people and the design of the physical environment. Students actively participate in this dialogue, providing their comments or posing questions. Some general comments on the methodology and the entire process are collected during this phase. After the discussion of results, students have to send a short text commenting the experience. Their feedback is relevant to refine the entire procedure. A process of systematization of results allows to organize some relevant key points emerged from the interaction with students, as described below. The closing of the course is devoted to the fine tuning of the design project considering the survey outcomes. Ideally, it should be a recursive process applying the testing procedure with final users until the results are satisfying; of course, due to time constraints this last process is hardly achieved.

As mentioned above, with the aim of deepening the teaching/educational potential of the procedure, students were asked to provide detailed feedback on the impact of the tools used on their ability to think/rethink the project and on the application of tools and procedure with the participants involved in the evaluation process. The collected observations highlight some emerging issues that can be organized in three macro-categories, discussed in the following paragraphs.

1.1 Impacts on design thinking

Under this umbrella, it is possible to collect all feedback related to the different ways in which students' design thinking has been "triggered" by the proposed methodology. In fact, students agree with each other in stating that the process had a significant impact on their design thinking, by enriching and improving their usual approach and contributing to the final solution. They affirm that this improvement process is twofold, as the increase in creativity is connected both with the quality and the quantity of information. On the one hand, through the questionnaire they gained access to information already taken into account during the design phase but represented and communicated in a more extensive way thanks to the visualization of questionnaire outcomes (Fig. 01). On the other hand, they appreciate the emergence of data/information not originally foreseen, that lead to a general increase of the design quality. To encourage their creativity, the direct confrontation with future users is perceived as crucial, since it has been an absolute novelty for them. Finally, among the elements that brought them to modify or fine tune their design project students mentioned:

- a more systemic perspective, given by the environmental perception as a psychological totality by the users, that empowered their abilities in thinking about the general relationships between spatial elements;
- the acquisition of a further professional point of view on space, mainly focused on the emotional and experiential aspects;

- a new practical ability related to the stronger and more evident connections between the physical and psychological environmental dimensions, that produced a more conscious and adequate response to users' needs;
- possibility of "measuring" the symbolic and iconic strength of their original idea.

1.2 Practical ways for implementing the project and interacting with the psychological data

This category collects the main feedback related to how the students have modified their design projects after receiving the feedback from the teachers, both architects and psychologists. Obviously, different intervention modalities were applied depending on the feedback. A basic modality to explain students the gap between their desiderata and users' perception was based on the direct comparison between these data: this procedure enables us to highlight relevant mismatches. From this shared starting point, different perspectives in redefining/redesigning the design project have been adopted. These can be roughly summed up as follows:

- extensive redesign of the project;
- modification of some variables, not only when evaluated as critical by users but also when not consistent with the general concept;
- addition of some visual elements or furniture to reinforce some spatial features/connotations;
- implementation of the simulation's technical quality, since some misunderstanding by users was ascribable to the lack or low quality of the simulation features (e.g. colors, sharpness, people etc.).

1.3 Features characterizing the tool itself in interacting with the audience

Students state that using a VR CardBoard had a direct impact on the audience, who was immediately involved and interested in interacting. HMDs seems to be an ideal tool to encourage a playful approach that support citizens' engagement in the participation process. They also stressed that this may be a risk to the results, as people tend to be more interested in the medium than in the design content. On the contrary, many participants stressed that the survey questions appear boring and repetitive.

Fig. 1. Bird's eye view of the preliminary design proposal "City Park" by one group of students (above); and one example of outcomes of the psychological assessment: the activities with the red square are the ones. The final design project at the end of the process can be view at: https://youtu.be/QG9t1R6j-Is (students credits: M. Ainsa, H. Erdin, M. Ghedamsi, A. Ginter, E. Hjorstam, I. Kovacevic, P. Kwiatkowska, E. Le Helloco).

2 Conclusions

In conclusion, it is possible to highlight that the entire process acts as a learning tool for students. In general, it is possible to understand from their comments that the method enables them to better envision and focus the experiential outcomes of their design project, and therefore its civic relevance. At the same time, the process allows students to deal with an interdisciplinary approach and to experiment first-hand the relevance of disciplinary collaboration. The role of representation, both on the side of simulation for design thinking and on the side of communication of design project, that permit to collect users' feedback, is crucial for the entire process. The use of HMDs, and specifically CardBoards, was relevant for three main reasons: (i) these provide a one-on-one scale interaction with the design project, that was traditionally unusual in the architectural field, enabling to virtually test the environment in a subjective way; (ii) participants to the survey are generally intrigued by the tool and engaged by the experience, and this is generally facilitating designers and interviewers in running the process; (iii) even if the quality of the simulation is not that high using CardBoards and standard smartphones, due to the affordable cost of the devices it is possible to run several virtual navigations in parallel; this last point contributes to the replicability of the setting in university courses or in the professional practice, even in small offices. Of course, the support of professionals for training the interviewers, performing the analysis, and interpreting the outcomes is in any case necessary. Regarding the trustability of such devices for these types of investigations, a recent research developed by the author (under publication) demonstrates that even this low-cost device enables a well-balanced quivalence of responses between the simulated and the real environment.

References

1. Piga, B. E. A. (2017). Experiential Simulation for Urban Design: From Design Thinking to Final Presentation. In B. E. A. Piga & R. Salerno (Eds.), *Urban Design and Representation. A Multidisciplinary and Multisensory Approach* (pagg. 23–36). Springer International Publishing.
2. Boffi, M., & Rainisio, N. (2017). To Be There, Or Not To Be. Designing Subjective Urban Experiences. In B. E. A. Piga & R. Salerno (Eds.), *Urban Design and Representation. A Multidisciplinary and Multisensory Approach* (pagg. 37–53). Springer International Publishing.
3. Piga, B. E. A., Morello, E., Boffi, M., Rainisio, N., & Di Liberto, G. (2017). Sharing the experience of urban design schemes through immersive simulation. In T. Maver, P. Chapman, C. Platt, V. Portela, & D. Eaton (Eds.) (pagg. 244–251). *Envisioning Architecture: Space / Time / Meaning*, 6-9/09/2017, Glasgow (GB): Mackintosh School of Architecture and the School of Simulation and Visualization at the Glasgow School of Art.
4. Piga, B. E. A., Boffi, M., & Rainisio, N. (2017). Experiential Simulation and Environmental Psychology for Anticipating Users' Reactions to Design Projects: An Application in Architectural Higher Education. *Proceedings of the International and Interdisciplinary Conference IMMAGINI? Image and Imagination between Representation, Communication, Education and Psychology*, 1(9:968), 1–9
5. Rasmussen, S. E. (1964). *Experiencing Architecture* (originally published in 1959). Cambridge MA: MIT Press.
6. Bosselmann, P. (1998). *Representation of Places: Reality and Realism in City Design*. Berkeley, Los Angeles, London: University of California Press.
7. Carmona, M., Tiesdell, S., Heath, T., & Oc, T. (2010). *Public Places, Urban Spaces: The Dimensions of Urban Design*. Routledge.
8. Kaplan, R., & Kaplan, S. (1989). *The Experience of Nature: A Psychological Perspective.* Cambridge University Press: New York, NY, USA.
9. Csikszentmihalyi, M. (1975). *Beyond Boredom and Anxiety.* Jossey-Bass: San Francisco, CA, USA.
10. Schön, D. A. (1987). *Educating the Reflective Practitioner: Toward a New Design for Teaching and Learning in the Professions*. Wiley.

Educational XRealities

Martijn Stellingwerff [1]

[1]Form & Modelling Studies, Faculty of Architecture, TU-Delft, the Netherlands

Abstract. XRealities is a recently approved pedagogic fellowship project that will run for two years, during 2019 and 2020. This paper will introduce the project plan, some early experiments and expectations. XRealities relates to the many real and digital situations that we encounter in our digitally expanded life experiences. In contemporary education, we notice a problematic situation with abundant screens. People sit next to each other and stare at screens, looking for contact and content. We describe the current dichotomy between latent online learning materials and potential real-world experiences in campus education. The XRealities project is focused on finding alternative approaches for educational settings in which both real-world social interactions and virtual content provide new forms of intertwined learning experiences. We assume there is huge educational potential in the right juxtaposition of visual digital and real-world experiences and the embodied interaction within this augmented reality.

1 Introduction – screens in education

The essence of education is bringing the world to the student. Educators assist in the exploration of world-aspects and challenge the learner to find and train the right responses to that world. In education, we make use of collections, descriptions, recordings, images, schemes, models, simulations and prototypes to represent selections of the world in the best possible ways, although real experiences, such as excursions, practical work and creative production of full-scale objects also belong to the educational curriculum, screens, in different sizes, take an increasingly dominant role in classes, lecture rooms, design studios and even in workshops.

With this French year of the Architectural Envisioning conference in mind, we could refer to the famous Lascaux Cave as an ultimate educational representation. The beautiful ancient cave drawings could be considered an early screen that depicts hunting strategies and characteristics of the most impressive animals in the field. This 360-degree cave experience is still as impressive as contemporary screens, with for example a 360-degree video about "Swimming with Giants" [1]. Huge buffalos and majestic deer, impressive whales and gigantic squids … they all have to compete with a massive amount of cute cat pictures on mobile phones.

Between the cave drawings and a spherical video experience on YouTube, we can consider a whole range of innovations in screen technology. Moses, for example, used mobile stone tablets to teach biblical laws. Flintstone was a thinner update that was also supporting the use of chalk pencils, which brought us 'easy-undo' features. The modern blackboard can still be found in school classes. A parallel technology focused on very thin foldable and bendable material, it can still be found and is known as … paper. Paper works with a real ink-based stylus, also known as pen. Eventually, with the efforts of clerks, human computers, and monks, a lot of knowledge was pencilled on paper and stored in libraries (physical repositories, comparable to cloud server buildings). The invention of printing technologies helped spreading our knowledge. The rest is prehistoric.

In contemporary education, we find screens in abundance. The ubiquitous screens are our predominant digital windows to the world. Four types of screens can be distinguished: screens with beamers and TV-sets for group presentations, laptop screens for sitting people, and mobile device screens to fill in all the time we cannot use the first two types of screen. VR goggles can be seen as a form of ultimate screen, close to the eyes, for full immersion.

2 Problem statement

Screens dominate and distract from valuable learning experiences involving collaboration, prototypes, simulations, tests and topical conversations. Contemporary physical and digital learning environments exist side by side but lack integration.

The lack of integrated physical and digital learning environments becomes apparent in social aspects, education spaces, learning materials and the deliverables (case studies, designs, proposals, models, prototypes, etc.):

- In the social behaviour of teachers and students, we can see the constant demand for attentiveness between real-near and virtual-remote collaborators. Screens and headphones isolate individuals from a group. Phone-screens often interrupt a good conversation. At the same time, there are global social ties that can come about, completely independent of place and time.
- Physical and digital education spaces offer totally different experiences. Lecture rooms, classrooms and studios bring uniquely focused events while online learning platforms (e.g. Blackboard, Brightspace and edX) support information exchange in a durable and accessible way.
- A third lack of integration can be found in educational collections, such as physical repositories for publications, samples, exhibition objects, tools and their digital counter parts with collections of open textbooks, images, 3D objects, code and all sorts of digital assets. Online learners and campus learners have different experiences.
- The lack of integration is also apparent between, on the one hand, physical objects in education laboratories and workshops and, on the other hand, virtual prototypes, digital performance testing, simulation, AI and VR.

The dichotomy of the physical and digital environments is already somewhat solved by deliberate disconnecting them in time and place, for example by introducing Blended Learning. The idea of blended learning is that the student prepares at home, using online learning materials, while contact hours are used more effectively for direct conversations, collaborations and exercises with peers and tutors. It should be noted that the word blended in blended learning indicates a combination of two more or less separate modes of learning, while the XRealities project aims to truly intertwine the experience of context and peers with digital content within a digitally augmented environment.

Other approaches partially integrate digital interactions and integration via screen sharing, e.g. using Clickshare [2], and classroom voting systems, e.g. Mentimeter [3], Kahoot

[4] and Feedback Fruits [5]. A good tool that truly integrates group collaborations with interactive screens is e.g. Nureva [6]. By combining physical actions of participants, using huge touch screens, Nureva brings the notion of shared idea creation to a visually consistent group experience.

3 The Educational XRealities project

Despite aforementioned good exceptions, we notice a huge mismatch between the abundant screens, online content and the real-world aspects of active, contextual, socially situated education. One new and ultimate type of see-through screen technology could solve this dichotomy between the real world and the digital world: Augmented Reality (AR). Although AR consists of screen technologies, the screens more or less disappear and give way to shared, digitally augmented, spatial experiences.

At the moment, we see rapid changes in the ways students use data from 'the cloud' and in the ways they form alternative networks for meeting and assisting each other. Soon, the changes will bring 'goggles and googles' (devices and data clouds), enhanced by artificial intelligence, that will have unprecedented impact on many details in our life. The setting of our education will change drastically.

In education, we have individuals and communities of people with different roles. They meet online and in social spaces such as classrooms, labs and studios. They make use of real and digital learning materials, including tools, instruments, data sets, models, prototypes, algorithms and simulations. Each gathering of teaching and learning people, in a composition of virtual or physical spaces, with a specific choice of learning materials, can be called an 'educational setting'. In the XRealities project, we will describe educational settings that allow us to investigate a range of future scenarios.

Different forms of Augmented Reality (AR) can be used to superimpose physical environments and digital content, to integrate prototypes and human actions. The AR environments create new educational settings that will be valued relative to existing solutions.

XR is a collective term for a range of key enabling technologies that will soon change the realm of education in exciting and challenging ways. XR encompasses VR and AR, but also the 'Real World' and the world of our imagination. The terms AR, VR and XR are currently adopted in different ways and the definitions vary. Sometimes XR is being disputed and disliked, while recently the term 'spatial computing' becomes more popular.

In our project, XR stands for XRealities. This will be a number of eXperimental education settings that range from digitally augmented university campuses to full-immersive VR settings that allow collaborations between participants from all over the world. Although technology is an enabling factor in the experiments, the outcomes will especially regard pedagogic insights for the organization of future campus- and online education.

3.1 Project plan

During the first year, we would like to make an ongoing and expanding augmented campus exhibition. This would start with examples of persistent AR content within the Teaching Lab. Based on those examples, other teachers would be invited and supported to build their own topical campus exhibition and eventually their own AR learning materials. At a certain moment, the campus will be full of exciting examples in line with the Open Science and Open Education policy [7, 8] of our university. Partial exhibitions could be built with students from different faculties or related to events, such as the International Festival of Technology, a conference, or they could play a role during the welcoming of new students at the start of the academic year.

During the second year, we would like to focus on meetings in VR and AR on top of the augmented exhibition content. This would involve making personalized avatars for teachers and students. An online XR conference would complete the project. Such a conference would offer local and global exchange of experiences, without the footprint of travel and without scheduling problems and absence.

The field of XR is developing rapidly. Developer tools are available and specific experiments can be created relatively easily. This implies that there is not a large risk in the actual technical challenge. The experiments depend more on the proper description of educational objectives and the right choice of content and interaction.

3.2 Experiments

Experiments are key in the XRealities project. We already started finding potential qualities of AR in relation to simulated education settings within the interior building context of our Teaching Lab. This laboratory opened in September 2017 [9], and it offers a creative meeting and working space for all educators and support staff at the TU Delft. The open, warm yellow interior of the Teaching Lab provides a unique spatial décor for meeting colleagues, and it is also the place for many sorts of training, sharing and experimentation. An additional and unexpected benefit comes from the wood chips in the yellow coated Oriented Strand Board (OSB), which is used all over the place. The subtle but unique OSB pattern in the floor and wall finish turns out to be ideal as a tracking pattern for AR applications.

We will get started by experimenting with an iOS app that allows to directly compose and test AR content. After an in-depth study of the available AR authoring platforms, we did not choose one of the main players, like Unity or Unreal Engine, nor did we choose to dive into programming challenges, using ARkit or ARcore. Instead, we chose to experiment with a new app and platform that allows to directly create and explore interactive AR experiences using mobile devices such as iPhones and iPads. The app 'Torch' [10] allows to quickly place and combine different 3D models and then add interactions and changes in scenes. This frees us from programming efforts and therefore makes it possible to be playful, quick, responsive and iterative in building and evaluating different educational AR prototypes. Torch comes with many improvements in each new app update and their roadmap reassures that many extensions to new and existing platforms and possibilities to export to the aforementioned Unity or Unreal Engine authoring tools will be added.

The experiments will include content from some of the various disciplines in the faculties of our university. We will connect the content to different learning goals, as described and schematised in Bloom's revised taxonomy [11, 12]. The range of learning objectives will be systematically explored during the experiments. In that way, we will explore how well the objectives in Bloom's taxonomy relate to the medium and user experiences offered by AR.

To illustrate the project, we could already describe one of our first experiments. This will be a small exhibition about changeable furniture. We will digitize chairs and one specific chaise longue. The real pieces of furniture belong to our large faculty collection of design chairs. Most of these chairs are designed by architects. The selected chairs will be augmented by 3D representations that demonstrate the state changes of the chairs. Some chairs can be changed by hinges, others can be decomposed and fit in another way around. The real chairs cannot constantly be touched. Therefore, the AR experiment would be successful if it teaches the different options. Referring to Bloom's learning objectives, the augmented chair exhibition should allow the visitor of the exhibition to *analyse* the chair positions and *apply* the understanding by moving or rotating the digital model representation in order to bring it in the alternative state.

3.3 Expectations of the project

We expect that Bloom's taxonomy in the 'Cognitive Process Dimension' allows us to come up with AR prototypes that facilitate and support each of Bloom's six learning objectives: *remember, understand, apply, analyse, evaluate, create*. We expect to be able to implement these learning objectives in a number of case exhibitions. Follow-up research would be needed to further understand the workings and effects of AR as an educational tool or medium. Given the predominant visual nature of most AR experiences, we assume that the successfulness of the learning objectives depends to a large extent on the subject matter of the education. In that regard, we will select cases that relate to design and understanding of complex spatial subjects. The tools within the AR environment should relate to the objectives. For example, if a learner needs to *analyse* a situation, this brings different prerequisites, compared to if he or she needs to learn to *create* something.

Within the 'Knowledge Dimension' of Bloom's taxonomy, we expect that conceptual and procedural knowledge relates better to AR than the factual and metacognitive knowledge. Facts can also be learned in less advanced ways, using descriptions, schemes, and illustrations. Metacognitive knowledge combines several aspects, and we assume that this type of knowledge needs more distance and reflection; we assume that will not be available in the direct, active modes of learning that belong to an AR experience.

Although we confront ourselves with a complex taxonomy with many aspects and interpretations, it is expected that this will not harm the richness of findings for this project. We are aware that within the given time, only the surface of this new medium can be explored. The width of scope, open to collaborations, experimentation, twists and turns excites. A new medium with so much potential should be explored in open and playful ways, while the findings should be noted, communicated and exhibited using evocative examples.

3.4 Deliverables

The educational fellowship within which this project will be carried out requires dissemination of knowledge and direct application in education. An added request is that the explored techniques should preferably suit the education of large numbers of learners, online and on campus. These requirements can be fulfilled if the expectations raised by the AR industry will also be fulfilled. This notion should be clear, because technical research is not proof of knowledge, instead it is explorative and creative and leads to new insights. The decision to focus on Blooms taxonomy while using available authoring tools is a practical choice with positive opportunistic sense for what might become our new way of teaching.

Therefore, the deliverables are open to surprise and influences from colleagues, students and anyone involved. We aim to make several AR exhibitions within the Teaching Lab and on the streets and squares of the TU-Delft campus. In the process towards the exhibition, we will record AR videos to disseminate early findings and we will invite teaching staff to workshops, and we will initiate special sessions to focus on their educational needs. Early results will be shown during the 2019 EAEA conference in Nantes.

Apart from the collaborations and experiments, we expect new insights regarding spatial user interfaces, new insights regarding juxtaposition of context and digital visualisations, and new insights regarding perception and the dynamic perspective within AR.

Finally, we expect new insights based on the creation of a large number of experiments. Those insights will be communicated in a new online course regarding the workflow and best practices for finding, making, using, and sharing 3D assets for education.

References

1. American Museum of Natural History, Swimming With Giants, published October 20th, 2018, https://youtu.be/wOsq7fQ9Z3A
2. Clickshare, https://www.barco.com/en/clickshare
3. Mentimeter, https://www.mentimeter.com
4. Kahoot, https://kahoot.com
5. Feedback Fruits, https://feedbackfruits.com/home
6. Nureva, https://www.nureva.com
7. Open Science: https://www.tudelft.nl/en/library/current-topics/tu-delft-open-science/
8. Open Education: https://ocw.tudelft.nl
9. Valkenburg, W. van; http://www.e-learn.nl/2017/09/28/opening-tu-delft-teaching-lab
10. Torch, https://www.torch.app/
11. Anderson, Lorin W.; Krathwohl, David R.; Bloom, Benjamin S; 2001, A taxonomy for learning, teaching, and assessing: a revision of Bloom's taxonomy of educational objectives, New York: Longman.
12. Revised Bloom's Taxonomy, http://www.celt.iastate.edu/teaching/effective-teaching-practices/revised-blooms-taxonomy/

SHS Web of Conferences **64**, 02015 (2019) https://doi.org/10.1051/shsconf/20196402015
EAEA14 2019

360° VR for Qualifying Daylight Design

Anette Kreutzberg[1,*] and *Emanuele* Naboni[2]

[1]Institute of Architecture and Design, The Royal Danish Academy of Fine Arts, School of Architecture (KADK), Copenhagen, Denmark.

[2]Institute of Architecture and Technology, The Royal Danish Academy of Fine Arts, School of Architecture (KADK), Copenhagen, Denmark.

Abstract This paper describes the initial findings in an ongoing project aimed at bridging the gap between quantitative daylight simulations and visually perceived daylight quality, using 360° rendered panoramas and animations displayed in virtual reality. A daylight studio equipped with a simple façade pattern for a simultaneous Thermal Delight study was used as case study and test room. The test room was recorded with a 360° camera in sequential image series on days with different weather conditions. The resulting 360° VR time-lapse recordings were proposed for visual diurnal daylight analysis as supplement to thermal measurements used for calibrating and varying the façade pattern on site and in a corresponding thermal simulation model. A comparative experiment was set up to calibrate the perceived visual qualities and ambiance of daylight in 360° photographic panoramas viewed in VR, compared to the perceived visual qualities and ambiance of the real world site. Subjective visual evaluations of the virtual as well as the real space were recorded based on 15 people answering to a questionnaire. Results from the comparative experiment indicate a variety in perception of daylight quality and ambiance but a rather uniform perception of daylight brightness in 360° photographs that can be transferred to 360° rendered panoramas.

* Corresponding author: anette.kreutzberg@kadk.dk

1 Introduction and background

Thermal delight is an important factor within architectural design [1]. Daylight influences thermal parameters like brightness and temperature and is important for thermal delight, but daylight can also cause thermal discomfort with glare and overheating if poorly designed or not properly regulated. Simulation tools can help in the design process to meet building regulation requirements with established metrics to quantify daylighting from a quantitative stand while the visually perceived daylight quality is purely subjective and challenging to envisage.

There has been a broad focus on light simulation, e.g. illuminance studies or glare studies, but the separation between quantitative daylighting studies and qualitative perceptual studies is evident in the influential Lighting Handbook of the Illuminating Engineering of North America [2], with only one chapter about Quality of the visual Environment and 28 chapters about Quantitative aspects. It is true that a quantitative assessment yields a fixed level, whereas a qualitative assessment of perceived daylight qualities reflects subjective interpretations, difficult for creating reproducible outcome, but it is never the less important to include these assessments in the design process.

Full-scale mock-ups are well-proven tools in architectural design to assess the aesthetic qualities of surfaces as well as to study daylight distribution and reflection. Virtual Reality can act as a digital substitute or supplement to full-scale mock-ups, simulating a bodily presence in space and is widely used in evaluating and communicating the perception of architectural space [3]. For evaluating thermal comfort, VR cannot simulate all climatic behaviour, but can provide a supplementary visual representation of daylight and its visual qualities in space based on several studies [4] indicating the potential of using VR in experimental daylight studies.

The present contribution focuses on qualitative daylight perception as material for daylighting design by evaluating perception of daylight quality and ambiance in 360° photographed panoramas and time-lapse videos viewed in VR compared to a real world scenario. The final aim in this ongoing project is to transfer the findings to 360° renderings of daylighting design, establishing a matrix for daylight quality and ambiance assessment in the digital model at any phase in the design process.

2 Methods

2.1 Mock-up

The test room used in this research is an existing 1:1 daylight studio used as mock-up equipped with a paper façade pattern for the Thermal Delight study.

The daylight studio is part of the Architectural Lighting Lab located in the campus of KADK (Royal Danish Academy of Fine Arts), Copenhagen (Fig. 1). The daylight studio faces East-Southeast 115°, and is located at top floor in a three storeys building hosting various types of research laboratories with specific functions, as opposed to standard design studios.

Fig. 1. KADK Campus and studio location (Image from Google Earth) (left), View from the daylight studio (middle) and view of the balcony (on the right).

The studio dimensions are W5,5 x L8,0 x H2,7 meters, and the window façade face East-Southeast and has an area of 14,85 m² with a small balcony in galvanized metal placed outside the window.

The studio has a grey linoleum floor and walls and ceiling are coated with white paint. The textile curtain covering the West-Northwest wall of the room is also white. The windows have glossy painted, white metal frames. The studio room is for this experiment set up with only one daylight opening facing East-Southeast. The studies are based on daylight only; no additional artificial light was used.

In the ongoing stage of the Thermal Delight research, the paper was chosen as a working material for implementation of design patterns in the façade. The modules designed are derived from a basic A4 paper, with the margins in the size of 210 mm x 297 mm. The element is half of an A4 paper, rotated in order to have the longest side (which is of 364m, the size of the A4 diagonal) on the bottom.

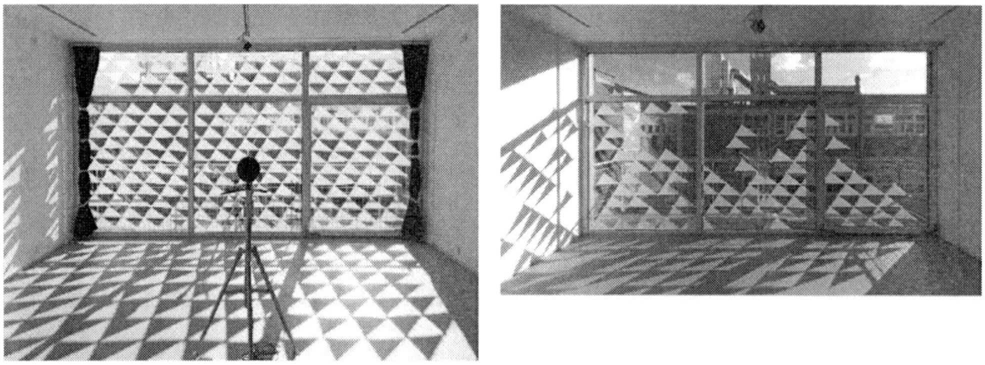

Fig. 2. First pattern - Full Grid (left) and third iteration (right).

The first iteration represents the introduction of the triangle pattern. As a first trial, the facade was covered with a grid of triangles, all of the same size (Fig. 2). A simple culling occurred, using a "true-false" logic, always removing every second module. The proposal took into consideration no optimisation strategy, but only aimed at physically simulating the behaviour of the decided-upon module on the architectural premises. However, the layout contributed to the observation of light patterns, shadow patterns, maximum glare points and, most important, human perception.
Four consecutive iterations of the façade pattern were recorded.

2.2 360° Photographic recordings

Sequential 360° panorama recordings were made with a ThetaS [5] camera and compiled to 360° Time-lapse videos [6] to supply the Thermal Delight study with a bodily and visual representation of the changing daylight.

The 360° recordings were captured over a period of one month between the 27th of September and the 23rd of October 2018. The initial recording was set up with five-second intervals (smallest possible interval) and an image size of 2048 x 1024 pixels (largest possible with 5 sec. intervals). AUTO shunter speed and ISO value was chosen since the ThetaS camera has a fixed aperture of F-stop 2.

Ten hours of recording compiled into a 1 min 22 sec time-lapse video, 25 fps (frames per second) with H265 encoding [7] and was uploaded to Vimeo [8] for sharing and displaying via the Vimeo App [9].

The first recording was made during a windy and partly cloudy weather condition, resulting in heavy flickering in the compiled time-lapse. It was evident that the daylighting condition in the room changed dramatically over time and at high speed (Fig. 3). Watching the time-lapse in mobile VR with the headset was somewhat uncomfortable due to the heavy flickering. Watching the time-lapse in 2D on mobile VR without the headset though was comfortable and still very informative.

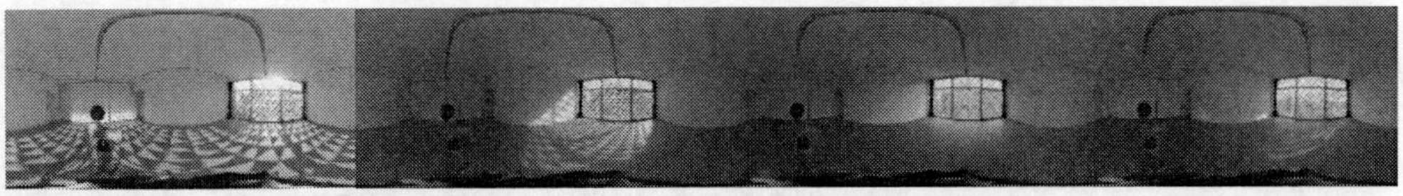

Fig. 3. Extracts from sequential 360° recordings revealing direct sunlight patterns, diffuse reflected skylight and reflections from nearby buildings all within a diurnal loop.

Based on findings from the first time-lapse recording, new time-lapse recordings were set up testing various exposure parameters. With extensive light variation due to natural skylight typologies during a diurnal recording, it was necessary to vary shutter speed or ISO values to balance the brightness levels with the fixed aperture of F-stop 2. The maximum shutter speed of 1/6400 provided the best results together with ISO set to AUTO.

The initial five-second intervals were increased to 10 seconds and eventually 20 seconds with the benefit of being able to capture images at an increased size of 5376 x 2688 pixels and at the same time not exceeding the internal storage capacity of 7GB within a 10-hour recording period.

2.3 360° Test renderings and time-lapse animation rendering

The digital Rhino [10] model of the studio room used for thermal simulations was refined and optimised for rendering with V-Ray [11] in 3dsMax [12].

A 360° diurnal time-lapse animation of the 3D test room model was rendered for comparison with the recorded time-lapse. The animation revealed a slight rotation mismatch between the digital model and the real test room; the light patterns did not match!

Several output variations were tested concerning level of abstraction as well as render quality and image size to minimize the total render time for time-lapse animations. Further studies will focus on the abstraction and aesthetics of 360° daylight renderings.

An alternative POV was used in a top/down 2D animation for a diagrammatic overview of the daylight movement in the room showing all walls at the same time (Fig. 4).

Fig. 4. Diagrammatic top/down 2D animation.

360° test renderings with varying level of abstractions like separated sunlight and reflected skylight were continuously made to explore new ways to utilise the digital model (Fig. 5).

Fig. 5. 360° panorama renderings. Combined Daylight (left), Separated Direct Sunlight (middle), Separated Reflected Skylight (right).

2.4 360° Exposure tests

Extensive exposure tests were made prior to the comparative experiment. The ThetaS camera has a fixed aperture F-Stop 2 with ISO ranging from 100 to 1600 in 13 steps and shutter speed ranging from 60 to 1/6400. Three test series with shutter speed 1/1600, 1/3200 and 1/6400 and all 13 ISO levels were captured and from these, four exposure combinations were selected for each Point of View: Shutter 1/6400 – ISO 100, 200, 400 & 800 (Fig. 6, 10, 11).

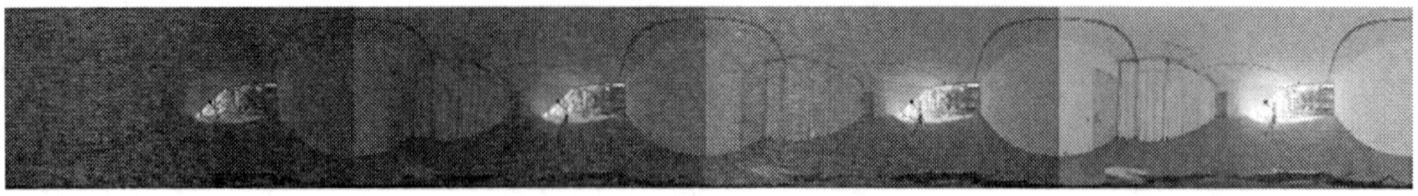

Fig. 6. 360° photos Position A, Sunny weather, ISO 100, 200, 400 & 800 from left to right.

Fig. 7. 360° photos Position A, Sunny weather converted for 2D perspective view (view direction B).

2.5 Screen and VR headset

For the comparative experiment a Samsung Galaxy S8 [13] mobile phone with Homido Mini VR Glasses (clip-on) (Fig. 8)[14] were used.

Fig. 8. Homido Mini.

The ideal VR headset for light evaluation is one fully covering the eyes letting no false light interfere with the screen display. A not-so-perfect solution with real-world light interference was chosen because of the compact design and portability – and in the realisation of its potential use in daily studio work. Testing with real-world light interference might also provide valuable unforeseen input to the study.

2.6 Comparative Experiment

Comparative experiments with a small sample of participants were conducted on three occasions (12th, 16th and 19th of October 2018) in the test room and once at another location. The motivation for the comparative experiments was to make initial calibration of the perception of daylight quality in 360° photographic recordings viewed in VR compared to a real-world scenario, with the purpose of transferring the findings to 360° panorama renderings in further studies.

Two Point of Views (POV) with three view directions (Fig. 9) covered the main differences in direct sunlight and shadow in the test room.

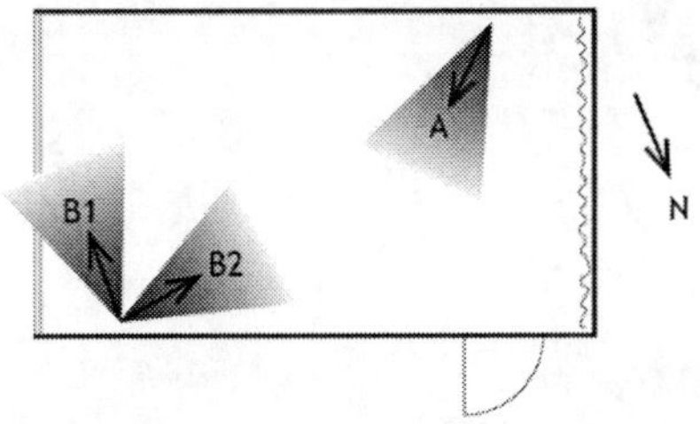

Fig. 9. Test room with two Point of Views and three view directions.

- Standing in the shaded area A looking in the direction of the directly lit area B.
- Standing in the directly lit area B looking in the direction of the shaded area A (B2).
- Standing in the directly lit area B looking in the direction of the light source (B1 the window).

A total of 15 participants, age 23 to 60, students and employees at KADK, viewed the 360° photos in VR with four different Exposure settings displayed in random order for each POV and compared them to the real world scenario. Four different combinations of real-time weather condition and recordings were tested in the experiment.

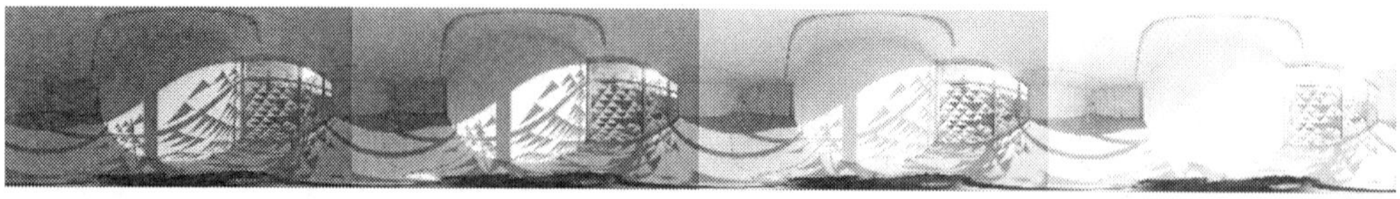

Fig. 10. 360° photos Position B, Sunny weather, ISO 100, 200, 400 & 800 from left to right.

Fig. 11. 360° photos Position B, Overcast weather, ISO 100, 200, 400 & 800 from left to right.

On location, sunny sky compared to 360° recorded sunny sky
On location, overcast compared to 360° recorded overcast
Different time and location compared to 360° recorded sunny sky
Different time and location compared to 360° recorded overcast

2.7 Survey

A questionnaire distributed on paper at the comparative experiment was designed to map the perceived visual qualities and ambiance of a real-world site at a given time and date compared to perceived visual qualities and ambiance of daylight in 360° photographic panoramas of the same location with identical weather condition, sunny sky or overcast, viewed in VR. Subjective visual evaluations of the virtual as well as the real space were recorded based on 15 people answering to a questionnaire. The participants at a different location tested both weather conditions.

Not all the participants in the experiment answered questions concerning the general quality and ambiance in the test room.

Participants were experiencing two series of 360° photographic panoramas in four different exposures in random order. After each series, participants filled out the corresponding questions in the questionnaire.

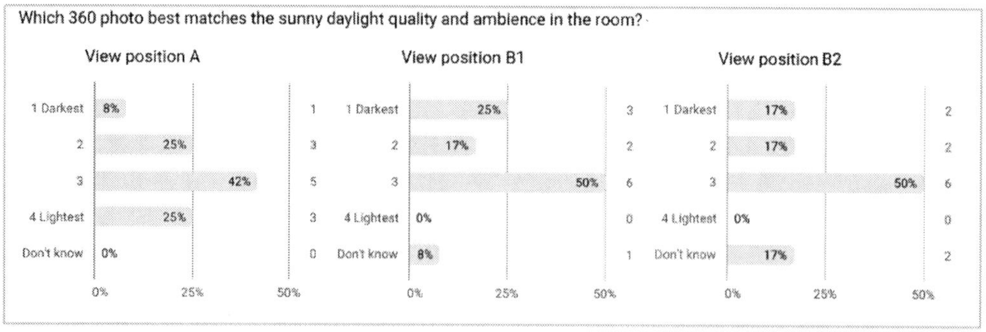

Fig. 12. Answers from participants in sunny or partly sunny weather.

3 Results and discussion

The time-lapse videos based on 360° photographed panoramas were very informative showing the impact of the façade patterns on direct sunlight during the day as well as the variation of luminous conditions in various weather conditions, clarifying the dynamics and complexity of daylight.

The corresponding 360° rendered animations were likewise informative with a stable skylight eliminating flickering. Render time was an issue though, even when lowering the image quality to a render time of five minutes per frame – a one-minute animation with only 25 fps would render for 125 hours! For practical use in the design studio, a low-resolution "preview" quality is proposed for time-lapse animations with the supplement of single frame panoramas of select times rendered in high quality and high resolution.

Only few of the participants in the experiment answered questions concerning the general quality and ambiance in the test room but they indicate a variety in perception of daylight probably caused by the ever changing and very dynamic light during the survey, which the comment sections of the questionnaire reflected. A larger sample in future surveys will provide more sufficient data.

While this study is still on-going, analysis of the questionnaires show a reasonable uniform perception of daylight brightness in the 360° photos compared with the real site. A small sample of 15 persons in total participated in the comparative experiment. Nine persons on site in sunny weather, three persons on site in partly sunny weather and three persons on a different site outside the studio in overcast weather. The daylight brightness was primarily matched with the second lightest exposure value in both sunny and overcast weather, on and off site. Position B standing in the direct sunlight was secondarily matched with darker exposures whereas position A in sunny weather as well as both positions A and B in overcast weather were secondly matched with the lightest exposures.

Although the results can be transferred to tone mapping of 360° renderings for assessment of luminous conditions in a virtual model, further studies with a larger sample must be done. Extreme brightness values from the sun must correspond to varying maximum brightness levels on different mobile screens – which influences the displayable contrast range. Contrast is important in understanding light, since the eye automatically adapts to different brightness levels. Further experiments with display on various mobile screens might result in findings of alternative tone mappings and should include image analysis to address problems with large variation in brightness level displayed in a single 360° panorama.

In further studies, the results from the survey will be compared to thermal measurements made by the Thermal Delight Project investigating the impact temperature might have on the perception of daylight brightness as well as quality and ambiance.

Results from the comparative experiment will also be transferred to tone mapping of rendered 360° panoramas and animations with various abstractions in further studies, and thus continue to unfold various ways rendered 360° VR output might inform daylighting design from conceptual to final phases in the design process.

References

1. L. Herschong, *Thermal Delight in Architecture*, MIT (1979).
2. *The Lighting Handbook*, 10th Edition, Illuminating Engineering Society (2011)
3. A. Heydarian, J. P. Carneiro, D. Gerber, B. Becerik-Gerber, T. Hayes & W. Wood. *Immersive virtual environments versus physical built environments: A benchmarking study for building design and user-built environment explorations*. Automation in Construction, **54**, 116-126 (2015)
4. K. Chamilothori, J. Wienold & M. Andersen. *Adequacy of Immersive Virtual Reality for the Perception of Daylit Spaces: Comparison of Real and Virtual Environments*. Leukos, 1-24 (2018)
5. *Ricoh Theta*, https://theta360.com/en/
6. *Adobe Premiere Pro CC*, https://www.adobe.com/dk/products/premiere.html
7. *HEVC/H.265*, https://www.divx.com/en/software/technologies/hevc/
8. *Vimeo*, https://vimeo.com
9. *Vimeo app*, https://play.google.com/store/apps/details?id=com.vimeo.android.videoapp&hl=en_US
10. *Rhino 6 for Windows*, https://www.rhino3d.com/ [Accessed].
11. *V-Ray for 3ds Max*, https://www.chaosgroup.com/vray/3ds-max
12. *3ds Max*, https://www.autodesk.dk/products/3ds-max/overview
13. *Samsung Galaxy S8*, //www.samsung.com/dk/smartphones/galaxy-s8/
14. *Homido mini*, https://homido.com/shop/homido-mini

Geometry, Pattern, Kinetics: Beginning Studio Pedagogy from 2D to 3D

M. Saleh Uddin[1]

[1]Department of Architecture, Kennesaw State University, Georgia, USA

Abstract. Architecture has the strongest educational ties to geometric organization because of the necessity for order and efficiency in construction, three-dimensional space composition, and the desire to create aesthetically pleasing structures. Kimberly Elam, the author of "Geometry of Design" mentions, *"Too often as a design professional and educator I have seen excellent conceptual ideas suffer during the process of realization, in large part because of the designer did not understand the visual principles of geometric composition. These principles include an understanding of classic proportioning systems such as the golden section and root rectangles, as well as ratios and proportion, interrelationships of form, and regulating lines."* With the primary objective of linking art and architecture through observation, analysis, and composition, this paper demonstrates beginning studio projects where geometry, pattern and kinetics are exercised to enhance understanding of two-dimensional pattern and three-dimensional construction. The objective is to enhance three-dimensional design thinking that has become a challenge in recent trends of questionable dependency on computer modeling applications for design decisions.

1 Introduction

Visual thinking and the understanding of geometric organization in both 2D and 3D are essential in architectural education. This is truer now more than ever before as students tend to depend on digital applications more and lean towards creating surfaces. With the primary objective of linking art and architecture through observation, analysis, and composition, this paper demonstrates beginning studio projects where geometry, pattern and kinetics are exercised to enhance understanding of two-dimensional pattern and three-dimensional construction. This paper promotes the concept of a beginning studio that starts with basic spatial composition of 2D and 3D in design thinking, linking with 3D physical construction.

The assigned projects had four specific goals:

- Observation of random geometry and pattern in nature and man-made objects
- Extraction and modification of geometry to formalize a pattern or a module for pattern
- Design thinking in three dimensions and kinetics, and
- Use of digital software and hardware to fabricate modified pattern in to three-dimensional kinetic element for architectural use.

2 Pedagogical Context

In architectural education, the foundation of understanding properties of geometry, composition, scale and proportion in 2D and 3D had been the essential ingredients for design process. They are also complementary to the notion of linking art with architecture. With recent dependency on digital tools in architectural studios there has been a noticeable change in overall understanding of design thinking, three-dimensional exploration, design iteration, and notion of scale and proportion. The confusion of scale in virtual environments and physical prints in this regard is notable.

In this context, it is relevant to note that the formal architectural education was established at the turn of the seventeenth century by the "Beaux-Arts" in France. It emerged in response to the value system at that time and continued to be the only model for two hundred years. Towards the end of the nineteenth century, the value system started to change. A new development of formal design education strongly appeared before World War I in Germany. The Bauhaus model of design education emerged in response to the technological development and industrial revolution. The word "Bauhaus" itself means "house of construction" or "school of building". Bauhaus was founded with the idea of creating a "total" work of art in which all arts, including architecture, would eventually be brought together. The Bauhaus style later became one of the most influential currents in modern design and architectural education.

The general world trend in the late twentieth century caused an emphasis towards scientific, mathematical, and quantitative approach to education, and art education is often refuted because it is based on perception. It is qualitative and subjective which makes it difficult to measure and evaluate. Current inception of digital media and personal computers in architectural education inherently seem to negate the role of art in architecture.

With current dynamic changes in the learning system, and evolving digital means, the design studio pedagogy is in the verge of reformation again after a century. From the academic perspective, Bauhaus principles still seem to be relevant because architecture is always seen as a fine art where principles of formal composition are considered being of greatest importance. Based on this belief, the author is convinced that a beginning studio should start with basic spatial composition of 2D and 3D in design thinking, understanding of geometry, graphic representation, and physical construction.

3 Geometry, Pattern, Module, Kinetics

Geometry, pattern, and module are interrelated in their definition, physical characteristics, and architectural interpretation. Discovery of geometry in our surrounding environment is an important learning for all visual designers. This learning includes observation, analysis, and synthesis of geometry for adaptation and reformation. This section of paper looks at definition of Geometry, Pattern, Module, and Kinetics for understanding and their cross relationships.

Geometry (from the Ancient Greek: geo- "earth", -metron "measurement") is a branch of mathematics concerned with questions of shape, size, relative position of figures, and the properties of space. In its simplest form, geometry is the mathematical study of shapes and space. Geometry can deal with flat, two-dimensional shapes, such as squares and circles, or three-dimensional shapes with depth, such as cubes and spheres. Studying geometry is important to develop skills such as logic, problem solving and spatial understanding. Since digital applications for drawing and models are based on geometry then it is important to study geometry and graphics of descriptive geometry.

Pattern is a regularity in the world, man-made design, or abstract ideas that repeats in a predictable way. Geometry, form, and shape are primary ingredients of a pattern. Although for architecture pattern is considered visual and structural, it can be behavioral for other fields. Several of our senses may directly be able to observe patterns. The visual element of pattern is constructed by repeating or echoing the elements of an artwork or drawing to communicate a sense of balance, harmony, contrast, rhythm or movement. There are two basic types of pattern in visual studies: Natural Pattern and Man-Made Pattern. Both natural and man-made patterns can be regular or irregular, organic or geometric, structural or decorative, positive or negative and repeating or random.

Natural pattern in visual field is often based on the inspiration we get from observing the natural patterns that occur in nature. We can see these in the shape of a leaf and the branches of a tree, the structure of a crystal, the spiral of a shell, the symmetry of a snowflake and the patterns on animals, fish and insects. Man-made patterns are usually a compositional pattern of lines, shapes, and geometry.

The studio projects illustrated in this paper have been chosen because they all use pattern in an inspirational manner both from nature and man-made.

Module is a self-contained component of a system, often interchangeable, which has a well-defined interface to the other components. Modularity in design is an approach that subdivides a system into smaller parts of modules that can be independently created and then used in different systems. Modularity offers benefits such as reduction in customization with more flexibility in design augmentation.

Kinetics is a new word in architecture borrowed from other disciplines. Wikipedia defines it as a study of motion and its causes. The English Oxford Dictionary defines as, "the branch of chemistry or biochemistry concerned with measuring and studying the rates of reactions". Webster Dictionary defines as, "a branch of science that deals with the effects of forces upon the motions of material bodies or with changes in a physical or chemical system". Historically, architecture has been perceived as a stationary creation and thus the study of kinetics was beyond its scope. In last thirty years, architectural education has integrated various digital tools such as fabrication, analysis, simulation, etc. to replace manual graphics and tactile design thinking and thus integration of kinetics and robotics in beginning studios are not rare.

4 Projects

Considering the fact that the recent trend is to design spaces through surface-making in computer modeling program this paper advocates three-dimensional design thinking through pattern and physical kinetic construction. The pattern in this instance is to be observed and analyzed as arts from nature and creation of objects by humans. During this process, students document and analyze texture, composition, pattern, and kinetics to formulate design strategies to create a three-dimensional system with and without kinetics.
Recent inclination of pattern and surface making in digital software by students cannot be ignored. If the understanding of scale, perception of depth, and control of desired geometry can be integrated then results would be more guided than accidental. This paper looks at these following projects where composition is documented as 2D graphic texture first, and then

https://doi.org/10.1051/shsconf/20196402016

explored in to pattern and 3D construction. Photographs, hand drawings, Photoshop filters, Illustrator outlines, AutoCAD drawing, CAD Laser cut, Rhino 3D model, and STL 3D print are the primary means and techniques of exploration. The final transformation is in three-dimensional kinetic construction. The assigned project had their respective criteria and objectives.

4.1 Project: Observation-Analysis-Synthesis

All assigned projects started with finding geometry, pattern, and texture from nature and man-made. Figure 1 illustrates student work highlighting observation and documentation from nature before expanding in to a system of geometric network or a structural system in three-dimensional construction.

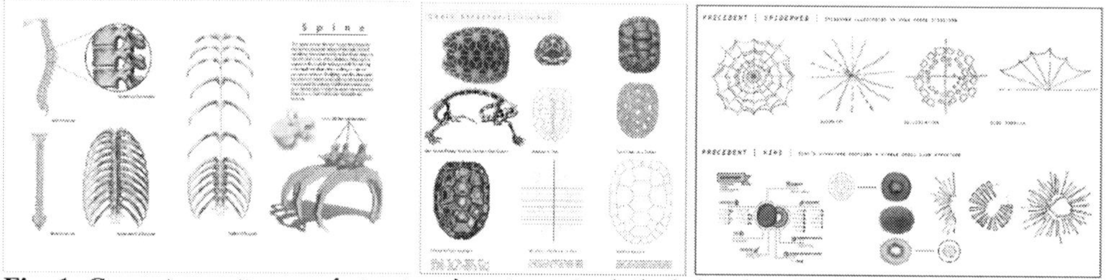

Fig. 1. Geometry, pattern, and structure in nature. Analytical drawing. Student Work from author's taught studio.

4.2 Project: 2D Image to Solid-Void Tactile Pattern

The highlight of this project was to observe texture from nature and man-made and manipulate in Photoshop to recreate that image in two or three shades that can be outlined and laser cut for tactile 3D model. The figure below shows original texture from grass blade and carpet transformed to two-dimensional tactile patterns. The primary objective was to observe, extract, manipulate, and recompose solid-void pattern. Techniques included Photoshop filter manipulation, outlining in Illustrator, and conversion to 2D AutoCAD for laser cut

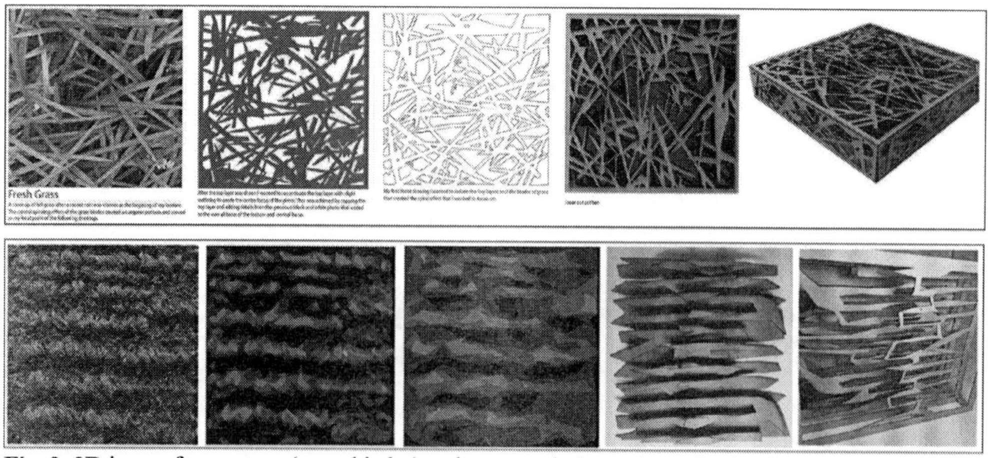

Fig. 2. 2D image from nature (grass blades) and man-made (carpet) manipulated and re-proportioned in to solid-void tactile fabrication pattern (author's 2nd year studio student work).

4.3 Project: Module and Pattern

The goal for this project was to understand the geometry of an individual element and its relationship with other elements to make a module that can be repeated as a tile. The start of this project was to manually draw individual components and to compose them in module for a desired pattern in solid-void. Techniques included hand drawing, Photoshop manipulation, outlining in Illustrator, and conversion to AutoCAD for laser cut.

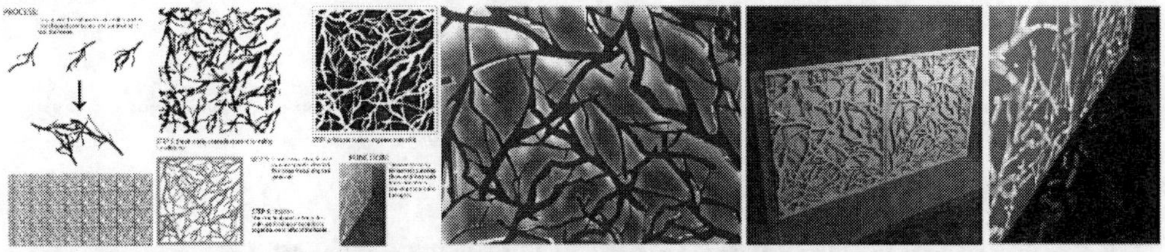

Fig. 3. The geometry of three tree branches overlapped and composed in hand drawing and then converted to CAD drawing for laser cut (Author's 2nd year studio student work)

4.4 Project: 2D Platonic Geometry to 3D KineticPattern

Below are two projects. They demonstrate exploration of platonic geometry (hexagon and rectangle) to form expandable and retractable kinetic systems by reforming original geometries. Using additional elements of pivot points the modules were able to expand and contract in all directions. The working principle of this kinetic movement was entirely conceived by tectonic experimentation in physical pieces rather than digital models. Only AutoCAD and laser cutter were used to create these expandable modular systems. Developing a connection system became an important factor for these explorations.

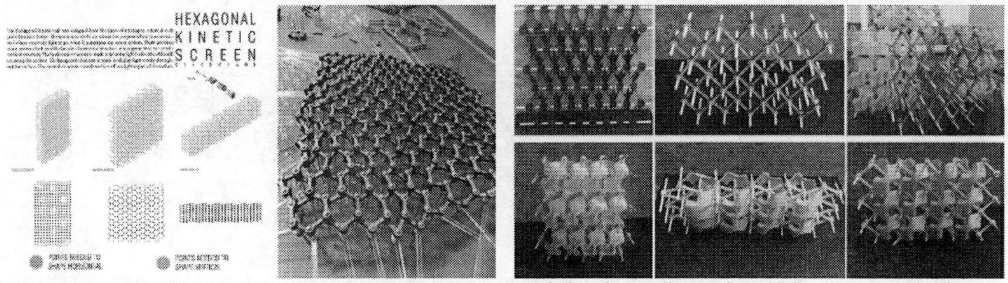

Fig. 4. Platonic 2D geometry transformed to create modular three-dimensional kinetic expandable systems through tactile exploration (author's 2nd year studio student work)

4.5 Project: 3D Geometry from Human Bone Tissue

This project takes lesson from microscopic human bone tissue formation and geometry formed by them. This exercise required detailed investigation, observation, and understanding in three dimensions. Since none of the surfaces are in a single plane the study had to be in Rhino 3D model and 3D print.

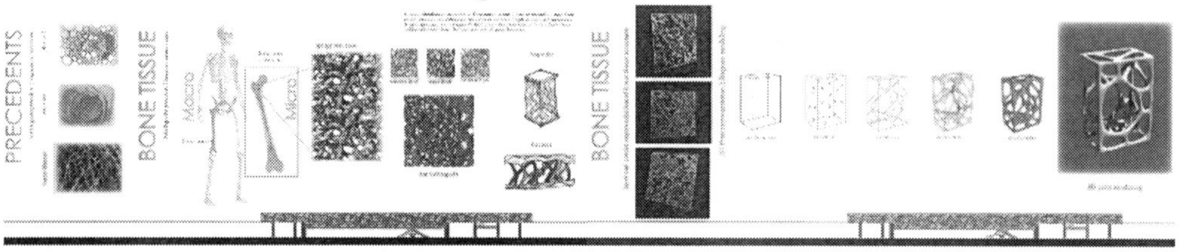

Fig. 5. Pattern extracted from bone tissue by 3D visualization and 3D print (author's 2nd year studio student work)

4.6 Project: Manual Mold Cast and Modular Wall System

Connecting the same unit in various ways to create a modular patterned wall was the objective of this short one-week project. Students first had to design a single unit in a hand-crafted model to create a female mold for casting. That mold was then used to cast several other copies of that same unit to construct the modular wall. The total process was executed entirely in manual techniques of mold casting. This is an effective substitute of 3D digital print.

Fig. 6. Mold casting technique to fabricate modular wall system from a single unit (author's 2nd year studio student work).

4.7 Project: Kinetics from Nature

Using inspiration from a Peacock's expandable feather system, this project analyzes the geometric pattern and constructs a 3D structural system that can expand and contract and span a large space. The fanned tail of a peacock presents geometry when diagrammed creates the mapping of a structural truss system. This system can be applied two-dimensionally as an expandable fan and three-dimensionally as an expandable dome. The method can be applied as a kinetic roofing or wall system that divides spaces according to program. It can also be applied as an expandable shell structure that can create temporary space. Analysis of geometry, kinetics, hand diagram and physical model iterations were the primary means for this exploration.

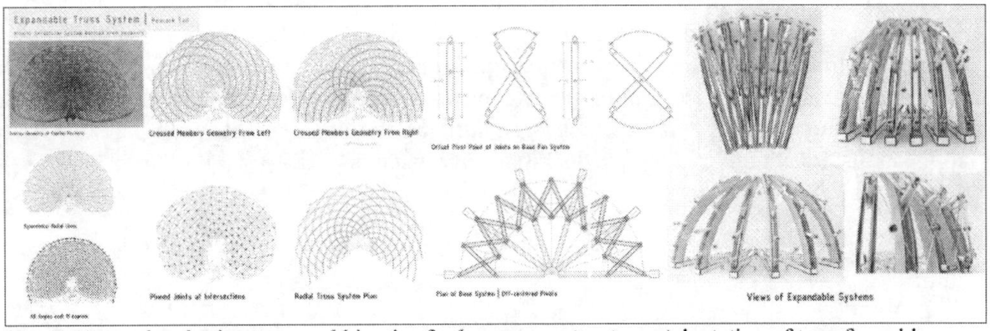

Fig. 7. Nature inspired pattern and kinetics for large span structure. Adaptation of transformable geometry in peacock's feather to design a structure with kinetic expandable truss system (author's 2nd year studio student work).

Conclusion

The lack of understanding between virtual and actual (digital and analog), in particular understanding of 'scale' seems to be alarming in all levels of architectural studios. In the author's studio half of the students had difficulties in constructing a laser cut model in 1/8th scale from a provided 3D digital file. Improper use of a 3D printer to cut simple pieces is becoming more common than before. These hints that the increasing use of digital tools has resulted in to decreasing the role of observation, analysis, and three-dimensional thinking in the beginning studios.

Fig. 8. Scale in print is vague and not within convention. Improper use of 3D printer demonstrates reluctance in exploration of other materials and techniques.

As mentioned before, architecture has the strongest educational ties to geometric organization because of the necessity for order and efficiency in construction. In that regard, understanding geometry and its order is of prime importance and thus should be studied before generating them in digital applications. Unfortunately, most architecture curriculums had taken out descriptive geometry courses in order to accommodate other courses primarily related to software learning.

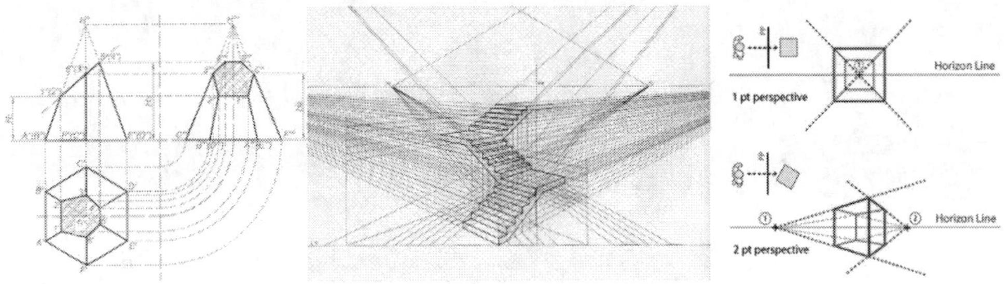

Fig. 9. Although descriptive geometry is no longer a part of architecture curriculum in most schools, it is the foundation of understanding geometry and three-dimensional thinking.

The current trends of design education are becoming dependent on the ability of digital application rather than incorporation for principles of design, understanding of geometry, and aspects of other design forces. This includes program, site, environment, etc. The projects in this paper are short exercises running between two to three weeks demonstrating an effort to incorporate three-dimensional tactile design thinking through geometry and pattern. Digital applications are not ignored but used strategically when needed.

SHS Web of Conferences 64, 02017 (2019) https://doi.org/10.1051/shsconf/20196402017
EAEA14 2019

A Missing Link: Thinking | making | presenting

Zamila Karimi[1,*] and *M Saleh* Uddin[1]

[1]Kennesaw State University, GA USA

Keywords: Studio Teaching, Architectural Representation, Drawing, Design Thinking

Abstract. *Let whoever may have attained to so much as to have the power of drawing know that he holds a great treasure* ... Michelangelo

The tradition of architectural drawings and making as a means of design thinking in a constant feedback loop results in discoveries that facilitate creative thinking in an iterative process. In the digital age, notion of drawing and making by hand as a cognitive process of thinking is fading. This trend is increasingly evident in upper-level architecture students who depend strictly on digital tools for design thinking, missing many critical decision-making steps. Concepts of scale, diagramming, composition, materiality are missing – part of the challenge is the computer screen and the lack of tactical autonomy with physical materials – pen, pencil, brush, architectural scale, materiality, construction, assemblage. How can we as educators assert that drawings are not just architectural representations, but a means to architectural inquiry? Why is it critical for our students to use hand drawings, sketching and diagramming when exploring ideas? Can small gestural models provide notions of scale, materiality, and construction? Teaching pedagogy has always engaged new modes of design thinking and communications as a way of design inquiry – a trait essential to architects. Historically, since the Renaissance, drawings have been the catalyst to advance architectural discourse. In the 20th century, different movements such as De Stijl, Constructivism and Bauhaus used a multi-disciplinary approach combining art and technology through the lens of drawing and making that led to a new wave of design pedagogy to push their imaginations into new territories which the digital augmented as technology advanced. We believe that today's students need to continue to develop both hand and digital skills in tandem to optimize design thinking, making and presentation techniques. In doing so, they can advance architectural pedagogy to new heights as those before did; critically in a material and physical sense: as an embodied spatial experience.

[*] zkarimi@kennesaw.edu : muddin3@kennesaw.edu

1 Introduction

And it's always interesting, I think, to see how the future, or rather the forward-looking form of any discipline, always carries within it the seeds of its own triteness. William Gibson

The tradition of architectural drawings and making as a means of design thinking using hand-mind-eye coordination in a constant feedback loop results in discoveries that facilitate creative thinking in an iterative process. However, in the digital age, this notion of drawing and making by hand as a cognitive process of thinking is fading. This trend is increasingly evident in upper-level architecture students who depend strictly on digital tools for design thinking without employing their hand skills to think with critically missing many important decision-making steps. Notions of scale, diagramming, composition, materiality are acutely lacking – part of the challenge is the computer screen and the lack of tactical autonomy with physical materials – pen, pencil, brush, architecture scale, materiality, construction, assemblage.

How can we as educators assert that drawings are not just architectural representations, but a means to architectural inquiry? Why is it critical than for our students to use hand drawings, sketching and diagramming when exploring ideas? Can small gestural models by hand provide notions of scale, materiality, and construction?

1.1 Pedagogy: Drawing as a mode of thinking

Teaching pedagogy has always engaged new modes of design thinking and communications as a way of design inquiry – a trait essential to architects. Historically, since the Renaissance, drawings have been the catalyst to advance architectural discourse. In the 20th century, different movements such as De Stijl, Constructivism and Bauhaus used a multi-disciplinary approach combining art and technology through the lens of drawing and making that led to a new wave of design pedagogy. Renewed interest in parallel projections/axonometry as a means of discovery and imagination was employed by both artists and architects to promote new vision of modernism. The impact of 20th-century avant-garde visualizing techniques was challenged by Russian Constructivist, which juxtaposed pure geometrical forms around a twisted frame as in Tatlin's monument of 1919 to announce a revolution in architecture. In the 1970s, these Constructivist drawings were explored by a group of rising architects at the Architectural Association in London as a means of design inquiry to find new meanings and relationships, Zaha Hadid's "The World (89 Degrees), 1984" is a critical commentary of the untested potential of Modernism through the lens of Russian Constructivist mode of drawings for a new lifestyle.

Even though modernism patronized straightforward and informative drawings, in many cases, the quality of architectural drawing stood out on those projects where getting the building constructed was not a serious consideration. While breaking away from the pragmatist tradition of the fifties, many new ideas became unveiled through visions for an architectural utopia. Archigram in England and the Metabolists in Japan and Vienna suggested a complete, attainable environment in their drawings rather than representing the structure to be built.

In the departure from modernism, the Post-Modern period introduced new intentions toward architectural drawing. Aldo Rossi in Milan, Oswald Mathis Ungers in Berlin, Raimund Abraham and Rob Krier in Vienna, and Leon Krier in London, all emphasized the relative independence of architectural drawing that led to a new aesthetic which eventually became recognized in its own right. Among American architects, Charles Moore, Michael Graves, John Hejduk, Robert Venturi, Steven Holl, and Peter Eisenmen are the pioneers.[2]

The MOMA "Deconstructivist Architecture" exhibition, curated by Philip Johnson and Mark Wigley (1988), featured drawings, models, paintings sketches of seven prominent contemporary architects—Frank Gehry, Daniel Libeskind, Rem Koolhaas, Peter Eisenman, Zaha Hadid, Coop Himmelblau, and Bernard Tschumi to showcase the revolutionary nature of their architectural drawings and models – even if it was on paper. These thinkers pushed the architectural discourse into new territories embracing digital technologies as is evident in the works of Frank Gehry, Zaha Hadid and the rest. This was a pivotal moment in history when "*... the imaginative spirit of drawing ... continues to be instrumental to the development of the field.*"[3] In retrospect, there is a common thread that runs amongst the thinkers who sketched, drafted and made models as a means to push their imaginations into new territories, which the digital augmented to catapult their careers. These thinkers were always in control of their craft like a true craftsman using both hand and digital in unison.

This type of criticality is lacking in today's emerging architects who use the digital not as a tool to think with but as a proxy to design with. There is a danger that comes with this dependency. They are as follows: a) lack of criticality in design thinking; b) spatial geometry and forms often is the result of some operations; c) notions of scale both figuratively and formerly is lost. The process is erratic and often missing many critical steps. Moreover, the iterative design process hand-mind-eye coordination in a constant feedback loop which often results in discoveries to facilitate creative thinking is compromised. Part of the challenge is the computer screen and the lack of tactical autonomy with physical materials – pen, pencil, brush, architecture scale, materiality, construction, assemblage.

Using student work from upper-level studios to illustrate the points, we will use three critical frameworks: *a) Design thinking: an iterative process; b) Analog|Digital: a hybrid approach; c) Presentation Techniques* to illustrate our hypothesis.

The lack of basic skills of drawing and making by hand as a cognitive process of thinking is fading amongst upper-level architecture students in the digital age. It is concerning to us as educators. We believe that our students need to continue to develop both hand and digital skills in tandem throughout their academic career thereby equipping them with proper skills to develop as critical thinkers.

2 A Theoretical Approach

I want to see things; I trust only this. I want to see, this is why I make drawings. I can see an image only if I draw it.[4]

Carlo Scarpa's profound statement about how an idea or an image comes to life and germinates when he physically draws is a concept that is foreign to today's students. Dependency on the computer as an all-knowing platform is a cause for grave alarm. Architecture is a complex multi-sensory tactical (haptic) phenomenon that requires hand/mind/making skills in a continuous loop as a cognitive process to generate ideas. Juhani Pallasma argues that "... the architecture of the eye detaches and controls, whereas haptic architecture engages and unites."[5] We as human encounter space via our seven senses: sight, hearing, taste, smell, touch, intuition, and equilibrium (body configuration/identification) as

[2] M. Uddin, Composite Drawing, Pg 4,5; McGraw-Hill, 1997
[3] A. Boyarsky, *Drawing Ambience: Alvin Boyarsky and the Architectural Association* http://www.kemperartmuseum.wustl.edu/exhibitions/9926
[4] F. Cocozza, *Axonometry: The Grip of Thought on Space - A Short Survey on the Relation between the Act of Planning and a Visionary Visualization Technique*. 2017
[5] J. Pallasmaa, *Hapticity and Time: Notes on Fragile Architecture*. 2000

an embodied experience. The Vitruvian concept of architectural values: utilitas | firmitas | venustas is still as valid today as it was during the Classical period, which suggests that successful architecture achieves its goal of function, form, and delight in a seamless assemblage(s) through skillful crafting of space formally, functionally, and is a delight to be in. Thus, such an endeavor requires us to viscerally engage our primal faculties to augment architectural thinking at this moment in time of advanced 3D software and virtual reality screens, which often is a form of detachment from the spatial experience.

2.1 Design thinking: an iterative process

Computers tend to dematerialize design thinking and results in abstractions far removed from the world of human sensory experience... Perhaps we need to defer the use of computer as a tool for design inquiry in the beginning stages to avoid notions of *dematerializtion and abstraction* so as to generate ... a *coherent design philosophy, good materials, craftsmanship and personal exploration.*[6]

Diagramming and Sketching as a mode of visceral design inquiry using the *innate capacities of the human brain for creative thinking*[7] in a constant feedback loop is missing in today's paperless studio environment. Research shows that for the brain to explore and incubate ideas one needs to be free flowing where physical, tactical gestures are activated critically to be imaginative and in a discovery mode to develop a design concept as a framework for an architectural project to germinate. During the design process, computers as design tools severely compromise the sense of scale, space, form, materiality as mere abstractions through a pixilated screen often lacking originality, which is a grave concern for us as educators. The sterile studio environment devoid of visual clues, hand sketches and physical three-dimensional explorations, suggest lack of passion and excitement for studio projects. Students are often scared to commit to making sketches and models, as is the case with one thesis student who is convinced that she can design digitally. Refusal to make physical models and sketches in an iterative process has resulted in her design lacking any sort of part/framework for design. Program requirements for urban farming, which is highly dependent on physical site conditions, context, sun path, and topography, is not totally understood via a digital model. Even when she hesitantly built a site-model with diagrammatic sketches, she did not quite know how to use it in an iterative manner to push design thinking viscerally. Architect Jacob Brillhart attests that since architecture is a spatial experience, "... observation and experience, translated through drawing, inevitably informs design ...hand drawing accesses the {intuitive} right side of the brain ... builds a foundation for visual thinking."[8] Student dependency almost entirely on digital as a means of design exploration is a cause for alarm as it detaches the student from spatial and formal considerations. Digital platforms run the risk of limiting the student's investigations especially during the early stages of concept development, as the mind/hand needs to follow a rational approach of sequential thinking via the flat-screen lacking subtle nuances of the hand/mind gestural drawings thus limiting ideation. This trend continues in many practices where new generations of architects are at a loss in client meetings without their digital tools and revert to senior architects to rescue them. Many practices are offering sketching classes to new architects.

[6] H. F. Mallgrave, *The Architect's Brain: Neuroscience, Creativity, and Architecture*; 216, Wiley-Blackwell. 2011
[7] H. F. Mallgrave, *The Architect's Brain: Neuroscience, Creativity, and Architecture*; 214, Wiley-Blackwell. 2011
[8] J. Brillhart, *Drawing Towards a More Creative Architecture: Mediating between the Digital and Analog.*

Figure 1. Desk Crits showing site on computer digitally prompting Professor to use hand sketches and physical materials to push design thinking tactically.

2.2 Analog | Digital: a hybrid approach

Modern brain scientists now know that your left brain is your verbal and rational brain; it thinks serially and reduces its thoughts to numbers, letters and words... Your right brain is your non-verbal and intuitive brain; it thinks in patterns or pictures composed of "whole things", and does not comprehend reductions, either numbers, letters or words.[9]

Historically, architects understood the idea of scale through hand-mind coordination via sketching and making on-site. The notion of light, shade, shadow, texture, materiality – had to be unraveled – rationalized and decoded to understand scale is all its dimensions. Relationships to the size of a reference object: human form or module of material formed the basis to understanding space with its three-dimensional form. However, that strategy gets compromised in today's pedagogical framework as digital tools get introduced to students at earlier levels. In such a context, some students find it difficult to leap between physical and digital models until a design is finalized. Students somehow believe that the sole use of digital will generate a thoughtful architectural outcome incorporating Pallasma's seven senses as spaces are organized compositionally.

Analog and digital forms of exploration have its own bias and a critical use of each media puts forward a mode of inquiry that enhances the design process. Neurological research shows that during the design development process an analog approach yields best results as the brain neurons and physical gestures tend to work together as an embodied experience to offer the most options/possibilities. Once the concept is achieved, then the digital mode of inquiry can be mobilized to now methodologically develop those concepts and ideas. The physical senses of the body – viscerally active – can never be replaced by the flat screen of the pixelated screen.

The notion of design thinking/making as a cognitive iterative process using both analog and digital in the current studio settings miss many critical steps thus compromising the outcome but more importantly highlighting deficiencies in our students. As students intersect/interweave the two platforms, there is utter confusion of rationalizing and making critical decisions in terms of scale, materiality, relationships, tectonics as they iterate site models and small diagrammatic design explorations, which require patience and commitment to the process in a methodological way. The data from google maps as a means of making physical models results in scalar confusions. Shifting scales between model space in digital drawings on 1 to 1 scale when juxtaposed onto a given site model scale with the physicality of material thickness or no thickness (foam) as contours lead to inaccurate physical models abstracted so drastically that their relationship to the existing context shows compromise. Mechanics of constructing with digital tools adds to the complexity of scale and relationships.

[9] H. F. Mallgrave, *The Architect's Brain: Neuroscience, Creativity, and Architecture*; 216, Wiley-Blackwell. 2011

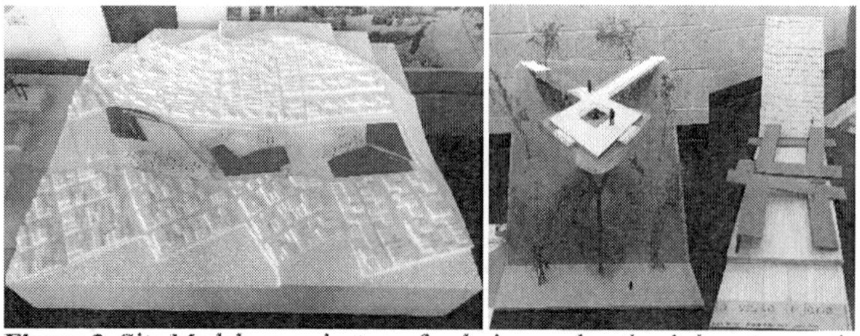

Figure 2. Site Models as caricatures for design exploration brings to question issues of scale | tools | materiality | relationships | context

Figure 3. Site diagramming using digital tools allow students to record site conditions abstractly which may lead to wrong assumptions and readings thus requiring additional strategies to complete diagramming.

2.3 Presentation Techniques: an opportunity

Clearly no single design direction dominates today, and while it is possible to map shifting intellectual agendas, the situation is not so much that one agenda supplants another as it is that one is layered over another, multiplying the possibilities and points of view. [10]

Compositional strategy in the digital age is an architect's license to be successful both in the academics and in practice. Learning and succeeding in architecture schools requires design thinking skills to ideate with and compositional and presentation skills to organize and curate in a manner that is both informational and provocative. Design schools today are inundated with architectural presentations' that all look similar printed on same glossy paper and digital small-scale models resulting in homogenization. How can students creatively curate and compose their presentations to avoid such challenges?[11] Studios traditionally have been a rite of passage to real-life practice where the students put their best effort forward both in terms of design intent and thought out presentation boards. However, today's students' dependency on digital tools as a primary driver of output may be an opportunity to rethink compositional strategies as a curatorial performance to optimize design thinking and presentation techniques. In doing so, today's students can advance architectural pedagogy to new heights as those before them such as Zaha Hadid, Frank Gehry, Thom Mayne.

[10] S. Allen, *The Future that is Now,* Places Journal, 2012

[11] H. F. Mallgrave, *The Architect's Brain: Neuroscience, Creativity, and Architecture*; 214, 215, Wiley-Blackwell. 2011

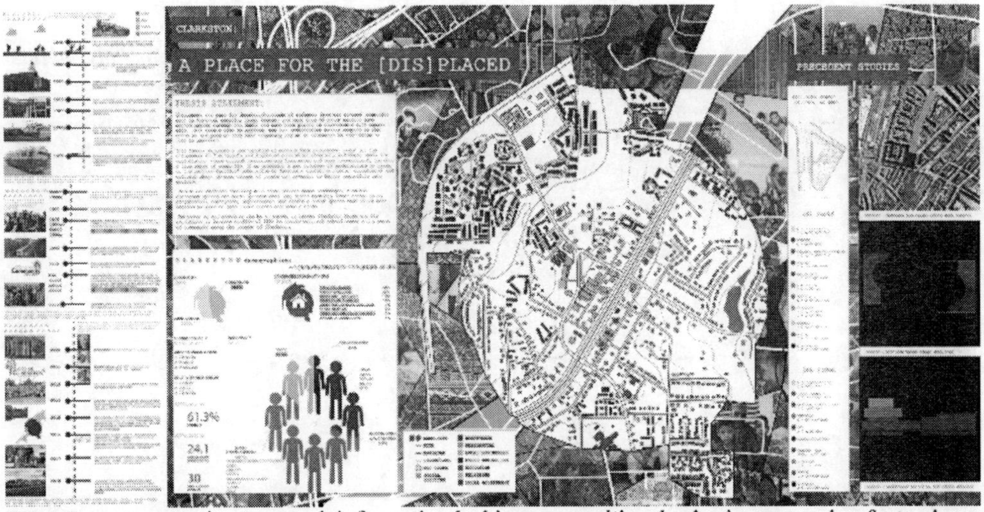

Fig 4. Posters cramming too much information lacking compositional cohesiveness: color, fonts, shapes resulting in boards hard to decipher.

Conclusion

The computer is an important tool – no one could do without it – but for me it's a tool and it doesn't replace thinking. It can make you disconnected and autistic, and that is why we always say, bring it out of the computer, print it up, use paper, use the physicality's and models to understand and anticipate what this thing will be in the end: something physical, something real, something for the people... Pierre de Meuron [12]

We as educators do not deny the importance of analog and digital to work in a complementary relationship to advance design inquiry responsibly so as not to compromise on the Vitruvian values of utilitas | firmitas | venustas and Pallasma's seven senses for a rich spatial experience. With the rise of digital technologies, we as architects must never give up our critical thinking skills of questioning and problem solving in a material and physical sense: as an embodied spatial experience. While we excitedly adopt and adapt digital tools to advance architectural pedagogy, it all begins with the *brain to draw upon all its specialized areas and potential strengths to foster creativity.* [13] To explore and incubate ideas one needs to be free flowing where physical, tactical gestures are activated critically to be imaginative and in a discovery mode to develop a design concept as a framework for an architectural project to germinate.

References

1. A. Boyarsky, *Drawing Ambience: Alvin Boyarsky and the Architectural Association* http://www.kemperartmuseum.wustl.edu/exhibitions/9926
2. F. Cocozza, *Axonometry: The Grip of Thought on Space - A Short Survey on the Relation between the Act of Planning and a Visionary Visualization Technique.* 2017

[12] H. F. Mallgrave, *The Architect's Brain: Neuroscience, Creativity, and Architecture*; 220, Wiley-Blackwell. 2011

[13] H. F. Mallgrave, *The Architect's Brain: Neuroscience, Creativity, and Architecture*; 220, Wiley-Blackwell. 2011

3. H. F. Mallgrave, *The Architect's Brain: Neuroscience, Creativity, and Architecture*; 216, Wiley-Blackwell. 2011
4. J. Brillhart, *Drawing Towards a More Creative Architecture: Mediating between the Digital and Analog.*
5. J. Pallasmaa, *Hapticity and Time: Notes on Fragile Architecture*. 2000
6. M. Uddin, Composite Drawing, Pg 4,5; McGraw-Hill, 1997
7. S. Allen, *The Future that is Now,* Places Journal, 2012

Ambiance of slowness. Brussels commercial gallery of the latter half of the 20th century

Claire Pelgrims[1,*]

[1] Aspirante F.R.S. FNRS, Faculté d'Architecture La Cambre Horta, Université libre de Bruxelles, 19 place Flagey, 1050 Bruxelles, Belgium

Abstract. In this paper, I study the sensory-motor effects of Brussels commercial galleries' ambiance in the latter half of the 20th century. The analysis of two case studies ("Deux Portes" networked galleries and Agora Gallery) reveals the different logics of slow mobility acceleration and immobilisation at stake in the emerging modernist grammar of slow mobility. This grammar – in arrangement with the grammar of fast automobility – structures and stabilises the design of spaces for slowness next to the roadscape in spatial segregation of transport modes. There are accelerating and decelerating sensory *dispositifs* that define galleries both as punctual destination spaces that capture passers-by and as alternative paths for pedestrians: logics of multifunctionality, fast mobility accessibility and setting of an ambiance on the one hand, and logics of securement, spatial and qualitative continuities, on the other hand. Accelerating and decelerating *dispositifs* and logics facilitate movement to better keep the consumer captive. Then, I discuss the possible contribution of iconographic archives in research about past ambiances. They effectively acknowledge sensory-motor effects of ambiance but do not constitute an autonomous corpus to grasp sensitivity and reshape past ambiances.

1 How to grasp sensitivity?

Commercial galleries may be considered as sensitive environments. As every other space in which one moves, it gives rise – as "active potentiality" [1–3] – to perceptive orientations and specific motor behaviours: specific sensory-motor ways of being. In their work at CRESSON, Thibaud and Chelkoff analysed urban micro-morphologies that structure visual, luminous and sound experience and reveal dynamics and located relations between built forms and social uses. They have constituted knowledge about built forms from discursive perception records [1] and field studies of spatial behaviours in specific micro-morphologies [4].

The sensory-motor effects result from the expressive power of surfaces and the motor power of visual orientations. First, the sensitive environment is expressive, in the sense that it gives rise to action and makes itself available as occurrence, thanks to the surface quality: texture, luminosity, reflection, contrast… Second, the attentional gestures of observation, contemplation, examination that underlie the perceptive activity of the passer-by engage his

* Corresponding author: claire.pelgrims@ulb.ac.be

full body, extending often what the sight outlines in movement [1]. Both steps and eye mobilities of the passer-by are considered in the concept of *"mise en vue"* (~exposure) that attempts to articulate spatial forms with social and perceptive forms [4]. It qualifies and characterises the different articulations of spatial behaviours and visual perceptions. Sensory-motor effects are described through the relation between the eye and the step, which dissociation is used as marker for urban environment quality [1]. Lower urban environment quality makes passers-by accelerate. In commercial galleries, "ambiance factors" – interactions between built form and physical signals (luminous, sound…) – engage and disengage, that way, motor activity of passers-by.

In the latter half of the 20th century, Brussels commercial galleries are defined both as alternative paths for pedestrians and as punctual destination spaces that capture the passer-by. As in the Louvre Museum's underground [1], passers-by become subject in modernist commercial galleries to a paradoxical demand to circulate and to stay. This demand manifests through programmatic, regulatory and discursive *dispositifs*. This ambivalence appears in the materiality of galleries as well: in an articulation of sensitive acceleration and deceleration *dispositifs* that defines the "locomotor efficiency" of the commercial gallery as a "space of movement". It operates on ambiance.

When dealing with ambiances of the past, there is a methodological challenge. Most research on ambiance is based on discursive perception reports. Enunciation is perceived as a key element in the very perception of ambiance. To study ambiance of the past and to construct a "history of action in built forms", Ben Hadj Salem, for example, crossed spatial materiality and archives sources describing the social world and perceptions, on the one hand, with the somewhat impressionistic literature sources describing experiences, on the other hand [5]. His methodology relates that way to the history of environment. To grasp sensitivity is about forms of writing and restitution. There is a tension between the use of the first person, the phenomenological "I", and the use of the third, the scientific and hermeneutic "she", "he" or "they". With regards to the knowledge produced the last decades, is it possible to discuss "ambiance factors" of the past without falling back on this methodology (the phenomenological "I")? Is it conceivable to discuss past accelerating and decelerating ambiance considering iconographic archives, and to do so without lapsing into spatial over-determinism? Iconographic archives effectively acknowledge sensory effects of light but they offer an overlooking point of view, less open to others' appreciation. It's somewhat already a hermeneutic restitution of the author's phenomenological perception.

The paper has multiple goals. First, I want to highlight the purpose of gaining insight in the meaning and the use of iconographic archives in research about accelerating and decelerating ambiances. Second, I aim for knowledge production about the "ambiance factors" that initiate or reduce motor activity in the Brussels commercial galleries of the latter half of the 20th century, and how iconographic archives analysis makes those effects visible. I analyse the spatiality of the ambiance documented through original plans archives, architect's drawings and photos. I study the light and material effects that appear in the representations of (1) the networked commercial galleries of Louise and Toison d'Or avenues, and (2) the commercial Gallery Agora in the city centre.

2 Next to the roadscape. Spatial segregation of speed and slowness

Between 1950 and 1989, the built environment of Brussels underwent large transformations to adapt to the increasing automobility, and became part of a roadscape. This new park-like infrastructural urban landscape enforced by the state administration embodies a strong and prevailing fast mobility imaginary articulating functionality and aestheticism in the materiality of the city [6]. The urban renovation of Brussels at the agglomeration scale thus relied mainly on this *imaginary* [7] of fast mobility, documented through plans, models and technical discourses. The roadscape was based on logics that were iteratively emerging during a process of fast mobility imaginary stabilisation – "grammaticalisation". Those logics regulated the articulation between fast and slow mobilities through different regulatory and material *dispositifs* making an acceleration of automobility possible.

In this articulation, urban renovation also created spaces of slow mobility that embodied the emerging slow mobility imaginary. There are punctual spaces of destinations (pedestrian or restricted areas, parks, commercial galleries) where ends fast mobility infrastructure and that tends to "immobilise", to "capture" slow mobility paths. Those separate spaces were bordering fast mobility infrastructure in modal segregation system. In the 1950s, the model of the commercial gallery was indeed re-appropriated by the urban renovation stakeholders to modernise the city, rationalise the uses of spaces and operate a stronger spatial segregation between fast and slow mobility. The model of the commercial gallery matched the logic of "vertical zoning urbanism": internalisation in the built environment of the slow mobility paths and immobility spots, public transport stops and car parking lots, multiplication of ground level and available spaces for mobility flows (tunnels, bridges…). In that way, the design of commercial galleries implements *dispositifs* that tend to "immobilise", to "capture" slow mobility, increasing the destination dimension of the commercial gallery (improved car accessibility, multifunctional venue and services, comfort).

However, commercial galleries also form part of a *passages* tradition. Rainy city, Brussels was already marked by a construction tradition of covered commercial galleries since the 19th century. It aimed at making the capital a "covered city" [8] emancipating pedestrians from the climate: Galerie du Commerce, Galerie de la Monnaie, Passage du Nord, Galerie du Parlement, Galeries Saint-Hubert… In that way, the design of galleries enforces *dispositifs* that, on the contrary, make slow mobility functional (spatial and qualitative continuity and securement of slow mobility paths).

3 A modern grammar of commercial slow mobility

The point here is to analyse the different logics of slow mobility acceleration and immobilisation at stake in the emerging modernist grammar of slow mobility. This grammar – in arrangement with the grammar of fast automobility – structures and stabilises the design of spaces for slowness next to the roadscape. It articulates accelerating and decelerating *dispositifs* and logics. I did so by analysing two case studies. The corpus is made of municipal archives (construction permit application and press reviews)[†], and of project publication in specialised journals. Both case studies lie next to the roadscape. The first is located along a portion of the "highway croissant" formed by the Léopold II Boulevard, the

[†] Archives de la Ville de Bruxelles : Fonds Marcel Lebouille 723, 724, 730, 737 ; DD542 ; DD529 ; Travaux Publics 91 341_1983, 92 091_1987, 82 066_1963. Archives d'Ixelles : Travaux Publics 389_1950,185_1986, 36_1986, 205_2007, 275_1953, 128_1983, 194_1987, 13_1962, 226_1961.

Petite Ceinture and the Louise Avenue. The "Deux Portes" neighbourhood is spread between two polarities from the upper city: the Porte de Namur and the Porte Louise. On one side, a night animated neighbourhood and, on the other side, a fancy commercial neighbourhood. The second case study is located at the heart of the Pentagon, next to the highway mini-ring around the Grand-Place designed in 1955 by Tekhné, along the boulevard of the railway Junction. It is bordering both the touristic and patrimonial neighbourhood of the Grand-Place and that of national institutions and mobility infrastructure (Mont des Arts, Royal Library, Central Station and Sabena terminus).

Commercial galleries as conceived in Brussels are widespread in the capital. Both cases are representatives of this proliferation and specific. The Louise Gallery (1952) constitutes the first modernist commercial gallery and the starting point for the most accomplished network of galleries in the capital: extension connecting Stéphanie square (1955, 1963), Porte Louise (1964), Espace Louise (1989), Toison d'Or Gallery (1964), Nouvelle Galerie d'Ixelles (1953), Bastion Tower Gallery (1964) (Fig.1). The Agora Gallery (1963), the 19th gallery located very close to the Grand-Place, is an interesting case study in terms of modernistic urbanism integration in the ancient urban fabric of the city centre.

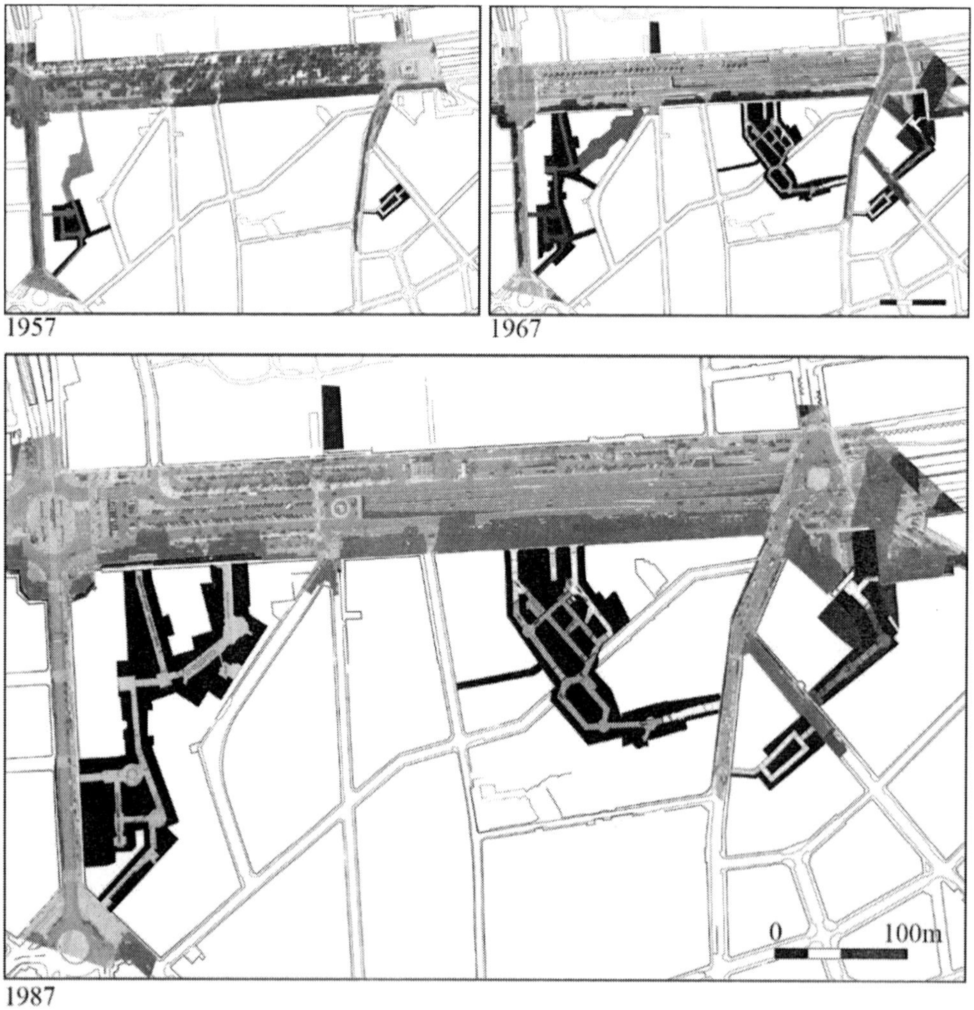

Fig. 1. The development of commercial galleries in the "Deux Portes" neighbourhood. Author's collage. Background map: Bruciel, OSM.

3.1 Capture of slow mobility

Inside the spaces for slowness next to the roadscape, the modernist grammar of slow mobility sets up *dispositifs* that are part of the deceleration and immobilisation logics capturing slow mobility paths. Those logics make slow mobility infrastructure become meeting point where everything is about staying, as in American shopping malls. The logics involve multifunctionality, accessibility and ambiance setting. Galleries are not pathways from point A to B, but the purpose of the journey.

The numerous adjacent functions of the commercial galleries exemplify the **logic of multifunctionality**: exhibition and conference centres, hotels, restaurants… Since the 1950s, galleries have related services. The first Gallery Louise, under the slogan "A city at the heart of the city" [9], already arranged shops, restaurants, private clubs, exhibitions and reception hall, parking lots, offices, shop storage and fifty "American flats" and apartments [10]. Those related services induce indoor spaces to withdraw into itself, as independent islands that accommodate visitors for a long time. This latter logic has strengthened since the 1970s with the slow paths' externalisation along outdoor pedestrian areas and another sidewalk extension. The Gallery Toison d'Or has been as a result transformed into a dead-end passage in the late 2000s. The length and continuity of the galleries became, notably with the third extension of the Louise Galleries [11], an argument for commercial attractiveness, even if it no longer constitutes a *passage* as such. One should go for a *promenade* but, above all, stay in this third extension.

The second logic that lets me define commercial galleries more as destination spaces than *passages* is the **logic of fast mobility accessibility**. Galleries are located next to pre-metro, expressways, tunnels and parking. The indoor and covered connections to those access means are further improved by lifts and escalators to reach destinations effortlessly. Parking issue is a key point in the development of each gallery. In the case of the Gallery Toison d'Or, the tunnelling of the pre-metro will allow the construction of access ramps for the underground parking in both directions of the Petite Ceinture highway. Everything guarantees ease, comfort and smoothness in the intermodal connection to the roadscape.

Third, the **setting of an ambiance** invites to *flânerie* and deceleration. On the one hand, the density of the public may alter the visibility of the passer-by and question the fluidity of build space. On the other hand, commercial galleries are specifically designed for slow paths in terms of materiality and dimension, and through some spatial configurations. The materials and the light create a calm atmosphere, a neutral environment. The atmosphere is comfortable, muffled and quiet. The standard musical background covers any noise.

Those complexes concentrate outside the urban fabric urban functions that prior participated in urbanity. The setting of an ambiance inside these complexes reproduces the reassuring setting of a traditional urban scenography. The model of the city – with arteries, small squares, terraces and stalls – is "recovered, absorbed, and, more importantly script-written, planned and managed" [12]. It creates fragments of *security city* and *managerial city*. The security city guarantees physical safety and encounters with socially close individuals in order to avoid conflicts with heterogeneous thoughts and groups. Following the community-based American example, it constitutes a "segmented city whose fragments have ceased to communicate" [12]. The managerial city relies on this segmentation to constitute "'communities' of consumers", target groups of commercial and spatial strategies [12]. Even if galleries seem to be open spaces for all, guardians, products adapted to the middle class' tastes and budgets, and more importantly, architecture guarantees the exclusion of poor people [13]. The ambiance controls and conditions both the material (air, light...) and the people. It transforms the urban *flâneur* and the passer-by in consumers [14].

The *linear spatial structuration* that imitates the traditional city centre is made of streets and squares. The latter ones are by nature sedentary places: rest area, exhibition area, reception, food court. The first give access to shops. Shop windows must be given all attention. "Paths are 'quaint', short, broken, avoiding tunnelling effect" [15] (that is to say the "effect of centration, slide and anticipation of sight" [1]). The design of the galleries is indeed based on the latest technical and scientific progress to make pedestrians decelerate, to invite them to a slow and "flowing" movement from one side of the gallery to the other. Sociological and psycho-commercial analysis were carried out to determine the best dimension of the hall so that the passer-by will not "be absorbed in a fast flow" [16]. The deceleration operates through the distraction that procure the environment, inviting to a sensory intensity within the modern and luxurious ambiance made of noble materials (blue stone, marble, mirror, vegetation). Those materials create rhythmic lateral appeals for sight along the path. Large halls with shop windows dilate the visual field, appeal lateral view and offer opportunities for breaks.

Furthermore, escalators generate spatial disorientation and confusion that tend to capture the visitor. "Escalator acts as a strategic tool to raise the complex and uncontrolled nature of flows" [17]. This confusion is backed by the trends of boundaries dematerialisation [4]. Visual field of galleries – flowing space with unclear limits – is characterised by crossfading and overprint. It requires passers-by to find other visual clues than that of a well-defined perspective space with clear boundaries and clear figure/ground ratio (as traditional public space made of full and empty spaces). The systematic use of glass shapes other boundaries. Either its transparency creates a visual dilation of public space as semi-public spaces behind the shop windows are made accessible for sight. Either its partial reflection back at passers-by deconstructs the space and articulates the two distinct spatial units by only suggesting spatial depth with a filtering effect. The escalators that link the different storeys allow an intimate relation with the built environment. The created space operates as a complex and progressive stretching deformation, connecting apparently incompatible spaces to create a large indoor landscape perceived in one continuous movement [17]. In their evolution, galleries evolve toward a more horizontal, *"landscaped" spatial structuration* (see the redesign of Toison d'Or Gallery in the late 2000s) that dilate the space even more.

On top of shop windows, some spatial configurations also offer opportunities for breaks: the show grip effect of the sky through roof windows, or of a diving view on other passers-by's overexposed activity. In the first configuration, the recurring sky contemplation behaviours are loopholes from feelings of lockdown as there is no visual horizon. "The sight compensates the impossibility for the body to go out."[1] This behaviour is a simultaneous stop of the step and the eye on the aerial landscape. After attracting the passers-by further in the gallery (see 2.2), the natural lighting captures them inside. In the second one, it's the "overexposure" [4] of the passer-by that invites for contemplation from a privileged perspective. The overexposure is caused by the spatio-visual *dispositif* of architecture itself: stylisation of build shapes and contrast with the light floor offering *panoramic views*. The indoor belvederes – balcony and stairs at the connection of the Porte Louise with Espace Louise and the double height of the restaurant area of Espace Louise – create a grip effect that captures the observer. It also creates tempting views from one level to the other.

3.2 Acceleration of slow mobility

Within commercial galleries, the logics of slow mobility deceleration create a tension with *dispositifs* that favour mobility of pedestrians. Those *dispositifs*, unfolding the notion of comfort, articulate different traditional, enduring *logics of slow mobility acceleration*: securement, spatial and qualitative continuities. According to Carmen Hass-Klau, the predecessors of both pedestrian areas and malls protected pedestrians in 19th century European congested city centres both from traffic and bad weather [18].

The **logic of securement** of slow paths expresses itself through the strong spatial segregation. It is materialised in externalised paths through commercial galleries and, later, also in external enlarges sidewalk protected by huge planter boxes from expressways (see Toison d'Or Avenue). In that way, it carries on the undertaking of the 19th century galleries that attempted to separate pedestrian and wheeled traffic. Twentieth century galleries are indeed still described as passages "safe from mechanical traffic" [19].

The **logic of spatial continuity**, through commercial galleries, solves the congestion problems at small scale: "the continuous congestion of sidewalks" caused by pedestrian crossing, traffic light and bus stop [20]. To build galleries is to construct "covered passages" that enable shortcuts in the urban fabric and avoidance of sidewalk congestion. It also connects commercial zones ("Deux Portes") or different neighbourhoods. The Agora Gallery is a "direct connection between the different attractive centres" [16] between the touristic and patrimonial city centre and the institutions neighbourhood. The indoor spaces of commercial galleries are that way fully considered as mobility infrastructure. It is not only an "analogous city" that supersedes the contemporary city with a simulacrum of the city with no connection to the urban surroundings [14,21]. It still reproduces the linear spatial structuration of the city centre. Galleries are based on a compilation of "streets", "squares" with fountains and benches. Architectural dimensions, transparency, even the planting, contribute to giving indoor spaces a public character. So does also the work of art: the sculpture in Porte Louise distract the visitor from commercial aspects of the venue. The "public space" model trivialise the gallery, make it become familiar and common [15], and therefore used as what it pretends to be: mobility infrastructure. The use of the street-related vocabulary ("roundabout", "artery", "square", "forecourt") supports this conceptual – and semantic – transfer and extension of public space into the built environment.

The notion of spatial continuity is closely linked to that of comfort. The **logic of qualitative continuity** relates both to functionality and aestheticism. Besides protecting from "traffic trepidation", galleries keep pedestrians "safe from bad weather conditions" [22]. First, this is made by climatic comfort: heating system, artificial and natural lighting, covered passages that allow "to go for a covered stroll (*se rendre en promenade et à couvert*)" [23]. Second, by specific and high standard design: "both measured and luxurious modernism" [10]. The design of commercial galleries helps to keep passer-by moving. Ironically, escalators also provide a way to speed up movement because of its mechanics of movement. The dimension of indoor squares also challenges the possibility to stay without consuming.

At a sensitive level, the demand to circulate also operates through slick and slippery quality of finishing materials. At the entrance of the Toison d'Or Gallery or at the connection between the Porte Louise and Espace Louise galleries, the passer-by is invited to progress into the building by an "effect of luminous drive" (*entraînement du lumineux*) [1]. An emerging and prevailing luminous occurrence as a roof window attracts the gaze and drive the full body. The transition from artificial to natural lighting is normally perceived as a way out but when it isn't, it projects the body forwards into the gallery. An accelerant "effect of sight centration, slide and anticipation" also operates in spaces designed with a strong directionality and a narrow visual field framed by opaque surfaces as corridors. "The sight can only escape by focussing on the exit way, the steps only repeat what has already been done by the sight."[1] For instance, this sensory-motor effect operates in some connection spaces as the narrow corridor linking the first Gallery Louise to Stéphanie Square between 1955 and 1963.

Conclusion

Unlike other urban spaces where walking is considered as a residual mobility vanishing with the rate of progress, commercial galleries constitute spaces where subsists a "right to speed" by walking, to efficient (small scale) slow mobility paths. This demand to circulate also relates to the definition of walking as a resource for the market (24). Property value and rental value are directly linked to the density of pedestrian flows. Therefore, material and regulatory *dispositifs* mostly aim at regulating walking by eliminating obstacles to traffic. This paradoxical demand to circulate and to stay is then due to the function itself of the gallery: it aims at attracting loads of people while preventing congestion threats that crowd generates. It facilitates movement to better keep consumers captive, to let them "legitimately stay, that is to say to consume" [15]. In that sense, commercial galleries prefigure the design of contemporary pedestrian areas, where walking is enforced and naturalised as the only modality for accessing this public space [24]. Historical European city centres will implement designs and forms of "living together" developed by their simulacrum [14,25].

To conclude, I wanted to draw attention to the relevance of iconographic archives for academic research on ambiance. The analysis reveals how the paradoxical demand to circulate and to stay is supplemented by different sensory-motor effects appearing in iconographic archives about those galleries. However, the analysis still relies to a large extent on other – discursive – sources. Iconographic archives are not sufficient. Dealing with iconographic archives also only take into account the visual dimension of modernist ambiance, neglecting the other senses: hearing, touch, smell and taste that have a greater importance in the concept of ambiance. Iconographic archives nevertheless confirm the paradoxical status of the commercial gallery of the latter half of the 20th century as a space of (functional) slow mobility and as a space of destination. The research gives that way a better understanding of the tension evolution, in mobility infrastructure development, between the fast and the slow mobility imaginary in Brussels.

References

1. J-P. Thibaud, Annales de la Recherche Urbaine **71**(1), 145-52 (1996)
2. JF. Augoyard, *Pas à pas* (Seuil, 1979)
3. JJ. Gibson, *The Ecological Approach To Visual Perception* (Houghton Mifflin, 1979)
4. G. Chelkoff, J-P. Thibaud, Annales de la Recherche Urbaine **57**(1), 7-16 (1992)
5. M. Ben Hadj Salem, *Les effets sensibles comme outils d'analyse et d'aide à la conception dans les gares du XIXe siècle* (Université P. Mendès, Laboratoire CRESSON, 2009)
6. G. Leloutre, C. Pelgrims, *L'entrée en ville Aménager, expérimenter, représenter*, 43-62 (Éditions de l'Université de Bruxelles, 2017)
7. C. Castoriadis, *L'institution imaginaire de la société* (Seuil, 1975)
8. Bruxelles, ville couverte. Face à main (15/01/1955)
9. M.Culot, *L'Immeuble et la parcelle* (AAM, 1982)
10. H. AM. La galerie Louise. Haut-lieu bruxellois. Reflets du tourisme **2** (1956)
11. F.R. Espace Louise-Sofitel : une galerie, 4 étoiles (27/12/1987)
12. P. Burniat, J-L. Genard, *Enclaves ou la ville privatisée*, 4-12 (La Lettre volée, 2002)
13. Y. Rouyet, J-M. Decroly, *Enclaves ou la ville privatisée*, 81-3 (La Lettre volée, 2002)
14. H. Lionnez, *Enclaves ou la ville privatisée*, 103-10 (La Lettre volée, 2002)
15. M. Martinez Jamart, *Enclaves ou la ville privatisée*, 84-102 (La Lettre volée, 2002)
16. En mars 1965 : cent boutiques, 4 niveaux de bureaux et un parking souterrain pour cent voitures occuperont l'emplacement du cinéma Agora incendié en 1959. Lanterne (10/05/1963)
17. G. Delalex, *Le génie de la marche*, 367-79 (Hermann, 2016)
18. C. Hass-Klau, *The pedestrian and city traffic* (Belhaven Press, 1990)
19. Goffaux R. Que veut la commune d'Ixelles dans l'affaire de l'O.T.A.N. ? L'installation du Conseil atlantique à Bruxelles (25/11/1966)
20. Les badauds pourront désormais déambuler dans la nouvelle galerie Agora, mais ils n'y seront pas encore chalands, Lanterne (3/05/1965)
21. T. Boddy, *Variations on a theme park*, 123-53 (Hill and Wang, 1992)
22. Prestige de Bruxelles... à la Nouvelle Galerie de la Porte Louise. Chambre Commerces Bruxelles, 734 (6/12/1963)
23. R. Detry, La nouvelle Galerie Louise ouverte au public en mars 1964. Prolongeant, sur une longueur de 150 mètres, celle qui existe déjà, Le Soir (7/11/1963)
24. S. Tonnelat, *Le génie de la marche*, 120-31 (Hermann, 2016)
25. C. Pelgrims, Espaces et Sociétés **175** (to be published in April 2019)

Ambiance through Spatial Organization in Vernacular architecture of hot and dry regions of India – The case of Ahmedabad and Jodhpur

Vibha Gajjar[1] and *Foram* Bhavsar[1]

[1]Associate professor, Institute of Architecture & Planning, Nirma University, India

Abstract : Most notable vernacular settlements of the past had a potential of generating stimulating ambiance because of their "organicness". The understanding of the "Nature of order" in this informality can identify the factors or parameters generating the definite spatial character appropriate for the time and place. The spatial character generates the spirit of the place, what we call the genius loci. This paper analyses the relationship of spatial organization and ambiance of the old city settlements in hot and dry climatic regions of two selected cities of India – Ahmedabad and Jodhpur at various levels. Both the selected cities are informal in nature and rich with its ambiance. Space syntax and digital simulations are used to decode the parameters of the spatial organization. The spatial morphology is analysed using tools such as justified graph analysis, visibility graph analysis, isovists analysis, agent movement analysis and sunlight study. The result of the study is summarised in tabular format and presented in a graphical manner for better understanding. This analysis is valuable in the derivation of the constants and variables to define the learning from the vernacular spatial organization.

1 Overview of Vernacular architecture

The sensibility and logic of vernacular architecture generate quintessential liveable built forms and ambiance of respective settlements. The vernacular has been widely understood as "the architectural language of the people" with its ethnic, regional and local dialects: the product of "non-experts" (Oliver, 2006).

The solutions inherited by vernacular architecture are largely based on necessity, efficiency and prevalent aesthetic sensibilities. The vernacular architecture of any place is based on natural and physical built environment enriched by the cultural heritage of a place and is developed over a long period of time. In the process of growth and evolution, it achieves "wholeness" not only in its totality but also in each fragment. Various building forms are generated in the process of learning from the existing natural environment and building suitably with climatic and social considerations. As a result, the settlements are synergetic with the social, cultural and climatic needs of the region.

1.1 Vernacular architecture of hot and dry regions of India

The Indian subcontinent's diverse range of landform and climatic conditions, such as gigantic Himalayan mountains in the north, Indian ocean in the south, hot sandy desert in the west and wetlands of Bengal in the east have generated varieties of built forms across the country. The natural landscape of the site and response to metaphysical concerns, climate, local proficiencies and construction techniques are the personification of the rich vernacular tradition of built form in India. In each location, the vernacular architecture has evolved in a unique way of response to the context.

Hot and dry climatic zones are experienced in areas far from sea coasts and usually have a low precipitation rate. The most important aspect to be addressed is protection from harsh sunlight and the usual answer is to develop a built form capable of reducing the radiation.

The most preferred solutions in a hot and dry climate are dense clustering of buildings, which allow for the minimum heat gain in buildings. The streets remain shaded in summer providing comfortable temperature throughout the day for outside activities. Hence even in harsh summers, the streets are walkable during the afternoon. The houses are constructed with thick insulation walls, tiny scale openings to keep the hot air outside, and compact clustering of built forms with mutual shading device and provision of the courtyard as a thermal regulator. The harmonious relationship between the thick walls, jalis (perforated openings), courtyards and transitional spaces of semi-private nature is an intelligent approach to deal with the climate.

The spatial character is at the heart of understanding the physical ambiance of any place. The initial step for understanding the spatial character is to understand the spatial organization and the experience attached with it. The interaction of this dense spatial organization with its culture generates the cohesive ambiance within the settlement. Amos Rapoport emphasized that climate, ecology, material, technology, and local economy act as controls rather than factors which affect the spatial ambiance which is articulated as "The built environment creates a specific microclimate for each site as an outcome of building configuration, its orientation, and arrangement in space. Further, the building element interacts with natural macroclimate to determine the factors affecting comfort: light, heat, wind, and humidity.

2 Introduction to Case study

Two cities in different geographical locations and cultural contexts in hot and dry climatic regions of India are selected to study the relationship of spatial organization with ambiance. The city of Ahmedabad located in the state of Gujarat and Jodhpur in the state of Rajasthan are selected for the study purpose as shown in Fig. 2.1.

Ahmedabad is a commercial capital city of the Gujarat state in the western part of India. The origin of Ahmedabad dates back to 1411 when the Nobel Ahmed Shah laid the foundation of the city on the east bank of river Sabarmati.

Jodhpur is the major city of the Marwar region of Rajasthan. Its roots of development are traced back to 1459 when king Jodha of the Rathod dynasty laid the foundation of the city. Both cities were once fortified towns with massive surrounding walls and bastions. In order to enter the city, Jodhpur had seven gateways, while Ahmedabad had twelve gateways safeguarding the city.

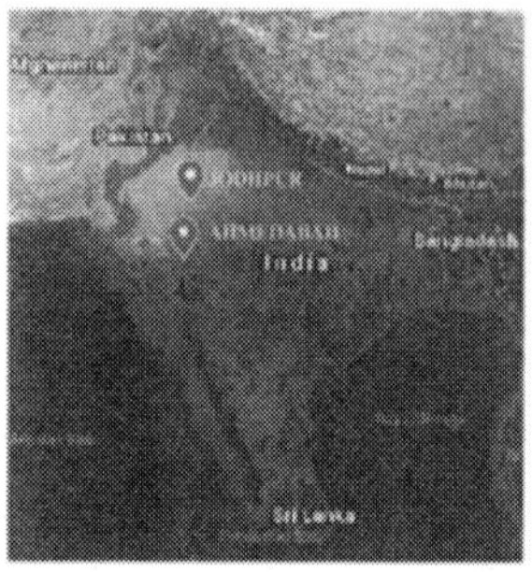

Fig. 2.1 cities on map of India, location of two surrounded by the Rocky Mountains of the Aravalli Hills

Fig.. 2.1.1 Street view of a Pol house in Ahmedabad

2.1 Ahmedabad

The vernacular settlement of the old city in Ahmedabad is known as 'Pols'. A pol is a gated community having one entrance where a similar group of people or community lives together. These pols are located on the eastern bank of river Sabarmati inside the walled city of Ahmedabad. The densely packed houses are arranged in a linear manner connecting to narrow streets and squares as shown in Fig. 2.1.1. In response to the situation of conflict between the communities and frequent events of riot, the city of Ahmedabad developed a gated community that ensures security and restricts singular access.

2.2 Jodhpur

The spatial organization in the old city of Jodhpur takes into consideration the issues of extreme temperature, solar radiation, glare, sand and dust storms in the development of houses. The houses have flat terraces, open courtyards and projecting balconies with perforated screens. The densely packed houses along the narrow streets breathe through their neighbourhood is formed on the basis of social stratification and occupations, like Brahmins live in Brahmpuri, goldsmiths in Sunaron Ki Ghati, and moneylenders in Bohron Ki Pol. Most of the courthouses are simple with only minimal entrance decoration. Private terraces are a multifunctional social space of the house, which is used for private functions, social interactions, drying the spices and for sleeping during the hot summer nights. In both cities, houses along the main streets have commercial activity on the ground floor with a residential unit above. Both cities have specific local terminology for various parts of the house as described in the table below.

Fig. 2.2.1 street view of jodhpur city

Table no.1 Local terminology of various parts of a traditional house

Local name	Description and meaning	Spatial characteristic	Spatial Quality
Khadki *or* ***Choubara***	Entrance or common space shared by a cluster of houses	open	public
Otla or Otta	The transitional space from outside to the inside of the house	semi-open	Semi-public
Chowk	Courtyard of varying scale for light and ventilation (At house & neighbourhood level)	open	Semi-private
Zarookha	The projected place for, shade, sun, light, and breeze usually on upper floors	closed	Semi-private
Parsal *or* ***Osri***	space inside the house usually surrounding the courtyard	semi-open	Semi-private
Ordo	A private space (rooms) usually closed with openings	closed	private
Rasodu	kitchen usually located near the courtyard as it gives a scope of providing the chimney like structure (Dhumadiyu) attached with fireplace (chulha) for the escape of smoke.	Semi-open	Semi-private

In both settlements, the ambiance is created through the hierarchy of built and open spaces in an unpredictable yet specific order. The spatial arrangement of narrow streets and open spaces, known as "chowk", act as social spaces for celebrations and social gatherings at a neighbourhood level. The "otla" outside the house acts as an interaction space contributing to the character at a micro level. Thus the physical and metaphysical behaviour of the spatial organization plays an important role in creating a lively ambiance.

2.3 The spatial organization of study areas

A house is an elementary unit in the spatial organization of a street. The morphology is a result of the interaction of a house with the street and the adjacent spaces at a cluster or neighbourhood level. Their spatial relations have an impact on the level of access and control over each other. The interaction of spaces within the house influences social behaviour. The visibility, permeability, movement and circulation patterns of the activity space within the house determines the social behaviour within the house. The built form of a house is the result of the individual needs and aspirations of each community. These structures reveal the understanding of climate, social structure, usage of locally available building materials, and craftsmanship. The courtyards play a significant role in the spatial organization for providing visibility, connectivity, light and ventilation. Further, the lifestyle adoption of users and multifunctional usage of spaces has greatly influenced building configuration and form. The size of family units, spatial usage, preparation of food, interaction of people and many other cultural considerations also affects the layout and size of dwellings (Rapaport, 1969).

In both case studies, the cluster formation is not following a rigid geometrical pattern allowing the degree of flexibility in spatial organization. The amalgamation of spatial organization and sensorial experiences generates the spirit of the place. This spirit of the space defines the special character of any particular place. In order to understand the ambiance of a place at a varying level from neighbourhood, cluster and house level the analysis of spatial organization plays a pivotal role. The symbiotic relationship of the forms of mass and space in architecture can be examined and found to exist at several scales (Ching, 2007). The detailed understanding of the spatial organization would be created using documented case studies for the city of Ahmedabad and Jodhpur.

3 Study area

3.1 Case study selection

The selected area for the study purpose is located in the old fabric of Ahmedabad and Jodhpur as shown in Fig. 3.1.1(a) & (b). Both areas belong to Hindu dominant community having a similar religious background and were built 100 to 200 years ago. The cultural background differs due to the social structure and regional differences. The analysis is done at three levels :

1. Neighbourhood of 200-300 houses
2. Cluster of 6-8 houses
3. Individual dwelling unit

The paper examines and analyses the physical spatial organization at various levels to understand the spatial character of a place. It examines important factors that play a vital role in creating a definite spatial organization. Each factor is examined using various analysis methods and derives the constants and variables in both the case studies. It further evaluates the reasons, which are socio-cultural or climatic for the variables that are specific.

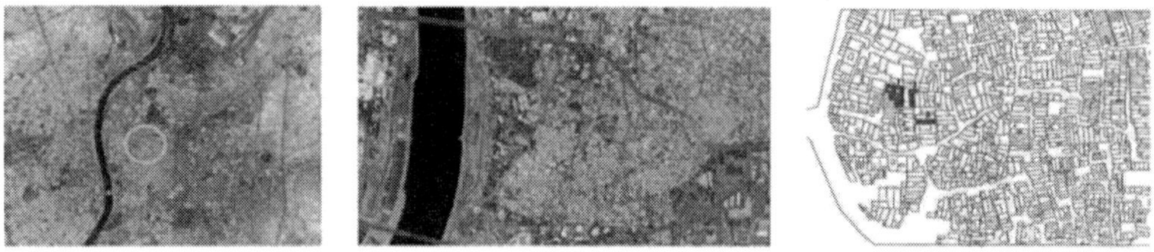

Fig. 3.1.1 (a) The figure shows the location of the study area in Ahmedabad at three levels City (Ahmedabad) – Area (Khadia) – Neighbourhood (Dhal ni pol)

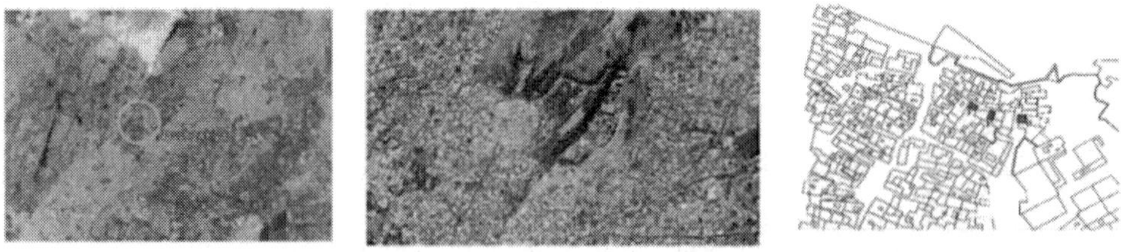

Fig. 3.1.1 (b) The figure shows the location of the study area in Jodhpur at three levels City(Jodhpur) – Area (Brahmpuri)- Neighbourhood (Fatehpol) to the site.

3.2 Method for spatial analysis

The space syntax analysis method is used for the purpose of spatial organization analysis. Space syntax tool is simply a way of transforming buildings into dimensionless forms to interpret spatial organization using graphical representation. Along with space syntax, the climatic analysis is done for the available sun hours and shadow patterns in summer and winter months. The analysis is done using a three-dimensional model to examine the results on the building surface as well as on the streets. The analysis generates the graphs and data indicating the behaviour of the space and behaviour of a user at a particular space.

3.2.1 Justified graph analysis

The justified graph is a tool to find out the relationships between the spaces. This method is used for understanding the adjacency of spaces at various scales such as adjacency of streets, clusters, houses and activity spaces within the house. The graph generated shows space as nodes, and line as connected spaces or edges. The dark thick lines in the images represent the main axis and the lighter and thinner lines represent the secondary, as well as the tertiary axis. The bigger dots in the diagram denote the nodes/convex spaces (the spaces between the houses) and the smaller dots denote the houses. The lines indicate the connectivity between the nodes and the houses. The figures below from 3.2.1(a) and (b) show the graphical presentation of Axis and J-Graph analysis at the neighbourhood to house level.

3.2.2 Visibility Graph, Agent movement and isovist analysis

The visibility graph analysis (VGA) is used to check the visibility and permeability as a dynamic component in spatial organization. This analysis will help in the revelation of the hierarchy of spaces from public to private and knowing the locations which remain directly visible or invisible through a colour-coding diagram. The colours from cyan to blue suggest low connectivity, and yellow to red suggest high connectivity. In order to find the natural movement pattern in the building, the Agent's model developed by Alasdair Turner is used for analysis. This helps in understanding the natural path of movement in the way of finding. It also helps in getting insight into private and territorial behaviour. The spatial quality of architectural space can be analyzed using Isovists. The potential field of view from the identified location can be generated using isovists. The depth map creates polygons and helps to measure the depth of each identified location and its relation with each other. A well-integrated location is colored in red which implies the shallowness, means no turns are required to get from one location to the other. The poor-integrated location is coloured in blue which implies the turns and depth with respect to their locations.

3.2.3 Sunlight hours and Shadow analysis

Sun and wind are the most essential cosmic attributes, as discussed earlier. The sunlight hours analysis is generated using the ladybug component of grasshopper on Rhino. This component calculates the direct sunlight received by any geometry in any given climatic zone. It further helps to understand whether the indoor and outdoor spaces are comfortable to use or not. Both selected cities have a similar climate. The studies are done on both surface and base levels to understand the behaviour of streets and house surfaces with respect to the climate. Low rise and high-density fabric of both the cities have narrow streets, pedestrian pathways and open chowks (open gathering spaces). Instead of such similarities, the study suggests relatively different results for both cities. The graph below shows the indication of hours of sun during the day time. The cold colours from blue to dark red suggest the fewer hours of sunlight where the warm colours from red to yellow suggest more hours of sunlight.

	AHMEDABAD	JODHPUR
SITE (Axis diagram)	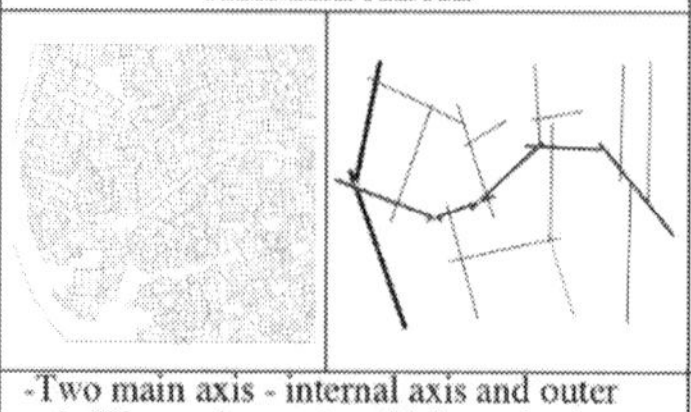	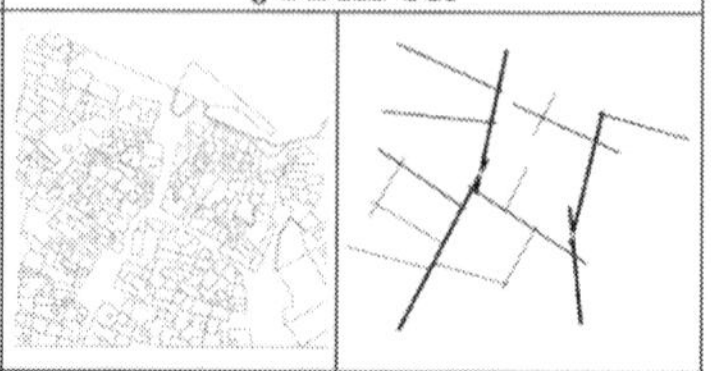
Axis	-Two main axis - internal axis and outer axis (Shown in varying thickness in the figure above) The outer axis is the main link between the east and west side of Ahmedabad - Internal primary axis is crooked - The relatively leveled ground around the primary axis	- Jodhpur has a single central primary axis which connects the neighbourhood to the other part of the city - Internal primary axis is continuous - The terrain is rising up towards east-west direction suggesting the primary axis at the lowest ground
Orientation	- The outer primary axis is in a north-south direction; the internal primary axis is in an east-west direction - The major number of secondary axis is in the north-south direction - The houses are oriented along the streets; mainly in an east-west direction to get better sunlight and ventilation within the house	- Primary axis is in the north-south direction - The major number of secondary axis is in the east-west direction along the terrain - The houses are oriented along the streets; In case of Jodhpur the houses face all four directions; the main source of light and ventilation is the courtyard
Adjacency	- All the houses open up on the shorter side of the house to get better accessibility to pathway and services	
CLUSTER (Axis and J-Graph diagram)	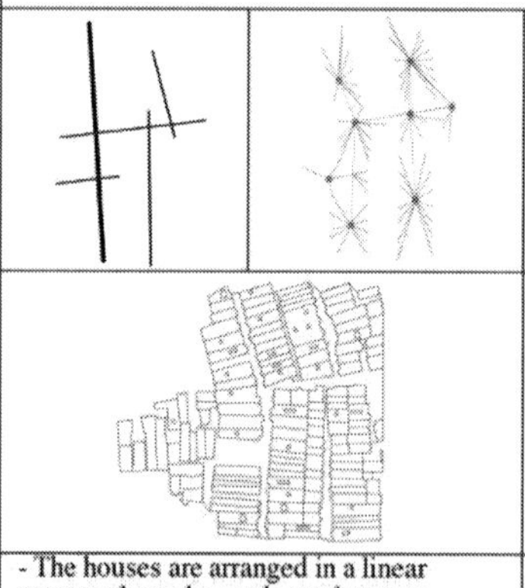	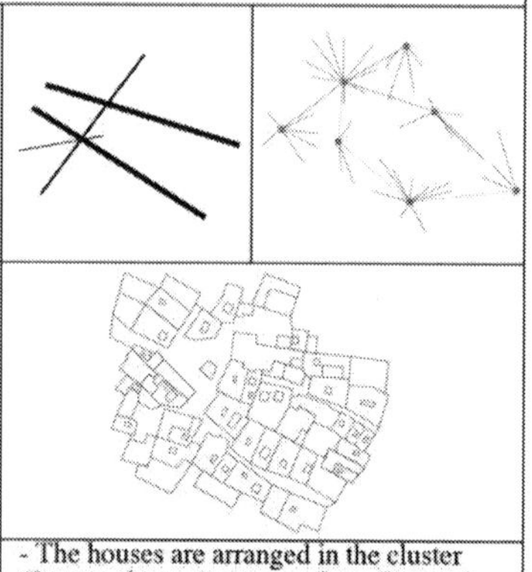
Axis	- The houses are arranged in a linear manner along the north-south streets - Linear but fragmented streets with more intersections between streets - Clusters are linear and smaller in size	- The houses are arranged in the cluster (Square clusters) manner along the north-south streets - Linear and less fragmented streets with fewer intersections between streets - The clusters are square in form and larger in size
Organization	- Suggests good connectivity and accessibility of houses along the street	
	- Linear organization of houses along the streets facing north-south direction - The houses are oriented along the streets; mainly in the east-west direction to get better sunlight and ventilation within the houses	- Cluster organization along the streets in all directions due to the availability of fairly flat land on the terrain - The houses face all four directions; the main source of light and ventilation is the courtyard
Connectivity	- Many nodes are connected along the axis - Many individual houses are connected to each node	- Relatively fewer number of nodes are connected along the axis - Fewer number of houses are connected to each node

Fig. 3.2.1(a) Analysis of Axis & J-Graph diagram at neighbourhood & cluster level

TYPICAL LAYOUTS (J-Graph analysis)

	AHMEDABAD		JODHPUR	
Large size house	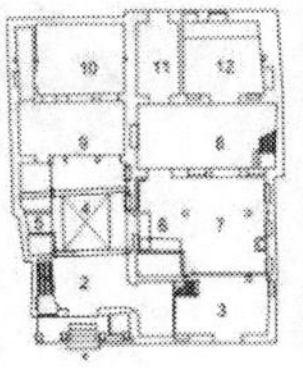	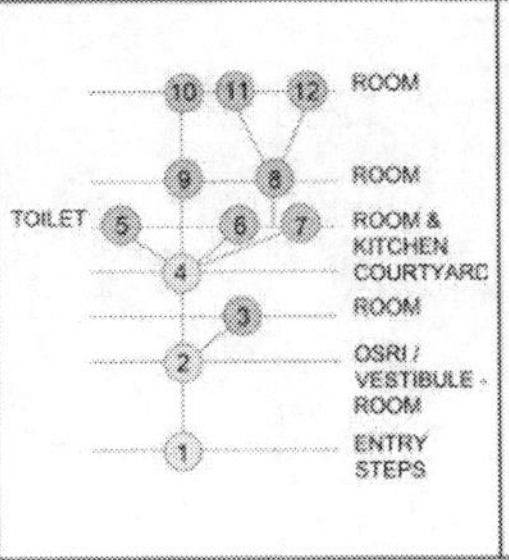	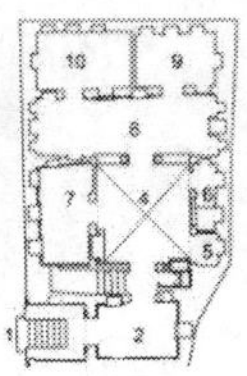	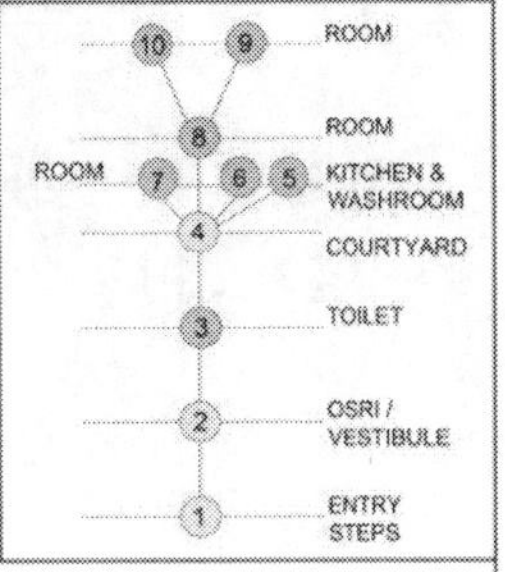
Medium size house	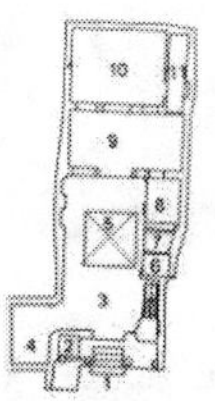	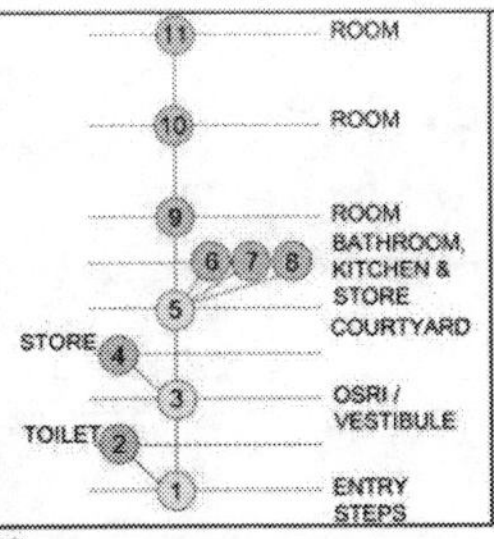	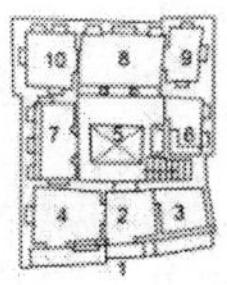	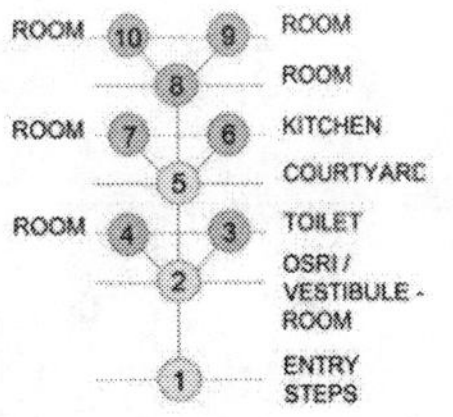
Medium size house	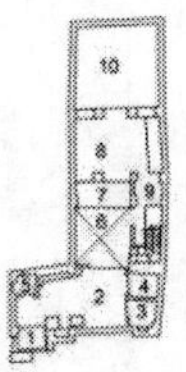	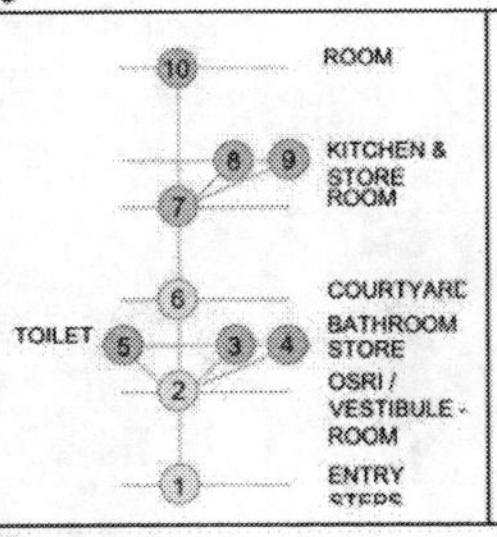	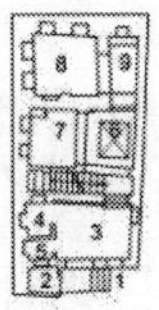	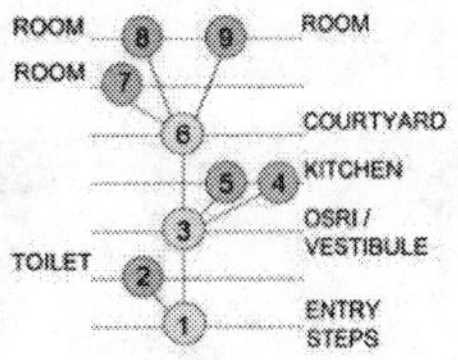

	AHMEDABAD	JODHPUR
Entry	- Direct entry from the street - Public space (otla) at the entry of a house - Toilet at the entry for easy access to the services (pipelines)	- Indirect entry - Use of change in direction and steps to disconnect from the street - The entry has high- steps which act as public space; Otla are visible only along the primary street - The toilet is located near entry – no pipelines but open drains along the terrain
Adjacency	- Linear adjacency of spaces; Branching of the spaces is taking places only along the semi-private spaces (Osri/vestibule and Courtyard) - Osri/vestibule or Courtyard is connected to services- Toilets and Kitchen - Access to private spaces (rooms) through the other private spaces	- Adjacency of the spaces suggest the branching at private and semi-private levels - The courtyard is connected to services- Toilets and Kitchen - Access to private spaces (rooms) through the passage around the courtyard
Centrality	- Courtyard 	- Courtyard 

Fig. 3.2.1(b) J-Graph diagram at a unit level **Private spaces** **Semi private spaces**

	AHMEDABAD		JODHPUR	
	Visibility	**Agent Movement**	**Visibility**	**Agent Movement**
SITE AND CLUSTER			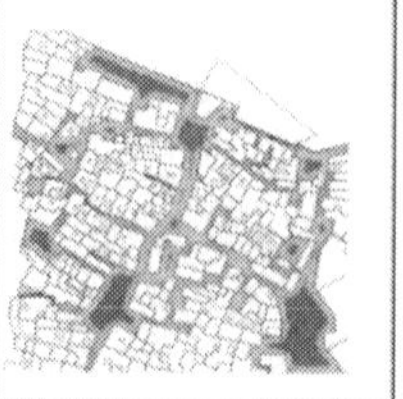	
	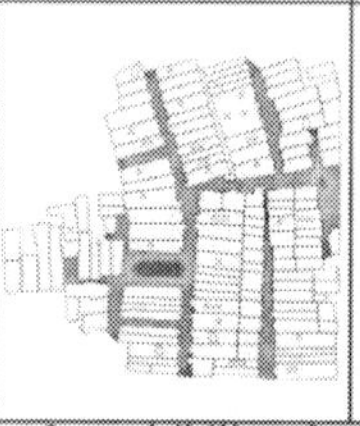	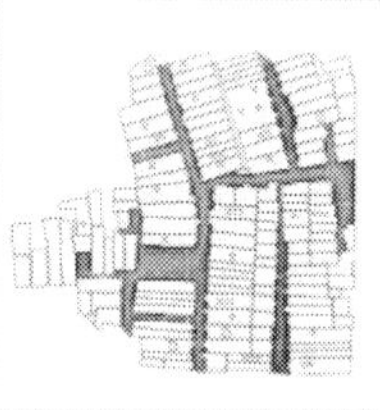	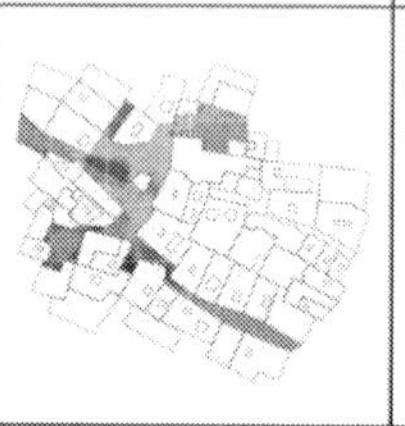	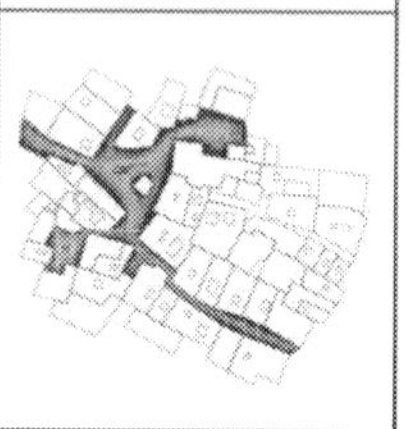
Visibility	- Less visibility of nodes - Movement and visibility along the periphery		- Higher visibility of nodes - Movement and visibility along the central axis	
Connectivity	- Nodes are higher in number and smaller in size		- Nodes are lesser in number and larger in size	
Privacy	- Privacy is attained by means of the organization of streets – Privacy is greater in the centre		- Privacy is attained by means of the terrain – Privacy is greater away from the centre along the terrain	
	Visibility	**Agent Movement**	**Visibility**	**Agent Movement**
TYPICAL LAYOUT	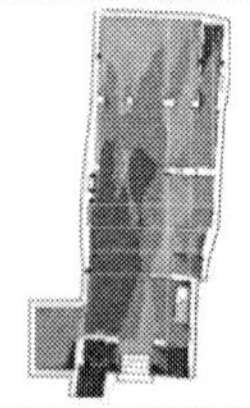	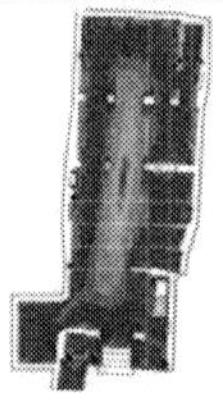	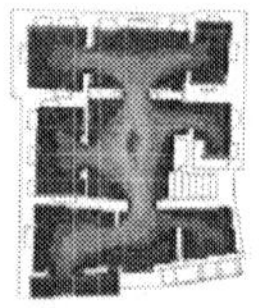	
	Isovist - Court	**isovist - Entry**	**Isovist - Court**	**isovist - Entry**
		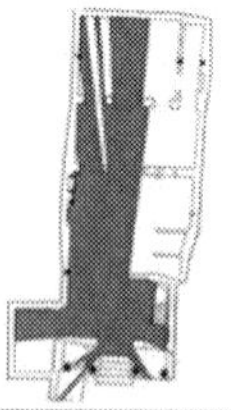	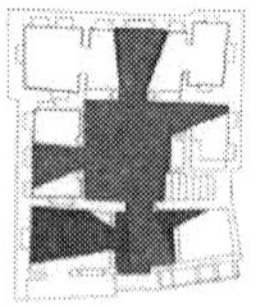	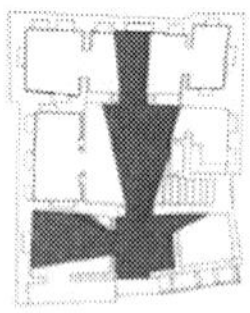
Visibility	- Linear visibility along the axis connecting all the semi-private spaces		- Visibility mainly in the central area along the courtyard	
Connectivity	- The private spaces (rooms) are partially connected with the semi-private areas		- The private spaces (rooms) are disconnected from the semi-private areas	
Privacy	- Streets are relatively more private in nature than Jodhpur - The house entry / otla work as a public space		- Streets are relatively public in nature – greater privacy along the terrain - Houses are very private in nature due to the socio-cultural reasons	

Fig. 3.2.2 Visibility graph analysis at neighbourhood, cluster and unit levels

	AHMEDABAD		JODHPUR	
	Winter	Summer	Winter	Summer
SITE AND CLUSTER (On ground)				
	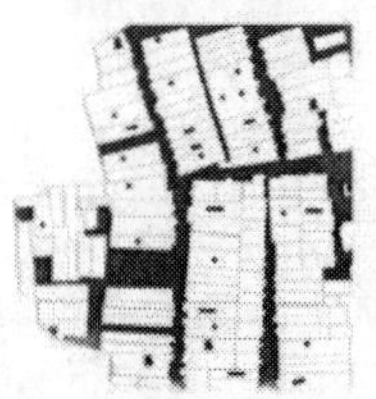	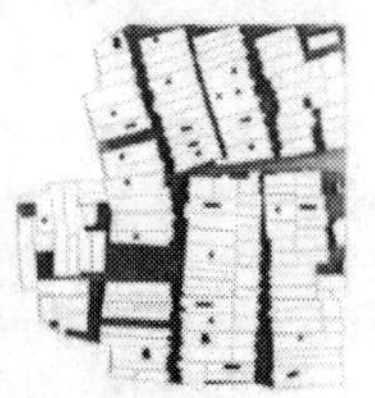	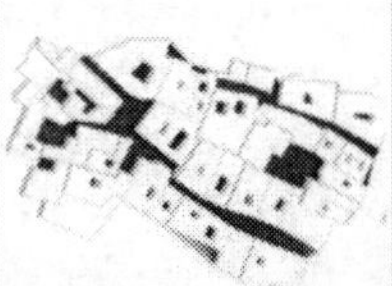	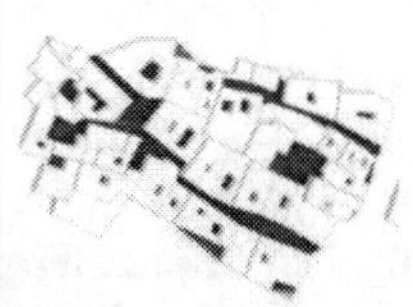
Sunlight	- Main and secondary streets are shaded in summer and winter; It allows a good comfort level in summer but remains cold in the winter time - Only larger nodes receive adequate sunlight in the winter time - Terraces receive good sunlight in summer and winter - Depending on the adjacent condition all the houses might not get enough sunlight throughout the day		- Main and secondary streets are relatively less shaded in winter and summer; works well in both winter and summer time - All the nodes receive reasonable sunlight during the winter time - Terraces receive very good sunlight in winter due to the terrain - All the houses get sunlight on the walls and terraces due to the terrain (The terrain generates stepped terraces)	
Daylight	- Longer hours of daylight in summers than winter time; the courtyard is the main source of energy within the house			
	- Relatively fewer hours of daylight in the house throughout the day due to the dense street configuration		- Relatively more hours of daylight in the house throughout the day due to the terrain	
	Winter	Summer	Winter	Summer
CLUSTER (On building surface) 1) Sun hours analysis 2) Shadow analysis				
	Shadow analysis			
				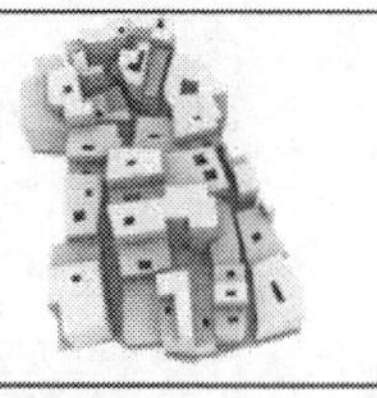
Shadow	- All the internal streets are shaded in the summer time			
	- Terraces are more shaded as compared to Jodhpur		- Terraces are less shaded and act as a good source of sunlight in the winter time	

Fig. 3.2.3 Sun hours and shadow analysis at neighbourhood and cluster levels

Conclusion

The comparative study of the selected case studies derives the constants and variables of the spatial organization that are responsible for the ambiance of the respective place. The constants are due to the hidden order in its organicness. The variables are due to the specific site conditions and dissimilar social structures. The results of the space syntax analysis and simulations derive the parameters affecting the spatial organization and ambiance generated with that organization. The constants are the result of climate responsiveness in both the settlements. The orientation of the houses promote better daylight and ventilation at the same time protects from the harsh sunlight during the summer season. The street orientation is a response to the existing landform, access and services. Both settlements provide a comfortable temperature in the summer season. The spatial organization of houses in the street responds to accessibility and privacy at the same time. Privacy here is understood both as physical and visual component. The physical component connects or disconnects various levels of accesses at thoroughfare, street or cluster respectively. The visual component is responsible for the sense of safety and connections between people. The city of Ahmedabad shows the character of the planned street as a decisive and conscious effort for achieving privacy. The character becomes more public and inviting at a cluster level. The neighbourhood street is a private entity hence the otla outside the house is the most public space used for interactions. This hierarchy of private to public at neighbourhood to house level generate the lively ambiance within the neighborhood. In Jodhpur, privacy is attained at a house level where the streets are public in nature hence the house edge is the threshold between public and private space. The social structure in Jodhpur necessitates the privacy from the public street. The house orientation and planning encouraged privacy. In both cities, the courtyard is the ultimate source of light and ventilation. This multifunctional space is the center of the house. Both the places have developed their own vernacular vocabulary appropriate for the time and place. The constants in both the case studies are relevent learning for any hot and dry climatic region with a similar background. The variables have more specific learning of the relationship between social behaviour and spatial organization. Altogether, they provide us with a concrete understanding of the vernacular spatial structure and specific ambiance they generate.

References

1. Oliver, Paul. Built to meet Needs Cultural Issues in Vernacular Architecture. Burlington ,MA: Architectural Press Elsevier Ltd., 2006.
2. Rapoport, Amos. house form and culture. N.J.: Prentice Hall,Inc., 1969.
3. Rudofsky, Bernard. Architecture without Architects. New York: Connecticut Printers, Inc., Hartford, Connecticut, 1965.
4. Fathy, Hassan. Vernacular Architecture . Ed. Walter Shearer and Abd-el-rahman Ahmed Sultan. Chicago: The University of Chicago Press, 1986.
5. Patel, Dr.Manek. Welcome to Ahmedabad: a complete city Guide. Ahmedabad: Gurjar Prakashan, 2010.
6. Hanson's, Hillier and. The Social Logic of Space. London: Cambridge University Press, 1984.
7. Dongre, Neeta Lambe & Alpana. "Analysing Social Relevance of Spatial Organisation: A Case Study of Traditional Pol Houses, Ahmedabad, India." Asian Social Science; Vol. 12.No. 9 (2016).

8. Mihir Vakharia, Rajan Rawal,Yash Shukla,Agam shah. "Summer in Ahmedabad:pol House performance." Indian Architect and Builder 1 October 2014: 10.

AMBIANTAL ARCHITECTURE – Defining the role of water in the aesthetic experience of sensitive architectural ambiances

Gregory Brais Sioui[1]

[1]Laval University, Architecture school, Québec, QC, Canada

Abstract. As part of an aesthetic approach, this design-driven research crosses two methodologies to determine what is common in the perception of atmospheres. First, a case study of the Vals' thermal baths led to the selection of five different atmospheres, which were then analyzed through Gaston Bachelard's lens, using his book *L'Eau et les rêves*. Bachelard's literary symbols echo in the water contained in the massive stone walls of Vals to identify common generators of sensitive atmospheres. Secondly, the construction of this dialogue between Bachelard and Zumthor leads to the elaboration of a conceptual architecture project which is voluntarily emotional. This project introduces the elements that generate five ambiences identified in the case study. This design-driven research is therefore based on Grégoire Chelkoff's theory of *formants* as vectors of transmission of atmosphere, pre-existing to the experienc of a place, of an ambiance, which itself is understood as a sensitive result of the perception of the space. The present work therefore questions the role of water as a sensitive vector, from the architecture to its visitor. The goal is to determine how water, in varying manifestations, can be used by architects to create a "*mise en scène*" for a voluntarily emotional architecture.

1 Architecture as art

Between artisans and artists, architects use the material and immaterial realms to set the *scènes* of reality, to set the backdrop of people's lives. Seen as the *first art*, architecture carries within it the potential to produce emotion: Le Corbusier said, "construction is made to hold, architecture, to move."[1] The individuality behind an emotive response to an environment makes it impossible and undesirable to establish an architectural recipe that would lead to a specific emotion. However, it's still possible and desirable to produce "an [architecture] work with emotion as a function"[1]. This goal was aimed at by Mathias Goeritz when conceiving the El Eco experimental museum. Not so much about generating a definite architecture which would produce a single predetermined emotion, the goal here is rather to offer various opportunities that might move a variety of visitors.

[1] Cited in: N. Gilsoul. "L'architecture émotionnelle, cadrage conceptuel." Chap. 1 In *Architecture émotionnelle : matière à penser*, pp.37-73. [Lormont]: Le Bord de l'Eau, (2011).

It's in this context of inherence between architecture and art that this work questions what is common in the perception of certain architectural atmospheres. The development of this thesis is constructed in two interdependent parts, the first building the second: a case study will lead to a conceptual architectural design exercise, with both parts to be interpreted through the same theoretical framework. Mainly, the objective of this work is to propose a creative approach based on the conclusions of an analysis that crosses a philosophical work and an architectural work, together generating a design-driven research.

More specifically, the aim is to define generic generators of atmosphere from different rooms of the Vals thermal baths, a project carried out by Peter Zumthor in 1996, by using the book *L'Eau et les rêves* by Gaston Bachelard, first published in 1942. The analysis then allows to produce an architectural project where emotion becomes a programmatic consideration, by staging and applying these generators of atmosphere in scotch tasting pavilions; it is, in other words, a matter of analyzing a contextual *aesthetic appreciation* in order to use it to nourish the production of an architectural design project. This design-driven research uses Gregoire Chelkoff's notion of "*formant*" as a common vector of transition from the atmosphere to the ambiance in a north-western cultural context.

2 Concepts' definitions

Four main concepts frame the ideational scheme underlying the reflection addressed in the two portions of this design-driven research. Aesthetics, emotional architecture, sensitive ambiance, and water constitute the themes discussed in this section.

2.1 The place of beauty

Etymologically, aesthetics, from the Greek word "*aisthēsis*" is defined as that which can be perceived by the senses. Since then, this definition has evolved to mean generally, in philosophy, a field studying the perception of nature and art (which includes architecture) regarding the emotions they evoke. The aesthetic experience is no longer an object-centered *result* perceived in a distant or outsider manner, but rather a multi-sensitive perceptive *engagement* with its source. In other words, architecture, as an object of art, has gone from an exterior object – to be contemplated – to an intrinsic one – to be lived. One's experience is therefore more engaging.

According to Arnold Berleant's article *Aesthetic Sensibility*[1], aesthetics is a theory of emotions defined by a developed, guided, focused and trained perceptual awareness. To evaluate the aesthetic functioning of an object, it must pass through the "aesthetic judgement" which is based on the combination of the "aesthetic appreciation" – the perceptive process – and the "cultivated sensibility" – the past experience and the knowledge that comes from it. Sonja Schürger echoes this point: "it is important to be aware of the potential indicator of hidden intentions."[2] According to her, the recurrence of this conscious action helps to develop a complexity in the perception of "ambiances' interactions".

2.2 To move, a goal for... architecture?

Even if the functionalist approach to architecture, which motored the modern movement, prevailed in its time, it was not universally accepted. It's in the after-war context that the German sculptor Mathias Goeritz, established in Mexico since the late forties, came with the idea of a perception and a feeling-oriented sensitive architecture: the *emotional architecture*. Proposed by the artist as a reaction to functionalism, which he considered dehumanizing, this concept became, for his friend and colleague Luis Barragàn and himself, an important consideration in their respective work. It then further evolved from a consideration to a full-blown conceptual approach, gaining popularity in Mexico. For Goeritz, the idea of an emotional architecture was defined as "a work of art whose main function is to provoke a "maximal emotion"[2] capable of transcending the modern socity"[3].

As a result, the production of an emotional architecture aims at the composition of a sensitive space which translates conceptual intentions such as the targeted atmosphere. The goal here is to make the visitor part of the architecture by involving them emotionally in it, because it includes a participatory dimension in its experience of space: "the aesthetic appreciation". Thus, the environment acts like a "receptacle of emotions.[...] The sensitive world becomes tangible and the space tunes with lived moments and feelings."[4]

2.3 The atmosphere precedes the experience

Atmosphere and ambiance are inherent terms. In architecture, they both share a notion of perception of space. Nevertheless, they don't have the same meaning. In order to separate them, here is a proposition for a definition which focuses on the border between those words' meaning. The atmosphere is the result of the work of an architect. With one's art, one can set up the conditions for the existence of an atmosphere. However, it precedes the experience. On the other hand, the ambiance is the *result* of a lived experience. The atmosphere is not self-sufficient, since an observer must experience it and therefore be in action in the atmospheric space in order for it to materialize into an ambiance - through the observer's perception.

The user experiences an atmosphere with all their senses. Its perception is multi-sensory. Reflecting a spirit-matter relationship, the ambiance is incorporated and embodied in the spatial dispositions, according to one's actions. A link is created between "the physical arrangement of an environment and our way of engaging ourselves within it"[5]. This condition makes the perception of an architectural atmosphere an "active or creative aesthetic experience"[6]. The ambiance is the lived result of these aesthetic experiences. For Gernot Böhme[7], it is defined by its "character". In other words, the way it affects our emotions.

Since "everyone reacts according to one's sensitivity, life experiences, sociocultural prerequisites, the performance of one's sensory receptors and the creativity of one's brain"[8], the ambiance is relative to each observer that experienced it. In consequence, there are as many ambiances as there are observers. Nevertheless, there is a certain uniformity in the perception, a common *ambiental* context. Thus, without speaking of universality, there is a collective part in the perception of an ambiance. Preconceptions arising from socio-cultural belonging form or distort the perception of a group of individuals. From the singular to the plural, the ambiance inscribed in a social context must be considered as an "inter-individual" experience[9]. Arnold Berleant[1] considers changes in movement and period in art as fundamental changes in the sensitivity of a community. This sensitivity is certainly a personal matter, since it derives from memories, experiences, etc., but also from culture. As a result, it includes a part of the collectivity that makes it "inter-individual"[9] .

Despite the uniqueness of the ambiance, it is still possible for the architect to influence the aesthetic experience and offer common opportunities to a variety of users. It is through the "*mise en scène*" that the architect will provide the necessary elements for the emergence of the ambiance.

---

[2] Originally cited in the text as: Mathias Goeritz. *Manifesto de la arquitectura emocional, (1954)*

2.3.1 The "mise en scène"

Scenographing "an ambiance is therefore equivalent to setting the conditions for its appropriation"[7]. For Gernot Böhme, the experience of architectural "scenic space" is "an intermediary between the subject[the observer] and the object[the architecture]"[13]. In order to study this experience, he proposes two approaches: the subject and the object. The subject approach is concerns with the aesthetic of reception, or the study of the sensitive effects of the aesthetic appreciation of ambiances. The object approach explores the aesthetic production: scenography, or rationalizing the ambiance because of the necessity to use "generators" of ambiences. It is this second approach that leads to the exploration of the available tools for the architect to move the visitors.

2.3.2 The "formants"

When questioning the perceivable qualities, transcendent the aesthetic object and its aesthetic appreciation, previously broached with Arnold Berleant, it resonates in both Gernot Böhme's and Grégoire Chelkoff's theories. First, Böhme argues that "[ambiance] production is not linked to the determination of objects, but to the way in which they irradiate space to serve as generators of ambiance"[7]. Then Chelkoff describes the ambiance as the lived phenomenon of a "sensitive network" *woven* between the environment and the user[2]. This brief definition lays the foundations of his theory parallel to Böhme's *generators*: the "*formants*". He first agrees that architecture is a manifestation of objects entering in relation in space. Through the perception of objects and spaces affecting various sensory registers, some *tensions* form a *sensitive network* between the multi-sensory sensitive being and the architectural space. As a result, the terms *object* and *space* give way to a representation of sensitive phenomena that he calls "*formants*". Natural or anthropogenic, tangible or intangible, these elements generate an interaction through time between the element that constitutes it and the sensory event it causes: "the aesthetic appreciation"[1]. The "*formant*" questioned by the present design-driven research is the material image of water.

2.4 The oneiric power of water

Water is an archetype of dichotomy. It is simultaneously a source of life, purity, freshness, inspiration, calming and meditation, as well as of strangeness, mysteries, unforeseen events, fears, vices and dangers. Water is at the same time matter, material and materiality in architecture. This tryptic of water falls in Chelkoff's categories to analyze an ambiance: "*form, "formant" and formality*"[5] respectively material, materiality and matter. These three characteristics of water would be as many considerations in the formation of a deliberately emotional architecture. In his book *L'Eau et les rêves*, Gaston Bacherlard explores water symbolism through what he calls material imagination : "direct images from the matter"[10]. These recurrent images in the important literary corpus evoked in his book form the body of reminiscences to which the "*formants*" should refer.

Density, which separates *superficial poetry* from *profound poetry*, is distinguished between the *sensitive values* of the art work – attributable to translation – and the *sensual values* of the art work – attributable to correspondence –, the latter providing a more refined and powerful image. The reverie is therefore deeper when it comes to matter than when it comes to form. This is what motivates Gaston Bachelard to restore "the individualizing power of matter"[19], often forgotten in aesthetic science which, in his time, focused too much on the notion of form. He develops, through hermeneutics of North-Western literature and poetry, a water psychoanalysis.

This whole design-driven research covers themes addressing the following symbols: the purity of clear water – a sacred purity –, the passage of time – the way of death –, the strongest composition of elements – water and fire –, the maternal evocations of water – in its actions – and the heavy water – the dark oppression it weighs. For the purpose of this article, two material images are presented. They both expose the oppositions of death and life images of water – which respectively refer to two chapter of *L'Eau et les rêves*: "*le complexe de Caron*"[3] and "*L'eau maternelle et l'eau féminine*"[4].

3 The crossover of two methods

As mentionned earlier, the present work comprises two sections. First, the case study allows us to explore some ambiantal "*formants*". Following this, the architectural design project explores the notion of *intent* in the creation of an emotional architecture and the notion of "*formant*" as a scenographic mechanism to achieve it.

The case study proposes an analysis of the five atmospheres "*mises en scenes*" by Peter Zumthor inside the thermal facilities of Vals. The analysis uses a philosophical work presenting a psychoanalysis of an important corpus of literary and poetic works which highlight different material images of water. In *L'Eau et les rêves*, Gaston Bachelard shows "that certain matters – in particular water – transport in us their oneiric power, a kind of poetic strength that gives a unity to real poems."[20] In other words, he demonstrates the coherence of the oneiric symbols offered by water in different aspects, as material images appear in a wide spectrum of works in north-western literature and poetry, going back to ancient Greece and stopping in the first half of the 20th century. The perception is no longer individual. It comes rather from the literary corpus studied by Bachelard, which demonstrates the inter-individuality in the analysis of the different "*mise en scène*" of atmospheres. It must be understood that the "*formants*" revealed below result from the ambiances perceived by the book, which brings us, of its own free will, its own experience in a narrative form.

The formants thus revealed under the bachelardian scope are then renewed in an architectural creation project leading to the completion of an original projectual process. The project is expressed in the form of five pavilions distributed along a 40km path which is part of the old General Wade military road in Scotland. This section of the road links two little town in the south of the Highlands: Fort Augustus and Laggan. Linking the sensitive experience of space to the one of tasting scotch whiskey, the project explores the aesthetic relationships to water represented in different forms, concrete or abstract, within an architectural space. The program and the site chosen for the project promote the aesthetic experience by focusing on contemplation and hedonism. Each of the pavilions serves to enhance a flavour and aroma of the scotch whiskey realm. In order, along the path, the presented flavors are: woody, smoky, floral/fruity, peaty and iodized. The two pavilions presented below start and end the journey: the woody and iodized ones.

4 From the case to the project

The present section exposes the design objectives as an extension of the above theoretical framework. It presents the analysis of two atmospheres "*mises en scène*" in two rooms of the Vals' thermal baths, which led to the respective identification of two "*formants*", themselves reinterpreted in the ambiantal scenography of the woody and iodized pavilions. In other words, the analysis is divided in two cases, each one presenting the complete process, from

[3] Chapter titles translated by the author: Charon's complex
[4] Chapter titles translated by the author: Maternal water and feminine water

the analysis itself to the creation of the pavilion. Those subsections present the following sequence of descriptions: the Vals' room' scenography; the perception from Bachelard's book; the atmosphere intended in the projected pavilion; and the expression of the "*formant*" reinterpreted in it.

4.1 The mark of time

In the baths' den, while five changing rooms follow one another on the left, a narrow lowering in the floor runs along the right wall and attracts the attention. Clear water flows through it, producing a light sound. This water which accompanies this first aesthetic experience guides the procession to the end of the corridor, from one door to the other. By looking up, one witnesses the source of this water. Synchronized with the rhythm of the changing rooms, five elementary gargoyles let out a continuous stream of water. Directly from the natural spring in Vals, the water marks the concrete with a layer of minerals that continues to thicken and widen over time.

For Gaston Bachelard, the poetic image linked to *Charon's complex* is hereby offered by the scenography of the stream of water and the trace that marks the passage of time. This trace, present long before the experiment described here, will also be present long after. While the aesthetic experience that is made of it is, all in all, ephemeral, the architecture that makes it possible is much more perennial. This controlled time mark underlines the difference in temporality scale between one's experience and what causes it. The patina, a testimony to this temporal gap, forces an awareness of the ephemeral reality of human life. Indeed, this first presence of water in the baths reminisces of a departure on water, which refers to the symbol of the *never made journey*, a symbol endorsed by the character of the ferryman, himself carrying a symbolic baggage allowing death to keep its sense of travel. This is how the formant of the ambiance is presented, referring the perceiving mind to the *Charon complex*. It is both the scenographic element that makes it possible to perceive the atmosphere and the sensitive result that imposes an awareness of death as a human condition.

The woody pavilion echoes this material image of the *Charon's complex* by using the patina. In order to give a plural meaning to this reverie and thus offer more opportunities for the visitor's reminiscence, the oneiric image poduced by the patina is used on different materials and over different time scales. Surrounded by a tightly rhythmic square colonnade, the upper floor is completely open to the outside. The tasting area is organized in two boxes on either side of the stairwell. Thin oak sections very close together at the perimeter of these boxes separate the tasting area from the rest of the pavilion. These two doorless rooms invite visitors to discover the woody flavour of Scotch whiskey (figure 1).

The choice of materials and their size in the tasting rooms (figure 2) results from this reflection based on the *Charon's complex*. From the most durable to the most ephemeral, stone, copper and oak each respect a temporal scale intelligible by the patina they present. In opposition on this scale, stone and oak define the main spaces. Copper, used in the roof and for the staircase, is mainly used to "*mettre en scène*" the patina.

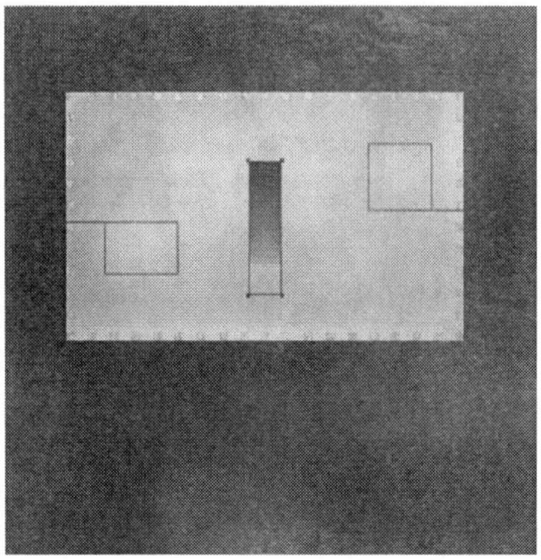

Fig. 1. Woody pavilion: 2nd floor plan

Fig. 2. Woody pavillion: interior perspective

4.2 Of water and other first liquids

In the heart of one of the stone monoliths that structure the space of Vals' installations, a particular smell attracts the attention. As one goes down into the pool, the smell is revealed. A small amount of essential oil gives the water a delicate aroma and texture. The basin contains thousands of yellow petals suspended in a cloudy, milky-looking water. The slight movement on its surface, combined with the essential oil mixed in, adds a cloudy appearance and a thick texture to the water, which is reminiscent of the appearance and texture of milk. Sitting on a submerged bench, the body floats between two waters. The perception of the body immersed and floating in complete safety refers to a level of comfort similar to the prenatal existence of the being. This sensation is here amplified by all the petals suspended in the basin. The scenography thus designed by Peter Zumthor helps to generate the maternal reverie of the petal basin.

In this basin, the water reveals its feminine character. Water is presented here as maternal, in its appearance as well as in its action. First, it diffuses a milky image through its slightly cloudy and thickened texture. As "all liquids are water"[21], the oneiric milk contained in the basin is reminiscent of filial love. Warm and thick, it envelops and cradles the body within it. Now the oneiric quality of water extends into the action. The experience of bathing, in this narrow and intimate room, connects the cradled body to the reverie of the prenatal bathing experience.

The iodized pavilion takes on this binary material image surrounding the reverie of maternal water – the appearance and the action. The oval-shaped pavilion floats in the center of a small lake along General Wade's road. Upon entry, the door closes behind, activating the light in a narrow space between two walls – the inner one being a scaled down copy of the outer one. In the heart of the inner wall is the tasting room. It is a brightly lit space but from which the light source remains a mystery (figure 3).

The ovoidal shape, in addition to referring directly to an egg, eliminates the edges while offering a relatively flat floor, therefore stable on water (figure 4). It is first of all this loss of visual limit, inside the tasting room, that evokes the image of the sea. It is filial love, the first love, that is the most similar to that of the sea: to love the infinite character of the sea "is to give a material meaning, an objective meaning to the infinity of love for a mother."[10] This characteristic of infinite water, specific to the sea, is coupled with its nourishing image. As an inexhaustible source of food, the sea plays a maternal role: it ensures a perennial abundance of food. In addition to these two reminiscences of filial love, which give the atmosphere of the pavilion a maternal character, the cradling action of the lake is also present. This action, an inherent quality of the matter, relaxes, rests and reassures. The rocking action is significant of the maternal reverie of water.

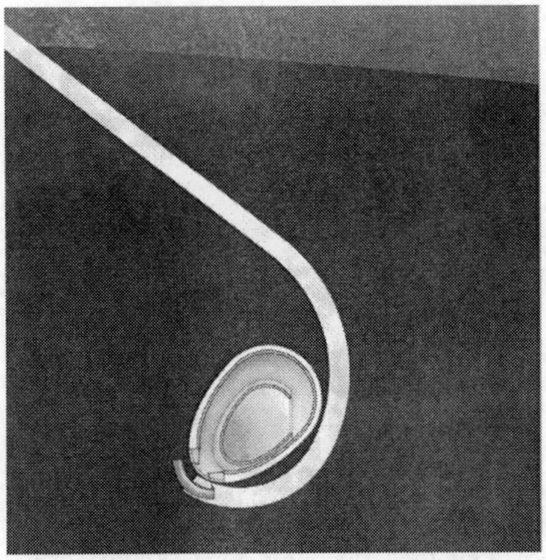

Fig. 3. Iodized pavilion: floor plan

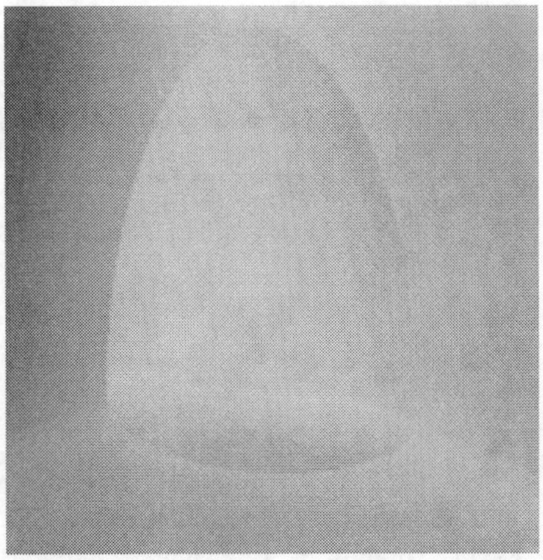

Fig. 3. Iodized pavillion: interior perspective

5 Retrospection

As a work of art, the architectural object conveys a sensitizing aesthetic, becoming a generator of emotions, themselves "a subjective translation of the ambiance of a place in its material as well as symbolic aspects"[11]. The method of reading the ambiances used in this design-driven research reveals a certain amount of objectivity to a northwestern literary context leading to the identification of "*formants*" of the sensitive ambiance. It also proposes a reappropriation of these "*formants*" by the "*mise en scène*" of a voluntarily emotional architecture.

In an interview with Mari Lending[12], Peter Zumthor explains the link between the memory of a place and the emotion that architecture can evoke. The loop is closed, the reminiscence is the key to the thesis. By its scope, the corpus explored by Bachelard contains an important collective memory to which everyone can refer. Whether direct or indirect, knowledge of this literary oneiric references refers to the same material image of water. Reminiscence refers to the collective memory giving access to the reading keys of common sensitive ambiances which lead to an architecture that touches a wide range of visitors.

Although they were as deep and meaningful, other symbols presented in the book were not applied in this design-driven research because they couldn't be linked to Zumthor's Vals' project. These symbols could be used and applied to some further analysis of other atmospheres.

References

[1] A. Berleant. "*Aesthetic Sensibility.*" Ambiances Enjeux - Arguments - Positions(2015). Published electronically 30 mars 2015. http://ambiances.revues.org/526 ; DOI : 10.4000/ambiances.526.
[2] S. Schürger. "*L'esthétique Du Paysage - Observer Et Créer Des Ambiances.*" Paper presented at the 1st International Congress on Ambiances, Grenoble 2008, Grenoble, France, (2008-09-10 2008).
[3] P. Ardenne, B. Polla, J.L. Genard, J. Le Maire, N. Gilsoul, P. Aebischer, L. Geninasca, et al. *Architecture Émotionnelle : Matière À Penser.* [in Français] [Lormont] : Le Bord de l'eau, (2011).
[4]Wolberg, Nathalie. 2008. "*Texture d'espaces - Territoires d'emotion : L'architecture comme objet intime.*" 1st International Congress on Ambiances, Grenoble 2008, Grenoble, France, 2008-09-10.
[5] P. Amphoux, J.P. Thibaud and G. Chelkoff. *Ambiances En Débats*. A la croisée, (2004).
[6] G. Chelkoff. "*L'ambiance Sensible À L'architecture : Paradoxes Et Empathies Contemporaines.*" Paper presented at the Ambiances in action / Ambiances en acte(s) - International Congress on Ambiances, Montreal 2012, Montreal, Canada, (2012-09-19 2012).
[7] G. Böhme. "*Un Paradigme Pour Une Esthétique Des Ambiances : L'art De La Scénographie.*" Paper presented at the 1st International Congress on Ambiances, Grenoble 2008, Grenoble, France, (2008-09-10 2008).
[8] N. Gilsoul. "*Jardiner L'architecture Émotionnelle - L'empirisme Comme Méthode Projectuelle De Luis Barragán.*" (2008). Published electronically 28-12-2008. http://www.projetsdepaysage.fr/fr/jardiner_l_architecture_emotionnelle.
[9] T. Ouard. "*Essai Sur Le « faire Une Ambiance » En Architecture.*" Paper presented at the faire une ambiance creating an atmosphere, Grenoble, France, (2008-09-10 2008).
[10] G. Bachelard. *L'Eau Et Les Rêves : Essai Sur L'imagination De La Matière.* [in Français] Paris: J. Corti, (2005). 1942.
[11] N. Audas. "*Le Rôle De L'affect Dans L'ambiance Ressentie.*" Paper presented at the 1st International Congress on Ambiances, Grenoble 2008, Grenoble, France, (2008-09-10 2008).
[12] P. Zumthor, H. Binet and M. *Lending. A Feeling of History.* Scheidegger & Spiess, (2018)

The Topological Reading of Ambiances in the Built Environment: The New Methodology for the Analysis of the Luminous Ambiance in the Museum Space

Selma Saraoui[1,*], *Azeddine* Belakehal[2], *Abdelghani* Attar[3] and *Amar* Bennadji[4]

[1] MAA at the Department of Architecture, University of Bejaia, Algeria
[2] Pr, Laboratory for Design and Modeling of Forms and Ambiences (LACOMOFA), the Department of Architecture, University of Biskra, Algeria
[3] MCB at the Department of Architecture, University of Bejaia, Algeria
[4] Dr, The Scott Sutherland School of Architecture and Built Environment, The Robert Gordon University, UK

Abstract. Daylight is currently at the centre of discourse on architectural space. The definition of architectural space takes essence from Euclidean geometry related to metric dimensions. The present study is an attempt to shed light on topology which is a non-Euclidean geometry. It can support non-metric components of space such as light to define architectural space. A corpus of six European museums has been chosen to study the immaterial or material relationships between form and daylight, since light is an essential element for the success of the exhibition. It also seeks to highlight discontinuity reports, and to confirm their existence through their software visualizations Therefore, the current study has taken into account an analysis model based on the notions of "route" and "sequence". The contemporary architectural project focused on taking into account human postures, both physical and psychological, within the architectural space. The results obtained show that light can release other spatial features for the museum space that can be highlighted by visualization with sequential analysis.

1 Introduction

Architectural space is one of the most crucial concepts and can offer a varied palette of definitions. It can be defined as a space organized by the construction, which often refers to the architectural object. Others define it as a hollow space that has meaning through function

*Corresponding author: saraoui.selma@gmail.com

and inner/outer relationship [01], or also as a three-dimensional space that includes man. This definition takes into account man's own internal space, which is often the space that surrounds us. [02]

Among other definitions that often appear in research work on architectural space, that of Jean Cousin [03] notes that the human body is surrounded by a bubble that constitutes its personal space. The latter is not only limited to the surface of the skin, but is also limited to the visual fields on spatial extensions (depending on what the human being sees around him). The definition of the architectural space presented by researchers is often an anthropological one. If we take architectural space outside its interaction with human beings, it becomes a mathematically geometric, three-dimensional space that has precise dimensions.

The evolution of theories on architectural space and its relationship to human experience, particularly those related to ambiances, has shown that there are two ways of studying space. A very rational way of studying it that excludes the human being, and focuses on the physical parameters of space called "quantitative", based on quantifiable measures and values of the environment. The second is related to the human being and his behaviour in space, and is based on the qualities of space.

The tools for representing space in our era do not only deal with geometric aspects, but also with the sensory aspects of architectural space. And so, we are witnessing the use of software that allows this new representation. This representation of space by software takes care of many aspects: dimensions to scale, climate files, context, orientations, materials, etc. These aspects make the representation very close to the reality of space. We are no longer talking about simple modelling, but rather about the visualization of architectural space.

The present research seeks to investigate the participation of the visualization of the lighting atmosphere, and the creation of a new definition of architectural space has become possible. The latter not only takes care of its geometric aspect, but also its sensory reality through the discontinuities of the lighting atmosphere noticed in the space. We will use a topological approach to space to see how topology can define architectural spatiality just by these discontinuity relationships.

1.1 The Approach of Ambiances Between Typologies and Visualization

The luminous ambiance is a critical subject nowadays because of its visible aspect. If the information comes from the different senses, our brain tends to decide for those that have a visible character (which comes to us from the eye); hence the absence of ambiguity in the luminous atmosphere.

J.R Pigeon [05] notes that it is only possible for man to visualize himself in his route if he becomes aware of his environment through his body. This awareness allows a better understanding of the architectural space. The first relationships between man and the route refer to visual capacities; man tends to understand the environment in which he finds himself and appropriate it through the decoding of information that space sends him.

Research on visualization of ambiances has been strongly developed; visualization converts information into images that can be used to explore and analyze. It is on the basis of human perception that we can build guidelines that make visualization very effective. The objective of the visualization was to improve teaching and learning procedures. [06]

What is to be noted is that despite the very high number of visualization software, they are not available to designers, and most have been tested by academics. Software simulation helps to create a modelling very close to reality, which would allow a visualization of the atmosphere or ambiances. This visualization supports the physical environment of the ambiance, but also makes the user an important element of the spatial reading. [07]

1.2 Topology and the Act of Displacement

Although it is old in architecture, topology appeared in the field of architecture as a response to an urban problem of the last century [08], then many works of classification of architectural buildings [09].

The birth of the topological trend in contemporary architectural thought has led to the emergence of two currents: i) The first one is that of deconstructivist architects; who wanted to express the differences and heterogeneity of the physical and cultural contexts of our age through formal discontinuity strategies, ii) The second one is that of architects of the new avant-garde who seek to create a link between architecture and mathematics; Through the topological theories inspired by mathematics.

No theory specific to topological architecture has yet been formulated. Nevertheless, we could talk about a topological trend that architects, both theorists and operational architects, i) talk about or ii) adapt to design their projects with pure reference to mathematics. [10]

In the light of what G. Di Cristina had presented as a definition, we were able to conclude that topology is defined as the act of displacement according to a temporal duration. The dynamic movement of the body crosses time creating points, successively, throughout a route (duration). This point once created becomes an event. [11,12] It is therefore considered that it is possible to define topology, from an etymological point of view, as a transformation into continuity.

1.3 The Luminous Ambiance in the Museum Route

Maria Saraiva mentions in her work two types of ambients found in the museum : i) the first one is objectal and aims to enhance the value of the exhibited object and it is the oldest[13] and highlights the relationship between visitor/object, ii) the second one is functional which is based in addition to the relationships between visitor/exhibited object, the route that enters the visitor's perceptive experience and becomes the basis for the visitor's sensory experience. [14]

The museum route is the fundamental unit of museum design and its success. The route is defined according to the dictionary [15] LAROUSSE (2009, p. 356): "the route or path followed to go from one point to another". Common sense shows the complexity of this term.

In the case of museums, we refer to the solutions to the principles of circulation that have been developed according to their form [16] (Mariani. R, 2000): i) The "linear" route: an obligatory scheme, ii) The "labyrinth" route: no traffic restrictions, iii) The "centred" route: The public has the freedom to choose the route of their visit.

For some researchers, the route represents the movement of the body, the movement in space. For others, it is described as a design-visit interaction. With the route, the simple act of moving begins to make sense. The lighting ambiance in the museum route is essential for the success of the exhibition. The architectural aesthetics of the route and the degree of visitor satisfaction depend on the daylight.

If we carry out an analysis of the interpretation of architectural space in the visual field, we will realize that the details of the space (working levels, floors and ceilings...) do not seem dominant, but the image of the room is often determined by the surrounding surfaces. This image is defined by the distribution of light and dark and is none other than the spatial deformation through light, which will influence the atmosphere and well-being. [17]

2 Materials and Method

Depending on the definition of topology cited in the section "Topology and the Act of Displacement" and projected into the spatial components and their ambient characteristics, we will conclude that from the environmental point of view the part of space in which the human body moves (duration) is none other than the route, or the dynamic movement of the body (displacement) creates throughout this route intersection points that we will call "sequences" that are considered purely light events. [18] In this section, we will explain the methodology adopted to verify this definition.

2.1 The Methodology

2.1.1 Sequential Analysis

In fact, it is a question of listing the cascade of parameters required to highlight the analysis model [19], which we summarize as follows:

- Step 01. Identification of the route: it is a question of observing the route of the visit route in the museum plans, according to the description accompanying the graphic documents. This route will be drawn by means of the drawing on the plans on a blue layer.
- Step 02. Identification of the bays by their axes: it is a question of identifying the bays which give on to the route, and on another layer draw the axes of the lateral bays and the bays in a discontinuous line of red color.
- Step 03. Delimitation of the sequences of the route: the limits of the sequences are at mid-distance between the axes of two consecutive bays. And here we return to the definition of the sequence that we have developed from the exploration of a literature review; the sequence is defined as the portion of the architectural space that is part of a route. A portion that is identified by the axis of a bay operated in the wall and whose boundaries are at the vicinity of the axes, the same route, located halfway between the bay of the sequence in question.
- Step 04. Description of the sequences: by establishing a table called the descriptor containing a description of all the variables of the bay and the conformation.

This sequential analysis model allowed us to highlight the light sequences and their characteristics as a function of the bay variables analysed. The qualifiers were chosen according to the rules of the European Handbook Of Daylight, which we quote here: i) - the type of lighting, ii) - the orientation, iii) - the position, iv) - the inclination) - the shape, vi) - the size, vii) - the glazing, viii) - the additional glazing. These variables are those that influence the lighting ambiance within a sequence of the studied route [20]

Then, according to the conformation, we noted our observations on the changes in conformation that are due to the presence of discontinuities according to the variables: i) -materials, ii) -colour, iii) -texture, iv) -morphology.

Once defined, the sequences can be analysed in several ways, namely:

- Statistical analysis: using STATESTICA 7 software to find: i) correspondences between the resulting discontinuities, ii) correspondences between the discontinuities and the design elements related to light.
- Simulation: through the ECOTECT 05 software for 3D modelling, and RADIANCE for surface illuminance and luminance, its role is to: i) confirm and visualize the discontinuities that exist in the museum route, ii) make a quantitative comparison between the results.

2.2 The Corpus of Study

2.2.1 The Selection Criteria and the Corpus of Study

To verify this hypothesis, we have chosen to work on permanent exhibition spaces in museums because the luminous ambiance is considered as a totalitarian ambiance, and permanent exhibition spaces provide a function that does not change over time. We chose six museums designed during the same historical period (from 1980 onwards and located in the European context), whose architects intended to work with natural light from the design stage, and on which we collected data going as far as describing the ambiances created by architects and journalists (In addition to photos, graphic files...).

Based on the definition of mathematical topology in architecture, we have classified our museums according to their architectural forms into two categories: i) a first where the volume is characterized by an absence of deformation or formal discontinuity, which includes three types of museums (the Departmental Museum of Prehistory, the Beyeler Riehen Museum, and the Acropolis Museum of Athens), ii) and a second where the formal deformations exist in the volume. This category also includes three museums (the Guggenheim Museum Bilbao, the Jewish Museum of Berlin, the Hergé Museum).

This choice of volume is also linked to the nature of the route, where we notice that for museums where there are no volumetric deformations (simple volume), the routes are unambiguous and symmetrical in most cases. In the case of museums with distorted volume distorted the route and mixed, or labyrinthine nature.

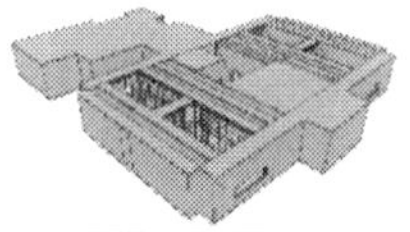	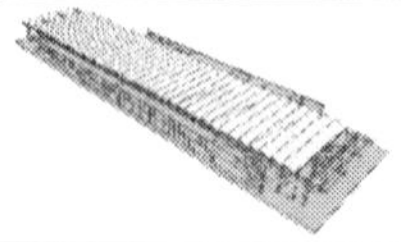	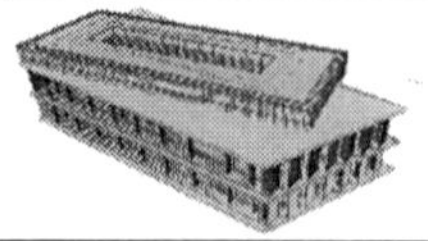
Prehistoric departmental museum Roland Simounet1981	Beyeler Riehen Museum Renzo Piano 1998	Acropolis Museum of Athens Bernard Tschumi 2007

Guggenheim Museum Bilbao Frank Gehry 1993	Jewish Museum Berlin Daniel Libeskind 1999	Hergé Museum Portzamparc Christian 2006

Table 01 : Presentation of the Volumes Made by Ecotect of the Case Studies

3 Results and Discussions

3.1 The Analysis of the Correspondences Between the Discontinuities Resulting from the Sequential Analysis

3.1.1 The Sequences

There is no precise typology to define the sequences, the nature of the volumetry, and the interior spaces or even the route. The existed sequences are very varied in nature. We have summarized the results of the sequential analysis in the following table where we find the number of typological variations and number of sequences, and finally the resulting discontinuities equal to or greater than 50% in all the study cases.

The architectural form	The case study	N° Typological variation of sequence	Total of sequences	Variable of discontinuity
without deformation	Prehistoric departmental museum	04	10	• Orientation
	Beyeler Riehen Museum Renzo	04	19	• Type of lighting • Orientation
	Acropolis Museum of Athens	09	80	• Type of lighting • Glazing complements • Orientation
With deformation	Jewish Museum Berlin	05	53	• Type of lighting • the size of the bay • orientation • bay position
	Guggenheim Museum Bilbao	09	31	• Type of lighting • wall inclination • shape of the bay
	Hergé Museum	15	15	• Type of lighting • the size of the bay • orientation • wall inclination

Table 02: Summary of the Results of the Sequential Analysis.

3.1.2 The correspondences between the discontinuities of the sequences and their visualizations

In this section, we will present the results of the sequential analysis, accompanied by some simulation to confirm the existence of the discontinuities. For the simulations, we selected the highest attendance periods and hours for each case study. We first performed our simulations by lighting curve, the curves are sometimes of a visible nature as in the case of the Hergé Museum. But in other cases, they are not and do not allow a good reading of the results.

What interests us in the simulation is not the results compared to the standards, but just the visualization that would highlight the discontinuities. To do this, we present the results as a synthetic image with some surface illumination values.

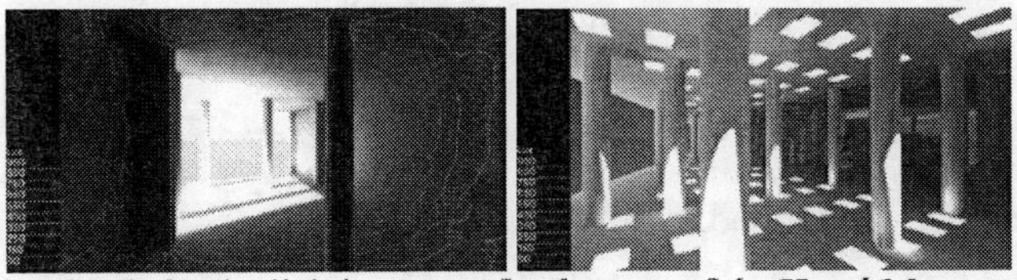

Fig. 01: results of the simulation by lighting curve for the case of the Hergé Museum and the Acropolis Museum.

3.1.3 Case-by-case study

a) For the cases of museums with undistorted volumes:

- The departmental museum of prehistory

Based on the analysis, it was found that there is no variation in the characteristics of the space in all sequences such as wall inclination.

Fig. 02: route visualized by radiance for one hour of high attendance at the departmental museum of prehistory.

There is a dominance of lateral lighting in the sequences (64%), logically distributed over the four orientations: i) a high percentage for EAST and WEST (39%), and ii) a low percentage for SOUTH and NORTH (11%). According to the analysis, there is a predominance of medium-sized side bays, followed by medium-sized zenithal bays, and large side bays.

- The Beyler Reihn Museum

It should be noted that the ceiling is completely open to the sky, consisting of zenithal bays. Space in all sequences retains these spatial characteristics. No inclination concerning the walls. We have visualized parts of the route showing this spatial stability for the time when the number of visitors is at its maximum at 2pm.

The museum is dominated by zenithal lighting in all sequences (61%). The zenithal bays occupy almost (100%) of the total ceiling surface, the lateral lighting is (39%), mainly EAST oriented (58%), NORTH oriented (25%), and SOUTH oriented at 17%. All the museum's bays are large.

The horizontal windows are protected by removable skylights that instantly regulate and manage the amount of natural light inside the museum. For side bays and whatever their orientation (SOUTH or NORTH), they are protected by a horizontal projection that the architect has designed as a decorative element.

- The Acropolis Museum of Athens

The space retains its characteristics for all sequences, with no inclination in the walls. All textures are smooth. The visualization allows us to see this spatial stability for part of the course, for the time when attendance is very important.

21 March **21 June** **21 December**

Fig. 03: route visualized by radiance for one hour of high attendance at the Acropolis Museum in Athens.

It can be seen that the dominant type of lighting in the museum is lateral. It is divided into several orientations: i) WEST (25%), ii) EAST (23%), iii) NORTH-(19%), iv) SOUTH-WEST (13%), v) NORTH and SOUTH (10%). Zenithal lighting represents 10% of all types of lighting.

The visualization of the lighting ambiance in the part of the linear route allowed us to draw the curve of the variation of the surface illumination variation, The graph represents a part of the route, where we can see that the surface illumination curves cross from one sequence to another and vary between 400 and 1900lux for the West orientation at 3pm, the curves are in an increasing order, this is due to the linear character of the route (time or attendance and very high in summer period).

b) In the case of museums with deformed volumes

- Jewish Museum, Berlin

There is no variation in the characteristics of the space. No wall inclination is observed. According to the analysis, it was found that the dominant type of lighting is zenithal (59%), followed by side lighting (45%). The latter is divided into several orientations: i) SOUTH-EAST (31%), ii) NORTH (23%), iii) SOUTH-WEST (31%), iv) EAST and NORTHEAST (12%), and finally v) WEST (8%).

The bi-varied analysis reveals that there is no dominance in terms of lighting type and bay size, but in terms of bay position and size, the middle bay in the middle of the wall is the most dominant for this case study.

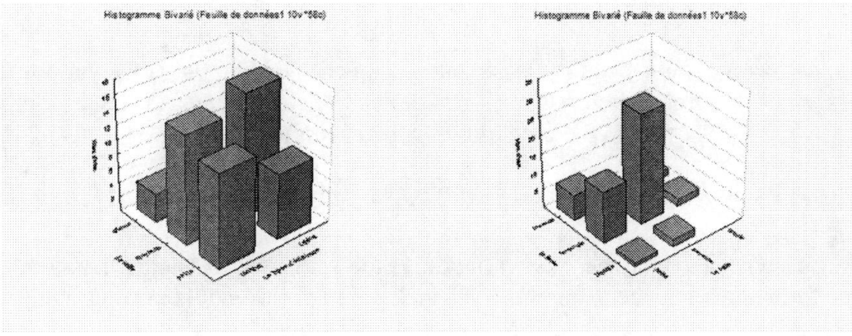

Fig. 04: Correspondence size - type of lighting and Correspondence Size - Position (Jewish Museum Berlin)

In the graph, the curves are all variable, which confirms the existence of discontinuity in the proportion of the bay to the wall in this part of the Jewish Museum in Berlin.

- Guggenheim Museum, Bilbao

We notice a discontinuity in the morphology of the walls, where the inclination of the walls in some parts of the route is done in a multidirectional way.

Fig.05: radiance visualized route for one hour of high attendance at the Guggenheim Museum in Bilbao.

The zenithal lighting constitutes 55% of the total lighting of the museum, we also find 45% of lateral lighting, several orientations with a dominance of the SOUTH orientation 38%, the NORTH orientation bays constitute 31%, that of the WEST orientation 15%, then an equality between the SOUTH-WEST and NORTH-WEST orientation with a percentage of 8%.

We also see a dominance of square bays for the medium size zenithal lighting type, followed by large rectangular side bays and a small number of large rectangular zenithal bays.

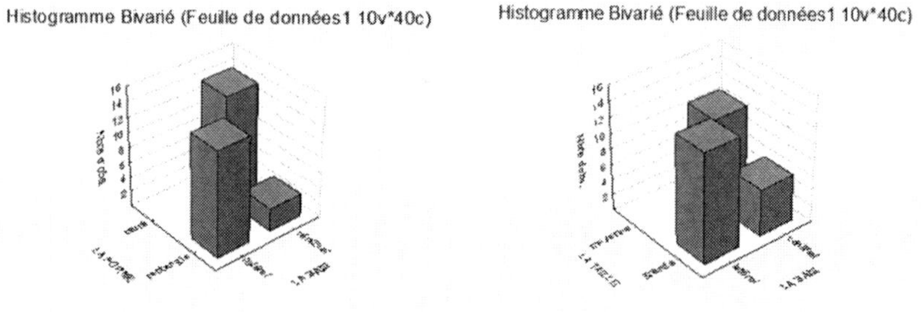

Fig.06: Correspondence Form-Type of lighting and Correspondence Size-Type of lighting (Guggenheim Museum Bilbao)

For the second graph, stability is present in sequence 03 and 04. However, for the rest of the sequences, the lighting curves are variable, and this characterizes the discontinuities in the proportions of the bay in the case of the Bilbao Museum.

- The Hergé Museum

The space is characterized by a variation of the walls, i.e. an inclination composing the passage from one sequence to another. The inclination of the wall is inward in one sequence, in the second it is absent, and vice versa for the entire route.

Side lighting is the majority (76%). Side lighting varies in orientation: a dominance of the EAST orientation (38%), equality between the SOUTH and NORTH orientation (23%), and finally a minority for the WEST orientation (19%).

It should also be noted that all the bays, regardless of their lighting types, are rectangular in shape and large in size overall.

3.1.4 The types of discontinuities according to the route in the museums of the study corpus

The sequential analysis has allowed us to identify the discontinuities that can characterize the lighting ambiance in our museums, the multivariate analysis between the different variables of the discontinuities to reveal correspondences between them. We summarize them as follows:

- For museums with a linear route, we have noticed the existence of three types of discontinuities: those of the proportion of the bay, those of its position, and its size.
- For museums with mixed or labyrinthine routes, we have noticed the existence of discontinuity in the complement of glazing, the shape of the bay, its orientation, and the type of lighting.

Conclusion

This research has allowed us to bring out in our museums a very varied range of sequence typology, each sequence is different from the other, thanks to the variables defined above that characterize it.

The corpus museums are formally varied, whether in terms of the volume of the building or the shape of the project's interior spaces, and depending on the characteristics of the route and its shape, we found that museums whose route is labyrinthine or mixed belong generally to museums whose volume is characterized by the existing deformation. On the other hand, museums with a linear or centred route are museums of the modern style or the new avant-garde, which is characterized by the absence of volumetric deformation of the project. And finally, thanks to the sequential analysis we were able to deduce that for each type of route, corresponds to a precise type of discontinuity.

The role of simulation in this research is to reproduce conditions of the exhibition spaces so that an accurate assessment and reading of the spatial characteristics as a whole and of the variations in illumination in the different sequences of the same route can be assessed through the different spatial visualizations.

This research has allowed us to confirm our hypothesis that light topology is defined as the discontinuities in the light sequences which characterize architectural space. The route in this research is composed of a sequence that can be defined as a succession of purely luminous events, where one passes from one sequence to another through a transition. This transition can be distinguished by the change of the lighting ambiance within the same space.

Topology can, through the relationship between sequences, create a new spatiality of architectural space other than geometric space. It will have a very important role if we treat each sequence separately in museums in order to improve the conditions for future exhibitions. The results can also contribute to the optimization of light comfort in exhibition spaces.

References

1. P. Von Meiss, *De la forme au lieu, une introduction à l'espace architectural,* (EDP polytechniques et universitaires romandes, Lausanne. 1993)
2. B. Zevi, *Apprendre à voir l'architecture* (1993)
3. J. Cousin, *L'Espace Vivant. Introduction à l'Espace* (Ed. Le Moniteur, Paris. 1980).
4. G. Bonnet, Rayons sonores et ondes de lumière ?cul,sciec,phys (2005)
5. J.R. Pigeon, *Essai (projet) soumis en vue de l'obtention du grade M.Arch. École d'architecture Université Laval* (2013)
6. A. D' Alba, *Analyzing Visitors' Discourse, Attitudes, Perceptions, and Knowledge Acquisition in an Art Museum Tour After Using a 3D Virtual Environment, doctorat thesis, university of north Texas.* (2012)
7. D.Siret , P. Woloszyn, *Ambiances architecturales et urbaines - Du complexe au simplexe, Les cahiers de la recherche architecturale 42/13 - Ambiances architecturales et urbaines* (ED Luc Adolphe Parenthèses, Paris,1998)
8. M.P.Corcuff, *Penser l'espace et les formes, Thèse de doctorat en géographie, le 26 novembre à l'Université de Rennes 2*(2007)
9. M. H. Ibrahim,A, J King Saud Univ.,9, Arch. &Planning, l-30. (1997)
10. G .Di Cristina, *Architecture and science 6-13,* (EDB wiley academy,Great Britain, 2001
11. S.Saraoui, A.Belakehal, *Actes de la conférence BASC, 156* (2011)
12. S.Saraoui, A.Belakehal, A.Attar, A.Bennadji, ZEMCH 2018, Melbourne, Australie, *92-94* (2018)
13. M. Saraiva, *L'environnement sensible dans les musées à caractère ethnologique Thèse de doctorat encadré par : J.F. Augoyard.* (2001)
14. S.Saraoui, A.Belakehal, A.Attar, A.Bennadji, ZEMCH 2018, Melbourne, Australie, *97* (2018)
15. Dictionary Le Petit Larousse 2009, p. 356 (2009)
16. S. Mariani-Roussset, *L'espace urbain en méthode,* 29-85 (1996)
17. S.Garry *The reasoning architect mathematics and science in design,* (EDB MC Graw Hill international ,edition Singapore,1990)
18. A.Sossinsky, *Noeuds Genèse d'une théorie mathématique*, (EDB Seuil, Paris, 1999)
19. S.Saraoui, A.Belakehal, *Actes de la conférence BASC, 159* (2011)

The space-time cube as an effective way of representing and analysing the streetscape along a pedestrian route in an urban environment

Thomas Leduc[1,*], *Vincent* Tourre[2], and *Myriam* Servières[2]

[1]AAU-CRENAU, CNRS, École Nationale Supérieure d'Architecture de Nantes, France
[2]AAU-CRENAU, École Centrale de Nantes, France

Abstract. The graphical representation of a complex polygonal object, such as an *isovist field* (i.e. a 2D *isovist* in its evolution over time along a pedestrian path in an urban environment) is a hard research issue that was stated in the late 1970s by Benedikt [1]. The generalized space-time cube from Bach and al. [2] is a conceptual framework allowing to propose innovative implementations for the analysis and the display of this complex object. The purpose of this article is to situate, within this formal framework, all the graphical representation solutions that have already been developed for *isovist fields*. It also aims to identify the problems raised by each of these solutions (spatial anchoring, arbitrary sampling, etc.) and to propose related solutions.

1 Introduction and state of the art

1.1 Urban space analysis and visibility issues

Because a strong trend at work today is to develop and design cities that allow people to adapt and mitigate phenomena related to major issues, such as climate change, several new criteria have emerged for the design of sustainable and healthy cities. These new criteria have to do with walkability, role of the streetscape in encouraging the development of pedestrian mobility, physical activity and well-being. As [3] points out, indicators related to walkability are not only the D variables (*design*, *density*, *diversity*, *destination accessibility*, or *distance to transit*) and must necessarily include "*the microscale streetscape features that affect the pedestrian experience*."

This attention to the landscape and its influence on pedestrian behaviour is part of a long line of work on the interactions between individuals and their built environment in terms of environmental psychology. It led to the identification of about fifty urban qualities likely to influence the behaviour of pedestrians moving in the street space [4].

* Corresponding author: thomas.leduc@crenau.archi.fr

The literature on visibility studies extends this questioning and provides definitions and operating procedures; tools that make it possible to formalize and to measure the essence of places in terms of some of the involved urban design qualities. In the state of the art, which aims to "*explore the spatial properties of the visual structure inherent in space*" [5], we can highlight the *isovist* (Figure 1a) and *isovist field* (Figure 1b) as pioneering methodological tools proposed by [1]. The *isovist* is indeed a tool that easily answers the questions raised by [6]: "*How far can we see, and how much can we see?*". All key questions according to the author "*in developing good design*".

Even if other tools exist, such as the *sky opening* studied by [7], to analyse and even design urban space, we will focus in the current article only on *isovists* for their ability to synthesize, at any generation point, the immediately surrounding open space.

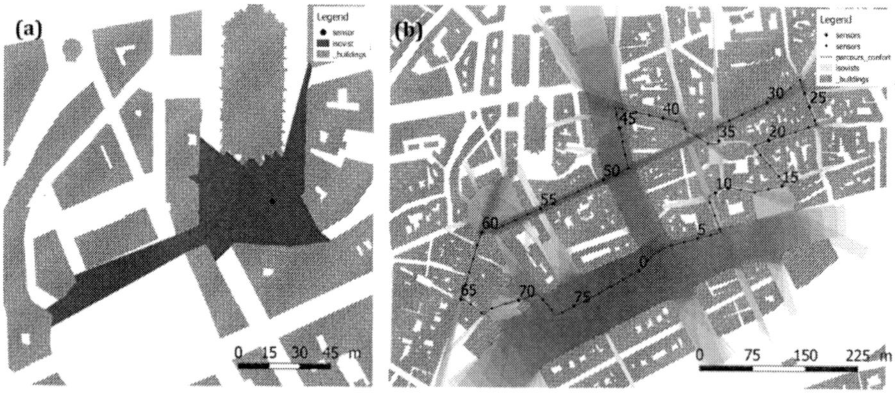

Fig. 1. (a) On the horizontal plane of the gaze, the isovist is the set of points seen from a given position (black disc); (b) The 1.6 km long pedestrian route is sampled every 20 m. An *isovist* is then calculated at each sampling point. The corresponding *isovist field* is drawn in a flat representation. The darker the colour, the more the corresponding portion of urban space is seen along the path.

1.2 Visibility representation issues along an urban route

According to [1]'s definition, in an environment composed of opaque facades, the *isovist* associated with a given vantage point corresponds to all the points seen from the corresponding position in all directions of space. In the case of a 2D planar environment, such as the one presented in [1]'s pioneering article, the *isovist* is a bounded polygon representing a horizontal slice of the pedestrian's immediate surrounding space. In the absence of any opaque surface, it may be useful to limit the scope of the gaze with an artificial horizon. This location-specific visual shape holds the same characteristics as a "visual pattern" as defined by [8] and therefore information extracted from the *isovist* provides a synthetic view of the studied terrain.

With reference to the work on Gibson's motion perspective, [1] proposes "*a cumulative understanding of the form of the environment [...] by perceiving variants and invariants in the transformation [...] caused by the observer's movement.*" Indeed, because "*the view of and visual exposure to the environment will gradually and sometimes suddenly change with [the] position*", the corresponding set of *isovists* and their shapes and sizes, but also the nature of their boundaries, vary along the way so as to create some sort of footprints' field specific to a given environment. To move beyond the study of single *isovist* and to provide an analytical tool to study a set of positions, he therefore proposes the *isovist field*.

However, this cumulative visual experience exposes the designer to a difficulty in terms of representation, as can be seen in Figure 1b. The juxtaposition of polygons in the same horizontal plane causes overlaps. To overcome this difficulty and exhibit the structure of the set of surrounding visual spaces, [1] proposes two strategies. The first consists in moving a panoptic light source along the path in an analogue mock-up. All the illuminated points of the mock-up correspond to the 3D *isovist* associated with the position of the light source. The second, described as more analytical by its author, consists in building a Minkowski model. This model is a sequential stack of 2D *isovists* where the Z-axis embeds the time dimension. Finally, the author proposes to apply a dimensional reduction to this *isovist field* and to study scalar fields (the *isovist* is abstracted at any point by a measure value such as its area). This simplified representation model makes it possible to derive iso-potential lines in a traditional way.

1.3 The space-time cube as a framework for temporal data visualization

A pedestrian's route through the city is a complex object that has a temporal dimension in addition to traditional spatial dimensions. As such, it is time-geography insofar as time and space are fundamental dimensions for the analysis of its dynamics. [9] proposed innovative visual objects such as the space-time cube to illustrate in a graphic way this type of phenomenon. More recently, [2] has developed a conceptual framework for the visualization of temporal data based on generalized space-time cubes. According to [10], "*maps are used to stimulate visual thinking about geospatial patterns, relationships, and trends*." The question is therefore whether the space-time cube is a relevant geovisual solution for the analysis of complex data such as an *isovist field*.

In this article, after a presentation of the conceptual framework developed by [2] and a review of some of the elementary operations resulting from their study, we will show that the representation techniques already implemented in the field of visibility analysis using *isovists*, fit perfectly into this formalism. With the help of this formalism, we will then propose innovative visualizations before opening up to extended visualization challenges.

2 Materials and method

2.1 Conceptual representation of spatial data based on the space-time cube

According to [2], "*a space-time object is a geometrical object within a space-time coordinate system*". The latter is a three-dimensional Euclidean space that adds a temporal dimension to a two-dimensional geographical space. In such a space, it is very common to represent the trajectory of a moving point object in motion on a plane map background by a static 3D curve. However, in the specific case of the representation of an *isovist field*, it is a matter of displaying not a point trajectory, but a collection of polygons as stated before. The issue lies in adapting a process for visualizing the evolution of a dimensionless object over time, to the visualization of the evolution and superimposition of a 2D shape over time. Regarding the issue of the representation device, the space-time cube is by definition a 3D object and its 2D visualization is part of the 3D rendering problems (with occlusion, depth ambiguity, perspective distortion, colour distortion and navigation difficulties [2]). To bypass the edge effects of a projection of a space-time cube on a 2D support as computer screens, various techniques exist to reduce its complexity and dimensionality.

2.2 Review of some elementary operations related to the space-time cube

According to [2], "*the ultimate goal of space-time cube operations is to transform a space-time cube into a space-time object whose shape is compatible with the shape of the media employed to convey the information*". With this constraint, they then develop a taxonomy of elementary operations that they group into four categories: extraction (selecting a subset of the space-time object), flattening (dimensional reduction by aggregation), geometry transformation, and content transformation.

In the extraction category, we distinguish the so-called *time cutting* operation. This operation consists in extracting a 2D subset of the space-time object that corresponds to a single time cut (slice without thickness). In the flattening category, we can mention the *time flattening* which is a dimensional reduction operation by aggregating time. In the category of geometric transformations, we can mention the *time scaling* operation, which aims to resize the space-time object along the temporal axis (an acceleration of the phenomenon under study is similar to a contraction and a slowing down to a dilation). Finally, in the category of content transformations, we can mention the *time colouring* operation, which aims to alter the colours of each time slice according to time.

2.3 Focus on compound operations formalism

The inherent complexity of certain space-time phenomena justifies the application of several elementary operations, in sequence, to the data set to be displayed. To formalize these complex processes, [2] introduced the notion of compound operation. Like the usual composition of functions in mathematics, a compound operation is a mechanism to combine elementary operations to build composite ones. To develop this formalism, the authors introduced in particular the operator + which corresponds to the composition and the operator * which refers to a repetition (operation applied several times).

Discrete time flattening, which is an operation that aims to reduce the complexity of the space-time cube by aggregating all data independently of the time-related dimension, is then written as follows: *(time cutting)* + time flattening*. It consists of starting with a first step of sampling the space-time cube as a function of time. Then to apply an aggregation operation on the resulting layers while preserving the spatial dimensions.

3 Case Study

3.1 Presentation of the study site

In order to test the few techniques for representing the above-mentioned complex geographical objects, we have identified an urban fabric that is emblematic of the city of Nantes (France). This 700 m by 500 m portion offers a diversity of urban morphologies, on a mainly pedestrian sector, flat ground of constant altimetry. To survey this urban fragment, we have chosen a 1.6 km pedestrian route (see Figure 1b) that winds its way from the medieval district in the east to a district of classical architecture dating back to the 18th and 19th centuries in the west. A research study conducted in 2010 on this same route [11] showed that its direction, if it is perfectly explained from a chronological point of view, is perfectible in terms of better visibility of heritage objects. For obvious reasons related to the constrained size of this article, the illustrations we present from now on are limited to a small portion of this path (from sampling point #30 to sampling point #40, see Figure 1b).

3.2 Review of techniques already implemented for 2D visualization of isovists

The *isovist* presented in Figure 1a is a flat representation of the space-time cube at a given time step. It results from a single elementary *time cutting* operation. The *isovist field* shown in Figure 1b is a flat representation of the initial space-time cube. It is the result of a compound operation namely *discrete time flattening* described in section 2.3. To be more precise, this compound operation embeds a content transformation operation (*translucent filter*) in order to produce an *alpha blending aggregation*. The resulting compound operation is written: (*time cutting + time colouring*)* + *time flattening*. The same *isovist field* could have been shown with several *isovists* side-by-side as in [1], thus corresponding precisely to the *discrete time juxtaposing* compound operation. It is indeed a matter of extracting "*multiple time slices and placing them side-by-side or on a grid*" [2].

The Minkowski model mentioned in [1]'s article is a 3D rendering resulting from a compound operation applied to the space-time cube of *isovist field*: (*time cutting + space shifting*)* + *3D Rendering*. In case this sequential stack of 2D *isovists* is coloured, as in Figure 2, an additional content transformation is applied and the corresponding compound operation is equal to (*time cutting + space shifting + time colouring*)* + *time scaling + 3D Rendering*. The *time scaling* operation that precedes the final 3D rendering step aims to facilitate the overall visualization of the shape by contracting it (by a factor of 2).

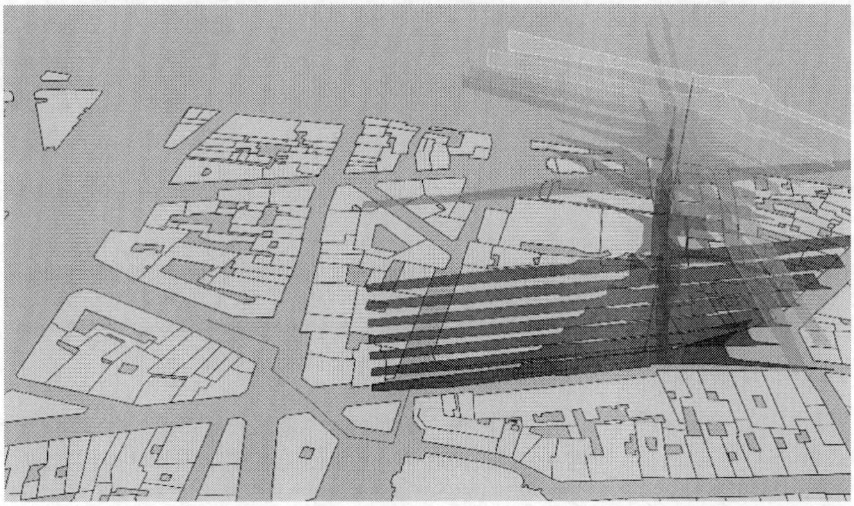

Fig. 2. Transposition of the "Minkowski model of the *isovists* along [the] given path" proposed by [1]. Sixteen *isovists* are stacked on the time axis, which is represented by the blue vertical axis. All the corresponding generation points are aligned on this axis. The more yellow the polygon is, the further away from the starting date.

The "*Analog production of isovists along a path by point-source illumination of a model*" stated in [1] is not a static visualization but a dynamic operation that consists of moving the cutting position over time. This is the compound operation known as *animated time cutting* according to the above-mentioned formalism.

3.3 Innovative 3D renderings

The stack of *isovists* presented in Figure 2 has the advantage of providing a synthetic view of the evolution of the pedestrian's *isovists* during his walk in the city. Its formal qualities make it possible to highlight the bifurcations and configuration changes in the pedestrian's surroundings over time. However, the *space shifting* operation that systematically accompanies *time cutting* results in a loss of all anchor points because it translates all *isovists* into the same vertical.

To counteract this loss of anchoring and avoid getting lost on the background map, we propose to remove the *space-shifting* step in the aforementioned compound operation. The space-time object presented in Figure 3 is a 3D rendering resulting from the following compound operation applied to the generalized space-time cube: (*time cutting + time colouring*)* + *time scaling + 3D Rendering*.

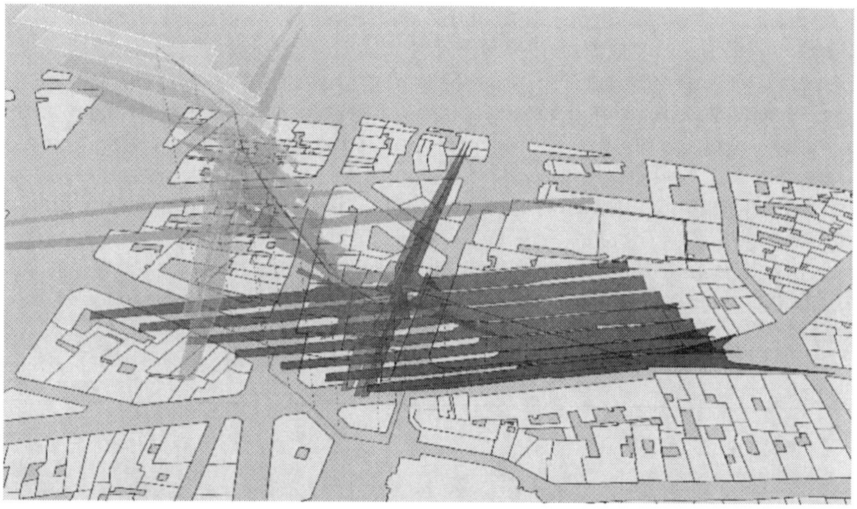

Fig. 3. By removing the *space shifting* operations in the Minkowski model from Figure 2, the *isovists* are relocated to match the path in spatial dimensions.

The shape presented in Figure 3 facilitates, through its stepped approach, a joint spatial and temporal interpretation of the *isovist field*. However, even if its sampling step is a little finer than that of Figure 1b (12 m vs. 20 m), one must recognize that this discretization remains coarse and that it may mask possible major visual events.

Figure 4 combines a response in terms of refinement of discretization, the sampling step is reduced to 2 m, with a thickness representation of the corresponding "visual volume". Indeed, instead of a stack of *isovist* slices, we represent a global volume by connecting the contour points from one time step to the next time step. This rendering also has the effect of guaranteeing a form of continuity of the entire shape, that could be associated to a single *3D Rendering* operation, and thus allowing to show the raw shape of the *isovist field*.

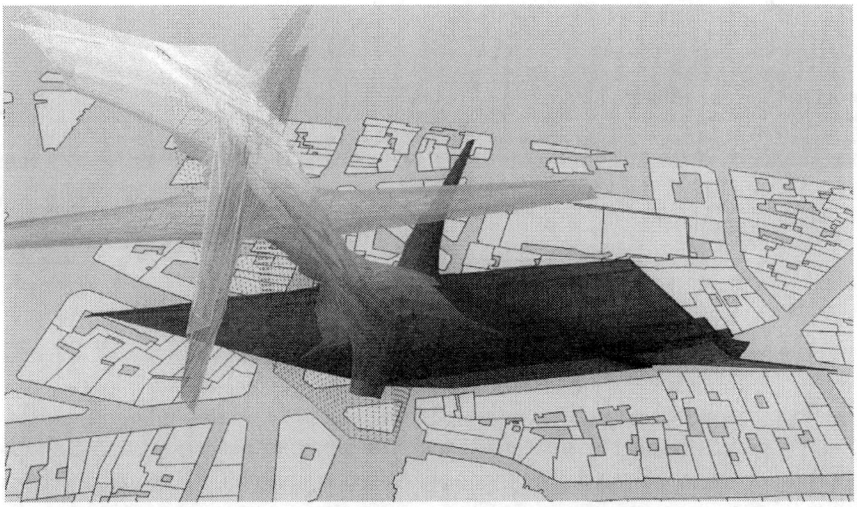

Fig. 4. Rendering in thickness of the space-time shape presented in Figure 3. In addition to refining the sampling step, the contour surfaces are drawn to shape a kind of "visual volume".

4 Discussion

4.1 Immersion

This article and the figures it embeds are intrinsically 2D objects. Being able to free ourselves from the 3D rendering step, by implementing stereographic rendering, would probably allow us a better understanding of an *isovist field* along a path (which is a typically 3D object). Indeed, by avoiding projection on a plane, we give free opportunity to immersion and navigation in the virtual space-time cube. We also allow to resize and adapt it to a human scale, as [12] points out: "*It can be seen that the scanned object* [...] *is systematically enlarged by students within the virtual environment in order to reach a height at least equal to the size of the user. The model is now 'inclusive' and in turn operates a form of immersion that we will call expansive*".

4.2 Increase dimensionality

As usual in the representation of the space-time cube, the background map in Figures 2, 3, and 4 is two-dimensional. To facilitate the location and identification of built volumes, we could imagine replacing this background plan with a three-dimensional urban model. However, this facility would have the disadvantage of introducing new masks and thus increasing occlusions. It would also introduce confusion in the use of the third dimension, which would be both spatial and temporal. In order to analyse the interstitial urban space in a volumetric perspective, we could imagine representing not two-dimensional *isovists* in their evolution over time but three-dimensional *isovists*. This dimensional increase of the item to be represented will consequently substantially increase the occlusion effects and depth ambiguities.

4.3 About adaptive sampling

The sampling distances used in this study are arbitrary and systematic. However, there are portions of the route that present more or less visual events. It is therefore entirely legitimate, if not desirable, to consider adapting the sampling step to the regularity or irregularity of the course at each point. The partitioning of the urban space of [13], as well as its extension by [14], who proposes to qualify the space according to the importance of changes in visual surroundings, could be very useful in this regard.

Conclusion

In this article, we have stated that attention to the visual landscape surrounding pedestrians as they walk around the city is a useful operating concept in urban planning processes. To this end, [1] proposed the *isovist field* as a solution allowing a dynamic analysis (spatial and temporal) of the space covered. The representation of a complex polygonal object, such as a 2D *isovist*, in its evolution over time along a pedestrian path, is however a hard research issue. From the generalized space-time cube proposed by [2], which is a conceptual framework adapted to the analysis and representation of the *isovist field*, we have resituated all the representation solutions that have already been proposed. In this article, we have also identified the problems raised by each of these solutions and proposed innovative evolutions. We now have to validate the usability of our proposals through a set of controlled experiments.

References

1. M. L. Benedikt, "To take hold of space: isovists and isovist fields," *Environ. Plan. B Plan. Des.*, vol. 6, no. 1, pp. 47–65, 1979
2. B. Bach, P. Dragicevic, D. Archambault, C. Hurter, and S. Carpendale, "A Descriptive Framework for Temporal Data Visualizations Based on Generalized Space-Time Cubes," *Comput. Graph. Forum*, vol. 36, no. 6, pp. 36–61, Sep. 2017
3. L. Yin, "Street level urban design qualities for walkability: Combining 2D and 3D GIS measures," *Comput. Environ. Urban Syst.*, vol. 64, pp. 288–296, Jul. 2017
4. R. Ewing, S. Handy, R. C. Brownson, O. Clemente, and E. Winston, "Identifying and Measuring Urban Design Qualities Related to Walkability," *J. Phys. Act. Heal.*, vol. 3, no. Suppl 1, pp. 223–240, 2006
5. M. Llobera, "Extending GIS-based visual analysis: the concept of visualscapes," *Int. J. Geogr. Inf. Sci.*, vol. 17, no. 1, pp. 25–48, Jan. 2003
6. M. Batty, "Exploring isovist fields: space and shape in architectural and urban morphology," *Environ. Plan. B Plan. Des.*, vol. 28, no. 1, pp. 123–150, 2001
7. J. Teller, "A spherical metric for the field-oriented analysis of complex urban open spaces," *Environ. Plan. B Plan. Des.*, vol. 30, no. 3, pp. 339–356, 2003
8. S. Steiniger, "Enabling pattern-aware automated map generalization," University of Zürich, 2007
9. T. Hägerstrand, "What about people in Regional Science?," *Pap. Reg. Sci. Assoc.*, vol. 24, no. 1, pp. 6–21, Dec. 1970
10. M. Kraak, "The Space-Time Cube Revisited from a Geovisualization Perspective,"

in *ICC2003: Proceedings of the 21st international cartographic conference: cartographic renaissance*, 2003, no. August, pp. 1988–1996

11. T. Leduc, F. Miguet, V. Tourre, and P. Woloszyn, "Towards a spatial semantics to analyze the visual dynamics of the pedestrian mobility in the urban fabric," in *Lecture Notes in Geoinformation and Cartography*, 2010
12. É. Bigot-Doll, "Échelles diffuses : Prospectives de conception via matière et VR," *SHS Web Conf.*, vol. 47, p. 01002, Oct. 2018
13. J. Peponis, J. Wineman, M. Rashid, S. Hong Kim, and S. Bafna, "On the description of shape and spatial configuration inside buildings: convex partitions and their local properties," *Environ. Plan. B Plan. Des.*, vol. 24, no. 5, pp. 761–781, 1997
14. T. Leduc, F. Miguet, and V. Tourre, "Motion Perspectives Integration in the Qualification of the Urban Spaces: Towards a 2D- and a 3D-Enrichment of the S-Partition Method," in *Urban Design and Representation*, B. E. A. Piga and R. Salerno, Eds. Cham: Springer International Publishing, 2017, pp. 69–82

A Performative understanding of spatial design, learning from exhibitions

Ayman Kassem[1]

[1]City University, Faculty of architecture and design, Tripoli, Lebanon

Abstract. The terms 'Performative' and 'Performance' are more and more emerging in the spatial design discourses, from exhibition, to interior, arriving to urban design. These notions are not clearly defined yet. They are characterized by a semantic width and multiple applicative possibilities. Between the different interpretations and uses of 'performance' and 'performative' in architectural discourses, this paper will focus on two main dimensions of a particular importance: The first refers to the concepts of the scenic, the narrative, and the theatrical qualities in architecture. The second relates to the 'event-character' of spatial interventions, and the relation between event, and soft intervention, which tackles the concept of the 'transformative power of the performative', which indicates to the capacity of architecture to activate spaces and processes.

Exhibitions, pavilions and installations are all concerned with changing the performance of the existing space, and therefore they are able to illustrate what a 'performative-oriented architecture' could be or mean. The Italian architect and thinker Andrea Branzi, in his essay "L'Allestimento come metafora di una nuova modernità" (exhibitions as a metaphor of a new modernity) (Lotus, n.115, December 2002), invites architecture to learn from the sub-categories of architecture. Architecture can learn from the practice of exhibition (displaying and installing). Architecture is invited to learn design approaches and strategies from performance-oriented spatial disciplines such as exhibition design, scenography, art-installations, and events, with a particular focus – in this paper – on: 1) Soft interventions and the transformative power of the event-intervention, 2) The aesthetic of experiencing and the narrative approach. Therefore, it is in the sub-disciplines of architecture that we can 'rehearse' a 'performative understanding' of the spatial project, in the sense of studying and designing spaces basing on performative criteria, and performance perspectives. The repertoire of the performative is large; it includes the concepts of the flexible, the event, the transformative, the theatrical, the narrative, etc. The aim of this paper is to explore and rehearse this kind of performative analytical tool for the architectural project (taking the case of Nevicata14, a project of refurbishment of the pedestrian Piazza Castello in Milan, 2015). A performative understanding of architecture is proposed as an alternative to a general theory, and as a new dramaturgy to analyse and design spaces.

1 The space as the performer

The terms 'Performance', 'event', 'to an audience' are key-concepts that can be directly exported to the architectural and the spatial discourse. 'Event' and 'Performance' also relate to the notions of the 'spectacular', and the 'temporary'. In the field of exhibitions, the 'spectacular' effect is exploited for communicative, commercial and entertainment motivations: 'Performing a spectacular experience', a 'memorable experience', are often the key- concepts for the so-called memory-design and experience-design [1]. To clarify how the terms 'performative' and 'performance' can be attributed to space; we definitely should study the space as the 'performer', and to differentiate between 'the performance of spaces', and 'the performance in the spaces'. In terms of terminology, many publications (in exhibitions, interior and urban design) used a performance-belonging lexicon, as metaphors and analogies with the world of performance, using all the related terminology. For example, in '*Mettere in scena, Mettere in mostra*' (2014) (in English: 'Exhibiting and displaying'), different contributors used a range of terms belonging to the world of performance to discuss architectural and spatial issues such as: architectural dramaturgy of the urban space, the event city, the urban scene, creative city, living architecture, media-tic space, urban dramaturgy, scenography of architects, media architecture scene, etc. [2]. Pier Luigi Salvadeo, one of the contributors, indicated that these kinds of analogies are necessary to understand the contemporary city, expressing that: "*A new performative reality, broad and inclusive, take on the contemporary city today. Space, time and action, the typical characters of the spectacle, contribute today to the formation of the spaces in the city, becoming dramaturgical places of constant change [...] a dramaturgy of the urban space that includes everything, where unlike the traditional dramaturgy, in close relation with the narrative theater, the scene is set as a condition of priorities and the dramaturgy becomes a peculiar feature of the space.*" [3] 'Performance' and 'Dramaturgy' can therefore be used as a metaphor, to illustrate and clarify some ideas, but they also function as analytical tools to study a spatial situation with all its spatial, social, and cultural dimensions.

2 Performance studies and the Performative turn

The notion of performance started to emerge in the humanities and social sciences and the philosophy of linguistics, in the mid-20th century. Theorizing about performance appeared in the 1940s and 1950s, when an intellectual movement known as The Performative Turn took shape. The performative turn in the humanities and social sciences consists on theorizing about performance as a social and cultural element [4]. Performance, as the central concept of this movement, was used as a metaphor and an analytical tool, employed to understand human behavior. Performance provides a perspective for framing and analyzing social and cultural phenomena [5]. Contributed to the movement the works of Victor Witter Turner, Erving Goffman and Richard Schechner, who focused on the elaboration of a dramaturgical paradigm to be applied to culture at large to view all culture as performance. Schechner understood performance as a continuum, where not everything is meant to be a performance, but everything, from performing arts to politics and economics, can be studied as performance [6]. In the same intellectual atmosphere, it could be mentioned also 'Dramaturgy' as a term that belongs to the world of performing arts, that also entered the world of humanities and social science, before entering the world of architecture. In relation with architecture, Pedro Gadanho understands by The Performative Turn, as the influence of performance over other artistic media and disciplines such as architecture. In the sense of "*reconnecting architecture with traditions of performance art*" [7], where architecture is inspiring and borrowing from the performative practices (and the world of performance) new dynamics and mechanisms. The performative aspects appear as: the participatory in and with architecture, ephemeral spatial practices, the use of flexible structures, the emphasis on the scenic and the spectacular, the emphasis on the experience, etc.

3 The Speech Act theories. Performative and transformative

"[...]Performance is intentional, if successful, it does cultural work in the world [...] As a transformational force, behavior performance has the power to restructure social order through the persuasive power of rhetoric and through the redefinition of both audience and context [...] " [8] (Carlson Marvin, 2008)

The philosophy of language, and with John Austin in the 1950s, introduced the concept of the 'performative utterance', arguing that 'to say something is to do something'. For Austin, the speech constitutes an active practice that can affect and transform realities therefore language is performative [9]. 'Performative' that refers to a situation in which a new reality is created [10]. Therefore, the performative has a transformative power. Judith Butler also considers that gestures and behavior are performative because they can generate the identities they express [11]. The 'speech act theory' is practically the central theory behind the notion of 'performance'. Performativity is the political aspect of performance and its manifestation of power [12]. What can transform a reality is therefore performative. In spatial design practice, the concept of the performative is relevant to analyze the constituents of a situation that can transform or be transformed with the design (the intervention, the architecture): communication, life forms, economy, the cultural, the architectonic, the social, the urban, etc. For Erika Fischer Lichte, the 'Transformative power' means also that architecture changes according to uses. The situation will influence and change the people involved. So even if spaces are created to serve certain purposes but that does not mean that they must only be used in this function [13]. The transformational force is in the power to restructure the social and the spatial order, through the transforming the audience, the context, and the users.

4 Spatial performances

4.1 Performative, narratives and the scenic potential

For Sophie Wolfrum, scenic, situationism, psychogeography and the psychosocial are the necessary concepts in the shaping of a city. She bases on Baudrillard's quote: *"[...] scenic space, without which [...] buildings would only be structures and the city merely an agglomeration." [14]*. 'Performative' refers to the execution of an action giving meaning to a situation in which a new reality is created or experienced. For Alban Janson, this new reality could be an experience lived on an internal individual level. Janson sees that scenic and theatrical situations could exist in architecture and urban spaces, and they can be intended or not by the planners or the designers. Architecture is a complex of intended and unintended scenic situations. He refers to the experience of ascending the Balthasar Neumann's staircase in the Bruchsal Palace, using it as an example to explain the performative as the scenic potential of a spatial situation. Janson narrates the spatial experience as below:

> *"[...]an initial approach to the staircase leads into darkness; at the same time ,we are confronted with an alternative to the cave-like atmosphere below in the form of ascending to the right or to the left, guided upwards by light, sweeping outward in an almost dance-like*

manner between dark depths and bright natural light , with no indication as yet of where the staircase ultimately lead[...]until we reach an oval platform that seems to be suspended in space, detached from the walls[...]vaulted over by the colorful , mysteriously illuminated depiction of the heavens on the ceiling, which serves as an antipode to the darkness below. Anyone ascending this staircase becomes a decisive part of a complete situation [...] essence of this architecture." [15]

It is a psychological, and perceptual experience, lived on an inner level, individually. The scenic, the theatrical and the narrative characters of the spatial situation were performed on the user while moving and experiencing the space (ascending a dramatic staircase). The user was part of the show, and that show is between him and his internal perceptual consciousness. Janson sees that *"Performances in architecture [...] surpass its mere practical use [...] The act of climbing a staircase, for instance, by a constructive arrangement according to a dramaturgical consideration into scenic experience without audience, so that we follow our movements through space as if it were a performance"* [16]. It is a self-reflexive experience. A special design/arrangement could enhance an experience of a pure functional movement, making it acquire more drama. And this when the user feels transformed into another reality. Such transformations should be called performative. It is the performative power of architectural situations [17]. Therefore, a performative understanding of architecture seeks to enhance these latent scenic potentials and narratives in spatial situation.

5 Exhibitions as performance-oriented architectures

Exhibitions are required to be more and more performative. Architect and exhibition designer Uwe Brueckner has attributed terms to spaces that are directly related to the world of performance and scenography, such as "choreographed space", "spatial potentials", "dramatized space" and "making spaces talk". He spoke about spatial potentials and parameters that allow the dramatization of space, creating choreographed spaces. Different strategies and tools can be employed to make a space perform. A performative space is a space stimulated to perform in different ways. To note that the expression "*making a space perform*" is in resonance with the title of the book "*Scenography- Making spaces talk, projects 2002-2010"*, Atelier Bruckner (2011). Bruckner underlines the design the space through scenography aspects [18]. For instance, creating immersive and suggestive atmospheres can be achieved through the manipulation of the interior environmental factors, in order to provoke a certain sensory experience. The communication of content is made through the lived spatial experience, and that approach underlines the shift from the design of space, toward the design of experience, where the aesthetic of perceiving becomes the aesthetic of experiencing. David Dernie differentiated describes a narrative space as the exhibition that is conceived as a story teller that adopts different styles of contextualization, rhythms, and levels of intensity and patterns of movements [19]. The approach of narrative design emphasizes the importance of visitors' experience, creating powerful spatial experiences, evoking emotional engagement, and creating memorable experiences [20].]. Unfolding narratives and sceneries are increasing interest in architecture in general, and in urban spaces in particular in order to enrich the experience of public architectures as not just containers of functions, sifting away from the pure functionalistic approaches.

6 Making spaces perform

'The performative understanding of spatial interventions' is proposed as a design approach and analytical tool. This understanding could be acquired through analogizing and studying the similarities between lightweight architectural interventions, exhibition design and art installations. A spatial intervention can be studied and analyzed in terms of performance/behavior, and in terms of interaction between the new (the inserted intervention) and the existing (the space). A spatial intervention aims at changing the performance of the existing. In this regard, we can define a gradient of strategies for spatial interventions, from solid and surgical interventions, to lightweight and even immaterial interventions. The architectural intervention can be studied in terms of theatricality, in this sense the architectural intervention can be seen as a sculpture, or as an installation that could be dramatic, narrative, scandalous, unnoticeable, spectacular, scenic, etc. Spatial interventions are also in a close relation with the notion of event. A spatial project can be studied and analyzed as 'as an event', 'provoker of an event' or 'hosting events'.

6.1 Art in spatial interventions. The collaboration art-architecture

The late version of modernism was incapable of providing successful domestic urban architecture, because of the pure functionalistic approaches. Art practices were called to participate in the formation of urban centers, to compensate for the cultural sterility of functionally determined solutions [21]. The use of art in buildings and public spaces became more and more popular as a response to tendencies and needs for the spectacular, the performative, the sophisticated, the entertaining, and the cultural image of the city or the institutions. The popularity of artistic interventions in architecture is increasing. Art-installations are used to experiment a manipulation of the environmental factors such as light, sound, and temperature. Altering the sensory experience in spaces induces different physiological and psychological reactions in people. Such spatial installations demonstrate how the physical characteristics of objects and environments act on people to influence the meaning making of spaces and objects [22]. Olafur Eliasson sees that architecture is more involved in problem-solving and nurturing collectivity and social values, and therefore art could do the same when collaborating with architecture [23]. Therefore, the collaboration serves the art by making it dedicated to the collective in real-life situations, outside museums, by calling artists for interventions in urban and architectural projects. In return, such collaborations give the architecture more sophisticated performative qualities, and enrich the perception of architecture as not just a shell of functions. Art in spatial sceneries is related to the interest in the unfolding narratives in space. Since the 1960s and 1970s, artists, writers and filmmakers were interested in this story-telling in motion [24]. In the space design disciplines, the attention is given to the engagement with places and objects, creating narrative paths and changing our common perception of daily spatial scenes. A kind of intrusion to break the order and continuity of the perception by creating new rhythms. Art is becoming more and more included in adaptive architectural projects, especially when we are dealing with public or semi-public spaces and institutions.

6.2 From solid, to soft, to immaterial interventions

In the field of preservation and spatial interventions, the tendencies are toward leaving the existing structures intact, and transforming the spaces through lighter and softer conceptual strategies, that are convenient to the cultural and economic limitations. Bryon Roberts believes the newer practices are more and more influenced by minimalism and conceptual art, operating between art, architecture, and preservation to make new spatial experiences [25]. In the same regard, Graeme Brooker and Sally Stone defined three categories of remodeling: Intervention, Insertion and installation: Intervention, is the most surgical and hardcore version of remodeling, while an insertion is defined as the introduction of a new element into, between or beside an existing structure. The softest version of interventions or remodeling is "the installation". It is defined when the elements of remodeling exist independently from the building, and therefore the two simply touch each other. Some installations can be ephemeral or temporary, and they generate a symbiotic relationship between a building and the new elements [26]. Pam Locker also spoke about the similarities between a display installation and an art-installation since art-installation and exhibition design are both site specific, usually interior and three-dimensional, and involve the transformation of the perception of space. They can use a variety of media including film, sound and light. They are both concerned with time and space, the sensory experiences of the audience and the idea of art as theatre [27]. An intervention, an exhibition display or an art-installation could perform in similar ways within an existing space and transform it. They are detectable as they stay independent from the hosting space, but at the same time, they have a transformative power (as they are able to induce momentary or even long-term changes in the spatial qualities of the exiting). Some kind of interventions can be described as immaterial as a very soft version of interventions, which aims to give the existing space a new performance with less architectural tools, where dramatic lighting, tactile and textured displays, changing temperatures, and low-tech audio environments are used in exhibitions to create emotional engagement between visitors and the event [28]. Making the space perform is achieved through a quasi-immaterial intervention.

6.3 The relation: Event, Performativity, Intervention

> *"[...] performativity is the capacity of architecture to become an event, to participate in a world which is more and more often defined in terms of occurrences rather than as a collection of objects and relations [...] What does it imply for architecture to be more and more often defined through performative criteria, from energy consumption to more qualitative characteristics like the capacity to generate affects?" [29] (A.Picon, 2012)*

As a performance, an exhibition, or art-installation, a spatial intervention can also be ephemeral, temporal and reversible. This temporality gives the intervention or the making of it, the character of 'event'. An 'Event' that can induce transformations in the space through time. The temporality, the scenic factor, the transformative power, the theatricality are the performative characteristics of an 'event'. 'Intervention as an event', 'architecture as event', or 'event-architecture', are expressions that are usually used to indicate to the kind of architecture that comes to life for a certain period of use, or as Sylvia Lavin stated *"when the lights are on"* [30]. Furthermore, some spatial situations hold in themselves some scenic potentials as spaces 'born' to host events. The term '*Eventmental*', a word that does not exist in the English language, was suggested by D. Leatherbarrow to discuss the performative character of architecture, as the character that unfolds beyond the scripted program, in the unpredictable singularity of situations. Leatherbarrow stresses on the relation between the '*eventmental*' character, the unpredictable and the unfolding: *"So the performance of a space is not revealed only in given occasion but also in 'indifferent emptiness' between various events [...] There are unexpected qualities as an unforeseen character [...] We could ask if this auditorium is working in certain occasion, if its 'generally adequate, but not until the event [...]"* [31]. A planned event in the space could unfold beyond the planned schedule of the event, revealing more about the performance of the space. So in other words, the space is seen as an event, and as an arrangement of factors and circumstances. Noting that the term 'unfold' is a keyword that is very present in many publications that discussed performativity. It is also used by artists who practice performance-art, where they prefer to describe their performances as '*an unfolding situation*' [32]. Therefore, beyond the performance of the physical qualities of the space, there is the performance of the setting, including the unpredictable factors. The event-character is also connected connects itself to the notions of openness, the participatory and the transformative. The event is open for participation, to transform and to be transformed; therefore, the event unfolds in space and time.

6.4 Installing and intervening as event, The before, the unfolding, the after

As in a performance, the installation animates this space. Sibylle Omlin reflected on the transformation of performance into spatial installation and vice versa, and on how a performance transformed into an installation can leave behind spatial relationships [33]. In the following case study (The case of Nevicata14, a project of refurbishment of the pedestrian Piazza Castello in Milan), we will examine the relation event-performance intervention. We will be analyzing and describing the design, the practice and the unfolding of the intervention. Therefore, the case-study will be dissected in three phases: the before, the during and the after. In other words, we will study the project as an event in space and time: the before, the during or the unfolding, and the after. This kind of dissected- analytical work, allows us to differentiate between: 1) Architecture as the event, 2) Architecture as a host for events, 3) Architecture as a generator of events. As well as the participation of architecture, and participation in architecture.

6.5 The case of Nevicata14, A project of refurbishment of the pedestrian Piazza Castello in Milan

Following the decision to pedestrianize the Piazza Castello in Milan, and after the construction of the Expo Gate (two temporary pavilions built for the Expo period), the Administration of Milan decided to try to solve the morphological arrangement of the Castello square, that became pedestrianized, but it was still left without a clear strategy, and without any specific criteria of use. Moreover, the decision for this project came also as a result of a series of pressures from the local residents, who complained about the not adequate use of this square, seeing its historical character, location and its urban quality. Under the name of 'Atelier Castello', and during the month of July 2014, Guidarini & Salvadeo associate architects (the winner firm), and other eleven architectural firms, were called for consultations, and to present for the citizens in the Expo Gate their proposals of possible future scenarios for the piazza. On behalf of the Municipality, the TRIENNALE of Milan was given the privilege in making decisions. The ultimate goal of the consultations was that, on one hand to propose a provisory solution to be implemented only for the period of the upcoming EXPO. On the other hand, this temporary accommodation/arrangement of the Piazza Castello had to be a testing ground to develop consciously the criteria for a final arrangement of the square, which will occur in the future, as a result of another international competition in two phases in progress [34].

6.6 A flexible proposal

From the initial competition call, the requests of the municipality and Triennale were not at all clear, and it was not yet clear the character that the project would have to have. All of these uncertainties had normally put the architects in a state of confusion. And thus, the winner studio claimed that at this point they decided to use this uncertainty as a positive condition for the project, from which to start. And so they had express their proposal through a different character of the project, that does not stop at its physical forms, but to imagine a flexible life of this future project, in which everything had to be fluid leaving behind any pretense definiteness. Therefore, it could be said that the confusion in the 'call' required a flexibility within the proposal to make [35].

6.7 Participation and Co-design

Through the public space of Piazza Castello, the architects wanted to think of a smart city, where the material hardware of urban territories could work in symbiosis with the immaterial software of social networks and the real city. A city where everyone is connected and where everything is linked together, as expressed by Salvadeo, basing on his reading of Carlo Ratti [36]. This 'participatory' concept is in the use and processes of formation of the city, which differentiates it from the classic notion of participation, as in the cases of the seventies, where it took the form of interviews, but where the architect was still fully controlling every stage of the design process till the final construction. And so maybe today 'participation' is more and more a kind of collective practice [37]. The inclusiveness in the design approach was expressed in the participatory mode of planning, the use of collaborative design software, and the resulting transparent use of city. The result should be hybrid spaces, material and digital. The Internet and the network not only changed the way in which people relate to one another, but this medium has allowed the emergence of practices like crowd-funding, capable of involving a large public, allowing each person to contribute to the definition of specific aspects of the project.
And therefore, if we go back to the notions of 'openness', 'performance-oriented', and 'participation', we will be able to differentiate-in architecture- between:
The participation <u>of</u> the project in its context: open, responsive to the context as a non-discreet architecture[39].
The participation <u>in</u> the architecture (of the users/the public): interaction and flexible uses.
The participation <u>in</u> the design process: co-design.

6.8 The proposal

As expressed by the architects of the project, the proposal consisted on a non-invasive project respecting the historical and monumental context, but without being totally intimidated by it. The Nevicata14 expresses a vision that can become reality. As during a snowfall, the space changes in an instant, and so how you see and use this space also change. The project proposal aims to empty space, rather than to add to it [38]. The project rearranges the space to make room for uses, therefore it avoids a delimitation of the space. The project thought about a geography of space's uses, offering a wide public pedestrian use of the space through providing public facilities that support the square (seating and benches of various types arranged freely in space, sun and rain shelters, distributors of fresh vapors, Wi-Fi, technology points, facilities for recreational activities, etc.). The idea was to continue with the calcite paving that was already present in the park next to the square, and to install under it a power/technological network to circulate and provide light and water. Nevicata14 seeks a hidden geography that is configured from time to time depending on the conditions of use. The vision regarding the castle is to be totally immersed in a vast new public space. And this public space will be an ideal continuation of the Sempione Park nearby [39].

6.9 The before

The use of social networks and the co-authorships were essential since the first moments of the project. Through an extensive use of social networks, the project has had for some of its phases, more than 400 authors. Experts and nonexperts, residents and citizens of Milan have contributed to the design process with their ideas, proposals and their actions online and on social networks, but also offline. More than 800,000 people have been intercepted in just over three months of work. During the consultation phases (the phases of 'Atelier Castello'), the presentations took different forms: from the exhibition of drawings, the organization of workshops, the involvement of citizens through debates and participatory activities, to the organization of demonstrative performances of possible new ideas [40]. All the works of consultations, presentations took place near the site at the Expo Gate. Later on, an exhibition of the design proposals was held at the Triennale. Various presentations were made for the council, the municipality. It is during that period that the debates about the destiny of the Castello were the most intensified. The Council also decided to launch an online consultation to check the approval ratings of the various project proposals. At this point the Triennale appointed a panel of experts, and asked the designers to submit a proposal at very short notice and a draft adapted to a low budget – finally declared – of 200,000€. The main requirement is that the project had to be completely and quickly reversible. The committee decided that the project to be realized is Nevicata14 [41]. Crowdfunding, social enterprise, and sharing economy, were also features of social innovative initiatives. As a result, imagining a new architecture of the square that will invert the usual logics of the construction of a space. And instead of moving from the space to the uses, it is the other way around, from the uses to the space, as noted by the architect. In other words, the uses and practices will participate in shaping the space [42].

6.10 The during

The project consisted of a series of circular areas in granular rubber of 1 cm thick, white and soft to the decking, which host a variety of structures of use, as expected from the earliest proposals. Metal structures/shelters for shade, facilities for punctual lighting, devices issuing fresh vapors, plants, signals poles that inform the public on how to use the square and its structures, etc. A team of 10 people was handling the social networks, and the communication part, to manage the processes of participations and proposed initiatives, and promoting the project, inviting the citizens to propose activities and events. Nevicata14, that began with '400 authors' now shifts its action from the co-design to the testing and the promotion of the uses of space. And this is how the concept of *Nevicata Lab* was born. Nevicata14 promoted a regulation of its uses, tried to define new models of coexistence and participation in a way that the public space could be self-sustaining [43].

6.11 The after

So a number of initiatives took place at Piazza Castello's Nevicata14, both profit and non-profit aimed at strengthening the civic fabric. The project also contributed to the creation of "community" and events, such as the 'Dinner in White', and 'the Critical mass bike' [44]. Therefore, it is a project that hosted events, encouraged events, but also generated events. It is a case of a performative intervention (urban scale), with emphasis on the character of event. *"Nevicata14 is an exportable model, not for its architectural forms, but for the fact that it triggers processes, procedures, of its conditions of use, and encouraging the formation of a community capable of producing proposals"*[45]. This statement of the architect, translates the concept of "transformative power of the performative". A performative spatial project, which embedded most of the repertoire of the performative: the event-character, the social, the flexible, the openness, the reversible, the participatory, the experimental aspect, the performativity of the process. Therefore, in this case of study, we were able to extract and detect the different 'dimensions of uses' of the notions 'performance' and 'performative'. The analysis, and the pure narrative description of this kind of projects, gave the possibility to dissect the 'before, after, during/unfolding' of the project in space and in time in which all the city was involved. We 'case studied' a project, basing on performance criteria and performative dramaturgy. Exhibitions, installations and interventions are spatial events that unfold in space and time. They all consist of a spatial intervention in which the designer or the artist employ different conceptual and nuanced strategies to make the existing space perform or to change its performance (temporary or permanent, short-term or long-term). This 'performative understanding of interventions' provides architecture with more nuanced, lightweight, performance-oriented strategies and curatorial attitudes towards urban design. Events/installations represent the lightest version of spatial interventions. They teach the architect how to intervene with almost an immaterial intervention.

7 Conclusion

A new dramaturgy for a new contemporaneity

Through our re-reading of Branzi's essay *'Exhibitions as a metaphor for a new modernity'* (Lotus 115, 2011) [46], we tried to demonstrate and propose what and how architecture can learn from performative practices such as exhibition design. We also tried to respond to another more recent essay by him, *'Towards a new dramaturgy of the project'* (2015) [47], in which he asks for a new 'dramaturgy' for the spatial project. 'Dramaturgy', which usually means the study of dramatic composition of the main elements that constitute a show. And therefore, we will use Dramaturgy as an 'understanding' of the series of internal/external connections that exist between the different agents that constitute the spatial situation when designing or studying a space. It is a metaphor and an analytical tool combined. Performative criteria and Performance perspectives come here to shape this proposed new dramaturgy for the architectural project. As stated by S. Lavin *"With the performative dimensions, architecture can try out new performative strategies developed by and for performance [...] to be distinct from the modern architecture"* [48]. Interior and exhibition design started emerging in parallel with the growth of commercial exhibitions, events, expos, and the tendencies for the performative, the spectacular, the polyfunctional, but since they are usually temporary, and reversible, these disciplines were considered as minor architecture as stated by Branzi who explains what and why the architecture of the city can learn from exhibitions:

"In the category of the design of displays [...] there are in fact all the genetic elements typical of a new modernity, which we shall call "light" and "diffuse" [...] (exhibition design) is superstructural and transitory, however, the sector is not a subcategory of architecture but on the contrary today plays a new and central disciplinary role in urban transformations. In other words, when an exhibition, a stand or a showroom is prepared, or when work or living spaces are fitted out, when a play or an opera or a ballet is staged, what is constructed is a piece of city. A piece of city [...] responding to the logic of reversibility, adaptation and re-functionalization of the constructed world. This evolutionary space, which can also be seasonal or promotional, is the one that most precisely differentiates the city of the twenty-first century [...]" [49]

Branzi, in his statement, used a vocabulary, or a lexicon, that belongs to theatre, exhibitions, and the world of performances, not just to 'metaphorize', nor to simply make an analogy between theatre and the city, between exhibition design and architecture, but he utilized this 'performance-belonging vocabulary' to expand in his analytical tool from the scale of installations and exhibitions towards the scale of architecture and the city. Architecture may borrow logics and strategies from the other minor subcategories of architecture, such as scenography, installations and exhibitions. New performative strategies and practices can be learned from these performative disciplines in order to handle a new complicated modernity and complex spatial conditions. Exhibition design is concerned with the transformation of existing spaces, in order to make them able to respond to new functions, and therefore architecture can learn a soft curatorial flexible approach to spatial interventions, in order to update the existing spaces with flexible spatial solutions and with nuanced performative interventions. The architect is challenged to choreograph spatial interventions that generate events, activate processes, and animate spaces, and lead to long term transformations. Architecture and urban design can also import from exhibition design the scenic, the narrative and the interactive approaches when creating spatial qualities and staging atmospheres. The design of the city is shifting from a pure sterile functional approach towards a performative approach that encourages the design of spatial experiences, urban narratives, ambiances.

Refurnishing, installing, displaying, intervening, or installing an event, are all practices that consist on rearranging systems of spatial objects, and making lightweight flexible interventions. Cultural, social, natural and economical contexts are in constant change, and thus prescribed dictated programs do not work; therefore, any new design or intervention should be adjustable and open for the continual changes in the modes of production and living [50]. Regarding this aspect of openness and flexibility, Valentina Signore defines the performative project as

"[...] spaces that are designed to be open to change: the design itself seeks to welcome and encourage transformations [...] The performative project lets things happen [...] It leaves room to the unexpected, to the indeterminacy [...] The idea of open form – distinctive to late modernity – not only accepts but even requires uncertainty , ambiguity, incompleteness, imprecision[...] Forms and functions are neither determined nor determining[...]"[51].

Architectural practices should be allowing themselves to be inspired from the performative spatial disciplines (exhibitions, installations, scenography, etc.) and take tools from the wide repertoire of the performative. Therefore, architectural practices had to reconnect with more performative spatial practices, and adopt a curatorial attitude towards the making of spaces. Performative-oriented approach to spatial design is the art of articulating spaces, curating experiences and ambiances, staging spatial events, designing spatial experiences, curating interventions and articulating them with events and positive urban transformations. This paper encourages deeper methodological reflections on architectural practices basing on a performative understanding of spatial design.

References

1. D. Dernie, *Exhibition design*, 23 (London, 2006)
2. L. B. Peressut., G. Bosoni, P. Salvadeo , *Mettere in scena Mettere in mostra* , (Siracusa : Lettera ventidue, 2014)
3. P. Salvadeo, "Le drammaturgie architettoniche delo spazio urbano", in Mettere in scena Mettere in mostra , ed. Basso Peressut L., Bosoni G. and Salvadeo, 25-6 (Siracusa : Lettera ventidue, 2014)
4. https://en.wikipedia.org/wiki/Performative_turn
5. 5-Ibid
6. https://en.wikipedia.org/wiki/Performance-studies
7. P. Gadanho, The Performative Turn, pedro gadanho's blog, Posted on 24/02/2012 https://shrapnelcontemporary.wordpress.com/2012/02/24/the-performative-turn/
8. C. Marvin, "Perspective on performance: Germany and America", in Erika Fischer-Lichte : *The Transformative Power of Performance: A New Aesthetics*, 9 (2008)
9. Carlson referred and was based on Turner,and Beeman .
10. S. Wolfrum, "Performativitvity and space", in Performative urbanism, generating and designing urban space, ed. S.Wolfrum and N. Frhr.v.Brandis,31 (Berlin :Jovis, , 2015)
11. S. Wolfrum, "Performativitve urbanism", in Performative urbanism, generating and designing urban space, ed. S.Wolfrum and N. Frhr.v.Brandis,(Berlin :Jovis, ,2015),13
12. Based on John L.Austin distinction between performance and performative,from Hempfer, Klaus, "Performance, Performanz, Performativität, 2011),13-43
13. E. Fischer-Lichter, "Performativitve urbanism", in Performative urbanism, generating and designing urban space, ed. Sophie Wolfrum and Nicolai Frhr.v.Brandis, 13,(Berlin: Jovis, 2015)
14. S. Wolfrum, "Performativitve urbanism", in *Performative urbanism, generating and designing urban space*, ed. Sophie Wolfrum and Nicolai
15. Frhr.v.Brandis, 13(Berlin :Jovis, 2015)
16. E. Fischer-Lichte, "Performativity and space", in in *Performative urbanism, generating and designing urban space*, ed. Sophie Wolfrum and Nicolai Frhr.v.Brandis, 31-32-33(Berlin: Jovis, 2015)
17. 14 -Sophie Wolfrum, "Performativitve urbanism", in *Performative urbanism, generating and designing urban space*, ed. Sophie Wolfrum and Nicolai Frhr.v.Brandis, 13(Berlin: Jovis, 2015)
18. A. Janson, "The performative power of architecture", in *Performative urbanism, generating and designing urban space*, ed. Sophie Wolfrum and Nicolai Frhr.v.Brandis, 164-5-6-7,(Berlin: Jovis, 2015)
19. Ibid

20. Ibid.163
21. http://www.atelier-brueckner.com , and Uwe R. Brüeckner , Scenography – Making spaces talk projects 2002–2010, ed .Atelier Bruckner, (US: Avedition, February 2011)
22. Ibid
23. D. Dernie, *Exhibition design*, 20 (London, 2006)
24. 20 http://www.atelier-brueckner.com And Uwe R. Brüeckner , *Scenography – Making spaces talk projects 2002–2010*, ed .Atelier Bruckner, (US: Avedition, February 2011)
25. D.Leatherbarrow, "architecture'sperformance unscripted", in *Performative Architecture: Beyond Instrumentality*, ed. Branko Kolarevic and Ali Malkawi , 9 (New York: Spon Press, 2005)
26. W. Griswold, G. Mangione and T. E. McDonnell, "Objects, Words, and Bodies in Space: Bringing Materiality into Cultural Analysis", Qualitative Sociology 36, no. 4, 343-364 (December 2013). http://link.springer.com/article/10.1007%2Fs11133-013-9264-6
27. B. Lanks, "An Interview With Olafur Eliasson, On Crossing Between Art And Architecture", fast company, (May 31, 2011) http://www.fastcodesign.com/1663951/an-interview-with-olafur-eliasson-on-crossing-between-art-and-architecture#4
28. "Publication Details: Jane Rendell, Art and Architecture: A Place Between", (London, September 2006)
29. http://www.janerendell.co.uk/books/art-architecture-a-place-between
30. B. Roberts, "Beyond Gordon Matta-Clark, Contemporary crossovers between art, architecture, and preservation", Architectural record, August 8, 2014, http://archrecord.construction.com/features/2014/1408-Beyond-Gordon-Matta-Clark.asp
31. G. Brooke and S. Stone, *Re-readings: Interior Architecture and the Design Principles of Remodelling Existing Buildings*, 102,104,105,127 (London: RIBA Enterprises, 2004)
32. P. Locker, *Basics Interior Design 02: Exhibition Design*, 31 (Lausanne: AVA Publishing, 2010)
33. BRÜCKNER ATELIER, Scenography – Making spaces talk, projects 2002–2010, Edited by ATELIER BRÜCKNER, With a preface by Frank den Oudsten, (Avedition, December 2010)
34. A. Picon, 'Architecture as Performative Art', in Y Grobman and E Neuman (eds), *Performalism: Form and Performance in Digital Architecture*, (Routledge, London , 2012)
35. S Lavin, "Performing the Contemporary, or: Towards an Even Newer Architecture", in Y Grobman and E Neuman (eds), *Performalism: Form and Performance in Digital Architecture*, 24-25(Routledge, London,2012)
36. D. Leatherbarrow, "architecture'sperformance unscripted", in *Performative Architecture: Beyond Instrumentality*, ed. Branko Kolarevic and Ali Malkawi, 11 (New York: Spon Press, 2005) it means "*Open or spread out from a folded*

position". no object (of events or information) gradually develop or be revealed. https://en.oxforddictionaries.com/definition/unfold

37. S. Omlin, "Perform the Space: Performance Art (Re)Conquers the Exhibition Space", in *Performing the exhibition*, ONCurating, no.15/12 , ed. Barnaby Drabble, Federica Martini and S. Omlin, 3-4, www.oncurating.org
38. http://nevicata14.tumblr.com/
39. http://milano.corriere.it/notizie/cronaca/15_febbraio_14/alberi-luci-rete-wi-fi-nevicata-trasforma-piazza-castello-milano-a6196b2a-b42411e4-9e87-eea8b5ef37a3.shtml
40. http://www.ananimation.com/nevicata-14/
41. http:// tro.milanotoday.it/piazza-castello-nevicata-14.html
42. a discussion with the architect.
43. http://www.domusweb.it/it/notizie/2014/12/16/_nevicata14.html
44. http://nevicata14.tumblr.com/
45. https://divisare.com/projects/274608-guidarini-salvadeo-architetti-associati-snark-space-making-atelier-castello-nevicata14
46. Lecture by the architect at the Politecnico di Milano
47. Lecture by the architect at the Politecnico di Milano
48. https://www.comune.milano.it/dseserver/nevicata14/index.html
49. https://divisare.com/projects/274608-guidarini-salvadeo-architetti-associati-snark-space-making-atelier-castello-nevicata14
50. A.Branzi, "Exhibition Design as Metaphor of a New Modernity", Lotus international, no.115, (December, 2002)
51. A. Branzi, "Towards a new dramaturgy of the project', in Mettere in scena Mettere in mostra , ed. Basso Peressut L., Bosoni G. and Salvadeo, (Siracusa : Lettera ventidue, 2015)
52. S. Lavin, "Performing the Contemporary, or: Towards an Even Newer Architecture", in Y Grobman and E Neuman (eds), Performalism: Form and Performance in Digital Architecture, 26-27 (Routledge, London, 2012)
53. A. Branzi, Exhibition Design as Metaphor of a New Modernity, 99, Lotus international, no.115, (December, 2002)
54. Ibid
55. V. Signore, "who said performative? Towards a critical posture", in Performative urbanism, generating and designing urban space, ed. Sophie Wolfrum and Nicolai Frhr.v.Brandis,173 (Berlin : Jovis, 2015)

Maintaining the ambiance of regenerated post-military complexes – current and emerging possibilities

Marta Małgorzata Rudnicka-Bogusz[1,*]

[1]Department of History of Architecture, Arts and Technology, Faculty of Architecture, Wrocław University of Science and Technology, Wrocław, Poland

Abstract. Ambiance provides scholarly value to historic architectural precincts and complexes. Through it, the built environment – a corpus of buildings – becomes an environment with its own unique narrative. It grants us the possibility of interacting with the atmosphere of the place as it was in the epoch. However, due to the nature of modern development, living historic cities are changing dynamically. To stay in use, and thus continue their scholastic mission, historic urban areas need to be integrated into local development through either within-use adaptation or conversion to new functions. For historic barrack complexes, this usually means extensive redevelopment. The difficulty comes from accommodating quotidian use while maintaining its historiographic values, especially with historically sensitive sites. This means the designers should take the legibility of the original complex into consideration in their designs and provide suitable aids to visitors. As it turns out there is not a universal approach, instead different approaches allow preservation of various degrees of the original ambiance.

1 Introduction: the ambiance

The need for preservation of historic architecture and urban typologies is self-explanatory. Cultural landscape carries embedded notions of characteristic local tradition, gives proof to local history, grants indigenous community a sense of belonging and makes various landscapes worth visiting for their uniqueness and genius loci. Customarily, it was believed (Charter of Athens 1931, Venice Charter) that the value of historic architecture derived from authenticity of original tissue: materials combined with traditional techniques. This reflected main value heritage held to scholars as a document. Since the birth of mass cultural tourism, people's relationship with heritage started to change gradually[1]. The majority of visitors had no tools to investigate the intricacies; they relied on the general impression for intellectual and spiritual impact. The material heritage became a mere vessel for associated meanings[2]: the tradition, social significance and sense of place, along with various other unquantifiable values. Combined, they constitute the ambiance. It develops in the perception of historic architecture held by people, based on features specific to a given location that make a place unique. In the face of the fact that even immaterial heritage is considered

* Corresponding author: marta.rudnicka@pwr.edu.pl

endowed with documentary and historiographic value (Convention for the Safeguarding of the Intangible Cultural Heritage 2003), it is safe to say that the authenticity of the actual building tissue is secondary to the experience. Scholars used to think that what validated the material aspect of heritage to viewers was patina[3], evoking the feel of antiquity. Still, recent history of heritage regeneration proves that even reconstructed objects, therefore those containing minimal amounts of authentic tissue (e.g. Dresden's *Frauenkirche*) induce veneration and incite historic curiosity. The relationship between historic architecture and its ambiance goes beyond the preservation of original components. Losing the authenticity of components does not necessarily mean loss of the authenticity of the whole. Integrity of the narration presented by the material heritage is the key factor to ambiance. It may well be argued that this was the reason behind anastylosis of ancient temples in Agrigento or reconstruction of the Old Town in Warsaw.

The aim of preserving or enhancing the ambiance within regenerated historic environments comes from the way a modern society functions – based on experiencing, rather than theoretical learning. A more approachable heritage becomes an active factor in fostering a conscious and evolved civil society. Therefore, it is the ambiance that provides scholarly value to historic architectural precincts and complexes. Through it, the built environment – a corpus of buildings – becomes an environment alive with a story to tell. This way ambiance creates personal sentiment, encouraging better understanding and prolonging the impact of an experience: from place to the sense of place[4]. For only an experience, which has the audience fully emotionally invested, can affect it in any lasting way: promote, dissuade etc. Only an emotional response triggered by a site can anchor historical consciousness in a dynamic relationship: site→ experience→ appreciation→ nourishment of the site[5]. Collective memory is bound up with the public realm in the sense that accessibility facilitates recognition. To stay in use, and thus continue its scholastic mission, historic urban areas need to be integrated into local development processes and urban planning, such as contemporary architecture and infrastructure development. This integration can be carried out by either within-use adaptation or conversion to new functions. Therefore, the story that the ambiance of a place tells can be a story of the past or present. It can reflect the original *genius loci*. The architectural makeover can also generate a new ambiance, based on current spatial development. Still, while telling what IS now, the story will inadvertently relate to what WAS there, in consequence in passing relating the origins of the complex.

2 Ambiance in post-military barrack complexes – case studies

All of the above relates as well to the bulk of historic post-military barrack complexes. Initially positioned on the outskirts of municipalities at the exit routes, they were eventually swallowed by swelling cities. When restructuring and modernization of the army rendered them redundant, they were relinquished. Throughout Europe, decommissioned military installations were either razed or adapted, due to dynamic changes brought upon living historic cities by modern spatial development. They were turned into multiple contemporary functions: housing, educational facilities, offices, mix-use, etc. In some cases, their layout and outer appearances were maintained as a front for new use. In others, spatial courtyards and cavernous interiors intended for swift movements of large troop of soldiers encouraged new owners to build upon and partition. Most of the historic post military facilities have not been recognized as heritage or even as real estate asset yet. Nonetheless, they were important to the local communities living in their immediate proximity. Establishing a garrison in a city meant influx of funds, jobs at construction sites and services, development of roads and railroads and last but not least – prestige. Local identity was therefore shaped by the military and benefits that came from mutual relations. This local tradition was embedded in historic military complexes. Loss of these objects would considerably distort local cultural landscape of garrison towns. Fortunately, post military barrack complexes can facilitate various functions. The difficulty comes from permitting quotidian use while sealing off the site against forgetting, especially with historically sensitive sites. As it turns out, there is not a universal methodology, instead different approaches allow preservation of various degrees of the original ambiance.

2.1 Traditional approach

The Collins Barracks in Dublin, Ireland (1704) have been preserved in their original layout and during adaptation for museum, hardly any outer appearance has been changed. Upon entering the complex, one instantaneously feels the mood of austere military life: granite-faced edifices encompass spacious empty muster yards. Both the original stylistics of the complex and its typology were preserved intact and can serve as scholastic asset. There are regular mock-up military parades and weapons reviews organized in the complex to uphold the military tradition. For those with enough imagination and information the ambiance of the place is close at hand.

Fig. 1. Minor infill in Collins Barracks does not thwart the original ambiance.

The Hyde Park Barracks in Sydney, Australia (1819) offer visitors a first-hand experience of barrack life. Originally, the complex was a penitentiary facility where convicts were rehabilitated under military supervision and subsequently a processing hub for female immigrants. Nowadays, it has been turned into a "living museum"[6]. Visitors are invited to don carefully replicated period attire and stroll amongst lime-washed bricks and floorboards, time worn roof timbers, doorways and observe layers of paint and plaster accumulated through the decades. Dormitories have been filled with replicas of period furnishings: hammocks, wash basins, a chiming clock. To immerse themselves in the experience, visitors are invited to use the props, for instance climb into the hammocks. The exhibition is completed by a searchable database of the names and identities of some of the barracks inmates. The experience is highly engaging and requires no prerequisite knowledge about the place. The complex is alive with performances that keep up the ambiance, giving out knowledge about the place as they unravel.

In Gdańsk, Poland, *Wrzeszcz* Garrison (1893) has been adapted to mix-use development, with schools, shops and office space now occupying what was once cavalry barracks, enriched with new-build housing district. Much of the original typological features were lost with the razing of stables and riding halls and the coming of the new-build that took up the space of *maneges* and drill grounds. However, preserved parts of the compound, even though adapted, still give a fairly accurate impression of how the property looked in the time of the military. The edifices were contemporarily adapted and expanded, yet with lavish materials like historic brick and COR-TEN steel. Despite that, muster yards encompassed by these dormitories were left unbuilt and retain their grandiose gauge. A passer-by can still experience original spatial relations between buildings and open spaces. It is hard to discern the original military feel of the place from the hip-hub vibe it possesses nowadays, but a well-informed visitor with enough knowledge will still find it possible to discover traces of the original ambiance.

Fig. 2. Preserved parts of *Wrzeszcz* Garrison strive to maintain their *genius loci* despite expansions.

In the 1970s, the *Bådsmandsstræde* Barracks (1840) in Copenhagen, Denmark, have been spontaneously turned into The Free Town Christiania – a unique social phenomenon of a self-governing city district. Drifters and squatters who first appropriated the compound adapted post-military buildings in the way that most suited them. They also built new houses where they saw fit, without supervision. What resulted was a distortion of the typology and stylistics of a historic military complex. However, a new ambiance has arisen – the free-spiritedness of cooperation with a hint of defiance and anarchy as the "free town" opposes government's efforts to subjugate it. It all happens in relation to the military origin of the complex, but it is only mentioned in passing, as something bygone, and there are hardly any spatial relations upheld to serve as proof of the original purpose of the compound. Therefore, the source ambiance is gone, replaced by a new one, which now becomes binding.

In all of the presented cases, there is information about the original use. However, in the first two spatial relations are visible to the naked eye, while in the latter – they are not. For the historic architecture to fulfil its purpose as the bearer of tradition and a part of cultural landscape its original spatial disposition etc. must remain visible or at least comprehensible with the right sort of narrative. In the first two cases, the ambiance was evoked with traditional means – patina, props, retaining a suggestive name, a non-invasive function allowing for retaining of original spatial relations. However, to allow for a more extensive change, facilitating a wider range of functions, the narrative has to be more explicit. In order to allow the eye of an unskilled tourist to appreciate the historic proof he or she is seeing, they must receive aid. Moreover, it would be dangerous to let any heritage become the sole vestige of memory, and consider it a self-sufficient repository of historical facts, able to trigger objective awareness of specific phenomena, especially the objects connected with the military power of the state, hence inevitably propagandistic. These objects are immersed in social context and thus require a commentary, an explanation, guiding the visitor through historic nuances, spelling out key points of the observed landscape.

2.2 Emerging technologies

Thus, while adapting historic architecture, aside from the physical space, the designers must take into consideration immaterial effects of their designs, such as the experience of the users, and account for them. It is even more challenging if the built environment is historic but serves contemporary use. The designers should program the ambiance to emanate the original *genius loci*, while at the same time leave enough space for the new ambiance – arising from the amalgam of the historic architecture and contemporary functions. The challenge comes from the ability of the designer to grasp the past sense of place and to display it effectively to the user. The solution to this problem might come with the emerging technologies e.g. the Augmented Reality – deeply immersive experience offered on site. However, the possibility of offering in-place simulation creates both new possibilities and new challenges for designers[7], such as petrification or stigmatization.

There already exist a few examples of AR and VR aids allowing for experiencing bygone architecture and urban environments. One of such experiences is an application granting the visitor the view of *Carnuntum*, the capital of the Roman province *Pannonia Superior* in modern-day Austria. The site of *Carnuntum* was never built upon, constituting today the largest preserved archaeological landscape in Central Europe. Using a combination of airborne scanners and ground penetrating radar archaeologists located the gladiator school, *castrum* (military camp) and barracks of the governor's guards. The data was then transferred to architects and computer graphics who created a precise 3D model and simulation of the gladiator school. This 3D model is currently used at the Petronell Visitor Center, providing visitors with a virtual animation of the site, and is fully integrated into the true-to-scale model of *Carnuntum* cohesive with geographic latitude and longitude points. Consecutive objects on the site are successively reconstructed, becoming a more substantial proof of what it was like in its prime. Such Augmented Reality environments, incorporating both the material substance of a given cultural landscape and elements of Virtual Reality are a way to enhance experiencing historic heritage sites' ambiance without limiting their potential for development.

Fig. 3. Some chapters in Šibenik AR guide include interactive avatars guiding through the fort.

Creating Augmented Reality is more demanding than Virtual Reality. In AR, digital characters and objects are linked to real-life space and appear as part of the surrounding world via smart device. This technology was used in the 17th century Croatian *Barone* Fortress of Šibenik. The AR features of the tour are available through glasses that can be rented at the entrance to the site as well as a personal smart devices via an application, allowing for selfies and snapshots e.g. with the avatars. The glasses have a built-in sightseeing tour scenario with five segments. Each of the parts lasts up to five minutes and gets activated approx. two meters from the point around which the chapter revolves. You can partake in a siege. Some chapters include interactive avatars, there is also a live-action movie imagining life in the streets of 17th century Šibenik. The fortress also houses various boutiques, shops and cafes. It facilitates space for cultural activity such as concerts and an open-air movie theatre. On the downside, the original layout of buildings inside the forte's walls is not mirrored by what was built there contemporarily, so the visitor has no way of recognizing the historic spatial relations between the elements of architecture. However, the new-build do not obscure the view inside nor the panoramas from the fortress and they allow for implementation of cultural program when the monument does not serve sightseeing. The modern functions and the monument-quality do not compete or impede each other as each of them happens in a separate slot of time. Contemporary additions were executed in an unobtrusive style that does not hinder the authenticity of the historic landscape. All in all, the ambiance was revived and the edifice was returned to active use within the society[8].

3 Ambiance in milieux de memoire – monumentalisation

An obvious problem with ambiance and urban recycling in barrack complexes begins when they had been used by a totalitarian regime. The ambiance there is one of gravitas and horror, nevertheless maintaining the original ambiance is essential to the scholarly value of the heritage. For only in a *milieu de memoire* can the ambiance of terror hold enough power to serve as a memento and preventive measure. However, it would seem, in cases involving sites of terror, the original ambiance inhibits any further investment other than a memorial and hinders any adaptation of the original tissue aiming to reintroduce active social use. On the other hand, if an adaptation were a priority, would it be even possible to downplay the ambiance to facilitate it. In addition, it does not seem ethical to put people in a position of day-to-day existence in an ambiance like that. At the same time, large homogenous historic complexes (such as post-military compounds) pose a challenge when it comes to integrated preservation efforts. Their sheer scale (tens of hectares of land) makes it virtually impossible to preserve them as monuments or turn into open-air museums. Due to the costs, such undertakings can only involve compounds that are exceptional within the genre. Moreover, preserving more of large-scale complexes as memorial sites might lead to petrification of historic towns.

3.1 Stigmatization

When a *milieu de memoire* is impossible to maintain in its entirety, a *lieu de memoire* may be created instead as a token. This was the case with the SS barracks in Oranienburg, built to accompany one of the first purposefully constructed death camps in Europe – *Sachsenhausen*. The compound consisted of a camp, barracks, single family housing district and various other facilities, laid out around exercise fields. While, after the war, the camp became a memorial, the barracks continued to be used during the GDR by the military. The single-family houses were turned over to private citizens. Some parts of the site were completely discarded, like muster yards that were overgrown with vegetation and a wooden mess-hall – dilapidated and rotten. The whole area was frozen in a state of transition, neither refurbished nor preserved, a constant reminder of the past. High status of the memorial site at the camp presented the local populace and politicians with contradictory incentives. On the one hand, there was the obligation to uphold the memory, on the other, the tourist potential of the site presented the local community with the opportunity to economically benefit and move on from the stigma[9]. The competition that ensued aimed to turn the underused post-military area of land into a new residential district with low-rise houses, parks, schools and sports facilities[10]. The idea of refurbishing post-military property as mix-use development proved most culturally suitable and economically lucrative in many cases[11]. It seemed the easiest way to reconnect the socially excluded area into the surrounding urban fabric. Rather than merging the parts of the compound and working out a joined way of dealing with the problematic inheritance, the 1992 competition sanctioned the existing spatial dissolution. If it had been carried out as planned, it would have ascribed a completely new meaning to different parts of the compound. The state competition only applied to the dishevelled SS barracks and office buildings, which were to be razed. The *Sachsenhausen* memorial, the single-family homes, as well as industrial facilities associated with the camp were excluded. The municipality decided that the memorial site is sufficient to do justice to the history of the concentration camp and that the barracks did not carry the ambiance of the camp. Daniel Libeskind submitted a proposal that stood in a stark opposition to the terms of the competition. Libeskind stipulated, that the *lieu de memoire* was not enough, that it was not possible to mourn the victims without condemning the persecutors[9]. Instead dividing the site into camp memorial and the "mundane" barracks viable for adaptation, he perceived them as opposite poles of the same horrific killing installation and designed an epic monument cultivating the original ambiance throughout the whole complex. At the same time, he managed to incorporate in it enough services and utilities, so that the monument could economically sustain itself and at the same time become the focal point of the local community thus maintaining its social viability. Libeskind divided this site into two parts. The western part would be flooded and the ruins of barracks would be left to rot in the water – symbolizing moral decay of their former inhabitants participating in an atrocity. Visitors would stroll on platforms and piers observing the decay. The eastern part was to be reforested. Between the trees, he placed a collection of newly designed buildings forming a linear pattern, housing workshops, private and public institutions, and common buildings. The architect called these buildings the Hope Incision, and placed it diagonally across the plot, so that it pointed to the *Sachsenhausen* camp. Project MoUrning managed to uphold the ambiance of the *Sachsenhausen* compound, while endowing it with new functions and altering its original appearance. Libeskind's design proved that heritage is less about tangible artefacts or material tissue than about the meanings placed upon them and the representations which are created from them[3] – the ambiance. Nonetheless, even this design could not solve the problem of how to include the single-family homes without petrifying them. Eventually, after much deliberation, the Oranienburg municipality abandoned the plan to raze the barracks and decided to repurpose them as the campus of the Brandenburg Police Academy, which opened in 2006. The buildings were fitted with information plaques, but the property is closed off and only occasional visitors are permitted inside.

3.2 Assimilation

Similarly, the La Marmora barracks in Via Asti (Turin, Italy) were a place of martyrdom of partisans during WWII. The complex was built in 1888, and in 1943 it was acquired by the fascist police and became a prison for political opponents. The walls of the complex still give proof to the executions that took place there. It is estimated that 400 people died on the premise in the moat adhering to barrack B, a fact remembered by a plaque placed there and sporadic commemorative events. However, only a portion of the historic complex (ground floors of barracks B and C) is set to become a museum honouring those grave events *per se*. The focus of the design is opening the previously inaccessible urban space and creating a green *piazza* 100x60 m, a meeting place for members of a civil society and so a monument for the freedoms the fallen dissidents had fought for as well. The historic buildings are going to be preserved in their original outer appearance and there are going to be no infills that might distort spatial relations between barracks in the original array – thus maintaining educational value. The modularity of post military architecture was useful in re-designing the barracks into open-source solutions – co-living and co-working modules. Hence, with very little intervention the complex is going to be turned into a monument for freedom of civic expression with all its scholarly value preserved, albeit not petrified[12]. It is hard to assess before its completion, but from the look of the design, it seems that the original ambiance and the newly created ambiance have the chance to coexist. The museum will serve as the aide for the original ambiance (providing narrative and guidance), but not curb the sites potential for assimilation in the city scape.

Conclusion

Saying that heritage conservation is a sum of economic, cultural and social factors is stating the obvious. Today "heritage" is a very fluid concept rather than a body of historic objects with a single fixed connotation. Still, we continue to concentrate the best part of conservation efforts and discourse on matters concerning physical condition and countermeasures for deterioration. But heritage shapes society not by its physical presence, but with the set of ideas embedded in it. Therefore, heritage is a social construct: its value results from social processes specific to present moment in time and space. It is used as an instrument against social change or just the opposite – facilitating a change. However, at the same time heritage possesses universal scholastic and historiographic values stemming from the possibility of reliving the past, actively experiencing "what it was like" for one's self. This possibility is granted by the place's ambiance. Ambiance does not depend on the amount of the original tissue – there may be hardly any left and still with the right tools, the character and atmosphere can be evoked. It is a positive thing, as it allows for extensive adaptation of historic architecture without forsaking its scholastic values. The original setting – the historic architecture – triggers ambiance, but is not enough to be self-evident, self-explanatory. From plaques, props, staged parades to HMD, VR and AR designers have tried to revive the *genius loci* of post-military barrack complexes while adapting them to new functions, granting their designs various degrees of legibility of the original ambiance. There is not a universal approach. Instead, it is a compromise between various factors: economic, functional etc. It would be most desired if there was a unified policy of regeneration of post-military barrack complexes based on an international know-how, to assure that heritage or not, it will remain a valued part of local cultural landscape.

References

1. Harvey D.C., *Heritage Pasts and Heritage Presents: temporality, meaning and the scope of heritage studies*, International Journal of Heritage Studies, **7:4**, 319-338, (2001)
2. Czerner O., *Wartość autentyzmu w zabytkach*, Ochrona Zabytków, **27/3(106)**, 180-3, (1974)
3. Graham B., Ashworth G.J., Tunbridge J.E., *A Geography of Heritage: Power, Culture and Economy*, (2000)
4. Piga B., Morello E., *Environmental design studies on perception and simulation: an urban design approach*, Ambiances, **1/2015**
5. Nora P., *Realms of Memory: Rethinking the French Past*, **1:1**, (1996-98)
6. http://sydneylivingmuseums.com.au/stories/mint-transformed
7. Morello E., Piga B., *Experiential simulation in architecture and urban space*, Ambiances, **1/2015**

8. http://www.barone.hr/stranice/augmented-reality/12/en.html
9. Sandler D., *Counterpreservation: Architectural Decay in Berlin since 1989*; 141-2, (2016)
10. Libeskind D., *Radix-Matrix: Architecture and Writings*, 102, (1997)
11. Rudnicka-Bogusz M., *Mix-use adaptations in sustainable regeneration of barrack complexes*, VII Conference SOLINA Sustainable Development, E3S Web Conf., **49 (2018)**
12. https://carloratti.com/project/caserma-lamarmora/

Architectural character in conservation design projects

Salim Elwazani[1,*] and *Pooja* Katara[2]

[1]Department of Architecture and Environmental Design, Bowling Green State University, Bowling Green, Ohio, USA
[2]SENSEcity, Glasgow, Scotland

Abstract. Embracing "character" as a form of spatial ambiance, this study has examined architectural character and the character's relationship with the cultural values, building configurations, and building integrity. Considered sequentially, cultural values, building configurations, and building integrity are the three conservation cognitive agents that need to be satisfied before and during the intervention process in the first place, and in a post-intervention sense if to keep the heritage quality of the preserved property intact. Therefore, the study deciphered the character interface with each of the three agents and reflected on these interfaces in completed conservation projects. The study culminated with that the building character plays the central role in moderating the workings of the three agents towards successful conservation projects. The study further pointed out to the challenges that conservation professionals and historic architects face in accommodating character, and by extension, cultural values, building configurations, and building integrity in conservation projects. There is a need for defining and "operationalizing" character—and each of the three agents—to move the conservation process forward.

1 Overview

This study embraces architectural "character" of historic buildings as a form of the spatial ambiance. The study defines architectural character and examines the character's interface with the cultural values, the building configurations, and building integrity within the context of heritage conservation design projects.

Character is an ethereal term in architecture and the built environment discourses. The Oxford Living Dictionary describes character as the "distinctive nature of something." [1]. Other lexical sources associate character with the mood, quality, and atmosphere particularly as related to an environment. Architectural works dealt with character—or its synonyms—in diverse veins and at various scales of the built environment. The 1969 Character of Towns by Roy Worskett, [2] a classic on the subject, advanced some groundbreaking ideas. A clear

* Corresponding author: selwaza@bgsu.edu

undertone ties the character construct with the *Genius Loci*, the Spirit of Place notion espoused by Christian Norberg-Schulz [3] and with the Image of the City scheme laid out by Kevin Lynch [4]. Lending itself to historic milieus, character has an eminent place in the historic preservation/heritage conservation field. The ICOMOS 2008 Québec Declaration on the Preservation of the Spirit of Place defined the spirit of place as "...the physical and the spiritual elements that give meaning, value, emotion and mystery to place" [5] and brought out the vulnerable and transformative dimensions of the term.

Stubbornly rooted in the ethereal realm of the place, character still holds a clear potential for worldly uses. The concept is particularly useful in historic structure conservation and in new creations within historic districts. The influence of character on the historic building conservation threads through the phases of the conservation process. On one side, character links with the cultural values that people attach to heritage and heritage resources. On the other side, character links with the physical integrity of heritage resources. In between, resides the building configurations—the building body, if you will—that function as the physical reference upon which the cultural values are placed and the building integrity is assessed. Considered sequentially, cultural values, building configurations, and building integrity are the three cognitive agents that need to be considered from early on in the conservation process (Figure 1). They are the bases upon which the practical steps of conservation projects should proceed.

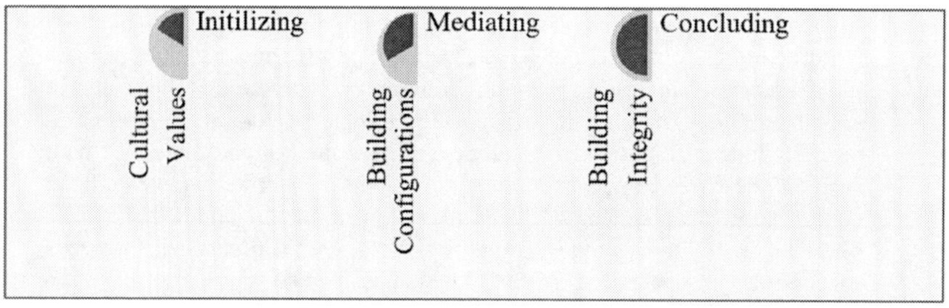

Fig. 1. Cultural values, building configurations, and building integrity interface with building character throughout the conservation process

How different jurisdictions approach character as a moderator of building conservation is a wide-open question. Character conceivably partakes in the conservation game in one form or another across the multiplicity and diversity of international, national, and local conservation programs. However, concrete trends in the interpretation and operationalization of the character in service of conservation are difficult to map. Explaining the strenuous relationship of the architectural character with the triad agents of cultural values, building configurations, and building integrity will help alleviate the difficulty.

This study examines architectural character and the character's relationship with the cultural values, the building configurations, and the building integrity in historic conservation design projects. This central objective gives rise for the discussion to proceed under the following headings:

- Architectural character and cultural values
- Architectural character and building configurations
- Architectural character and building integrity

2 Architectural character and cultural values

The raison d'être for the heritage conservation activity, the cultural values of communities, cities, and nations grounds the requirements for conservation at any level and for any type of resources. The cultural values and conservation actions figure in a cause and effect relationship.

Cultural values spring into what many conservation circles normalized as cultural significance (importance, worth), rendering "values" and "significance" two sides of the same coin. The variations in the methodologies for evaluating cultural significance across the globe are, unsurprisingly, striking. The criteria for significance evaluation generally recognize the economic, social, scientific, political, and aesthetic achievements associated with the resources. Conservation agencies work to cater for heritage resource geographies and stakeholders that come under their jurisdiction resulting in significance evaluation and recognition systems with, assumedly, commensurately responsive evaluation criteria. While countries tend to have nation-wide criteria, some may, in addition, have provincial, State, or city criteria for recognizing heritage resources at those jurisdictional levels. Yet, we can see criteria to recognize resources that are of value to the world community, such as those of UNESCO World Heritage List, or to a continental union, such as those of the European Heritage Label.

By the virtue of their weight and precedence, the cultural values, expressed in a more operational sense as the evaluated significance of a historic building, dictate the attention to the building character—and by the same token, dictate the understanding of the perceived building physical entity features (configurations) and the compliance with the building integrity safeguarding measures during the ensuing conservation process. To this point, the cultural values spell into the evaluated significance of the building; the evaluated significance leads to the safeguarding of the character of the building. To safeguard the character, there is a need to identify it more tangibly first. The sensuous character construct can be understood in terms of the building's physical, spatial, and polychromatic dimensions. The roof configurations, the plan layout, and the exterior paint color are all contributor examples of the building character. Because of the heritage embodiment of time, time comes to play as a domain of the building character and of the possibilities for change in the character. To draw on the examples above, if the roof configurations, the plan layout, or the exterior paint color had changed from what they were during the historic "period of significance," the building character is said to be compromised. Hence the requirements for safeguarding the character of buildings based on the premise that such character, spelled in terms of "protected" building configurations and features, is a credible framework for conveying significance.

3 Architectural character and building configurations

Character is a function of visual perception. Seeing a building is seeing its constituent parts and features which impart the sense of a unique ambiance or, said otherwise, the observed character. While psychology, neurology, and other disciplines delve into the inner workings of the perception phenomenon, architecture and the spatial fields take an abbreviated, yet rather subjective, expression of the perception of buildings and places, that is, the term character.

What are the spatial and physical attributes of a building character? What are the circumstantial variables that might modify the perceived character of a building? As the circumstances under which perception occurs vary, the circumstantial variables of perception, and by extension of character, beg for better definition. As subjective as they may be, the circumstances of a perceptual experience of a building, a plaza, or a sculptural piece modify the perception in congruity of the observer's savviness, frequency of observation, and duration of the activity. Taken further, perception—and the ensuing characterization activity—is influenced by the observer's cultural grounding.

The heritage conservation discourse keeps character in high regard. Hence the need to maintain character as intact as possible during the preservation or rehabilitation intervention processes. However, international heritage programs and organizations do not sufficiently describe the character defining strategies and techniques that lend themselves to technical expediency. Of the few that do, the Technical Preservation Services Division of the U.S. National Park Service defines a set of observable attributes for the visual character of buildings such as the building roof configurations, the façade shape, and finish materials [6]. The Division's further disseminates a guidance for reconfiguring the building character. The NPS visual character scheme elevates the subjective perception of buildings to a situational experience by setting up three vantage points from which the perception of the building takes place. Each vantage point serves as a reference for a set of observations from which the character is presumably visually being constructed. The aggregate of the observations tagged by the three vantage points weaves the collective effect of the building character. The vantage points are described by location as follows: a) at the exterior: from afar to identify the overall visual aspects; b) at the exterior: at close range to observe the building's "materials, craftsmanship and surface finishes;" and c) in the building interior to identify "spaces, features and finishes" (Figure 2).

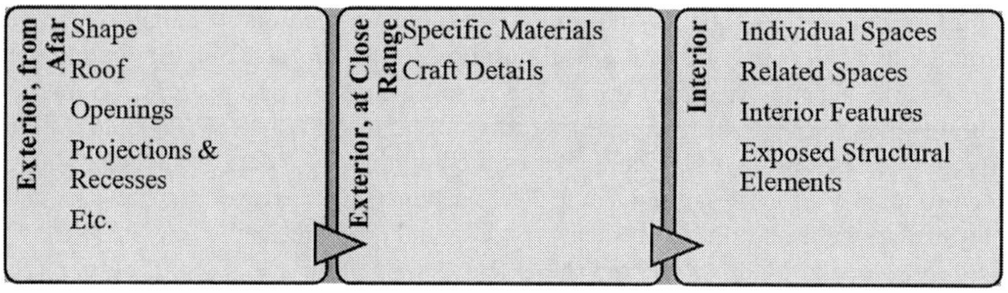

Fig. 2. The three observation vantage points and the building attributes captured

4 Architectural character and building integrity

In addition to the role of cultural significance and building configurations in the conservation process, building integrity peeks to corroborate or disqualify the maintenance of the building character. More elucidated, these relationships stand as follows: articulating cultural significance grounds the rationale for maintaining the architectural character; defining the building configurations provides a base for measuring the building integrity; assessing building integrity assures or refutes the safeguarding of the building character.

Building integrity, to recall, denotes the wholeness and completeness of the building's physical entity over time. The International Council on Monuments and Sites (ICOMOS) advance a definition for integrity: "it is a measure of the overall coherence and the wholeness and intactness of the property and its attributes" [7]. ICOMOS further echoes the building integrity in the closely intertwined concept of authenticity, defined in "an authentic property expresses its cultural values in a truthful and credible way through a variety of attributes such as its form, materials, function, management system, location, spirit, etc." [8]

 https://doi.org/10.1051/shsconf/20196403008

To determine whether the building has integrity or not, one would ask: to what degree the current building configurations (features, materials, and immediate context) are similar to or different from what they were during the building historical period of significance? A negative assessment of the building conditions would indicate that the building character has been disturbed. A disturbed character of the building does not honestly impart the significance of the building.

Where does the building integrity come to play in the conservation process? Evaluating integrity becomes critical at two distinct phases of the process, before and after the start of the intervention into the building—to preserve, rehabilitate, or restore. The aim in each of the "before and after" phases is to corroborate the harmonious relationship of the members of the triad: cultural significance, building configurations, and building character. This, in other words, underscores the need for keeping the relationship consistent from the pre-intervention phase into the treatment phase of the conservation process. The mechanism that shores the relationship is "integrity". By extension, a third, post-treatment phase can be envisioned in which the harmonious relationship between the members of the triad must be kept (Figure 3). Stated otherwise, if the appropriately treated building subsequently witnesses physical and spatial changes that disturb the triad harmonious relationship, the building character—measured through the mechanism of integrity—will suffer. This, in other words, underscores the need for keeping the relationship consistent throughout the three phase continuum in order to claim not only a compliant conservation, but also to claim the maintenance of building integrity as a cultural edifice.

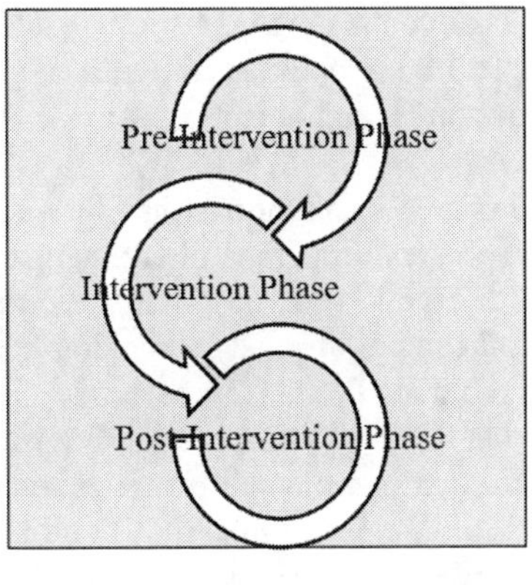

Fig. 3. Phase continuum of intervention

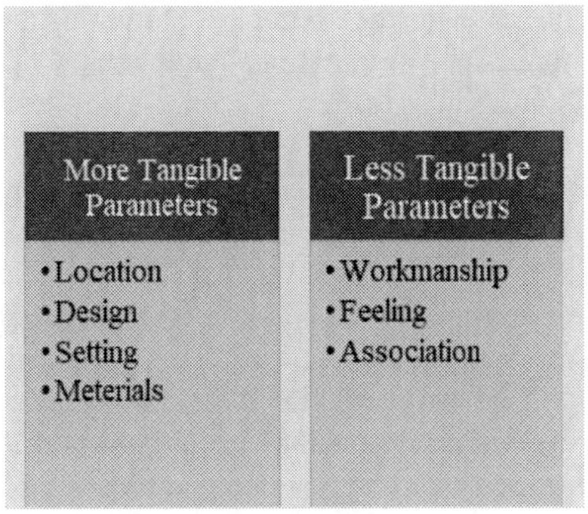

Fig. 4. Integrity parameters

Building integrity as a mechanism for checking on the building character and on the maintenance of the building significance works by triggering a number of parameters. This is clearly spelled out in the National Register Bulletin 15, How to Apply the National Register Criteria for Evaluation [9]. The Bulletin's list of integrity measurement parameters includes more tangible parameters and less tangible parameters (Figure 4).

5 Applied aspects of building character

Conservation principles and practices paid attention to the character attribute of heritage resources universally. This attention crosses the pre-intervention, the intervention (treatment), and the post-intervention stages, and is usually nuanced with one or more of the character-related attributes of cultural values (significance), building configurations, and building integrity. Examples related to buildings and other spatial environments abound. The "president of the Royal Society of Ulster Architects (RSUA) said it is important that the BBC retains the historical character of the listed building when it undergoes its Pounds 77m revamp" [10]. Kelm aimed to "analyze the interiors of four historic mills and determine how historic character and integrity was maintained during the rehabilitation process" [11].

Character is particularly recalled when considering threatened resources. In *Heritage at Risk*, ICOMOS recalled character in at least eleven places. For example, the Kathmandu Valley resources (Nibal), have been identified with their character: "The traditional settlements within the Kathmandu Valley are being lost to development. The earthquake has further decimated the settlements that have managed to retain at least a semblance of the historic character" [12]. Character can be a backdrop for reconciling the vulnerable existing with the unrelenting new. This is apparent in case of the Historical Museum, Sarajevo, a product of the socialist Yugoslavia's architecture and an example of the 20th century European heritage: "the expressionist character of the building's architecture opens up a dialogue with the tendencies of international contemporary architecture, which, in turn, is not formally unilateral" [13].

Yet, character is made more meaningful when its maintenance is set by the jurisdictional authorities as a prerequisite for both the historic structure inventory listing and the certified completion of the intervention into the structure, including rehabilitation and restoration. In the process of nominating a property to the National Register of Historic Places, character serves as a common thread for substantiating cultural significance, elucidating the building configurations, and corroborating the overall integrity for making the case for the "heritage" stature of the property. This postulate holds true in each and all successful nominations—now stand at well over "93,000 properties… [with] 1.8 million contributing resources - buildings, sites, districts, structures, and objects" [14].

The requirement for retaining the property's character is common to the four main types of treatment sanctioned by the National Park Service: preservation, rehabilitation, restoration, and reconstruction. Standard 2 of the Standards for Preservation states: "the historic character of a property will be retained and preserved. The replacement of intact or repairable historic materials or alteration of features, spaces and spatial relationships that characterize a property will be avoided" [15]. Standard 2 of the Standards for Rehabilitation is slightly different to accommodate for the very nature of the rehabilitation treatment: "the historic character of a property will be retained and preserved. The removal of distinctive materials or alteration of features, spaces and spatial relationships that characterize a property will be avoided" [16].

Concluding remarks

The three cognitive agents of cultural values (property significance), building configurations, and building integrity are not only intertwined in their workings but also inherently instrumental in guiding the conservation process. These agents are checked against the requirement for retaining the building character throughout the stage continuum of conservation: before the intervention, during the intervention, and after the intervention. In a way, therefore, character moderates the role of each member in the triad to shoulder the conservation goals.

Except for the building configurations—which for some still remains a conundrum—the remaining agents are apt to continue to pose challenges for preservation professionals and historic architects to capitalize on their potential for shoring conservation efforts. This, for one, stems from the ethereal, subjective nature of the cultural values, building integrity, and of course, building character. This has not slowed down conservation agencies and building professionals from undertaking preservation, rehabilitation, and restoration ventures with varied degrees of success in meeting the proper conservation requirements. The more these agents are defined and "operationalized," the more they will move the conservation practice forward, public and private.

References

1. Oxford English Dictionaries, https://en.oxforddictionaries.com/definition/character (accessed October 19, 2018).
2. R. Worskett, *The Character of Towns*, London: Architectural Press, (1969).
3. C. Norberg-Schulz, *Genius Loci: Towards a Phenomenology of Architecture*, New York: Rizzoli, (1980).
4. Kevin Lynch, *The Image of the City*, Cambridge, MA: The MIT Press, (1960).
5. UNESCO, World Heritage Committee, "The ICOMOS 2008 Québec Declaration on the Preservation of the Spirit of Place," https://whc.unesco.org/uploads/activities/documents/activity-646-2.pdf (accessed October 19, 2018).
6. National Park Service, Technical Preservation Services, "The Walk Through," https://www.nps.gov/tps/education/walkthrough/ (accessed October 19, 2018).
7. ICOMOS, "Glossary," https://www.icomos.org/en/2016-11-10-13-53-13/icomos-and-the-world-heritage-convention-5/glossary3#authenticity (accessed January 17, 2019).
8. Ibid.
9. National Park service, "National Register, Bulletin 15: How to Apply the National Register Criteria for Evaluation," https://www.nps.gov/nr/publications/bulletins/nrb15/ (accessed January 17, 2019).
10. S. "BBC Urged to Keep the Historical Character of Building during Revamp," *Belfast Telegraph,* Apr 04, 2018, https://search.proquest.com/docview/2021239365?accountid=26417 (accessed January 21, 2019).
11. K, Megan J., "North Carolina mill rehabilitation: An analysis of historic building character and design strategies," The University of North Carolina at Greensboro, ProQuest Dissertations Publishing, 2012. 1511826.

12. ICOMOS, *Heritage at Risk: World Report 2014-2015 on Monuments and Sites in Danger*, Ed. Christoph Machat and John Ziesemer, Berlin: hendrik Bäßler verlag, (2017), https://www.icomos.org/images/HR2014_2015_final.pdf (accessed January 18, 2019)
13. Ibid.
14. National Park Service, National Register of Historic Places, "What is the National Register of Historic Places?" https://www.nps.gov/subjects/nationalregister/what-is-the-national-register.htm (accessed January 18, 2019)
15. National Park Service, Technical Preservation Services Division, "Standards for Preservation," http://www.scc.ca.gov/webmaster/ftp/pdf/sccbb/2017/1709/20170928Board03B_Matthew_Turner_Ex4.pdf (accessed January 18, 2019)
16. National Park Service, Technical Preservation Services Division, "Standards for Rehabilitation," (accessed January 18, 2019)

Heritage interpretation through an architectural design

Bartosz Marek Walczak[1,*]

[1] Lodz University of Technology, Institute of Architecture and Urban Planning, Department of Heritage Conservation, Revitalisation and History of Architecture, al. Politechniki 6, 90-924 Lodz, Poland

Abstract. The paper presents results of a heritage conservation studio at the Institute of Architecture and Urban Planning, Lodz University of Technology. Students were challenged to propose an architectural intervention aiming at storytelling of the palimpsestic past, as well as upkeeping the unique ambiance of the place. Furthermore, students were encouraged to play with the use of the building and perception of the site. Submitted projects allowed to make a number of interesting observations.

1 Introduction and background

The paper presents results of a heritage conservation studio at the Institute of Architecture and Urban Planning, Lodz University of Technology. Participating students were challenged to propose an architectural intervention at the site with a rich history.

This place already in Medieval times was used as a location of a water powered flour mill (recorded for the first time in the late 14th century). This was the property of the church leased to mayors of Łódź, which was reflected by the name of this site: Wójtowski Młyn. The whole area was not subject to investment and development until late 18th century, when all church properties – including this one – were secularised. This opened new perspectives. In the early 19th century, by means of governmental decision, the town of Łódź and nearby villages were designated as a major textile centre production. Among the reasons there were ample water resources and well-developed milling infrastructure [1], which could be easily transformed to the needs of new textile factories.

The site was acquired by Christian Friedrich Wendisch – one of the Saxon entrepreneurs invited to Łódź to establish industrial production. As a consequence, in 1824, the existing all-timber structure was adjoined by the very first purposely-built cotton-spinning mill in Łódź. The location of a water wheel lade was adopted. The new industrial building was timber-framed, on the masonry basement walls, according to its description and measured drawings hold in the archives [2] (fig. 1.). The enterprise was thriving over a very short time. The factory master died unexpectedly in 1830 and the building itself remained abandoned for several years. The mill was re-opened and enlarged by the new owner – Friedrich Moes.

* Corresponding author: bartosz.walczak@p.lodz.pl

Production continued until 1863, when the factory was badly damaged by a fire. Finally, the site was overtaken by another successful industrialist – Karl Scheibler – who transformed this area into a farm providing food for his workforce and some organic components necessary for dyeing and fixing processes. The foundation walls of previously existing building ware used to build a house for a manager of the farm (fig. 1.). The building survived until present time. However, its uses changed several times and the structure was adjusted to the subsequent needs such as car repair workshop. Finally, in 2002 it was given to a local scout team as storage for boats and yachts.

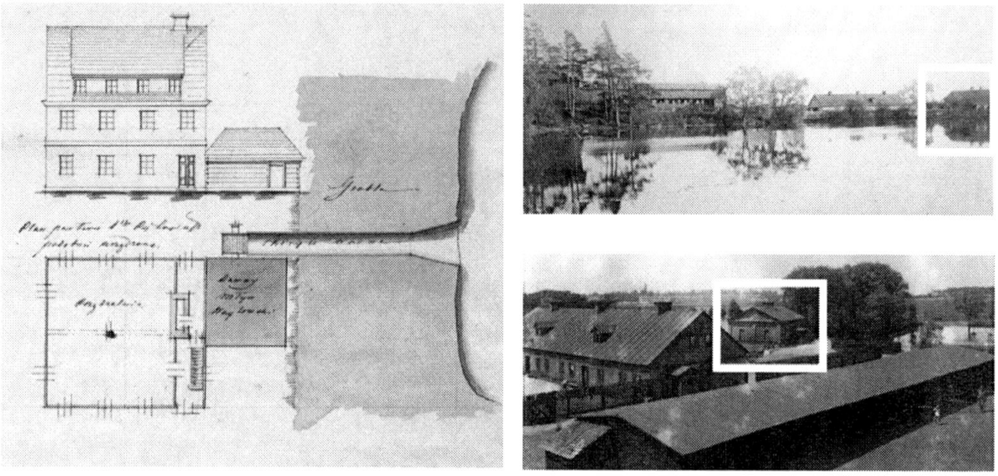

Fig. 1. Archival records of the buildings on the site in question: The Wendisch spinning mill with adjoining old mill (left); images of the Scheibler farm in the 1880s with existing building framed in white (right). Sources: State Archives in Warsaw; Central Museum of Textiles in Lodz.

In 2008, the site drew attention of Zbigniew Rybacki, a local archaeologist. With the support from the state heritage conservation office in Łódź, archaeological research was launched and continued until 2011. The excavations proved to be a valuable source of information on the site history and its palimpsestic nature. First, the exact location of the very first mill was identified. Unfortunately, only a ground beam and couple of boards were excavated. It was, however, enough to recognise a structural type as "post and plank" and indicate the time-frame of its construction as the 16th century. The shape, volume and appearance had to remain unknown (fig. 2.).

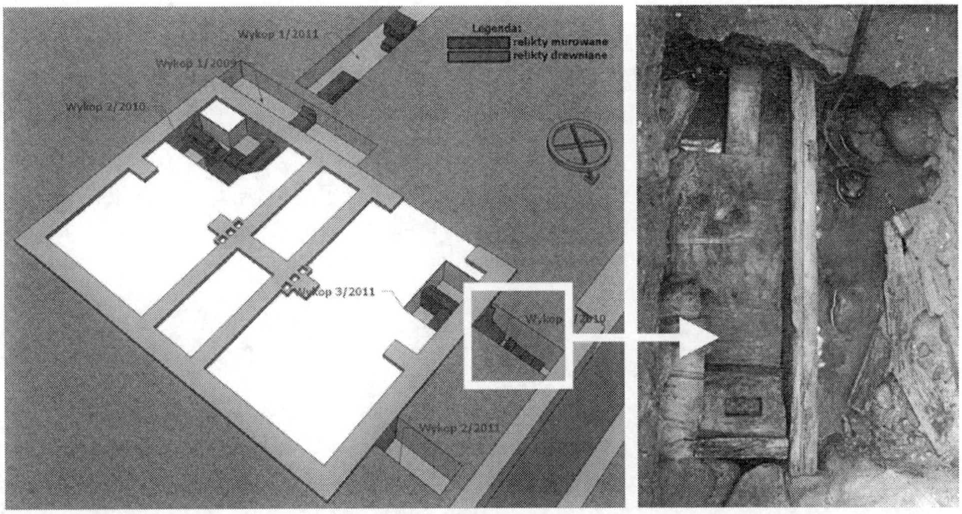

Fig. 2. Archaeological research report. 3D representation of excavations (left) and relics of the old water mill (right). Sources: image and photo provided by Zbigniew Rybacki and Maciej Milczarek.

Furthermore, the location of the Wendisch mill of the 1820s was proved. Additional investigations of the existing building walls allowed to identify relics of the Moes mill. As a result, it was possible to reconstruct the history of the site and locate all the stages of its development. In order to communicate the research outcomes, the team of archaeologists invited experts in virtual reality. Unfortunately, the proposed visualisations left no place for further interpretation, as they were so obvious and complete, even there had been not enough evidence. But it did not prevent the visualisation team from proposing the full reconstruction of a mill and of other associated buildings of which we know nothing (fig. 3.). The result is in strong opposition to the requirement of the London Charter [3], an important document, based on the longstanding (and still ongoing) debate on how to transmit knowledge to a larger audience and to avoid any misleading of members of the public. Sadly, authors of the VR reconstruction were, most likely, not aware of this Charter existence. Fortunately, the virtual reconstruction was shown only during a few presentations of the archaeological research on a forum of local history association. For most of the local society, the site remains just another uninteresting old building, like many similar sites in the city.

Fig. 3. Archaeological research report. Stages of building development (left); virtual reconstruction of the old water mill (right). Sources: images provided by Zbigniew Rybacki and Maciej Milczarek.

2 Project description

Without doubts, there is a discrepancy between the rich historic content of the site and its current condition. Despite being used by scouts, the building is semi-derelict, and entrance to the site is concealed due to open-air storage of the boats. Only a small board on the front façade provides basic information on the site history. But it is not enough to make an effort to visit the site. Picturesque location on a former mill pond, on opposite site of which, there is a public park on the edge of a large housing estate, which seems, however, to create some opportunities (fig. 4.). But apparently, the use and the appearance have to be improved significantly to attract local community and other city dwellers.

Fig. 4. Existing building – former farm manager's house (left); site in the context of allotments and parklands (right). Sources: photo by Filip Tomaszewski; Google Maps.

This situation inspired the author to conduct an experiment with a group of master-degree students. The project was expected to take 7-8 weeks (half a semester). At the beginning, students were given initial information on the history of the site. The package included a map, measured drawings of the building, copies of archival documents and images, and other secondary resources. Then, the site visit was arranged. Students could see the building and meet the scout leader who reported major maintenance problems. This was enhanced with a lecture by Mr. Zbigniew Rybacki – the archaeologist responsible for the mill excavations. On this basis, students were to analyse the site and its context and to formulate their own design guidelines/principles. On this basis, they were challenged to propose an architectural intervention aiming at:

- storytelling of the palimpsestic past, including different strategies matching the variable knowledge (and even lack of knowledge) on particular stages of the site development;
- upkeeping the unique ambiance – feeling of being in a remote agricultural location, while the site just outside the city centre;
- adoption of matching building materials;
- proposing a new use attracting members of contemporary society.

In other words, the project was expected to investigate opportunities of providing real architecture instead of virtual reconstruction to make the site more meaningful [4]. This means that storytelling has to be embedded – to some extent – in the proposed architectural interventions. Furthermore, students were encouraged to play with the use of the building and perception of the site. Thus, locality was expected to become another important strategy in the designing process.

The new use was to be decided by students. They were, however, informed that the site would be considered as one of the pilot projects related to the EXPO Horticultural, which would be organised in Łódź in 2024 [5].

3 Project outcomes

It is necessary to stress that the project was preceded by another task, aiming at the commemoration of a protestant church destroyed during World War II. The students gained, therefore, some experience in heritage interpretation. They learned how to tackle ruins and erect new structures above/aside. The studio was accompanied by a series of lectures on heritage conservation theory and good practices in this domain. The students formed two-person teams. More than 30 designs were submitted as a result of this project. This was an ample material to make some interesting observations in respect of the heritage interpretation by means of architectural solutions.

A large group of students focused on the oldest structure – viz. the water-powered flour mill. Łódź, as a city has Medieval origins, but its actual development took place throughout the 19th century. Therefore, the existence of a structure older than 200 years was apparently inspirational. Since there was not enough information on the mill appearance, many students decided to manifest its existence and primary use in the most obvious way – by using a waterwheel as an icon. Some of them tried to play also with a platform (usually made of wooden boards) depicting the building location and outline. Surprisingly, the most of such proposals disregarded presence of running water and – as a result – these wheels remained merely static sculptures – conceived as either free-standing objects or incorporated in a new structure (fig. 5.).

In this respect, it is necessary to mention one proposal, which included a "reconstruction" of the mill. Unexpectedly, authors of this design decided to follow the VR presentation ordered by archaeologists, even all its drawbacks had been underlined during the introductory lecture. For some reason, the students did not use an outline of the mill seen on the archival document (fig. 1.). Instead, they repeated literally the shape and volume shown in the aforementioned presentation (fig. 6.). Without doubt, this case proves the delusive nature of virtual models and their potential malfunction in the heritage interpretation. It seems that such representations of hypothetical appearance of no longer existing structures have to be less obvious and stimulating imagination of visitors – like, for example, Venturi's famous Benjamin Franklin house in Philadelphia.

Fig. 5. The waterwheel as a prime source of inspiration for new architectural interventions: a café by Malgorzata Szczesna and Patrycja Wawrzyniak (left) and a sculpture by Marta Canert and Maja Ciszewska (right). Sources: the Department of Heritage Conservation, Revitalisation and History of Architecture, TUL.

The latter idea was successfully deployed by another team of students, who proposed simple, white-painted steel frames. Their geometry and distribution suggest the existence of the mill and, what is more interesting, location of other buildings, including the Wendisch factory, so important for the early industrialisation of Łódź. The concept was enhanced by a kinetic installation (a rotating waterwheel linked with cogwheels by a transmission belt) referring to the both stages of site development: milling and spinning (fig. 7.).

Fig. 6. A small pavilion by Adrian Smigielski and Jakub Wlostek (right) shaped on the basis of the virtual model commissioned by archaeologists (left). Sources: images provided by Zbigniew Rybacki and Maciej Milczarek; the Department of Heritage Conservation, Revitalisation and History of Architecture, TUL.

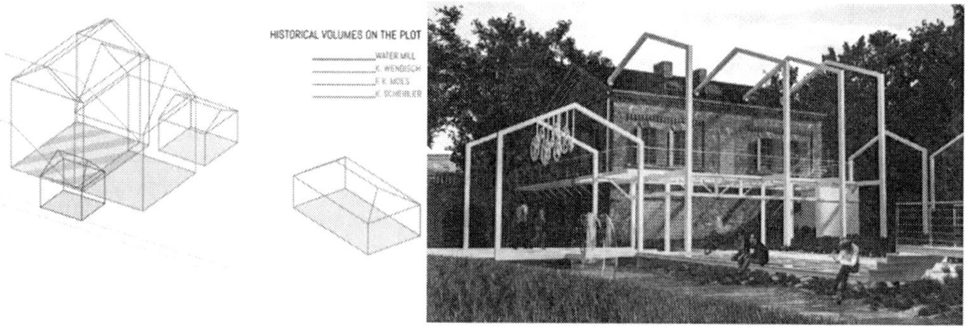

Fig. 7. Three-dimensional representations of palimpsestic past: volumes analysis by Joanna Sobczak and Paula Popczak (left); frames and platforms reflecting no longer existing structures by Patryk Jankowski and Julia Mejer (right). Sources: the Department of Heritage Conservation, Revitalisation and History of Architecture, TUL.

This was not the only design, which explored palimpsestic history through superimposed volumes of buildings existing now and in the past. Some students studied this topic during the analytical stage (fig. 7.), while their creation for some reasons did not reflected the outcomes. Only a few teams decided to take the risk and experiment not only with volumes and shapes but also with textures and materials to make their proposal more meaningful. One of the most comprehensive proposals included a minimalistic platform on the site of the Medieval mill, and a timber-framed pavilion on the site of the Wendisch cotton-spinning factory. The pavilion had glazed curtain walls instead of masonry fillings, while its roof was proposed as a textile membrane. These decisions aimed at the creation of a structure which would be recognised as a contemporary even by laymen. Furthermore, some small additions (like an exhibition of actual archaeological relics) were designed as simple as possible, without any reference to the past. This was, again, underpinned by careful usage of cladding material. Steel mesh was chosen for the purpose. The students also eagerly responded to the idea of EXPO Horticultural. They proposed to transform surrounding allotments into vineyards and the mill site into a winery. Their idea was to draw attention of visitors to climate change, allowing successful winemaking in such unexpected location (fig. 8).

Fig. 8. The comprehensive approach to the issue of the past: a winery by Piotr Mrowczynski and Anna Rawa. Sources: the Department of Heritage Conservation, Revitalisation and History of Architecture, TUL.

Apart from storytelling, students attending the studio were encouraged to focus on the particular atmosphere of the site. In this respect, a self-restriction proved to be the most fruitful design strategy. There were several proposals using very modest means to represent history of the site in order to upkeep peaceful ambiance of the area. Simple decking, pathways, and greenery matched a semi-agricultural appearance of the neighbourhood. Also proposed new uses reflected this unique location: the building could serve as a community centre with a charming outdoor dancing area, workshop area, or significantly improved scouts meeting place (fig. 9.).

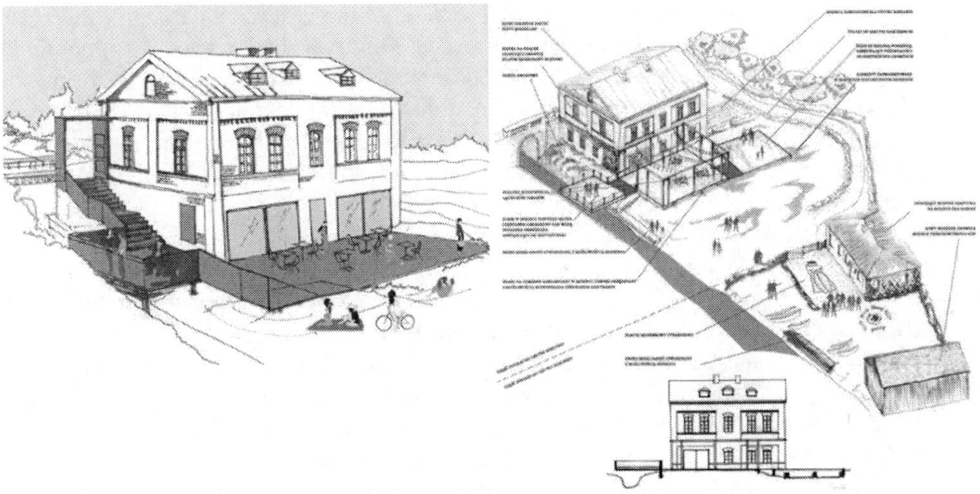

Fig. 9. Simplicity and self-restriction as a design strategy: by Patrycja Stelmach and Marta Zielinska (left); by Anna Helena Sluzewska and Justyna Zareba (right). Sources: the Department of Heritage Conservation, Revitalisation and History of Architecture, TUL.

It is impossible not to mention intangible factors affecting perception of the site when the ambiance issues are being discussed. The students usually properly recognised the value of peaceful location in the middle of the large city. Some of them – as was mentioned above –

tried to upkeep this ambiance. Very few, however, made a step further and tried to represent the site in a sensitive way. Proposed visualisations were typically focused on shapes and volumes of buildings. Lightning conditions, colours of the sky, greenery varied significantly and were far from the reality. Even when the effort to create particular nostalgic atmosphere was made, the result was rather disappointing (fig. 10.). Ironically, those teams, who used analogue methods of representation achieved better results in this respect. Previous experiments and research conducted by author proved however that students would be capable of proper representing the "genius loci" of the place when instructed properly [6]. Yet, the representation issues were beyond the scope of the heritage conservation studio.

Fig. 10. Atmospheric vision created by Aleksandra Durak and Katarzyna Gulink – nostalgic but far from the reality. Sources: the Department of Heritage Conservation, Revitalisation and History of Architecture, TUL.

Conclusions

Tutors and students alike were attracted by the topic of the described heritage conservation studio. Academics underlined that the Wójtowski Młyn site had a good size, possible to analyse in a limited time frame and offered a number of heritage issues to be tackled. Students' feedback revealed that this was something new and therefore interesting. Some of them indicated also that from one hand they enjoyed experimenting but from another they had not enough confidence/knowledge/experience to understand the limits of acceptable intervention. Even so, most of the submitted proposals were of high quality and the studio goals were achieved to a large extent :

- Architectural solutions proved to be an effective tool for the storytelling of the palimpsestic past, but only the minority of participating students exploited the full potential of shapes, volumes and materials to make place more meaningful; it is also significant that, apart from the waterwheel, technology was a disregarded part of heritage, which is however problematic worldwide as recent research revealed [7].
- The task of upkeeping the unique ambiance was approached in a less successful manner – feeling of being in a remote agricultural location, while the site just outside the city centre was properly identified, but most of students did not include it in their design strategies; only a few proposal revealed a proper sensitivity to the ambiance of the neighbourhood both in terms of design and representation.
- The adoption of matching building materials appeared challenging for many students; probably they are not enough knowledgeable in this matter, which is surprising since

usually they used to be familiar with works of Peter Zumthor as well as his writings [8]; nevertheless, a particular attention should be paid in the future to draw their attention to opportunities offered by various materials in terms of texture, feeling, etc.

Despite some reservations and minor mistakes, the outcome of the heritage conservation studio may become valuable material for a discussion on the future of the site. Some ideas will be presented to the city officials responsible for the EXPO Horticultural along with some recommendation on possible scenarios of re-using the site in a manner matching its importance for the local heritage.

Acknowledgements

Author would like to thank all the individuals involved in the Wójtowski Młyn project: Dr. Magdalena Bednarkiewicz, Dr. Maria Dankowska, Prof. Jan Salm, and Mr. Filip Tomaszewski who were tutoring students; Mr. Zbigniew Rybacki who provided the results of archaeological investigations; Mrs. Joanna Filipiak-Markiewicz who allowed access to the site.

References

1. A. Bartnik, D. Bartnik, Rozmieszczenie młynów wodnych na obszarze Łodzi w świetle wykorzystanych historycznych źródeł kartograficznych, Z Dziejów Kartografii, **XXI**, 11-40 (2017)
2. The Central Archives of Historical Records in Warsaw, AGAD PL 1 402 30-4 003
3. http://www.londoncharter.org/ access: 19.10.2018
4. UCL, *Making spaces meaningful through architectural storytelling* (2017)
5. https://expo2024.uml.lodz.pl/en/ access: 04.02.2019
6. A. Kepczynska-Walczak, B.M. Walczak, Built heritage perception through representation of its atmosphere, Ambiances Review (2015)
7. J. Douet (ed.), *Industrial Heritage Re-tooled: The TICCIH Guide to Industrial Heritage Conservation* (2013)
8. P. Zumthor, *Thinking architecture* (1998)

Design of new daylight simulators on architectural scale models

Thierry Blandet[2], *Levin* Erbilgin[1,*], *Nicolas* Vergnes[2], *Nicolas* Hoerter[2], and *Olivier* Lehmann[1]

[1]National Strasbourg School of Architecture, Research Department, 67068 Strasbourg Cedex, France
[2]Telecom Physique Strasbourg, ICube laboratory (IGG team), Strasbourg University, France

Abstract. To enable students to evaluate the impact of light on the scale models of their architectural projects, two physical simulation systems have been built: a mirror box and a sun and sky simulator. Both devices integrate LED-based light sources. We have integrated several models for the computation of the sky image: luminance models and colour appearance models. The integration of the recent and evolving LED technologies explains how our simulators are innovative.

1 Introduction

At the School of Architecture of Strasbourg (ENSAS), teachers developed a lighting simulation activity with the objective of placing the students under experimental conditions and considering the issues of light ambiences [1,2]. This is a way to directly enter into the architectural space and its conception by integrating physical and sensory phenomena. The teaching is based on the use of scale models under controlled light conditions and computer simulations.

Until now, experiments were conducted at the Belgian Building Research Institute (BBRI) in Louvain-la-Neuve, in collaboration with the Climat & Architecture laboratory of the Louvain-la-Neuve School of Architecture [3]. This pedagogical agreement allowed teachers to define their educational needs and to gain experience with the use of these simulators. The teaching feedback and the relevant exchanges between BBRI and ENSAS technical teams led to the definition of the design and the technology involved in our simulators.

The Strasbourg School of Architecture decided to take advantage of a recent building extension to develop its own laboratory made by two high-tech devices designed for experimentation, research and teaching purposes: a mirror box and an artificial sun and sky. Both are installed on the ground floor of one of the two buildings that make up ENSAS. This educational and academic initiative required the development of modular systems to ensure the optimization and maintenance of the simulators. We will detail the design and the main

*Thierry Blandet: thierry.blandet@unistra.fr; Levin Erbilgin: levin.erbilgin@strasbourg.archi.fr

characteristics of the sun & sky simulator by presenting the several prototypes required for its conception.

2 Theoretical approach

2.1 State of the art of daylight simulators

Daylight simulators applied to architecture models have been developed since the middle of the twentieth century. Each laboratory develops its own simulators according to the current technologies and depending on the requirements expected on the results [4,5,6].

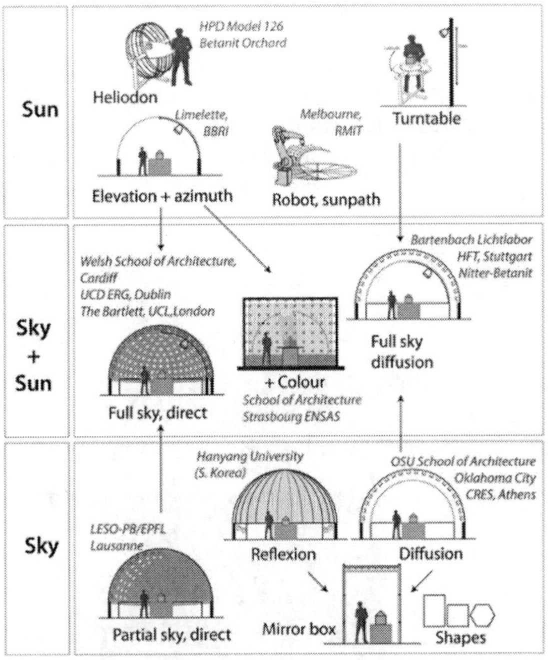

Fig. 1. Daylight simulators, a state of the art

- The figure on the left presents all the daylight simulators by organizing them in three groups depending on the lighting emission they reproduce (only the sunlight, only the light emitted by the sky or the light emitted by both the sun and the sky).
- An artificial sun reproduces the sun path defined by a sequence of azimuth and elevation values. This path is relative to the model, so the artificial sun can move and illuminate the model (heliodon), the model can rotate under a static light source, or the sun handles the elevation while the model consider the azimuth. The mechanical movement is mostly linear even if a 6-axis robot can also reproduce the sun path [7]. The sun can be simulated using a spotlight or a parabola to produce a parallel beam.

The colour of emission of an artificial sun is usually white, and the intensity is possibly dimmable depending on the period of the day simulated.

Fig. 2. Sky domes by direct lighting (Cardiff, Welsh School of Architecture ©), by reflection (Hanyang University, South Korea ©) and diffusion (HFT, Stuttgart ©).

The sky vault can be partially [8] or fully reproduced. There are three ways to reproduce sky light: by direct lighting, by reflection, or by diffusion. By reflection, spotlights disposed below the model illuminate a white dome, the reflections on the dome illuminate the model.

Considering that the real sky emits a diffuse light, the most realistic principle is a sky working by diffusion. Light sources are disposed behind a diffusing material. Usually artificial skies have spherical shapes [9]. The most realistic systems combine an artificial sky and a mechanical sun.

3 Experimentation, study, research and prototypes

Considering the state of the art, we decided to develop a simulator that combines an artificial sky working by diffusion and an artificial sun. Regarding the current lighting technologies, we decided to work only with LED sources which allow the control of the luminous flux and the colour with an excellent energetic performance. For the artificial sky, we chose RGB sources individually controllable and high-power RGBW sources for the sun.

To obtain a Lambertian diffusion and reach a realistic sky, it is necessary to find the optimal diffusing material and the right distance between the light sources and this material. Because LED technology offers a possible control of each source, it is no longer necessary to build the sky with a dome shape, therefore we chose a cubic shape. After calibration, each source will have the same contribution to the centre of the experimental platform.

3.1 Diffusing panels scattering characterization

3.1.1 Experimental setup

To characterize the diffusion of a material, it is necessary to determine its transmission factor depending on the angle of observation. To achieve these measurements, a directional source is more appropriate than an extended source. The source used is a He-Ne continuous laser.

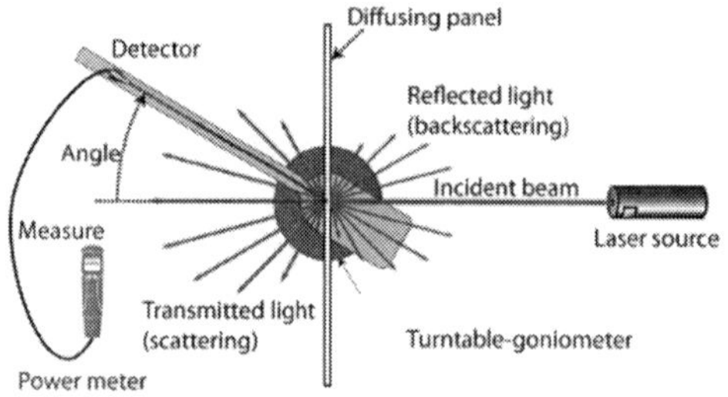

Fig. 3. Experimental set-up for scattering characterization

The laser beam is perpendicular to the sample, and a power meter is on a computer controlled mobile arm. Scattering is a balance between light transmission at 0° and transmission for higher angle values. To limit the light losses, the selected material should not be too absorbent. Theoretically a perfect diffuser commonly qualified as Lambertian is a material through which its illuminance is angularly uniform. The intensity value I depending on the angle of observation θ for a Lambertian diffuser is defined by the following equation:

$$I = I_0 \cos\theta \tag{1}$$

The objective of these measures is to identify which sample (fire resistant) is the closest to the luminous intensity defined by equation (1).

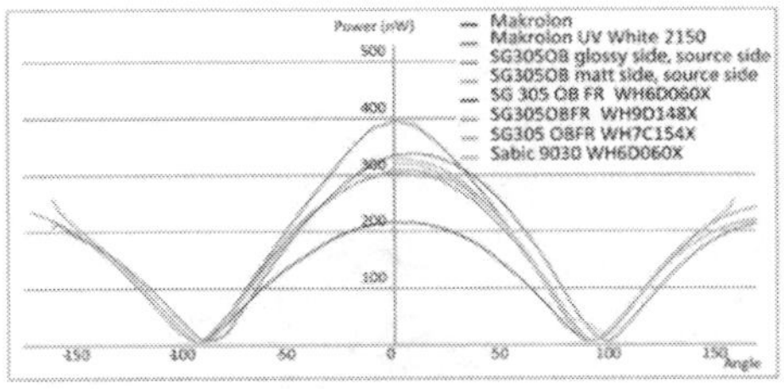

Fig. 4. Samples scattering characterization

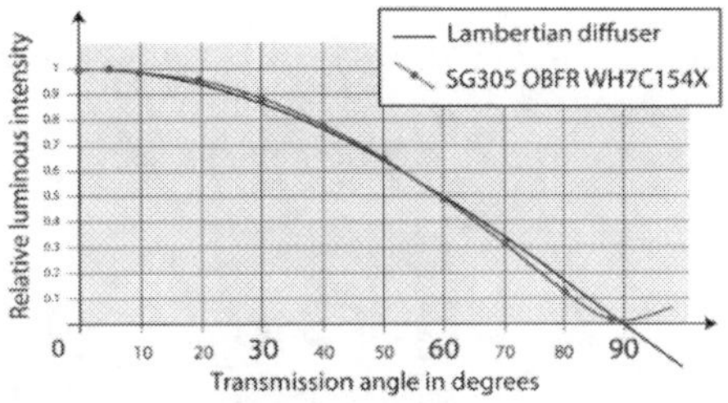

Fig. 5. Lambertian diffuser vs SG305OB

Finally, the results led to select the SG3050B FR Lexan polycarbonate diffusing panel with colour reference WH7C154X.

3.1.2 Diffusing panel junction design

The standard dimension of the selected product is 3.05 m x 2.05 m. To cover a large surface several diffusing panels are arranged next to each other.

To avoid "light leaks" the following junction has been designed. The two millimetres gap between each panel allows to take into account the polycarbonate thermal dilatation and the tolerance required by the structure.

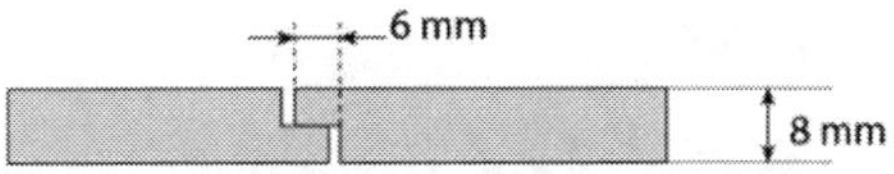

Fig. 6. Junction between two diffusing panels

3.2 Sky prototype: presentation & details

3.2.1 LEDs distribution and distance between sources and diffusing panels

The purpose is now to define the LEDs disposition on the wall of the artificial sky, the optimal distance between the sources and the diffusing panel.

Fig. 7. Artificial sky prototype in a staggered arrangement and visual impact validation

One-hundred-pixel balls lights are disposed in a staggered arrangement. The 10 cm distance between the sources and the diffuser permits to obtain a homogenous scattering. The sources used in this prototype are screwed on the back and their dome is not necessary. Therefore, LED sources have been specially manufactured to meet our specifications (12 RGB LED sources on a 67 mm flat square on an aluminum plate drivable through the WS2811 protocol).

3.2.2 Pixel mapping

In the simulator, each LED source composing the sky is individually controlled via a pixel mapping process. To fit with our specific needs, we decided to develop our own LED driving system using Arduino devices.

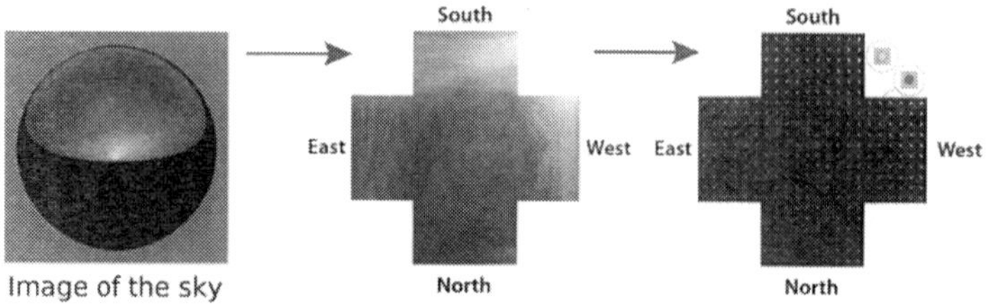

Fig. 8. Pixel mapping principle

We developed software able to compute the images of the sky, then, knowing the position and the address of each source, it is possible to consider our artificial sky as the display system for these images. To validate our approach, the final structure of the artificial sky was reproduced on a small-scale prototype (580 x 580 x 460 mm). We disposed in a staggered arrangement one hundred RGB light sources inside the model.

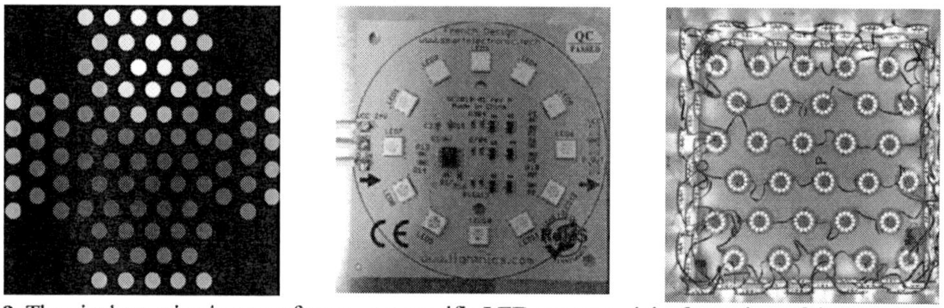

Fig. 9. The pixel mapping in our software, our specific LED source and the front view of the prototype

In addition to the pixel mapping, this prototype served as a support to control a fish-eye camera that will be needed for the final sky calibration and the visual feedback of the inside of the studied architectural model.

Because of the cubic shape, the calibration of the final system consists of assigning to each LED the right intensity so that each one has the same contribution to the centre of the experimental platform.

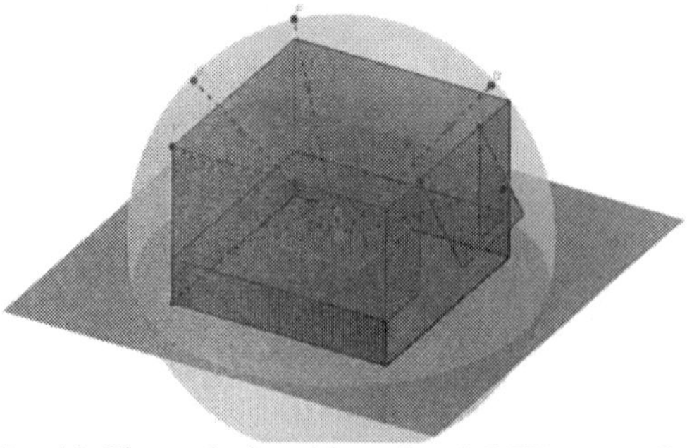

Fig. 10. Theoretical position of each LED source for calibration process

4 Our sun and sky simulator

4.1 General presentation

In our system, the model is placed on a circular technical platform. Our artificial sky has a cubic shape with the following inside dimensions: 4780 mm x 4780 mm x 3780 mm (width x length x height). The user defines the geographical position of the desired solar path as well as the day. During a typical three minutes simulation, the sun goes from sunrise to sunset, follows its trajectory in the sky [10], the colour and intensity of the artificial sun and sky will evolve in the simulated day as in reality. The sun and sky simulator has to be considered mainly as a qualitative device.

The experiments can be observed directly by the users from the outside of the simulator through a large opening at north. A bright awning compensates the loss of light due to this opening. The simulation is recorded by cameras positioned inside and around the architectural model.

Fig. 11. Mechanical sun within the artificial sky and a computed simulation of a model within the cubic shape of our sun-sky simulator

8,000 RGB LED sources are distributed on the walls and ceiling of the structure behind polycarbonate 8 mm thick diffusing panels. Our software calculates in real-time the image of the sky, through the pixel mapping process we can simulate several sky models.

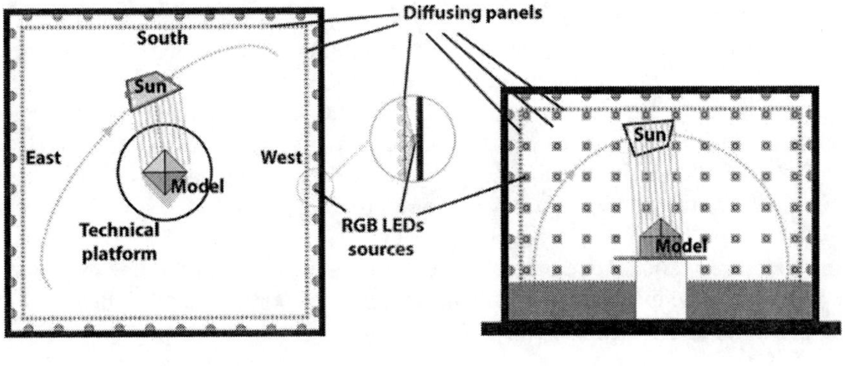

Fig. 12. General principle of the sun & sky simulator

The entire sun path is reproduced by the mechanical sun moving around the scale model inside the artificial sky. The mechanical sun follows its trajectory on a semi-sphere of a two-meter radius. RGBW LED sources which simulate the sun are disposed on a one-meter square plate. The artificial sun allows to study the contribution of light inside the model and its shadows. To produce parallel rays, RGBW LED sources are associated with their collimation optics.

Sky models

The sky models implemented in our software are of two different types: luminance models defined by Perez [11] and the CIE [12] and appearance models defined by Nishita, Haber or Preetham.

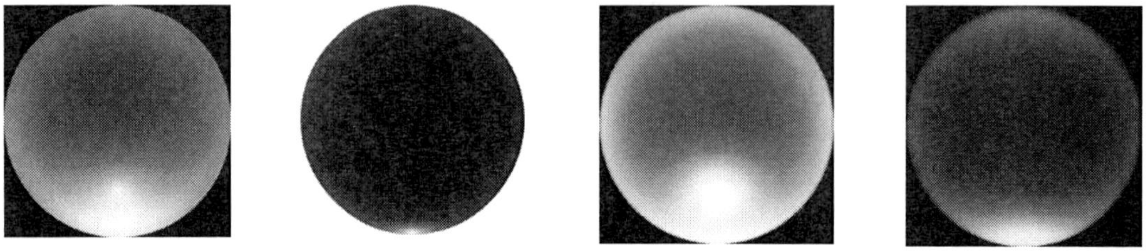

Fig. 13. Overcast sky defined by the CIE at midday, the clear sky sunset defined by Perez, sky image at midday by Haber and sunset image by Nishita.

4.1.1 Luminance models

Luminance models can be used to determine the relative luminance of any point in the sky or sky element defined by its height and azimuth. To calculate the luminance of a sky element, three main parameters are considered:

- The sun position which is defined precisely via the Sun Path Algorithm program [13]
- The sky conditions (clear sky, overcast sky…)
- The position of the sky element in the celestial vault
-

Perez proposes to calculate sky images for 8 ranges of sky clarity from clear to overcast sky. The CIE has standardised equations which define 16 different sky types, including the "standard overcast sky" and the "clear sky with polluted atmosphere".

4.1.2 Appearance models

Appearance models are in colour and based on the results of the Perez luminance models. Appearance models appear to be the most interesting to use because the colour increases the realism of the simulation. In order to compare the different sky models an implementation of the main appearance models were made by Bruneton. Its code being freely accessible, modifications were applied to produce the data required for our needs. The models implemented are those of Bruneton, Elek, Haber, Hosek, two models of Nishita (one defined in 1993, and the other in 1996), of O'Neal and Preetham [14-19].

Some are less realistic than others because they have not been defined the same way. For example, some reproduce sunsets better than others. The best models will however be defined during the calibration of the system.

4.2 User interface details

Two computers are used to control and synchronize all the components of the simulator: one sends all the information needed to achieve the experiment, the other one collects the experiment data and gives a visual feedback in real time.

The software handles the computation of the images of the sky, the communication with the Arduino to achieve the pixel mapping, the communication with the motors of the mechanical sun, the DMX communication to control the light sources of the artificial sun, the different video and photo acquisition cameras.

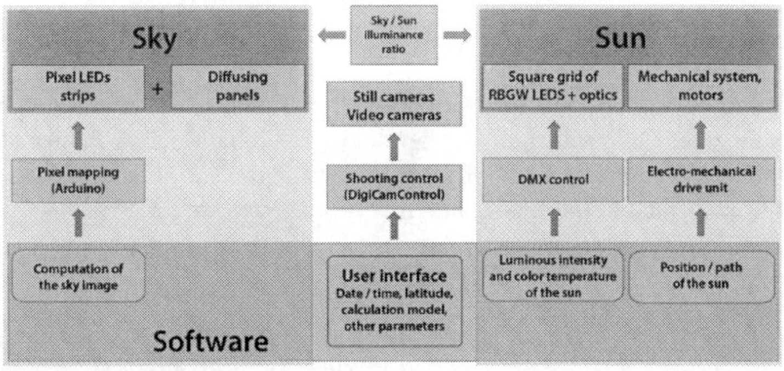

Fig. 14. Software functional chart

The software allows to define the parameters of the experiment, the geographical location, the day, the experiment duration, the sky model, atmosphere conditions (turbidity, albedo), etc.

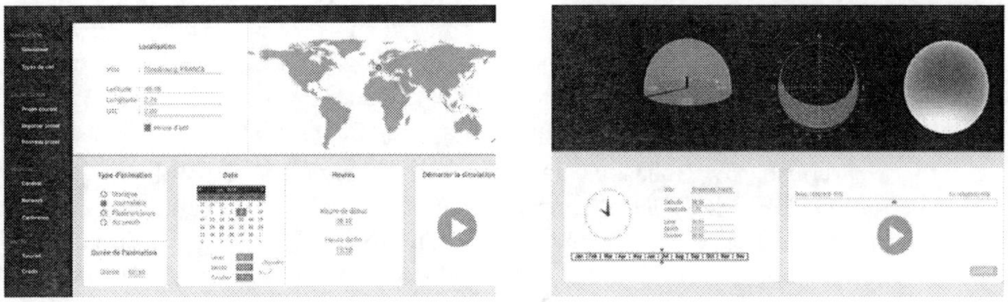

Fig. 15. User interface for simulation settings, solar diagram and other feedback

Once the simulation is started, the interface shows the evolution of the sun on a solar diagram, the current image of the sky, etc.

5 Conclusion

As each laboratory develops its own simulators, a state of the art of daylight simulators led us to define the main features of our systems, considering the use of current and emerging technologies. Several prototypes allowed us to study and validate our technological choices.

In our sun and sky simulator, a mechanical sun is integrated inside the artificial sky, moving around the scale model. LED lighting technology allowed us to produce a non-homogeneous colour system with calculated images displayed on the walls and the ceiling of a cubic shape sky. As each LED source composing the sky is individually controlled via a pixel mapping process, it is possible to display in real-time simulations using luminance or appearance sky models. High-power RGBW LED sources associated with their collimation optics on one-meter square aluminium reproduce the sun carried by the mechanics. A sun & sky simulator is considered as a qualitative device. As we master the pixel mapping software, the calibration processs, sky models, and the visual feedback, our simulator can evolve into a quantitative tool.

In a next step, in-situ photometric measurements, daylight computed simulations and measurements under our simulators will be compared to identify the interest and the limits of each method to produce the most relevant results.

References

1. Dominique Laburte, Jacques Rizzotti. Camera obscura, sensing spaces - Pour un enseignement expérimental des ambiances. Thibaud, Jean-Paul and Siret, Daniel. *Ambiances in action / Ambiances en acte(s) - International Congress on Ambiances*, Montreal, Canada. International Ambiances Network, pp.589-594, (2012).
2. D. Laburte, J. Rizzotti, S. Varano, JP. Wetzel De Lumine 7 ENSAS Evaluation des performances de la lumière du jour dans les musées. Rapport de synthèse (2013-2014)
3. M. Bodart, A. Deneyer. A guide for the building of daylight scale models. In PLEA2006 - The 23rd Conference on Passive and Low Energy Architecture, Geneva, Switzerland, 6-8 September (2006)
4. D. Croghan, Daylighting by Design: 2. Some early examples of the use of an artificial sky, (2015) (unpublished paper, retrieved on ResearchGate)
5. Nawab, M. (1981). DEVELOPMENT AND USE OF A HEMISPHERICAL SKY SIMULATOR. Lawrence Berkeley National Laboratory. Report #: LBL-13400.
6. Okado, Masaaki & Nakamura, Hiroshi & Koga, Yasuko & Goto, Koichi & Fujii, Shun'ichi. (1997). Development of an Artificial Sky: part 1: Various Artificial Skies in the World. Journal of the Illuminating Engineering Institute of Japan.
7. Mehrnoush, L.,Prohasky, D.,Burry, J.,Akbarzadeh, A.,Williams, N. (2014). ROBOTHERMODON: An artificial sun study lab with a robot arm and advanced model platform -A thermal heliodon (STEVE: Solar Thermal Evaluation Experiment) In: Proceedings of the 34th Annual Conference of the Association for Computer Aided Design in Architecture (ACADIA) 2014, Los Angeles, United States, 23-25 October (2014)
8. Performance of a new sky scanning simulator / Michel L., Roecker C., Scartezzini, J.-L. – In Lighting Research and Technology, vol. 27, num. 4 (1995), p. 197-208
9. Designing and building a geodesic dome as a bearing structure for an 'artificial sky' lighting installation Maria K. VRONTISSI Proceedings of the International Association for Shell and Spatial Structures (IASS) Symposium 2009, Valencia
10. Richard Kittler, Stanislav Darula, Determination of time and sun position system, Solar Energy, Volume 93, 2013, Pages 72-79, ISSN 0038-092X
11. Perez, R., Seals, R., & Michalsky, J. (1993). All-weather model for sky luminance distribution—preliminary configuration and validation. Solar energy, 50(3), 235-245.

12. Darula, S., & Kittler, R. (2002). CIE general sky standard defining luminance distributions. Proceedings eSim, 11-13.
13. Reda, I.; Andreas, A. (2003). Solar Position Algorithm for Solar Radiation Applications. 55 pp.; NREL Report No. TP-560-34302, Revised January 2008.
14. Nishita, T., Sirai, T., Tadamura, K., & Nakamae, E. (1993, September). Display of the earth taking into account atmospheric scattering. In Proceedings of the 20th annual conference on Computer graphics and interactive techniques (pp. 175-182). ACM.
15. Nishita, T., Dobashi, Y., Kaneda, K., & Yamashita, H. (1996, August). Display method of the sky color taking into account multiple scattering. In Pacific Graphics (Vol. 96, pp. 117-132).
16. Preetham, A. J., Shirley, P., & Smits, B. (1999, July). A practical analytic model for daylight. In Proceedings of the 26th annual conference on Computer graphics and interactive techniques (pp. 91-100). ACM Press/Addison-Wesley Publishing Co..
17. Bruneton, E., & Neyret, F. (2008, June). Precomputed atmospheric scattering. In Computer Graphics Forum (Vol. 27, No. 4, pp. 1079-1086). Blackwell Publishing Ltd.
18. Elek, O., & Kmoch, P. (2010, May). Real-time spectral scattering in large-scale natural participating media. In Proceedings of the 26th Spring Conference on Computer Graphics (pp. 77-84). ACM.
19. Hosek, L., & Wilkie, A. (2012). An analytic model for full spectral sky-dome radiance. ACM Transactions on Graphics (TOG), 31(4), 95.

Light as an ambient material in transformative urban "night" spaces

Meliti D. Dikeos[1]

[1] Department of Architecture, Wentworth Institute of Technology, Boston, USA

Abstract. This study explores the subject of design and community engagement with regard to light as a transformative material in urban "night" spaces. Four unique case studies, each of which originated under the auspices of meeting specific site purposes or needs, nevertheless all represent the successful coming together of design, local suitability and ultimately, community support and participation.

1 Introduction

The mysterious and spiritual qualities of *natural* light and shadow have long been studied by artists and architects alike. Louis I. Kahn poetically wrote about "silence and light" and how "Greek architecture taught [him] that the column is where the light is not, and the space between is where the light is. It is a matter of no-light, light, no-light, light." [1] Whether in the temples of the Greco-Roman world or the religious buildings belonging to faiths around the globe, natural light has always been understood as having transformative spiritual qualities that designers hope to attain in secular structures as well. However, what happens *in darkness*, *in the city*, when the sun can not affect the aura of a place? At night, light is often seen as an *artificial* necessity, rather than as the carrier of a more powerful, experiential message. One can argue that design creativity is limited because of the understandable concern and focus on nocturnal, urban safety. The author of this paper proposes to investigate how architecture and the public, together, can create transformative urban "night" spaces through the innovative use of artificial light.

In her 1961 book, *The Death and Life of Great American Cities*, writer Jane Jacobs argued that it is a city's role to ensure safety and that good public lighting is essential. She recognized the different personalities of a city during daytime versus nighttime hours and emphasized the importance of street life with people as the "eyes on the street," "unconscious[ly] policing." [2] As an activist herself, Jacobs believed in the power of community wisdom and engagement as a means toward urban improvement. Although her ideas may seem mainstream today, the public as a partner for design and community pride cannot be underestimated. As such, this study will seek to emphasize the active participation of local residents and business people in understanding nocturnal urban problems and insight in how to solve them. Case studies of atmospheric community and architect designed "night" projects with varying financial means will be reviewed as a base against which further exploration can be generated in both professional practice and academia.

2 Community developed and designed nightscapes

2.1 "Pittaki Street" | Athens, Greece

The gritty Psyrri neighborhood in downtown Athens, Greece is known for its night life, alternative Greek eateries, live music scene, offbeat shops, and more recently, for a creative light art installation. Since the economic crisis of 2008, Athens, and the country as a whole, has battled severe austerity measures that have weighed on both the financial and emotional livelihood of its citizens. In an effort to conserve money and energy to meet demanding taxes, many residents have turned the lights off at their building entries, darkening both their surroundings and their state of mind. Pride of place has diminished and some streets, like the more industrial Pittaki Street in the Psyrri neighborhood, had become inhospitable, run-down alleys of trash, criminal activity, and graffiti. Frustrated by their depressed urban condition, in 2013 residents took on the challenge of regenerating the life of the neighborhood by creatively re-illuminating this darkened street, symbolic of many.

The Greek non-profit group Imagine the City, in partnership with the visual arts group Before Light, organized the collection of unwanted ceiling lights, sconces, old lanterns, and other creative shades, including personal hats, to fulfill a simple design vision. The motley donations were weatherproofed, wired for low consumption LED bulbs, and strung across the street from building to building (Fig. 1).

Fig. 1. Pittaki Street - Athens, Greece | © Anastasios71

What started off as a community experiment with a targeted six-month installation has since grown into a unique landmark that shows no signs of diminishing after six years. With sponsorship by Coca-Cola and its "Believe in a Better Tomorrow" campaign, this remarkably modest project of $40,500 has rejuvenated the community with not merely a sense of safety, but also one of pure joy. Athenians have been brought back outdoors in Psyrri with the help of numerous nighttime cultural activities held here. Although Project Manager Stephania

Xydia of Imagine the City recognizes the humble character of this installation, she also recognizes its influence: "This is very small scale; it's not a revolution...But since the (economic) crisis, more Greeks want to be involved in the revival of urban spaces." [3] This modest project can be seen as a "beacon of hope" and as a case study for further partnerships of communities, non-profits, and designers.

2.2 "Universal Lantern" | Cartagena, Columbia

A comparable case study to that of Pittaki Street can be found in Cartagena, Colombia, where a collaborative process with locals, including architecture students, and lighting specialists from the multinational Arup Group reactivated streets in the Getsemani neighborhood. Remarkably similar to the once rough Psyrri neighborhood, Getsemani was formerly a place of drugs, prostitution, and spotty crime. On the edge of Cartagena's historic district, it is now considered the trendy locale for restaurants, music, and mural art. However, as in Athens, while the public squares enjoyed a nightlife common to larger, thriving urban gathering areas, the side streets found themselves in the dark with limited social interaction.

Seen as untapped, yet promising areas for greater residential engagement, the design team – led by Don Slater, co-director of the Configuring Light research group at the London School of Economics, Arup lighting leader Leni Schwendinger, Universidad Jorge Tadeo Lozano and Despacio, PLANE-SITE managing director Andres Ramirez, Italian lighting manufacturer iGuzzini, and numerous other participants – hosted neighborhood gatherings and workshops to understand the specific characteristics of these streets and the inhabitants. Their process developed into a strategy of night lighting through what was called a "universal lantern," a concept based on the use of traditional lanterns. [4] (Fig. 2).

Fig. 2. "Universal Lantern" - Cartagena, Columbia | © Dr. Don Slater, LSE

Private residential building faces and public commercial storefronts were the backdrop against which these "universal lanterns" of varying colors and creative patterns were installed. They were arranged in a variety of ways – off balconies, on low stools, on tripod-like stands, along building faces, and stitched together with delicate string lights crossing the street. As in Athens, this spare installation in Cartagena proved to be an effective method for social interaction that originated from a process of collaborative community engagement and concluded with a sense of neighborhood ownership.

3 Architect designed nightscapes

3.1 "White Night, White Noise" | Athens, Greece

Interestingly, the Pittaki Street project became a physical reality in 2013, almost a decade after the 2004 Summer Olympics in Athens, an event that intensified Greece's economic difficulties. During the feverish preparation that preceded the Games and within walking distance of the Psyrri neighborhood, Boston-based architects Höweler + Yoon were selected by the Athens Olympic Committee for a temporary (18-day) art installation sited in the more distinguished and renowned Makrigianni neighborhood at the base of the Acropolis. Their celebrated design, titled *White Night, White Noise*, was a sensor-activated light installation on a platform with a field of fiber optic stalks. (Fig. 3).

Fig. 3. "White Noise, White Light" - Athens, Greece | © howeleryoon.com

The movement of pedestrians entering the field would trigger white light from the LEDs and noise from associated speakers below. Playful and mysterious qualities of flickering light and sound were generated during the dark hours of the night, under the steady glow of the Parthenon itself.

Höweler + Yoon's sophisticated technical detailing at this stately, tourist-filled site seems to share little in common with the less refined Pittaki or Gensemali Street projects. However, despite obvious differences, these three designs are urban experiments that share a similar message regarding the potential of nocturnal light as a generator of community and civic engagement. Architects Eric Höweler and Meejin Yoon described their Athens project in a manner that furthers the significance of public reciprocity in design ventures:

> "By acknowledging the unpredictability of the architectural encounter, designers can capitalize on the public realm as a fertile testing ground for interactivity, explicitly relying on public participation and play for the completion of various projects. These projects could be called 'open source': they invite multiple participants to co-produce unanticipated outcomes in an evolving scenario of scripted and unscripted behaviors." [5]

The body's interaction with this particular piece generated varying qualities of light and noise, from darkness and silence to steady and sound-filled. The uncertainty of what is to be experienced in this nocturnal scene are suggestive of Louis Kahn's words regarding daylight and its ability to inspire awe. While the quality of light in the *White Night, White Noise* installation embrace these more poetic qualities, the charm of the eclectic light in Athens' Pittaki and Cartagena's Getsemani neighborhoods are equally significant. Although the projects themselves were initiated from differing starting points, resources, and stakeholders, they all share a common thread in their value of "public participation and play."

3.2 "Luminous Drapes" | Kuwait City, Kuwait

Unlike the physically activated, playful installation of Höweler + Yoon's "White Noise, White Light," the architectural firm Studio Toggle designed a temporary installation for "The Human Capital 2018" cultural forum in Kuwait City, Kuwait as a place to socialize through rest, rather than movement. In contrast to gathering in a place of passage, whether through "lanterned streets" or a "lit field," this piece creates a place of both daytime and night-time rest through a series of minimal pavilion structures. The piece itself was made from a "kit of parts" that included construction scaffolding, tube lights, drapes, and benches (Fig. 4).

Fig. 4. "Luminous Drapes" – Kuwait City, Kuwait | © Gijo Paul George

Through Nuqat, a Kuwait nonprofit organization, the architects were challenged by the broad directive of designing an outdoor gathering place in the expansive plaza of the newly opened Sheikh Jaber Al-Ahmad Cultural Center. To do so, they strategized a series of rooms with drapes laser-cut and arranged in ways that would encourage the public to "collaborate," "lounge," "socialize," "dine," and "meet," all terms used in the actual designers' diagrams. At night, when the pavilions were lit, these outdoor rooms assumed a romantic, almost alluring quality, with the softness of the layered drapes creating a light, billowy mass. The name of the installation, "Luminous Drapes" appropriately defines this lit pavilion that, in the architects' own words, "changes its ambience and makes for a dynamic space," through the use of light as a "sculptural medium." [6] The simple detail of outlining the scaffolding with tube lights shows an understanding of how a shared, yet intimate, nighttime environment can originate through an inventively low-tech and low-cost approach. From natural light in daytime hours to artificial light at night, the character and atmosphere of the space changed.

Conclusion

While there are numerous case studies worthy of study regarding the nocturnal urban condition and the transformative qualities of artificial light, these four were selected for the thematic stories they share. In each of these urban locations – whether in Athens, Greece, Cartagena, Colombia, or Kuwait City, Kuwait – the nighttime environment was reinvented in ways that go beyond the status quo by putting artistic prowess at the forefront. In the Psyrri and Getsemani neighborhoods, urban weaknesses inspired activists, non-profit organizations, lighting manufacturers, architecture students and professors, to collaborate. Through limited financial means, people were brought together outside, on the streets, in a way that regenerated the community. In the more refined and touristy neighborhoods of Höweler + Yoon's Athens installation or Studio Toggle's Kuwait City installation, one also sees a responsive understanding for unique, social place-making at night. Regarding "White Noise, White Light," Meejin Yoon stated that:

> "The most amazing thing was how public this project was…One of the most incredible moments was the opening night when an elderly Greek woman came with roses for the project team and said, 'Thank you so much for bringing this to Athens. Thank you from the city.'" [7]

Despite the temporary nature of this installation, the project clearly brought not only a sense of beauty and pride to the community, but also a sense of local unity and grace in the midst of a larger global event such as the Olympics. In contrast, although the long-term effects of groundswell projects which come to meet less specific but nonetheless vital neighborhood challenges are more difficult to measure, they are no less important to a locale's vitality. This is even more significant when the results have a creative and artistic bent to them rather than solely fulfilling a utilitarian requirement. Although there is much debate regarding the negative effects of light pollution, each of these projects embodies a delicate balance between nocturnal light and all it represents in terms of safety, showmanship, and social gathering. The sensitive use of artificial light as an "ambient" material, along with an inspired design vision and public participation, becomes a powerful tool, which, as Jane Jacobs reminds us, can serve to revitalize night street life and community.

References

1. J. Lobell, L. Kahn. *Between Silence and Light : Spirit in the Architecture of Louis I. Kahn*. Distributed in the U.S. by Random House (2000, c1979) www.azquotes.com/quote/645512, accessed on 1/31/19
2. Jacobs, Jane. *The Death and Life of Great American Cities*. New York : Modern Library (1993)
3. T. Hsu, "Whimsical Project Puts Athens' Pittaki Street in a Better Light" (*Los Angeles Times*, October 11, 2013)
4. T. Forde, "Smart Everyday Nighttime Design Aims to Use Light as a Means to Build Better Communities" *(Archdaily*, September 28, 2017)
5. E. Höweler, M. Yoon, "Verify in Field" (*Perspecta*, **Vol. 42,** pp.163-167, 2010)
6. Studio Toggle, "Luminous Drapes: Experimental Urban Pavilion 2018" https://studiotoggle.com/portfolio/exhibition/luminous-drapes
7. E. Donoff, "White Noise, White Light" (*Architectural Lighting*, March 8, 2007)

Acoustic comfort in the living environment and its association with noise representation: A systematic review

Farrah Jahangeer[1,*]

[1]Ecole Nationale Supérieure d'Architecture de Nantes – Mauritius

Abstract. Noise is commonly known to have adverse health effects on residents in urban living environment. In the past, great focus has been placed on the impact of external traffic noise on residents' health and the development of urban noise mapping. On the other hand, neighbourhood noise – a combination of noise generated from outside residences and noise generated from within residences – are also considered to be as detrimental as external traffic noise and are yet to be widely investigated. This paper focuses on the systematic review of studies assessing neighbourhood noise with the intention of determining the presence of architectural variables and methods of noise representation for micro-scale areas. 153 articles are retrieved from specific research databases and out of which 12 were considered to be relevant to the study. The results indicate that architectural variables present in the studies were mostly related to the analysis of external traffic noise and that representation of noise in analyses were rarely devised which highlights the need for further investigation in this area.

1 Introduction

This paper is a review of studies assessing noise annoyance in residences with the intention of determining the association with representation of both architectural parameters and the aural environment. The rise in global urban population generates greater need for economically designed multi-storey houses providing sufficient comfort to residents. The increasing tendency in designing and building micro-houses [1] and multi-storeys further increases housing density which questions the intensity and impact of structure-borne and airborne-noise experienced by residents.

Over the decades, the impact of traffic noise on residents' health have been widely investigated and published. For instance, Belojevic et al. (1997) observed in a comparative study that residents in noisy traffic areas (72 dBA_{leq}) experienced more fatigue, headaches, nervousness and sleep disturbances in comparison to other residents exposed to less urban noise (56 dBA_{Leq}). A meta-analysis study showed that residents with constant noise exposure to traffic noise above 55 dBA had high risks of cardiovascular disease [2]. However, noise

* Corresponding author: farrah.jahangeer@nantes.archi

experienced by residents, also known as 'neighbour noise' or 'neighbourhood noise', are caused by multiple sources linked to neighbour activities (TV, music, voices, footsteps and so on), airborne noise from outside (traffic) and noise from building services. Neighbour noise could be considered a result of dynamic events occurring in residences and the perception of this type of noise is likely to be subjective to culture and type of built environment.

The literature in the field of room acoustics, psychoacoustics and built environment has very few rigorous studies assessing noise dissatisfaction in dwellings. In a recent review evaluating the relationship between acoustic field data and subjective responses in dwellings, Vardaxis et al. [3] found only 10 complete studies since 1985. While noise in residences may not be as critical as noise in the workplace determining work performance, associations between neighbour noise and health have been observed in several studies; Maschke & Niemann [19] highlighted that annoyance with neighbour noise was related to stress and increased health risks in cardio-vascular systems in a cross-sectional study across Europe; Rautio et al. [4] observed that noise (from airborne or structural) experienced by residents was significantly associated with depressive mood of residents.

In the field of room acoustics, physical aspects of rooms or spaces are important measures to be considered. Sabine's equation [5] for reverberation time established since the 20th century demonstrates the link between volume of room and absorptive surface area. In a laboratory experiment investigating the perception of distance of noise source in relation to reverberation time, Mershon et al. [6] reported that in spaces with longer reverberation time (1.36s) subjects tended to overestimate the distance of noise sources than those in settings with shorter reverberation time (0.36s). Similar results were reported by Cabrera & Gilfillan [7] where perceived distance of speech source was found to increase with reverberation time. Mommertz [8] also indicates the relationship between the shape of rooms and variations in sound propagation in the related space. The aural environment, is described as sounds being integrated in a space or built environment that contributes to the experiential identity of the space [9] and the graphic two-dimensional representation of such environment is yet to be sought. Despite the advances in auralization to virtually experience three dimension spaces, architectural designs at concept stages commence with two-dimensional sketches, which accentuates the ability to represent the aural environment in two dimensions. Alternatively, graphic noise representation on a large scale in the urban environment, known as 'noise mapping' is greatly being investigated in the recent years with the support of developing technology and programs thereby assisting in the monitoring and control of noise pollution.

This review evaluates past studies assessing noise dissatisfaction in residences with the aim of investigating the following research question: to what extent is architecture and the aural environment being represented when analysing neighbour noise? The following sections describes the methodology applied in selecting articles for the review and their evaluation.

2 Review methodology

A similar systematic review to that used in the study of Rautio et al. [4] was used here including three specific stages in the review. Articles used in this study were population-based studies with peer-reviews published in academic journals. In the first stage, the following journal databases – Sciencedirect, Sage, BIB CNRS and JSTOR – were researched manually for articles with titles including one of the three keywords to ensure that all articles addressing noise issues in the living environment were included: 'domestic noise', 'residential noise' and 'neighbour noise'. These three different keywords relate to different degrees of living proximities and the focus in this study is on the evaluation of neighbour noise experienced by indoor residents. In the second stage, articles assessing neighbour noise with objective and subjective measures were selected and their abstracts read. Acoustic studies focusing on aircraft and railway noise were excluded from the study because these specific noise sources were considered to be unrepresentative of typical central urban residential areas in cosmopolitan cities. In the third stage, the selected articles were fully read and the quality of the articles was rated based on the following inclusion of the following variables: architectural parameters, urban setting variables, subjective noise rating, objective acoustic measures, graphic noise representation and significant statistical results. One point was attributed to the presence of the aforementioned variables.

3 Results

3.1 Study selection

At the first stage, the keyword search in the aforementioned databases led to the retrieval of 153 articles– domestic noise: 33, residential noise: 86, neighbour noise: 34. The selected studies at that stage in relation to the time of publication, as shown on the graph in Figure 1, indicates the recent increasing awareness and concern with noise in the living environment. Studies focusing on 'residential noise' investigated mainly noise caused by vehicles and its impact on health and economy. Studies focusing on 'neighbour noise' analysed mainly the transmission of noise caused by neighbour activities in relation to construction systems and sound insulating materials. Studies investigating 'domestic noise' focused mainly on isolated noise sources, such as pets and air conditioning units. After the second stage of the selection process, only 12 complete studies met the above selection criteria and these were evaluated in the final stage.

3.2 Study characteristics

As seen in the Table 1, 6 studies were carried out in Europe, 6 in East Asia and 1 in USA. Most of the studies had cross-sectional research designs and only 3 with experimental laboratory research designs indicating the prevalence for population based survey at a single point in time which is most likely linked to the restricted access to dwellings. In all studies in this review, subjective ratings of annoyance with neighbour noise were recorded for varying neighbour noise sources. Acoustic measures present in some of the studies here were weighted sound levels (dBA) and sound insulation index. Architectural parameters such as spatial measurements were not present but other variables such as dwelling typology and perceived distances were present. Sample sizes in the review varied between 20 and 9000. Age of sample populations in most studies varied between 18 and 90 years. In some singular cases, Wang et al.[10] only considered the age group of 20 to 26 years because of the expected decline in hearing acuity. Both males and females were considered in all studies but Jeon et al. [11] highlighted that its sample contained mainly (73%) housewives.

 https://doi.org/10.1051/shsconf/20196403012

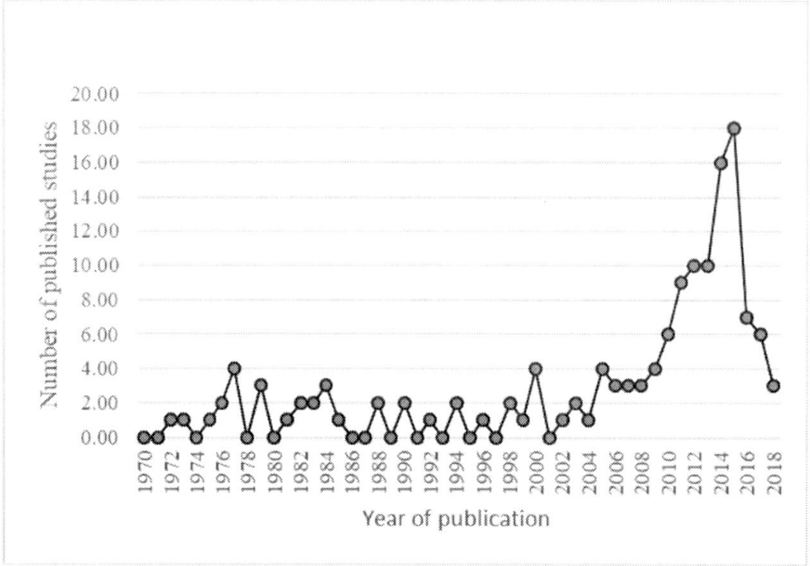

Figure 1. Number of studies published over the years after searching the databases for the specific keywords

https://doi.org/10.1051/shsconf/20196403012

Table 1. Description and evaluation of studies in the review

Study No.	Author	Study Type	Architectural parameter	Noise source investigated	Urban typology variable	Annoyance & disturbance rating	Objective acoustic measure	Measurement Location	Graphic noise representation	Significant statistical results	Rating
1	Namba et al. (1991)	cross-sectional	none	Neighbour, airborne, traffic, impact & building services	none	USA= 40% (neighbours' automobiles), Japan=48% (motorcycles), Germany=20%(drainage pipes), Turkey=40%(drainage pipes), China=20% (television & radios)	none	none	None	Univariate analysis - Neighbourhood noise annoying in all countries but with varying noise source selection. In all countries, except Germany, populations tended to endure noise rather than complain directly to neighbours.	2
	USA (n=261), Japan (n=375), Germany (n=296), Turkey (n=325), China (n=387)										
2	Grimwood (1997)	cross-sectional	Dwelling type: flat. Shared party walls	Neighbour, airborne & impact	none	Noise annoyance related to perceived poor insulation =98% Sleep disturbance related to perceived poor insulation= 83% Music, TV & radio = 98% Footfall on floor = 95%	47 - 63 dBA (airborne sound insulation)	Inside	None	Univariate analysis - Noise sources are music, television, radio and voice. Poor insulation is perceived as the cause of noise dissatisfaction.	4
	UK (n=40)										
3	Bluhm et al. (2004)	cross-sectional	Dwelling type: detached house and flat. Window orientation	Traffic	Urban transport netw	Annoyance in rooms with windows facing street=39.6% Sleep disturbance in rooms with windows facing street=27%	<45, between 45 and 55, >55 dBA from noise map	None	None	Univariate and bivariate analyses - Bedroom window orientation is found to be an important factor for the occurrence of disturbances linked to traffic noise.	5
	Sweden (n=657)										
4	Gidlöf-Gunnarsson & Öhrström (2007)	cross-sectional	Dwelling type: flat. Availability to green area (10-200m and 600m walking distance) and access quiet side in dwelling	Traffic	Distance to main road and to green spaces	Poor access to green areas and with quiet access in dwelling=43% Poor access to green areas and with no quiet access in dwelling= 51% Better access to green areas and with quiet access in dwelling=31%	60-68 dB	Outside	None	Multivariate and bivariate analyses - Degree of perceived availability to green space influenced residents' responses to traffic noise.	5
	Sweden (n=500)										
5	Maschke & Niemann (2007)	cross-sectional	Dwelling type recorded but not used in analysis	Neighbour, traffic, plane, train, commercial & road	none	Annoyed with traffic noise=39% Annoyed with noise from neighbours=36%	none	N/A	None	[illegible] annoyance is associated with risks for respiratory symptoms and cardiovascular symptoms	2
	8 European cities (n=8539)										
6	Jeon et al. (2010)	cross-sectional & lab	Dwelling type: flat	Neighbour, impact, airborne, drainage & traffic	none	Overall annoyance=50%(of 67% annoyed with floor impact noise by children)	[illegible] no measures in actual flats)	Prerecorded in two apartments	None	[illegible] with noise sources and proposed quantification model highly correlated with energy-equivalent model but predictive powers	4
	Korea (n=512 survey, n=109 lab)										
7	Li et al. (2010)	cross-sectional	Dwelling type: flat	Traffic	different types of residential estate layouts	Average score for home nois annoyance = 2.47/5.00. Noise annoyance increased with dust concern	60-70 dB(A) (predicted by CRTN method)	none	None	Logistic regression - Perception of nearby greenery can reduce noise annoyance in homes	5
	Hong Kong (n=688)										

Table 1. continued

8	Ryu & Jeon (2011) Korea (n=512 survey, n=109 lab) Housewives	cross-sectional & lab	Dwelling type: flat	Neighbour, impact, airborne, drainage & traffic	none	Annoyance with indoor noise greater than that of outdoor noise. >50% of high sensitivity people annoyed with indoor noise.	30-50 dB (in lab only, no measures in actual flats)	Inside	none	One way analysis of variance - noise sensitivity influenced reactions to indoor residential noise and outdoor noise.	4
9	Yu & Kang (2014) UK & Taiwan (n=80 to 300) Uk- pensioners,	cross-sectional	Dwelling type: Flat & detached	Neighbour, noise inside of dwelling & traffic	none	Traffic noise most annoying in Sheffield and two-wheelers most annoying in Taipei.	77.5 -42.5 dBA, Taiwan 67.5-37.5 dBA, UK	Outside	Noise Map	Correlation analysis - perception of living environment is affected by different cultural and social factors.	5
10	Hongisto et al. (2015) Finland (n=159)	Cross-sectional	Building type. Light weight and heavy weight floor and wall construction	Neighbour, airborne, from exterior, stairs & building services	none	Annoyance with neighbour noise almost similar in both light and heavy building(56%). No effect of noise on sleep disturbance. Airborne neighbour noise annoying in light and plumbing noise in heavy.	55-60 dBA (day) from noise map	None	None	Mann-Whitney test - airborne sound insulation lower in light building type and no difference between light and heavy building type for satisfaction with sound insulation. Airborne neighbour noise annoying in both building types.	4
11	Wang et al. (2015) Malaysia (n=171)	Cross-sectional	Dwelling type: detached, terraced, semi-detachd, flat	Neighbour, traffic, noise inside	Urban setting type - urban, suburban and rural	Annoyance lower in detached houses (50%) than in other house types. Same for loudness ratings. Traffic noise and neighbourhood noise most annoying noise sources.	None	None	None	One way analysis of variance & correlation analysis - Apartments have loudest neighbour noise than terrace house.	4
12	Park & Lee (2017) UK (n=21)	Laboratory	Test building with simulated Korean living room	Neighbour & impact	none	60% of subjects noticed noise when above 38,5 dBA and greater annoyance recorded when exposed to sounds by impact ball.	31.5 - 63 dBA	Inside (pre-recorded)	None	Bivariate, multivariate and correlation analysis - noticeability of noise increased with sound levels. Sound level was associated with physiological responses in participants	4

3.3 Objective acoustic measures

Noise sources being investigated tended to vary; airborne noise, impact noise, traffic noise, building services noise and other external types of noises (such as commercial noise). Noise levels of sources were considered in only nine studies and were determined by various methods. Study 3,4 and 7 in Table 1 focusing mainly on traffic noise had noise levels varying between 43 dBA and 70 dBA. Bluhm et al. [12] derived traffic noise levels from noise maps available while Gidlöf-Gunnarsson & Öhrström [13] and Li et al. [22] derived noise levels from CRTN (calculation of traffic road noise) method. Gidlöf-Gunnarsson & Öhrström [13] further referred to the speed of the vehicles which according to UK Noise Association [14] a reduction in 10 km/h can reduce noise level by 40%. Other parameters considered to influence traffic noise levels, such as speed zones of areas being investigated, road surface textures and types of vehicles in the investigated areas were not mentioned. Hongisto et al. [15], on the other hand, determined the building selection for their study according to noise maps highlighting areas with traffic noise levels between 50 to 60 dBA.

Other six studies (Study 2, 6, 8, 9, 10 & 12) investigated several noise sources and the range of noise levels varied considerably, between 30 dBA and 63 dBA. In the laboratory experiment by Jeon et al. [11] annoyance was evaluated in relation to each noise source separately. None of the studies included noise levels recorded within dwellings for a defined period to fully represent the soundscape in dwellings. Grimwood [16] and Hongisto et al. [15] measured the sound reduction index and sound insulation in relation to noise transmission through walls and floors. Reverberation time, the acoustic measure used to define quality of acoustics, was not used in the studies reviewed here.

3.4 Representation of acoustic measures

The requirements of the European Commission to produce noise maps every 5 years for specific areas by member states have further encouraged the use of noise maps in research (END 2002). A few studies in the review (Study 3, 9 & 10) have used noise maps to derive estimates of noise levels of the areas being investigated in relation to the built area and density (see Figure 2). The noise maps used were similar to those produced by Scotland's Environment and accessible by the public (see Figure 2). The noise maps are required to be updated every five years according to the Environmental Noise Directive. No other forms of two-dimensional graphic representation of noise, such as location points of noise sources in dwellings, were used in the studies to represent the aural environment and its variation with time.

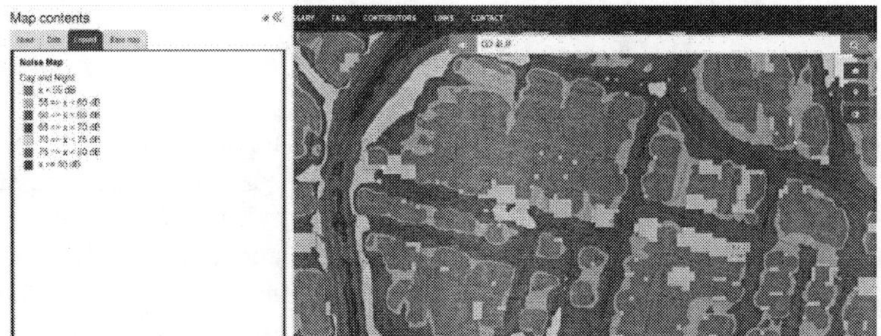

Figure 2. Noise maps available on Scotland's Environment.

3.5 Architectural parameters

The review indicates that a few architectural parameters were used in the analyses of noise annoyance in living environments, such as housing typologies, spatial urban context, construction type and orientation of windows in dwellings. Housing typologies were used in the analysis of studies 3, 9 and 11. Bluhm et al. [12] analysed both detached houses and flats and the window orientation in both dwelling type suggesting that annoyance with traffic noise was 16.3% less in dwellings with windows facing gardens than those facing the street. However, there was no mention of glazing ratio or technical details of windows. Yu & Kang [17] considered two types of housing; detached houses and flats but failed to compare the subjective data collected between the two. Wang et al. [10] performed a comparative analysis for subjective data collected in four types of dwellings; detached, terraced, semi-detached and flat and indicating that perceived noise annoyance was highest in flats.

Gidlöf-Gunnarsson & Öhrström [13] observed that flats with one side having average noise level less than 48dBA and located within walking distances to green areas (10m to 200m) had 20% less annoyance than flats with higher average noise level and with further walking distances to green areas. Hongisto et al. [15] investigated the relationship between building construction type –light and heavy weight – in relation to neighbour noise transmission and specified the construction period and the materials used for walls and floors. Light weight floor materials consisted of parquet, soft underlay, cast concrete, flexible isolation board, steel reinforced concrete and walls consisted of plasterboard, mineral wool steel studs and plasterboard. Heavy weight floors and walls consisted of steel reinforced concrete. This study also mentioned the number of floors in the buildings being surveyed. Park & Lee [18] referred to the simulation of living rooms in the laboratory experiment providing details of surface material finishes and construction materials used – volume of $38m^3$, wooden flooring and furniture and concrete structure. However, there were no further details on the layout of the space, openings or location of participants during study. Similarly, no other architectural elements were present in other studies reviewed here such as the floor plan of dwellings and architectural parameters that have been analysed so far could be considered sparse and superficial.

While urban context also contributes to the aural environment in which dwellings are located only few studies have considered parameters in the urban environment. Bluhm et al. [12] considered the type of urban network surrounding the residences and Gidlöf-Gunnarsson & Öhrström [13] analysed the distance between green spaces and buildings being surveyed. Li et al. [22] and Wang et al. [10] investigated the type of residential setting (e.g. urban, suburban or rural). There were no other mentions of urban form, urban pattern, building intensity or building heights which can be expected to influence noise propagation.

3.6 Subjective satisfaction measures

All of the studies in the review assessed perceived noise annoyance but the types of noise sources being evaluated differed. To begin with two different definitions of annoyance were observed in the review; Jeon et al. [11] defined annoyance as a feeling of displeasure associated with any agent or condition while Maschke & Niemann [19] defined annoyance as a feeling of discomfort which is related to adverse influence on an individual by any substance or circumstance. Traffic noise and noise arising from neighbour activities (such as TV, music, talking) were the most investigated noise sources in the review. Table 2 indicates the percentage rating reported in the studies here except of three studies (Study 4, 9 & 12) that were excluded because of the type of research study (laboratory based) and missing data. Based on the average noise annoyance ratings available in the studies, neighbour noise was found to have higher noise ratings than annoyance caused by traffic noise which highlights the need for further investigation into neighbour noise. Some studies, such as that of Maschke & Niemann [19], evaluated neighbour noise as one cluster while in other studies, such as that of Jeon et al. [11] each noise source was evaluated separately. This distinction of noise sources is likely to vary according to households and culture. For instance, in the study of Namba et al. [20] drainage sounds were found to be the most annoying in West Germany and Turkey whereas in China sound from television and radios were the most annoying. In Sheffield, traffic noise was rated as the most annoying noise source whereas in Taipei talking, music and TV were the most annoying noise sources (Yu & Kang 2014).

Table 2. Percentage ratings reported in studies reviewed here.

	Average percentage of annoyance ratings						
Study no.	traffic	voices	music	TV	impact noise	drainage	stairs/circulation
1	–	16.6	20.2	14.2			18.6
2	–	78	98		95	50	–
3	12.8	–	–	–	–	–	–
5	39	36				–	–
6	87	65		–	67	60	–
7	49.4	–	–	–	–	–	–
8	87	67				60	–
10	42	43					
11	49.1	57.5			–	–	–

Other independent variables present in dwellings that could possibly be linked with noise annoyance were also evaluated; Hongisto et al. [15] assessed noise sensitivity and extraversion of participants, Li et al. [22] evaluated the relationship between perceived foul odour from exterior and perceived annoyance with traffic noise, Gidlöf-Gunnarsson & Öhrström [13] assessed subjective psychosocial stress in relation to traffic noise annoyance and Yu & Kang [17] analysed the correlation between perception of quiet living environment and education level and age groups.

The methodology deployed for collection of subjective data varied considerably. Questionnaires were mostly administered in two ways - by post or verbally during interviews. The response rate was reported in several studies (Study 3,4,7,10 & 11) and was above 50% except for study number 11 where the response rate was 17.1%. The studies are in line with the recommended response rate between 40% and 80% as suggested by Baruch [21]. The evaluation scales also varied – 5-points scale, 7-points scale, 10-points scale and several studies such as that of Gidlöf-Gunnarsson & Öhrström [13] and Li et al. [22] used different scales thereby providing the possibility of counterchecking responses obtained. While correlation coefficients are indicative of the type of association present in the aforementioned studies, the causality is difficult to predict in such types of research designs.

3.7 Study evaluation

After evaluation of each study, the highest score achieved was 5 and with an average of 4. The majority of the studies scoring 4 highlighted the absence of acoustic measures and urban typology variables in the studies. The lowest score obtained was 2 in the study of Maschke & Niemann [19] where only subjective data was collected in a large cross-sectional study across Europe. Four cross-sectional studies scored 5 and out of these three focused entirely on annoyance with traffic noise and included either measured noise levels or noise levels estimated from noise maps.

4 Discussion

The review of studies evaluating noise annoyance in the living environment brings to light the complexities associated with such studies and the areas that could be further explored. The definition of 'neighbour noise' needs further precision because clusters of different noise sources in dwellings were given similar name – Maschke & Niemann [19] considered neighbour noise as the presence of voices, TV, music and impact noise while Jeon et al. [11] considered only voices and music as neighbour noise. The demography of studies does not indicate significant findings when assessing age but gender and occupation seem to be important factors given that in some studies housewives were main participants in surveys. The aural experience of housewives is likely to differ from those of their husbands and same goes for employees working from homes.

In general, great importance is given to the destructive impact of traffic noise on health and annoyance – WHO [23] recommends an average night exposure of 40 dBA away from busy streets. However, the findings here indicate greater annoyance with neighbour noise which is most likely because of its unpredictable and uncontrollable characteristics. Noise levels provide some indication of intensity of noise but does not necessarily reflect noise annoyance of inhabitants. Guski [24] reported that annoyance of inhabitants were only 33% dependent on acoustic measures and therefore other measures are required to fully characterize the aural environment of dwellings. It is considered that acoustics measures and architectural spatial parameters investigated in the studies were sporadic and do not fully describe the aural environment of dwellings Furthermore, the aural environment is likely to vary in different cultures and the relationship between cultural geographies of home [25] and noise perception is yet to be established. The difficulty of accessing dwellings for long periods of time could be considered as a limiting factor in studies here.

The Environmental Noise Directive in 2002 initiatives triggered the increased use of noise maps but these have certain shortcomings when assessing noise on microscales and temporality of noise. Alternative noise mapping strategies, such a smartphone-based noise mapping could be sought to obtain more frequent and accurate data whilst developing a participatory approach in field surveys. The two-dimensional graphic representation of aural environment is an essential element that could be integrated in early stages of architectural designs and studies here have not shown any association between any form of representation of the aural environment and noise annoyance in dwellings.

Conclusion

The review here assessed studies investigating noise annoyance in living environments and the findings indicate that the integration of architectural measures and the representation of architecture and the aural environment are sparse. Further investigation is required to fully comprehend the transient nature of aural environments in dwellings. There are insufficient data, be it subjective or objective, on neighbour noise and more investigations are required in the field to improve acoustic comfort in dwellings. The limitation of data collection caused by access difficulties in dwellings require the development of innovative methodologies and measures for data collection. The evaluation of subjective acoustic measures occurs in several countries and some even take into account the influence of culture when assessing noise perception in cross-sectional studies. The association between spatial organisation of dwellings across cultures and noise annoyance ratings could be further possible areas of investigations.

References

1. R. Post, "Are tiny houses and micro-apartments the future of urban homes? | Guardian Sustainable Business | The Guardian," 2014.
2. T. Münzel, T. Gori, W. Babisch, and M. Basner, "Cardiovascular effects of environmental noise exposure.," *Eur. Heart J.*, vol. 35, no. 13, pp. 829–36, Apr. 2014.
3. N.-G. Vardaxis, D. Bard, and K. Persson Waye, "Review of acoustic comfort evaluation in dwellings—part I: Associations of acoustic field data to subjective responses from building surveys," *Build. Acoust.*, vol. 25, no. 2, pp. 151–170, Jun. 2018.
4. N. Rautio, S. Filatova, H. Lehtiniemi, and J. Miettunen, "Living environment and its relationship to depressive mood: A systematic review," *Int. J. Soc. Psychiatry*, vol. 64, no. 1, pp. 92–103, Feb. 2018.
5. W. Furrer, *Room and Building Acoustics and Noise Abatement*. London: Butterworths, 1964.
6. D. Mershon, W. Ballenger, A. Little, P. McMurty, and J. Buchanan, "Effects of room reflectance and background noise on perceived auditory distance," *Perception*, vol. 18, pp. 403–416, 1989.
7. D. Cabrera and D. Gilfillan, "Auditory distance perception of speech in the presence of noise," in *Proceedings of the 2002 International Conference on Auditory Display, Kyoto, Japan*, 2002.
8. E. Mommertz, *Acoustics and sound insulation: principles, planning, examples*. Basel: Birkhäuser, 2009.
9. T. Sheridan, N. York, and K. Van Lengen, "Hearing Architecture Exploring and Designing the Aural Environment," ACSA, Inc, 2003.
10. C. Wang, Y. Si, H. Abdul-Rahman, and L. C. Wood, "Noise annoyance and loudness: Acoustic performance of residential buildings in tropics," *Build. Serv. Eng. Res. Technol.*, vol. 36, no. 6, pp. 680–700, Nov. 2015.
11. J. Y. Jeon, J. K. Ryu, and P. J. Lee, "A quantification model of overall dissatisfaction with indoor noise environment in residential buildings," *Appl. Acoust.*, vol. 71, no. 10, pp. 914–921, Oct. 2010.
12. G. Bluhm, E. Nordling, and N. Berglind, "Road traffic noise and annoyance--an increasing environmental health problem.," *Noise Health*, vol. 6, no. 24, pp. 43–9, 2004.
13. A. Gidlöf-Gunnarsson and E. Öhrström, "Noise and well-being in urban residential environments: The potential role of perceived availability to nearby green areas," *Landsc. Urban Plan.*, vol. 83, no. 2–3, pp. 115–126, Nov. 2007.

14. P. Mitchell, "Speed and Road Traffic Noise The role that lower speeds could play in cutting noise from traffic A report commissioned by the UK Noise Association. Researched and written by Paige Mitchell," 2009.
15. V. Hongisto, M. Mäkilä, and M. Suokas, "Satisfaction with sound insulation in residential dwellings – The effect of wall construction," *Build. Environ.*, vol. 85, pp. 309–320, Feb. 2015.
16. C. Grimwood, "Complaints about poor sound insulation between dwellings in england and wales," *Appl. Acoust.*, vol. 52, no. 3–4, pp. 211–223, Nov. 1997.
17. C.-J. Yu and J. Kang, "Soundscape in the sustainable living environment: A cross-cultural comparison between the UK and Taiwan," *Sci. Total Environ.*, vol. 482–483, pp. 501–509, Jun. 2014.
18. S. H. Park and P. J. Lee, "Effects of floor impact noise on psychophysiological responses," *Build. Environ.*, vol. 116, pp. 173–181, May 2017.
19. C. Maschke and H. Niemann, "Health effects of annoyance induced by neighbour noise," *Noise Control Eng. J.*, vol. 55, no. 3, p. 348, 2007.
20. S. Namba, S. Kuwano, A. Schick, A. Aclar, and M. Florentine, "A cross-cultural study on noise problems: comparison of the results obtained in Japan, West Germany, The U.S.A., China and Turkey," *J. Sound Vib.*, vol. 151, no. 3, pp. 471–477, 1991.
21. Y. Baruch, "Response Rate in Academic Studies - A Comparative Analysis," *Hum. Relations*, vol. 52, no. 4, pp. 421–438, 1999.
22. H. N. Li, C. K. Chau, and S. K. Tang, "Can surrounding greenery reduce noise annoyance at home?," *Sci. Total Environ.*, vol. 408, no. 20, pp. 4376–4384, Sep. 2010.
23. "WHO night noise guidelines for Europe," Mar. 2019.
24. R. Guski, "Psychological determinants of train noise annoyance," *Euro-Noise*, vol. 98, no. 1, 1998.
25. A. Blunt, "Cultural geography: cultural geographies of home," *Prog. Hum. Geogr.*, vol. 29, no. 4, pp. 505–515, Aug. 2005.

Issues of wind mode visualization and pedestrian comfort assessment when designing residential buildings on sloping territories in the Arctic zone

Ilya V. Dunichkin[1, 2,*], *Emanuele* Naboni[3], *Anna E.* Korobeinikova[2] and *Olga I.* Poddaeva[1]

[1]Educational, scientific and industrial laboratory for aerodynamic and aeroacoustic tests of building structures, Moscow State University of Civil Engineering (National Research University), Russia

[2]Urban planning Department, Moscow State University of Civil Engineering (National Research University), Russia

[3]Institute of Technology, School of Architecture, Royal Danish Academy of Fine Arts, Copenhagen, Denmark

Abstract. Subject of research: visualization of the wind regime of residential buildings on the slope area in the Arctic. Goals: the purpose of the study is to identify the problems of visualization of the wind regime on the slopes and the analysis of patterns of airflow around the sloping territories, affecting the comfort of pedestrians. Materials and methods: airflow patterns of slope areas with different characteristics and comfort assessment for humans are analyzed in the course of work. Geotechnical methods are presented in solving the problems of wind erosion and the stability of hillsides and complex terrain within the city limits. Results: The importance and novelty of the research in studying the relationship of slope geometry and environmental quality, as a decrease in comfort inevitably leads to a decrease in the development of nearby urban areas in settlements on the coast of the Arctic Ocean, problems with climate and local climate, as well as a decrease in the quality of life of people. The article considers the relationship of plastic relief with the aeration regime of the territory, the dependence of aerodynamic roughness on their height, features of aeration of the slope and hilly terrain, factors affecting the direction and speed of the wind and methods for studying the aeration regime of slope areas. Findings: The possibilities of applying existing approaches to research and visualization for slope areas are demonstrated. The direction of development of the technique for visualization of slope areas has been determined.

* Corresponding author: ecse@bk.ru

1 Introduction

Wind mode of buildings is one of the most important criteria for the comfort of a residential area, especially in conditions of sub-zero temperatures in winter. The quality of the environment of pedestrian paths and landscaping sites is greatly influenced by the architectural and planning organization of development and relief. Therefore, it is important to carry out a qualitative study based on a wind mode visualization before making design decisions. Naboni, E. said that this is particularly true in a design that is paying attention to to assess the comfort of pedestrians both the streets and in the courtyards [1]. It should be noted that most studies on thermal and visual comfort consider objects in hot and temperate climates, so it is important to determine the research methods and the planned type of results, taking into account the conditions of the cold climate.

The cities practically did not face the problems of mastering of complex reliefs until recently, as they had sufficient areas of free reserve areas. Up to date, architects and city planners have paid attention to the development of so-called inconvenient land, which include slope areas. Residential development in the port cities of the Arctic often has accommodation in slope areas. This is due to the fact that areas with flat terrain are used for port infrastructure, logistics and industrial facilities. Residential areas on the coast of the Arctic Ocean are usually located at a distance of a sanitary protection zone from the port, which forces them to develop in sloping areas. Such a situation takes place, for example, in Murmansk, both in the existing building facing the port and in new areas nearby the ports and shipyards of the Kola Peninsula of Russia that are under construction. The design of residential buildings on slope areas is necessary for the qualitative development of these strategic areas [9]. The design needs to be parametric, thus account for a comparison of multiple options, to account for the specific climatic features of the site, in the form of sub-zero temperatures, high humidity, icing of structures and large snow loads. All these factors depend on the wind regime and with a significant effect reduce the comfort of pedestrians and also pose a risk to human health due to the risk of frostbite [7].

In the framework of pre-project analysis, must be determined morphological and geometrical characteristics, as they will affect, fig. 1. It is necessary to choose flat, not steep, slopes in order to avoid the formation of leeward rotor and dynamic upstream, or to take measures for local relief changes in the process of geoplasty. This issue should be given special attention as the increased wind and vortex flows inevitably affect the microclimate and the local climate of the territory and as a consequence the comfort of being in this area, especially in regions with high humidity. At the same time, the building should have the correct aeration and be ventilated.

In addition to taking into account the morphology of the slope, affecting the wind regime in the process of studying and preparing the territory of the analysis, it is necessary to realize that the development of urbanized countryside will make significant changes in the local climate. Therefore, the designer should pay special attention to determining the shape, orientation and height of the building, taking into account all the climatic features of the region, as well as other equally important features. Moreover, Poddaeva, O., Churin, P., & Dunichkin, I. argue that the wind speed will be affected by the roughness of the underlying surface, which in the conditions of development of slope surfaces is an interesting task requiring a more detailed study of aerodynamics and wind protection of slopes and with potential urban development [14]. The need to take into account these factors and the features of the aeration regime of slopes associated with their geometric parameters pose a difficult task for urban planners and designers to develop the slope areas, taking into account bioclimatic comfort.

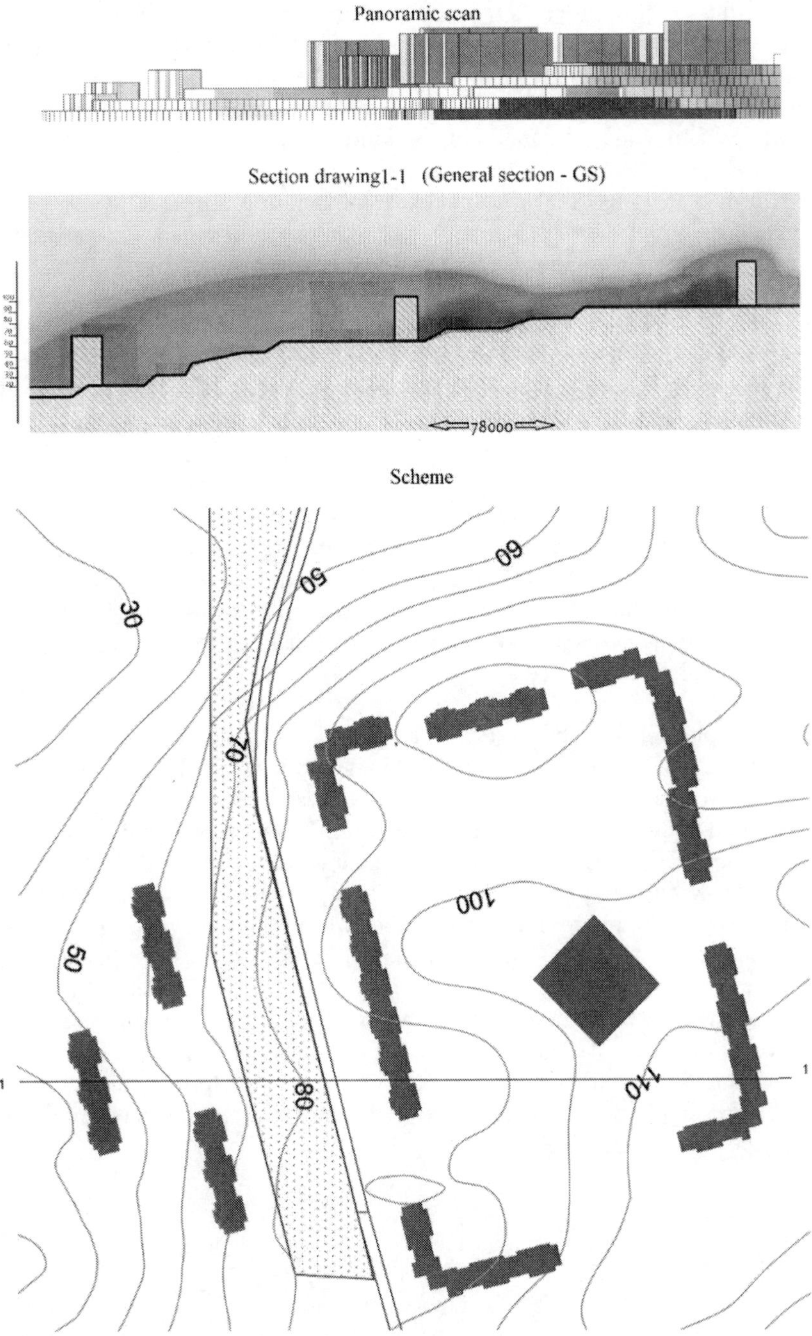

Fig. 1. Scheme, section drawing and panoramic scan of the layout of buildings on the slope area for the subsequent visualization of wind flows (Option 1).

2 Materials and methods

The study of the wind flow around the slope is very important both at the stage of pre-design analysis and in the design process. On the basis of the obtained regularities and quantitative characteristics of this process, recommendations are developed to improve the aeration regime in this territory by architectural, construction and engineering methods [1]. Determining the aerodynamic characteristics of the slope is possible by several methods – the method of full-scale laboratory studies, theoretical calculations, as well as methods for experimental and numerical simulation of wind effects

It is necessary to use the wind mode visualization in high-quality parametric design to solve the problem of the development of residential development on the slope areas in the Arctic [5]. The implementation requires solving the problem of drawing up a visualization of the wind regime on the masterplan of residential buildings, which is that the flow distribution fields must be parallel to the slope surface at the height of 2 meters or be horizontal and terraced with a step that allows one to get a reliable picture for evaluating comfort. Thus, there is a technical task to visualize the wind regime in sloping areas. This is a critical analysis since buildings, structures and green spaces can modify the speed and direction of air flow and, consequently, the nature of snow positions in winter. The initial approach involves the allocation of flat sections of the slope and terraces for consideration of their wind regime in numerical simulation and on physical tests in a wind tunnel [13]. This facilitates the visualization of the wind regime of the slope area. To solve the problems of pedestrian comfort assessment in the study, the method of physical testing of the location of snow deposits in the wind and the method of photo fixation of snow deposits on the model in the wind tunnel with subsequent digital processing to visualize the plan of distribution of snow deposits in height were applied.

For example, consider the project of residential development of the Kola Peninsula in the Arctic zone of the Russian Federation. Within the framework of the study, CFD-modeling of wind flows was carried out in order to identify areas that are uncomfortable for pedestrian safety [4]. The issue of visualization arose during the numerical CFD-modeling to identify comfort in terms of aeration of local areas. There was a task to investigate a wind mode. The problem of obtaining results consisted in that at reception of the map of speeds in the inclined territory where buildings are located at different levels, and it is necessary to receive data at the level of growth of the pedestrian to receive a field of speeds one plane it is not possible. In the case of the study of the development project for the Kola Peninsula were to be taken three horizontal planes for each group of houses at different levels of the slope, fig. 2. Also, for the accuracy of the data it was necessary to use vertical velocity fields. This method is longer in comparison with the study of development on flat terrain, but in this case, we see a direct relationship between the complexity of regulation and evaluation of aeration regime on the slope and its modeling.

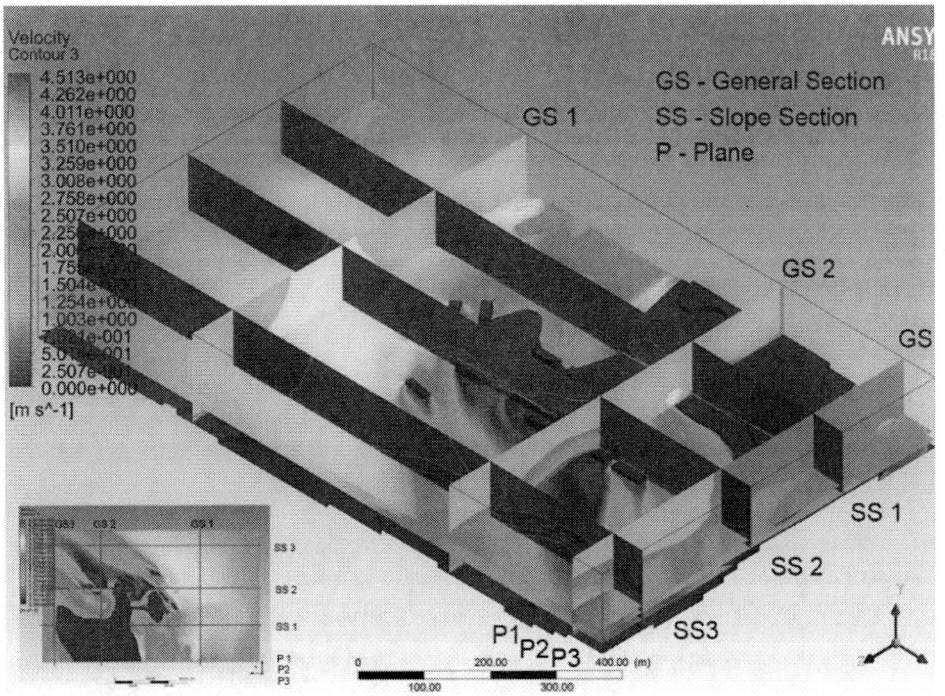

Fig. 2. 3D model of the method of visualization of the slope using two-dimensional planes, taking into account the parameters of the relief (Option 2).

3 Analysis and results

Parametric design of micro-districts in severe weather conditions, such as those prevailing in cold climatic zones, is one of the most critical tasks requiring a complex solution with BIM-technologies [10]. A multifactor analysis for determining the initial environmental parameters, which includes climatic analysis of the seasons, is required to develop a comprehensive solution of parametric design, taking into account the features of cold climatic zones according to the following parameters:

Average monthly and maximum wind speeds;
Average monthly and extreme air temperatures;
The most repetitive wind directions for a month;
The amount and state of aggregation of precipitation per month.

Cold climatic zones include the polar type of climate (arctic zones) and subpolar climate. In winter, temperatures can drop to -50°C due to the influence of arctic air masses, so winters are long and cold here, with lots of snow and significant snow transport in buildings.

The complex analysis includes the presence and influence of landscaping, orientation of the development of the neighborhood unit and the type of its composition [11]. Building parameters allow classifying the air permeability with the calculation of the perimeter density of the neighborhood unit (the long perimeter of the neighborhood unit to the sum of the lengths of the external facades) and the influence coefficient parameters and position of the slope relative to the building. Classification of permeability to air allows one to pre-determine the approximate ranges of the parameters of building and its location relative to the slopes. This makes possible the modeling of aeration, multivariate analysis and comparative analysis of a limited number of options for design solutions of construction, which reduces the work on the visualization of air flow and the time for evaluating the results by an average of 2-3 times.

The results of experimental studies allow one to determine the arrangements for the placement of a set of small architectural forms and various snow protection structures. Experimental studies conducted at the Aerodynamic and Aero Acoustic Testing of Building Structures at the Scientific and Production Laboratory of the National Research University MGSU indicate how snow transfer phenomena occur in the microdistrict territory, fig. 1, [2]. A verified simulation is carried out on the basis of physical experiments and the parameters of the development environment are determined for parametric design and optimization of design solutions in computer-aided design [12].

The method of assessing pedestrian comfort initially requires the allocation in the neighborhoods of building groups for visualization in different sections (Fig. 2) of wind speed, streamlines of air masses (Fig. 3), height of snow deposits in buildings and on a slope (Fig. 3 b).

These visualizations make it possible to adopt decisions on the optimization of the road network, the placement of landscaping, fences and small architectural forms, which can change the pedestrian comfort indicators for the better [6].

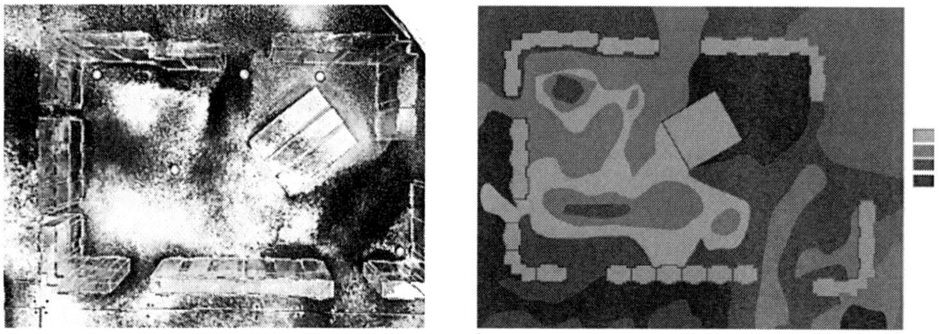

a) b)

Fig. 3. Photo fixation of a residential development layout on a terrace of a slope in a wind tunnel after a snow transfer test and snow drifts plan (Option 1).

Conclusions

It is important to note that the regulation of the aeration regime can be carried out by placing small architectural forms on the adjacent territory that also fulfill recreational functions in addition to the methods of geoplastics. Such a decision is rational for the organization of bioclimatic comfort in the construction of Northern cities, in which the creation of an artificial relief can be difficult because of severe climatic factors.

Thus, the study presents the possibility of developing design solutions for sloping areas that take into account the aeration regime and the dangerous effect of wind erosion, as well as the application possibilities for fixing slopes of geotechnical structures with biopositive structures affecting the parameters of aeration and the bioclimatic comfort of the environment, and hence the health indicators of the population. The development of the considered direction consists in integration of calculations of geotechnical structures and wind conditions of slope territories into the environment of complex parametric design with parallel modeling of engineering, geotechnical and socio-functional landscape systems that will allow to reach a qualitatively new level of design solutions [8]. Visualizations for assessing pedestrian comfort include a minimum of an additional six vertical sections and three horizontal section-plans on the slope for a complete picture, taking into account the relief considered.

It is possible to model and visualize the influence of small architectural forms and landscaping on the forms and types of snow deposits by conducting experimental studies [3]. Given the lack of existing standards for accounting snow transfer it is necessary to carry out simulation, physical testing, verification, visualization of the wind regime and snow transfer in the pedestrian zone [15]. Zones of snow deposits with visualization, located on motorways, pedestrian paths, landscaping sites, show places needed for snow protection and are the visualization basis for placing additional fences, landscaping and small architectural forms. These studies should refer to the terraces and flat slopes, and should then be conducted above surfaces with significant slope. This requires matching the areas of the visualization zones, which entails additional research and analysis of the wind regime in places where the slopes of the terrain change.

References

1. Naboni, E. (2014, December). Integration of outdoor thermal and visual comfort in parametric design. In *30th International PLEA Conference* (pp. 1-10)
2. Dolganov, A., & Kagan, P. (2018). On the design of high-rise buildings with a specified level of reliability. In E3S Web of Conferences (Vol. 33, p. 02061). EDP Sciences
3. Vigier, T., Siret, D., Moreau, G., & Lescop, L. (2013, August). Sensitive suggestion and perception of climatic effects in virtual urban environments. In Proceedings of the ACM Symposium on Applied Perception (pp. 139-139). ACM
4. Dunichkin, I. V., Poddaeva, O. I., & Golokhvast, K. S. Studies and evaluation of bioclimatic comfort of residential areas for improving the quality of environment. In Building Simulation (pp. 1-6). Tsinghua University Press
5. Pham, T., Musy, M., Siret, D., & Teller, J. (2007). Methodology for integrating and analyzing environmental and urban data in 3D GIS. In Proc. of 10th AGILE International Conference on Geographic Information Science
6. Blocken, B., Stathopoulos, T., & Van Beeck, J. P. A. J. (2016). Pedestrian-level wind conditions around buildings: Review of wind-tunnel and CFD techniques and their accuracy for wind comfort assessment. Building and Environment, 100, 50-81
7. Valger S., Fedorov A., & Fedorova N. (2013). Simulation of incompressible turbulent flows in the vicinity of low-flow bodies using PC ANSYS Fluent.. Computingtechnology, 18(5). 27-40
8. Cherkina, V., Shushunova, N., & Zubkova, J. (2018). Application of BIM-technologies in tasks of quality management and labour safety. In MATEC Web of Conferences (Vol. 251, p. 06004). EDP Sciences
9. Korol, O., Shushunova, N., Lopatkin, D., Zanin, A., & Shushunova, T. (2018). Application of High-tech Solutions in Ecodevelopment. In MATEC Web of Conferences (Vol. 251, p. 06002). EDP Sciences
10. Poddaeva, O., Dunichkin, I., & Gribach, J. (2018, June). Conducting calculating and experimental researches of the bioclimatic comfort of the residential area territory.

In IOP Conference Series: Materials Science and Engineering (Vol. 365, No. 2, p. 022029). IOP Publishing

11. Kagan, P. (2018). Monitoring of the development of urban areas with the use of information technology. In MATEC Web of Conferences (Vol. 193, p. 05031). EDP Sciences
12. Billen, R., Cutting-Decelle, A. F., Marina, O., De Almeida, J. P., Caglioni, M., Falquet, G., ... & Rabino, G. (2014). 3D City Models and urban information: Current issues and perspectives-European COST Action TU0801
13. Miguet, F. (2007). A further step in environment and bioclimatic analysis: the software tool Solene. IBPSA 2007, 3-6
14. Poddaeva, O., Churin, P., & Dunichkin, I. (2016). Experimental study of wind loads on unique buildings and structures in Russia. In MATEC Web of Conferences (Vol. 86, p. 02012). EDP Sciences
15. Antoniou, N., Montazeri, H., Wigo, H., Neophytou, M. K. A., Blocken, B., & Sandberg, M. (2017). CFD and wind-tunnel analysis of outdoor ventilation in a real compact heterogeneous urban area: Evaluation using "air delay". Building and Environment, 126, 355-372

Visual effects of colors on façades during rain

Takayuki Kumazawa[1,*]

[1] Assoc. Prof., Graduate School of Science and Engineering, Ibaraki University, 4-12-1 Nakanarusawa, Hitachi, Japan

Abstract. In this study, rain would be regarded as an important factor in landscape color planning, and the visual effects and impressions of colors on façades during rainfall were focused. Verification experiments in which the authors produced a rain simulation device that can control rainfall and illuminance were conducted. Experiments were conducted based on an experimental design with four factors of hue (5R / 5Y / 5G / 5PB), Chroma, lightness, and rainfall. In the rainfall simulation experiment, the visual effects of the color based on visual colorimetry were evaluated, and impression evaluations were extracted. As results, this study presented our findings about the relationship between rainfall and the visual effects and impressions of color, and determined that there are differences in the evaluation depending on the Hue and the amount of rainfall. In landscape color planning, considering the influence of rainfall is important for a fascinating landscape. This study will show basic important references of such new landscape color planning.

1 Background

There are many days when it will rain in Japan. Culture and custom on rain are deeply rooted in everyday daily life. Rain is an important factor in capturing the landscape, and in Japan, rainfall has long been a feature of landscape appreciation. Generally, depending on the strength of rain, the size and density of raindrops and the speed of falling greatly differ. As a result, landscape appearance will vary depending on the situation of rain. The appearance with the strength of rain is one of the important landscape features of rain. However, it is unknown how the appearance of the facade color changes in the rain. Besides, there are no studies about visual positive effects of colors on facades during rain. In Japan, as there is much rain season, landscape planning and landscape color plans should have been focusing on rainy landscapes.

In this study, rain should be taken into consideration as an important factor in landscape color planning, and the visual effects and impression of colors on facades during rainfall were investigated.

* Corresponding author: takayuki.kumazawa.1@vc.ibaraki.ac.jp

2 Method

2.1 First investigation

Firstly, the actual situation survey was conducted outdoors around Ibaraki University Hitachi campus. A building consisting of a facade color on the south side was selected. In figure 1, Facade color of the building was shown. And then, in table 1, Hue, Value and Chroma by color difference meter was shown. These were surveyed by visual colorimetric evaluation during daylight (10 o'clock - 14 o'clock) and evening (17 o'clock - 19 o'clock) during fine weather and rain. In table 2, results by visual colorimetry on sunny or rainfall in the noon were sugested. And in table 3, results by visual colorimetry on sunny or rainfall in the evening were shown. As results, the recognition for value of color was changed for medium value of color at the time of fine weather and at the time of rain, and the recognition for Chroma of the color was changed at high Chroma color at the time of rain. Therefore, it turned out that a tendency of the appearance of the facade color at the time of rain depends on lightness, value and Chroma of color.

Fig.1. Facade colour of the building

Table 1. Hue, Value and Chroma by color difference meter

Object	Hue	Value	Chroma
Object 1	8.5G	6	4
Object 2	4.2B	3.9	1.3
Object 3	1.8Y	6.8	0.9
Object 4	5.2GY	8.9	1.5
Object 5	8.3BG	7.6	1.6
Object 6	6.9Y	7.4	9
Object 7	8.0R	4.2	5.2
Object 8	8.5GY	7.9	1

Table 2. Results by Visual colorimetry on sunny or rainfall in the noon

	Sunny day (Noon)		Rainfall day (Noon)				
Value	M	SD	M	SD	t	p	t-test
Object 1	7.55	1.32	7.05	0.78	-1.425	0.162	-
Object 2	3.70	0.86	3.89	1.10	0.616	0.541	-
Object 3	8.90	0.31	8.74	0.45	-1.323	0.194	-
Object 4	9.00	0.00	9.00	0.00	-	-	-
Object 5	7.80	0.95	8.58	0.51	3.165	0.003	***
Object 6	7.85	0.37	8.05	0.23	2.057	0.047	**
Object 7	5.20	0.70	5.16	0.96	-0.158	0.876	-
Object 8	8.95	0.22	8.89	0.32	-0.634	0.530	-
Chroma	M	SD	M	SD	t	p	t-test
Object 1	4.30	1.81	5.00	1.25	1.399	0.170	-
Object 2	4.80	1.82	4.21	1.27	-1.165	0.252	-
Object 3	1.90	0.55	1.58	0.84	-1.420	0.164	-
Object 4	1.10	0.55	1.21	1.23	0.366	0.717	-
Object 5	2.85	2.03	2.05	0.62	-1.637	0.110	-
Object 6	11.50	1.43	11.79	0.92	0.747	0.460	-
Object 7	8.95	2.24	8.79	3.28	-0.180	0.858	-
Object 8	0.55	0.51	0.58	0.61	0.162	0.873	-

***,p<0.01 **,p<0.05

Table 3. Results by Visual colorimetry on sunny or rainfall in the evening

	Sunny day (Evening)		Rainfall day (Evening)				
Value	M	SD	M	SD	t	p	t-test
Object 1	7.00	0.918	6.39	0.698	-2.290	0.028	**
Object 2	4.45	0.945	3.61	1.092	-2.539	0.016	**
Object 3	8.55	0.605	8.28	0.461	-1.547	0.131	-
Object 4	9.00	0.000	8.94	0.236	-1.056	0.298	-
Object 5	8.25	0.910	8.22	0.732	-0.103	0.919	-
Object 6	8.10	0.308	8.17	0.383	0.594	0.556	-
Object 7	5.35	0.587	4.78	1.060	-2.086	0.044	**
Object 8	9.00	0.000	9.00	0.000	-	-	-
Chroma	M	SD	M	SD	t	p	t-test
Object 1	5.25	1.251	5.67	1.237	1.031	0.310	-
Object 2	4.30	1.689	5.39	1.720	1.967	0.057	-
Object 3	2.10	0.912	2.11	0.832	0.039	0.969	-
Object 4	1.45	0.686	1.22	0.428	-1.211	0.234	-
Object 5	2.65	1.348	2.50	0.786	-0.413	0.682	-
Object 6	9.40	2.836	12.78	1.215	4.677	0.000	***
Object 7	9.70	2.080	10.33	1.414	1.085	0.285	-
Object 8	0.55	0.510	0.39	0.502	-0.979	0.334	-

***,p<0.01 **,p<0.05

2.2 Second investigation

Secondly, verification experiments were conducted in a rain simulation device that can control rainfall and illuminance. In figure 2 and 3, a plan of experiment device was shown. Experiments were conducted based on an experimental design with four factors of hue (5R / 5Y / 5G / 5PB), Chroma, lightness, and rainfall. In the rainfall simulation experiment, the visual effect of the color based on visual colorimetry, and conducted impression evaluations. In particular, the amount of rain and illuminance as conditions of the visual environment were changed. Subjects underwent the visual stimulus colorimetric evaluation of the stimulus. In figure 4, the levels of hue and Chroma in these experiments were shown. In table 4, factors and levels of visual effects were shown. Factors of the visual effect were

Hue, Chroma, Illuminance, and Rainfall. In the experiments, visual colorimetric evaluation using the Mansell color sheet at hand for the presented stimulus was conducted with varying rainfall. This evaluation experiments were also conducted at each illuminance. The experimental pattern is a total of 64 patterns of stimulus (hue: 4 levels / saturation: 2 levels) × rainfall (4 levels) × illuminance (2 levels).

In addition, impression for participants was extracted by indicators for impression evaluation in table 5. After participants had an experience of experiment device, they answered their impression by using these indicators.

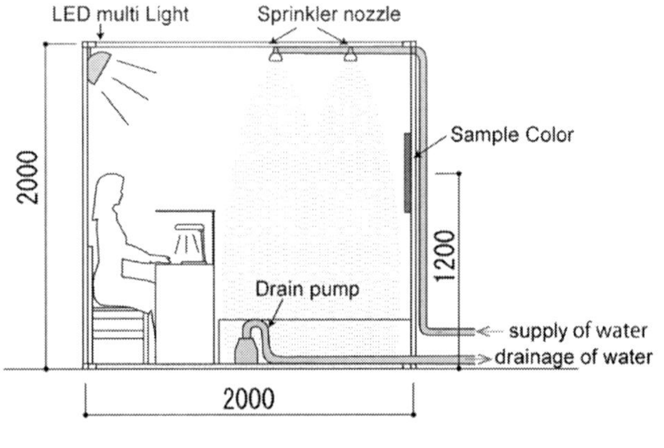

Fig.2. Laboratory sectional view

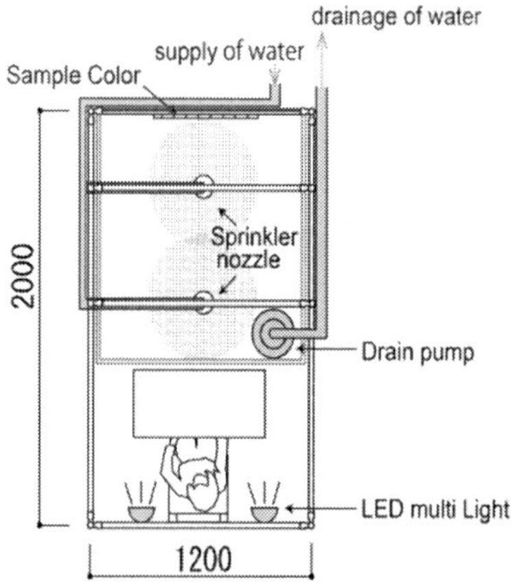

Fig.3. Plans of experiment device

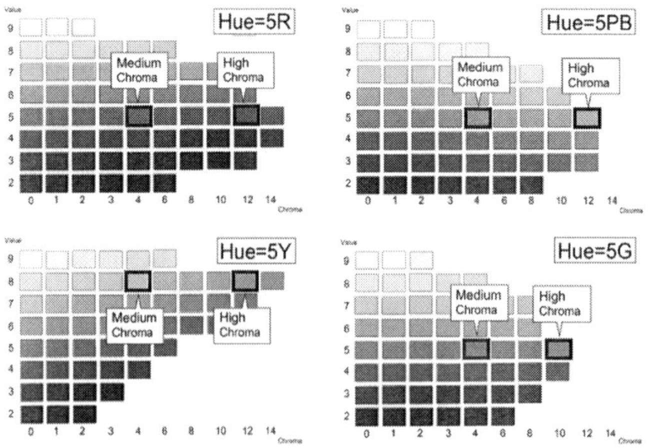

Fig.4. The levels of hue and Chroma in these experiments

Table 4. Factors and levels of visual effects

Factors	Level				Pattern	Total
Hue	5R	5Y	5G	5PB	4	64
Chroma	Medium Chroma (C=4)		High Chroma (C=10～12)		2	
Illuminance	Low illumination (100lx)		High illumination (1000lx)		2	
Amount of rainfall (mm/h)	No rain 0	Normal rain 4	Slightly stronger rain 11	Strong rain 21	4	

Table 5. Indicators for Impression Evaluation

Evaluation	Evaluation scale →	7	6	5	4	3	2	1
【Texture】 質感	Soft — Hard やわらかい — かたい	Very soft	Soft	Slightly soft	No soft, no hard	Slightly hard	Hard	Very hard
【Lively】 生き生き感	Be lively — Be lifeless 生き生きとした — 活気のない	Very be lively	Be lively	Slightly be lively	No be lively, no be lifeless	Slightly be lifeless	Be lifeless	Very be lifeless
【Aesthetic】 美的感	Beautiful — Ugly 美しい — 醜い	Very beautiful	Beautiful	Slightly beautiful	No beautiful, no ugly	Slightly ugly	Ugly	Very ugly

3 Results and discussion

The data obtained by the subject's visual sense colorimetric evaluation was tabulated for each of Value and Chroma of the color. After aggregation, 4-way analysis of variance was carried out based on factors of hue, Chroma, illuminance and rainfall to show the effect on color. As results, many interactions were detected in hue. For this reason, analysis of variance was performed separately for each stimulus hue (5R / 5Y / 5G / 5PB). That is, a three-way configuration analysis of variance based on three factors of Chroma of color, illuminance and rainfall was carried out. In addition, it was judged that there was a significant difference when the significance level was 0.05 or less.

After verification experiments, the findings obtained by the verification are as follows. In Table 6, Visual effects on Value (Hue=5R/5Y/5G/5PB) were shown. First, in order to examine the color lightness of the visual effect, visual colorimetry by Value of the color was conducted. It was found that the Value of the color got higher during rainfall. The color Hue (5R/5G/5PB) showed that the lightness effect of the color Value was at its maximum in rainfall 11mm/h, and above that the lightness effect of the color Value began decreasing. At the same time, the color of Hue 5Y showed that the lightness effect of the color Value increased as rainfall increased.

In table 7, visual effects on Chroma were shown. In order to examine the color saturation of the visual effect, we conducted visual colorimetry by Chroma of the color. Then, in figure 5, transitions of Chroma by rainfall difference (Hue=5R/5Y/5G/5PB) were shown. It was found that the Chroma of the color got lower at rainfall. The color of Hue (5Y/5G/5PB) showed that the somber effect of the color Chroma increased with increasing rainfall. Other hand, the color of Hue 5R showed that the somber effect of the color Chroma decreased with increasing rainfall.

Also, in order to examine the texture as an impression of color, we evaluated impression. It was found that the color impression became soft during rainfall. It is thought that the lightness of the color increased because of the rainfall, hence softening the texture of the color. Impression was evaluated in order to examine liveliness as an impression of color. By analysis only the color of Hue 5R took on a lively impression at rainfall, with the effect reaching its maximum when rainfall was at 11mm/h. It is thought that the movement of rain changed because of the rainfall, and the color took on a lively impression through its influence. Finally, impression was evaluated to examine the aesthetic sense of colors. It was demonstrated that only the warm color of Hue (5R/5Y) took on a beautiful impression at rainfall. We believe that the movement of rain was changed by the rainfall, and the color took on a beautiful impression by its influence.

Table 6-1 Visual effects on 【*Value*】 (Hue=5R)

Factors	Level	Ave.	Effect	-1.0 -0.5 0.0 0.5 1.0	F-value	p-value	Analysis of variance
Chroma	Medium	6.01	0.24		21.97	0.000	***
	High	5.53	-0.24				
Illuminance (lx)	100	5.56	-0.21		16.67	0.000	***
	1000	5.98	0.21				
Amount of rainfall (mm/h)	0	5.06	-0.71		21.79	0.000	***
	4	5.96	0.19				
	11	6.12	0.35				
	21	5.95	0.18				

***,p<0.01 **,p<0.05

Table 6-2 Visual effects on 【*Value*】 (Hue=5Y)

Factors	Level	Ave.	Effect	-1.0 -0.5 0.0 0.5 1.0	F-value	p-value	Analysis of variance
Chroma	Medium	8.54	0.26		87.03	0.000	***
	High	8.02	-0.26				
Illuminance (lx)	100	8.08	-0.20		48.95	0.000	***
	1000	8.48	0.20				
Amount of rainfall (mm/h)	0	8.06	-0.22		7.37	0.000	***
	4	8.29	0.02				
	11	8.34	0.06				
	21	8.42	0.14				

***,p<0.01 **,p<0.05

Table 6-3 Visual effects on 【*Value*】 (Hue=5G)

Factors	Level	Ave.	Effect	-1.0 -0.5 0.0 0.5 1.0	F-value	p-value	Analysis of variance
Chroma	Medium	5.12	-0.01		0.05	0.823	-
	High	5.14	0.01				
Illuminance (lx)	100	4.58	-0.55		124.88	0.000	***
	1000	5.67	0.55				
Amount of rainfall (mm/h)	0	4.55	-0.58		16.11	0.000	***
	4	5.21	0.09				
	11	5.38	0.25				
	21	5.36	0.24				

***,p<0.01 **,p<0.05

Table 6-4 Visual effects on 【*Value*】 (Hue=5PB)

Factors	Level	Ave.	Effect	-1.0 -0.5 0.0 0.5 1.0	F-value	p-value	Analysis of variance
Chroma	Medium	5.69	0.29		22.88	0.000	***
	High	5.10	-0.29				
Illuminance (lx)	100	4.88	-0.52		71.67	0.000	***
	1000	5.91	0.52				
Amount of rainfall (mm/h)	0	4.91	-0.48		7.26	0.000	***
	4	5.58	0.19				
	11	5.63	0.23				
	21	5.46	0.07				

***,p<0.01 **,p<0.05

Table 7-1 Visual effects on 【*Chroma*】 (Hue=5R)

Factors	Level	Ave.	Effect	-1.0 -0.5 0.0 0.5 1.0	F-value	p-value	Analysis of variance
Chroma	Medium	2.93	-3.23		899.51	0.000	***
	High	9.38	3.23				
Illuminance (lx)	100	6.12	-0.03		0.08	0.772	-
	1000	6.18	0.03				
Amount of rainfall (mm/h)	0	7.03	0.87		8.29	0.000	***
	4	6.15	0.00				
	11	5.69	-0.46				
	21	5.74	-0.41				

***,p<0.01 **,p<0.05

Table 7-2 Visual effects on 【*Chroma*】 (Hue=5Y)

Factors	Level	Ave.	Effect	-1.0 -0.5 0.0 0.5 1.0	F-value	p-value	Analysis of variance
Chroma	Medium	2.96	-3.49		1237.12	0.000	***
	High	9.93	3.49				
Illuminance (lx)	100	6.41	-0.04		0.16	0.694	-
	1000	6.48	0.04				
Amount of rainfall (mm/h)	0	8.01	1.56		30.72	0.000	***
	4	6.43	-0.02				
	11	5.72	-0.73				
	21	5.63	-0.81				

***,p<0.01 **,p<0.05

Table 7-3 Visual effects on 【*Chroma*】 (Hue=5G)

Factors	Level	Ave.	Effect	-1.0 -0.5 0.0 0.5 1.0	F-value	p-value	Analysis of variance
Chroma	Medium	4.79	-1.34		222.93	0.000	***
	High	7.47	1.34				
Illuminance (lx)	100	5.66	-0.47		27.50	0.000	***
	1000	6.60	0.47				
Amount of rainfall (mm/h)	0	7.38	1.25		27.60	0.000	***
	4	6.33	0.20				
	11	5.50	-0.63				
	21	5.31	-0.82				

***,p<0.01 **,p<0.05

Table 7-4 Visual effects on 【*Chroma*】 (Hue=5PB)

Factors	Level	Ave.	Effect	-1.0 -0.5 0.0 0.5 1.0	F-value	p-value	Analysis of variance
Chroma	Medium	3.57	-2.51		697.56	0.000	***
	High	8.60	2.51				
Illuminance (lx)	100	5.56	-0.52		30.46	0.000	***
	1000	6.61	0.53				
Amount of rainfall (mm/h)	0	7.31	1.23		22.15	0.000	***
	4	6.16	0.08				
	11	5.57	-0.52				
	21	5.29	-0.79				

***,p<0.01 **,p<0.05

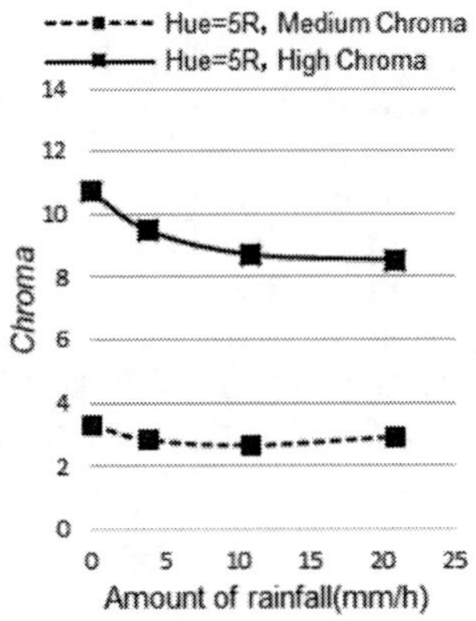

Fig.5-1 【Chroma】 by rainfall difference(Hue=5R)

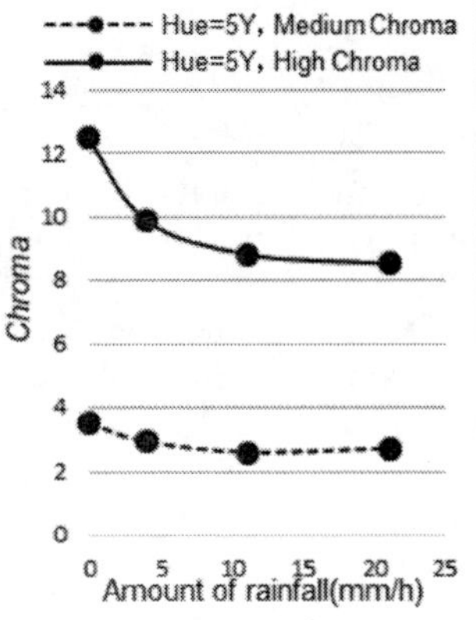

Fig.5-2 【Chroma】 by rainfall difference(Hue=5Y)

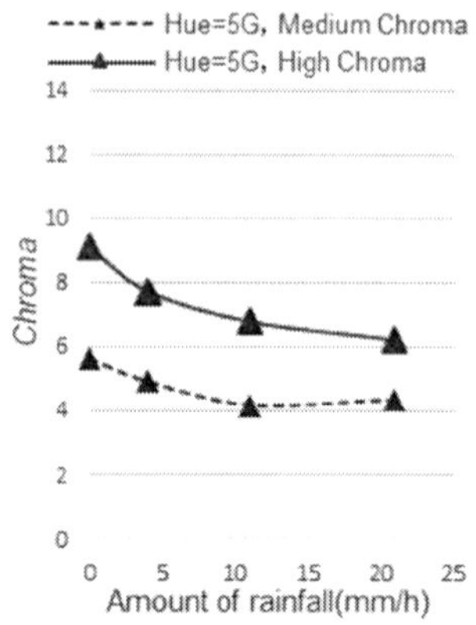

Fig.5-3 【Chroma】 by rainfall difference(Hue=5G)

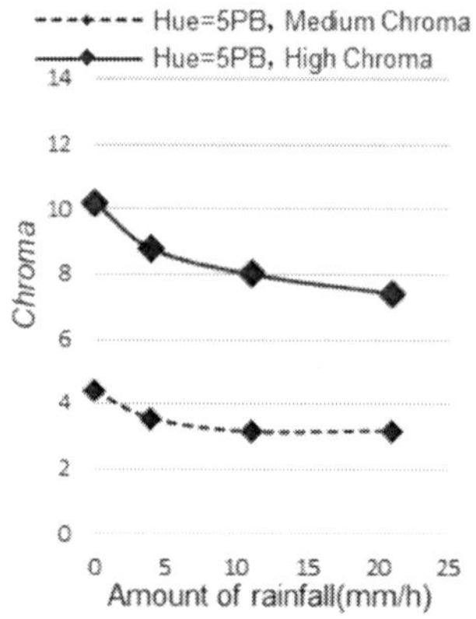

Fig.5-4 【Chroma】 by rainfall difference(Hue=5PB)

4 Conclusion

In this study, the relationship between rainfalls and the visual effects and impressions of color were confirmed, and there were differences in the evaluation depending on the Hue and the amount of rainfall. In landscape color planning, it was to consider the influence of rainfall is important for a fascinating landscape. This study will show basic important references of new landscape color plan.

This study was summarized as results of graduation research by Mika Kawakatsu who was a fourth grade student at the Ibaraki University. The author thanks his student for his work on this research.

References

1. Kobayashi, T. ：A Study on the Effects of the Changing Scenery, journal of the Japanese Institute of Landscape Architecture (5), pp.279-282, 1987.3
2. Fukushima, H. and Horikoshi, T. : Realization of Climatic Landscape by fluctuation with weather and time for appearance of architecture, Summaries of Technical Papers of Annual Meeting, Architectural Institute of Japan, Urban Planning, pp.529-530, 2014.9
3. Takamatsu, M and Nakashima, Y and Yoshiwara, S. ：Apparent Color of 10 Standard-color-cards in Dense Fog, Journal of the Color Science Association of Japan, pp.207-214, 1999.12
4. Kuwabara, Y and Takamatsu, M and Nakashima, Y and Terakawa, H and Tada, K and Iwane, H. ：Visual Characteristics of Colored LED Lights In Dense Fog, Journal of the Color Science Association of Japan, pp. 86-87, 2012.5
5. Kawakatsu, M and Kumazawa, T：A study on changes in color appearance on facade during rain, Architectual Institute of Japan, pp739-740, 2017.7
6. Inoue, M and Kobayashi, T. ：The Research Domain and Scale Construction of Adjective-Pairs In A Semantic Differential Method In Japan, The Japanese Association of Educational Psychology, pp. 253-260, 1985

AUTHOR INDEX